PROBLEMS AND SOLUTIONS ON ATOMIC, NUCLEAR AND PARTICLE PHYSICS

Major American Universities Ph.D.
Qualifying Questions and Solutions

PROBLEMS AND SOLUTIONS ON ATOMIC, NUCLEAR AND PARTICLE PHYSICS

Compiled by

**The Physics Coaching Class
University of Science and
Technology of China**

Edited by

Yung-Kuo Lim

World Scientific

NEW JERSEY • LONDON • SINGAPORE • BEIJING • SHANGHAI • HONG KONG • TAIPEI • CHENNAI

Published by

World Scientific Publishing Co. Pte. Ltd.

5 Toh Tuck Link, Singapore 596224

USA office: 27 Warren Street, Suite 401-402, Hackensack, NJ 07601

UK office: 57 Shelton Street, Covent Garden, London WC2H 9HE

Library of Congress Cataloging-in-Publication Data
Problems and solutions on atomic, nuclear and particle physics / compiled by the
Physics Coaching Class, University of Science and Technology of China ; edited by
Yung-Kuo Lim.
ix, 717 p. ; ill. ; 22 cm. -- (Major American universities Ph.D. qualifying questions and solutions)
Includes index.
ISBN-13 978-981-02-3917-6
ISBN-10 981-02-3917-3
ISBN-13 978-981-02-3918-3 (pbk)
ISBN-10 981-02-3918-1 (pbk)
1. Physics--Problems, exercises, etc. 2. Physics--Study and teaching. I. Lim, Yung-kuo.
II. Zhongguo ke xue ji shu da xue. Physics Coaching Class. III. Title
QC32.P76 2000
539.7076--dc22

2005277578

British Library Cataloguing-in-Publication Data
A catalogue record for this book is available from the British Library.

First published 2000
Reprinted 2003, 2007, 2008

Printed in Singapore by World Scientific Printers

PREFACE

This series of physics problems and solutions, which consists of seven volumes — Mechanics, Electromagnetism, Optics, Atomic, Nuclear and Particle Physics, Thermodynamics and Statistical Physics, Quantum Mechanics, Solid State Physics and Relativity, contains a selection of 2550 problems from the graduate-school entrance and qualifying examination papers of seven U.S. universities — California University Berkeley Campus, Columbia University, Chicago University, Massachusetts Institute of Technology, New York State University Buffalo Campus, Princeton University, Wisconsin University — as well as the CUSPEA and C.C. Ting's papers for selection of Chinese students for further studies in U.S.A., and their solutions which represent the effort of more than 70 Chinese physicists, plus some 20 more who checked the solutions.

The series is remarkable for its comprehensive coverage. In each area the problems span a wide spectrum of topics, while many problems overlap several areas. The problems themselves are remarkable for their versatility in applying the physical laws and principles, their uptodate realistic situations, and their scanty demand on mathematical skills. Many of the problems involve order-of-magnitude calculations which one often requires in an experimental situation for estimating a quantity from a simple model. In short, the exercises blend together the objectives of enhancement of one's understanding of physical principles and ability of practical application.

The solutions as presented generally just provide a guidance to solving the problems, rather than step-by-step manipulation, and leave much to the students to work out for themselves, of whom much is demanded of the basic knowledge in physics. Thus the series would provide an invaluable complement to the textbooks.

The present volume consists of 483 problems. It covers practically the whole of the usual undergraduate syllabus in atomic, nuclear and particle physics, but in substance and sophistication goes much beyond. Some problems on experimental methodology have also been included.

In editing, no attempt has been made to unify the physical terms, units and symbols. Rather, they are left to the setters' and solvers' own preference so as to reflect the realistic situation of the usage today. Great pains has been taken to trace the logical steps from the first principles to the final solution, frequently even to the extent of rewriting the entire solution.

In addition, a subject index to problems has been included to facilitate the location of topics. These editorial efforts hopefully will enhance the value of the volume to the students and teachers alike.

Yung-Kuo Lim
Editor

INTRODUCTION

Solving problems in course work is an exercise of the mental facilities, and examination problems are usually chosen, or set similar to such problems. Working out problems is thus an essential and important aspect of the study of physics.

The series *Major American University Ph.D. Qualifying Questions and Solutions* comprises seven volumes and is the result of months of work of a number of Chinese physicists. The subjects of the volumes and the respective coordinators are as follows:

1. Mechanics (Qiang Yan-qi, Gu En-pu, Cheng Jia-fu, Li Ze-hua, Yang De-tian)

2. Electromagnetism (Zhao Shu-ping, You Jun-han, Zhu Jun-jie)

3. Optics (Bai Gui-ru, Guo Guang-can)

4. Atomic, Nuclear and Particle Physics (Jin Huai-cheng, Yang Bao-zhong, Fan Yang-mei)

5. Thermodynamics and Statistical Physics (Zheng Jiu-ren)

6. Quantum Mechanics (Zhang Yong-de, Zhu Dong-pei, Fan Hong-yi)

7. Solid State Physics and Miscellaneous Topics (Zhang Jia-lu, Zhou You-yuan, Zhang Shi-ling).

These volumes, which cover almost all aspects of university physics, contain 2550 problems, mostly solved in detail.

The problems have been carefully chosen from a total of 3100 problems, collected from the China-U.S.A. Physics Examination and Application Program, the Ph.D. Qualifying Examination on Experimental High Energy Physics sponsored by Chao-Chong Ting, and the graduate qualifying examinations of seven world-renowned American universities: Columbia University, the University of California at Berkeley, Massachusetts Institute of Technology, the University of Wisconsin, the University of Chicago, Princeton University, and the State University of New York at Buffalo.

Generally speaking, examination problems in physics in American universities do not require too much mathematics. They can be characterized to a large extent as follows. Many problems are concerned with the various frontier subjects and overlapping domains of topics, having been selected from the setters own research encounters. These problems show a "modern" flavor. Some problems involve a wide field and require a sharp mind for their analysis, while others require simple and practical methods demanding a fine "touch of physics". Indeed, we believe that these problems, as a whole, reflect to some extent the characteristics of American science and culture, as well as give a glimpse of the philosophy underlying American education.

That being so, we considered it worthwhile to collect and solve these problems, and introduce them to students and teachers everywhere, even though the work was both tedious and strenuous. About a hundred teachers and graduate students took part in this time-consuming task.

This volume on Atomic, Nuclear and Particle Physics which contains 483 problems is divided into four parts: Atomic and Molecular Physics (142), Nuclear Physics (120), Particle Physics (90), Experimental Methods and Miscellaneous topics (131).

In scope and depth, most of the problems conform to the usual undergraduate syllabi for atomic, nuclear and particle physics in most universities. Some of them, however, are rather profound, sophisticated, and broad-based. In particular they demonstrate the use of fundamental principles in the latest research activities. It is hoped that the problems would help the reader not only in enhancing understanding of the basic principles, but also in cultivating the ability to solve practical problems in a realistic environment.

This volume was the result of the collective efforts of forty physicists involved in working out and checking of the solutions, notably Ren Yong, Qian Jian-ming, Chen Tao, Cui Ning-zhuo, Mo Hai-ding, Gong Zhu-fang and Yang Bao-zhong.

CONTENTS

Preface   v

Introduction   vii

Part I. Atomic and Molecular Physics   **1**

1. Atomic Physics (1001–1122)   3
2. Molecular Physics (1123–1142)   173

Part II. Nuclear Physics   **205**

1. Basic Nuclear Properties (2001–2023)   207
2. Nuclear Binding Energy, Fission and Fusion (2024–2047)   239
3. The Deuteron and Nuclear forces (2048–2058)   269
4. Nuclear Models (2059–2075)   289
5. Nuclear Decays (2076–2107)   323
6. Nuclear Reactions (2108–2120)   382

Part III. Particle Physics   **401**

1. Interactions and Symmetries (3001–3037)   403
2. Weak and Electroweak Interactions, Grand Unification
   Theories (3038–3071)   459
3. Structure of Hadrons and the Quark Model (3072–3090)   524

Part IV. Experimental Methods and Miscellaneous Topics   **565**

1. Kinematics of High-Energy Particles (4001–4061)   567
2. Interactions between Radiation and Matter (4062–4085)   646
3. Detection Techniques and Experimental Methods (4086–4105)   664
4. Error Estimation and Statistics (4106–4118)   678
5. Particle Beams and Accelerators (4119–4131)   690

Index to Problems   709

PART I

ATOMIC AND MOLECULAR PHYSICS

1. ATOMIC PHYSICS (1001–1122)

1001

Assume that there is an announcement of a fantastic process capable of putting the contents of physics library on a very smooth postcard. Will it be readable with an electron microscope? Explain.

(Columbia)

Solution:

Suppose there are 10^6 books in the library, 500 pages in each book, and each page is as large as two postcards. For the postcard to be readable, the planar magnification should be $2 \times 500 \times 10^6 \approx 10^9$, corresponding to a linear magnification of $10^{4.5}$. As the linear magnification of an electron microscope is of the order of 800,000, its planar magnification is as large as 10^{11}, which is sufficient to make the postcard readable.

1002

At 10^{10} K the black body radiation weighs (1 ton, 1 g, 10^{-6} g, 10^{-16} g) per cm^3.

(Columbia)

Solution:

The answer is nearest to 1 ton per cm^3.

The radiant energy density is given by $u = 4\sigma T^4/c$, where $\sigma = 5.67 \times 10^{-8}$ Wm^{-2} K^{-4} is the Stefan–Boltzmann constant. From Einstein's mass-energy relation, we get the mass of black body radiation per unit volume as $u = 4\sigma T^4/c^3 = 4 \times 5.67 \times 10^{-8} \times 10^{40}/(3 \times 10^8)^3 \approx 10^8$ kg/m^3 = 0.1 ton/cm^3.

1003

Compared to the electron Compton wavelength, the Bohr radius of the hydrogen atom is approximately

(a) 100 times larger.

(b) 1000 times larger.

(c) about the same.

(CCT)

Solution:

The Bohr radius of the hydrogen atom and the Compton wavelength of electron are given by $a = \frac{\hbar^2}{me^2}$ and $\lambda_c = \frac{h}{mc}$ respectively. Hence $\frac{a}{\lambda_c} = \frac{1}{2\pi}(\frac{e^2}{\hbar c})^{-1} = \frac{137}{2\pi} = 22$, where $e^2/\hbar c$ is the fine-structure constant. Hence the answer is (a).

1004

Estimate the electric field needed to pull an electron out of an atom in a time comparable to that for the electron to go around the nucleus.

(Columbia)

Solution:

Consider a hydrogen-like atom of nuclear charge Ze. The ionization energy (or the energy needed to eject the electron) is $13.6Z^2$ eV. The orbiting electron has an average distance from the nucleus of $a = a_0/Z$, where $a_0 = 0.53 \times 10^{-8}$ cm is the Bohr radius. The electron in going around the nucleus in electric field E can in half a cycle acquire an energy eEa. Thus to eject the electron we require

$$eEa \gtrsim 13.6 \ Z^2 \ \text{eV}\,,$$

or

$$E \gtrsim \frac{13.6 \ Z^3}{0.53 \times 10^{-8}} \approx 2 \times 10^9 \ Z^3 \ \text{V/cm}\,.$$

1005

As one goes away from the center of an atom, the electron density

(a) decreases like a Gaussian.

(b) decreases exponentially.

(c) oscillates with slowly decreasing amplitude.

(CCT)

Solution:

The answer is (c).

1006

An electronic transition in ions of ^{12}C leads to photon emission near $\lambda = 500$ nm ($h\nu = 2.5$ eV). The ions are in thermal equilibrium at an ion temperature $kT = 20$ eV, a density $n = 10^{24}$ m^{-3}, and a non-uniform magnetic field which ranges up to $B = 1$ Tesla.

(a) Briefly discuss broadening mechanisms which might cause the transition to have an observed width $\Delta\lambda$ greater than that obtained for very small values of T, n and B.

(b) For one of these mechanisms calculate the broadened width $\Delta\lambda$ using order-of-magnitude estimates of needed parameters.

(Wisconsin)

Solution:

(a) A spectral line always has an inherent width produced by uncertainty in atomic energy levels, which arises from the finite length of time involved in the radiation process, through Heisenberg's uncertainty principle. The observed broadening may also be caused by instrumental limitations such as those due to lens aberration, diffraction, etc. In addition the main causes of broadening are the following.

Doppler effect: Atoms or molecules are in constant thermal motion at $T > 0$ K. The observed frequency of a spectral line may be slightly changed if the motion of the radiating atom has a component in the line of sight, due to Doppler effect. As the atoms or molecules have a distribution of velocity a line that is emitted by the atoms will comprise a range of frequencies symmetrically distributed about the natural frequency, contributing to the observed width.

Collisions: An atomic system may be disturbed by external influences such as electric and magnetic fields due to outside sources or neighboring atoms. But these usually cause a shift in the energy levels rather than broadening them. Broadening, however, can result from atomic collisions which cause phase changes in the emitted radiation and consequently a spread in the energy.

(b) *Doppler broadening:* The first order Doppler frequency shift is given by $\Delta\nu = \frac{\nu_0 v_x}{c}$, taking the x-axis along the line of sight. Maxwell's velocity distribution law then gives

$$dn \propto \exp\left(-\frac{Mv_x^2}{2kT}\right)dv_x = \exp\left[-\frac{Mc^2}{2kT}\left(\frac{\Delta\nu}{\nu_0}\right)^2\right]dv_x,$$

where M is the mass of the radiating atom. The frequency-distribution of the radiation intensity follows the same relationship. At half the maximum intensity, we have

$$\Delta\nu = \nu_0\sqrt{\frac{(\ln 2)2kT}{Mc^2}}.$$

Hence the line width at half the maximum intensity is

$$2\Delta\nu = \frac{1.67c}{\lambda_0}\sqrt{\frac{2kT}{Mc^2}}.$$

In terms of wave number $\tilde{\nu} = \frac{1}{\lambda} = \frac{\nu}{c}$ we have

$$\Gamma_D = 2\Delta\tilde{\nu} = \frac{1.67}{\lambda_0}\sqrt{\frac{2kT}{Mc^2}}.$$

With $kT = 20$ eV, $Mc^2 = 12 \times 938$ MeV, $\lambda_0 = 5 \times 10^{-7}$ m,

$$\Gamma_D = \frac{1.67}{5 \times 10^{-7}}\sqrt{\frac{2 \times 20}{12 \times 938 \times 10^6}} = 199\ \text{m}^{-1} \approx 2\ \text{cm}^{-1}.$$

Collision broadening: The mean free path for collision l is defined by $nl\pi d^2 = 1$, where d is the effective atomic diameter for a collision close enough to affect the radiation process. The mean velocity $\bar{v}$ of an atom can be approximated by its root-mean-square velocity given by $\frac{1}{2}M\overline{v^2} = \frac{3}{2}kT$. Hence

$$\bar{v} \approx \sqrt{\frac{3kT}{M}}.$$

Then the mean time between successive collisions is

$$t = \frac{l}{\bar{v}} = \frac{1}{n\pi d^2}\sqrt{\frac{M}{3kT}}.$$

The uncertainty in energy because of collisions, ΔE, can be estimated from the uncertainty principle $\Delta E \cdot t \approx \hbar$, which gives

$$\Delta \nu_c \approx \frac{1}{2\pi t},$$

or, in terms of wave number,

$$\Gamma_c = \frac{1}{2} n d^2 \sqrt{\frac{3kT}{Mc^2}} \sim \frac{3 \times 10^{-3}}{\lambda_0} \sqrt{\frac{2kT}{Mc^2}},$$

if we take $d \approx 2a_0 \sim 10^{-10}$ m, a_0 being the Bohr radius. This is much smaller than Doppler broadening at the given ion density.

1007

(I) The ionization energy E_I of the first three elements are

Z	Element	E_I
1	H	13.6 eV
2	He	24.6 eV
3	Li	5.4 eV

(a) Explain qualitatively the change in E_I from H to He to Li.

(b) What is the second ionization energy of He, that is the energy required to remove the second electron after the first one is removed?

(c) The energy levels of the $n = 3$ states of the valence electron of sodium (neglecting intrinsic spin) are shown in Fig. 1.1.

Why do the energies depend on the quantum number l?

(*SUNY, Buffalo*)

```
3d (l = 2) ─────────────── - 1.5 eV
3p (l = 1) ─────────────── - 3.0 eV
3s (l = 0) ─────────────── - 5.1 eV
```

Fig. 1.1

Solution:

(a) The table shows that the ionization energy of He is much larger than that of H. The main reason is that the nuclear charge of He is twice than that of H while all their electrons are in the first shell, which means that the potential energy of the electrons are much lower in the case of He. The very low ionization energy of Li is due to the screening of the nuclear charge by the electrons in the inner shell. Thus for the electron in the outer shell, the effective nuclear charge becomes small and accordingly its potential energy becomes higher, which means that the energy required for its removal is smaller.

(b) The energy levels of a hydrogen-like atom are given by

$$E_n = -\frac{Z^2}{n^2} \times 13.6 \text{ eV}.$$

For $Z = 2$, $n = 1$ we have

$$E_I = 4 \times 13.6 = 54.4 \text{ eV}.$$

(c) For the $n = 3$ states the smaller l the valence electron has, the larger is the eccentricity of its orbit, which tends to make the atomic nucleus more polarized. Furthermore, the smaller l is, the larger is the effect of orbital penetration. These effects make the potential energy of the electron decrease with decreasing l.

1008

Describe briefly each of the following effects or, in the case of rules, state the rule:

(a) Auger effect
(b) Anomalous Zeeman effect
(c) Lamb shift
(d) Landé interval rule
(e) Hund's rules for atomic levels

(Wisconsin)

Solution:

(a) Auger effect: When an electron in the inner shell (say K shell) of an atom is ejected, a less energetically bound electron (say an L electron) may jump into the hole left by the ejected electron, emitting a photon. If the process takes place without radiating a photon but, instead, a higher-energy shell (say L shell) is ionized by ejecting an electron, the process is called Auger effect and the electron so ejected is called Auger electron. The atom becomes doubly ionized and the process is known as a nonradiative transition.

(b) Anomalous Zeeman effect: It was observed by Zeeman in 1896 that, when an excited atom is placed in an external magnetic field, the spectral line emitted in the de-excitation process splits into three lines with equal spacings. This is called normal Zeeman effect as such a splitting could be understood on the basis of a classical theory developed by Lorentz. However it was soon found that more commonly the number of splitting of a spectral line is quite different, usually greater than three. Such a splitting could not be explained until the introduction of electron spin, thus the name 'anomalous Zeeman effect'.

In the modern quantum theory, both effects can be readily understood: When an atom is placed in a weak magnetic field, on account of the interaction between the total magnetic dipole moment of the atom and the external magnetic field, both the initial and final energy levels are split into several components. The optical transitions between the two multiplets then give rise to several lines. The normal Zeeman effect is actually only a special case where the transitions are between singlet states in an atom with an even number of optically active electrons.

(c) Lamb shift: In the absence of hyperfine structure, the $2^2 S_{1/2}$ and $2^2 P_{1/2}$ states of hydrogen atom would be degenerate for orbital quantum number l as they correspond to the same total angular momentum $j = 1/2$. However, Lamb observed experimentally that the energy of $2^2 S_{1/2}$ is 0.035 cm^{-1} higher than that of $2^2 P_{1/2}$. This phenomenon is called Lamb shift. It is caused by the interaction between the electron and an electromagnetic radiation field.

(d) Landé interval rule: For LS coupling, the energy difference between two adjacent J levels is proportional, in a given LS term, to the larger of the two values of J.

(e) Hund's rules for atomic levels are as follows:

(1) If an electronic configuration has more than one spectroscopic notation, the one with the maximum total spin S has the lowest energy.

(2) If the maximum total spin S corresponds to several spectroscopic notations, the one with the maximum L has the lowest energy.

(3) If the outer shell of the atom is less than half full, the spectroscopic notation with the minimum total angular momentum J has the lowest energy. However, if the shell is more than half full the spectroscopic notation with the maximum J has the lowest energy. This rule only holds for LS coupling.

1009

Give expressions for the following quantities in terms of $e, \hbar, c, k, m_e$ and m_p.

(a) The energy needed to ionize a hydrogen atom.

(b) The difference in frequency of the Lyman alpha line in hydrogen and deuterium atoms.

(c) The magnetic moment of the electron.

(d) The spread in measurement of the π^0 mass, given that the π^0 lifetime is τ.

(e) The magnetic field B at which there is a 10^{-4} excess of free protons in one spin direction at a temperature T.

(f) Fine structure splitting in the $n = 2$ state of hydrogen.

(*Columbia*)

Solution:

(a)

$$E_I = \left(\frac{e^2}{4\pi\varepsilon_0}\right)^2 \frac{m_e}{2\hbar^2},$$

ε_0 being the permittivity of free space.

(b) The difference of frequency is caused by the Rydberg constant changing with the mass of the nucleus. The wave number of the α line of hydrogen atom is

$$\tilde{\nu}_H = R_H \left(1 - \frac{1}{4}\right) = \frac{3}{4} R_H ,$$

and that of the α line of deuterium atom is

$$\tilde{\nu}_D = \frac{3}{4} R_D .$$

The Rydberg constant is given by

$$R = \left(\frac{e^2}{4\pi\varepsilon_0}\right)^2 \frac{m_r}{m_e} = \frac{m_r}{m_e} R_\infty,$$

where m_r is the reduced mass of the orbiting electron in the atomic system, and

$$R_\infty = \left(\frac{e^2}{4\pi\varepsilon_0}\right)^2 \frac{m_e}{4\pi\hbar^3 c}.$$

As for H atom, $m_r = \frac{m_p m_e}{m_p + m_e}$, and for D atom,

$$m_r = \frac{2m_p m_e}{2m_p + m_e},$$

m_p being the nucleon mass, we have

$$\Delta\nu = c\Delta\tilde{\nu} = \frac{3}{4}c(R_D - R_H) = \frac{3}{4}cR_\infty\left(\frac{1}{1 + \dfrac{m_e}{2m_p}} - \frac{1}{1 + \dfrac{m_e}{m_p}}\right)$$

$$\approx \frac{3}{4}cR_\infty\frac{m_e}{2m_p} = \frac{3}{4}\left(\frac{e^2}{4\pi\varepsilon_0}\right)^2\frac{\pi^2}{\hbar^3}\frac{m_e^2}{m_p}.$$

(c) The magnetic moment associated with the electron spin is

$$\mu_e = \frac{he}{4\pi m_e} = \mu_B,$$

μ_B being the Bohr magneton.

(d) The spread in the measured mass (in energy units) is related to the lifetime τ through the uncertainty principle

$$\Delta E \cdot \tau \gtrsim \hbar,$$

which gives

$$\Delta E \gtrsim \frac{\hbar}{\tau}.$$

(e) Consider the free protons as an ideal gas in which the proton spins have two quantized directions: parallel to B with energy $E_p = -\mu_p B$ and antiparallel to B with energy $E_p = \mu_p B$, where $\mu_p = \frac{\hbar e}{2m_p}$ is the magnetic moment of proton. As the number density $n \propto \exp(\frac{-E_p}{kT})$, we have

$$\frac{\exp\left(\dfrac{\mu_p B}{kT}\right) - \exp\left(\dfrac{-\mu_p B}{kT}\right)}{\exp\left(\dfrac{\mu_p B}{kT}\right) + \exp\left(\dfrac{-\mu_p B}{kT}\right)} = 10^{-4} ,$$

or

$$\exp\left(\frac{2\mu_p B}{kT}\right) = \frac{1 + 10^{-4}}{1 - 10^{-4}} ,$$

giving

$$\frac{2\mu_p B}{kT} \approx 2 \times 10^{-4} ,$$

i.e.

$$B = \frac{kT}{\mu_p} \times 10^{-4} .$$

(f) The quantum numbers of $n = 2$ states are: $n = 2$, $l = 1$, $j_1 = 3/2$, $j_2 = 1/2$ (the $l = 0$ state does not split and so need not be considered here). From the expression for the fine-structure energy levels of hydrogen, we get

$$\Delta E = -\frac{2\pi Rhc\alpha^2}{n^3}\left(\frac{1}{j_1 + \dfrac{1}{2}} - \frac{1}{j_2 + \dfrac{1}{2}}\right) = \frac{\pi Rhc\alpha^2}{8} ,$$

where

$$\alpha = \frac{e^2}{4\pi\varepsilon_0\hbar c}$$

is the fine structure constant,

$$R = \left(\frac{e^2}{4\pi\varepsilon_0}\right)^2 \frac{m_e}{4\pi\hbar^3 c}$$

is the Rydberg constant.

1010

As shown in Fig. 1.2, light shines on sodium atoms. Estimate the cross-section on resonance for excitation of the atoms from the ground to the

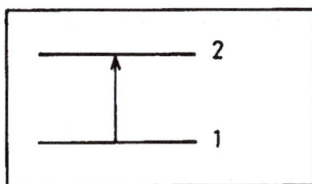

Fig. 1.2

first excited state (corresponding to the familiar yellow line). Estimate the width of the resonance. You need not derive these results from first principles if you remember the appropriate heuristic arguments.

(*Princeton*)

Solution:

The cross-section is defined by $\sigma_A = P_\omega / I_\omega$, where $P_\omega d\omega$ is the energy in the frequency range ω to $\omega + d\omega$ absorbed by the atoms in unit time, $I_\omega d\omega$ is the incident energy per unit area per unit time in the same frequency range. By definition,

$$\int P_\omega d\omega = B_{12} \hbar \omega N_\omega ,$$

where B_{12} is Einstein's B-coefficient giving the probability of an atom in state 1 absorbing a quantum $\hbar\omega$ per unit time and $N_\omega d\omega$ is the energy density in the frequency range ω to $\omega + d\omega$. Einstein's relation

$$B_{12} = \frac{\pi^2 c^3}{\hbar \omega^3} \cdot \frac{g_1}{g_2} A_{21}$$

gives

$$B_{12} = \frac{\pi^2 c^3}{\hbar \omega^3} \cdot \frac{g_1}{g_2} \cdot \frac{1}{\tau} = \frac{\pi^2 c^3}{\hbar^2 \omega^3} \cdot \frac{g_1}{g_2} \Gamma ,$$

where τ is the lifetime of excited state 2, whose natural line width is $\Gamma \approx \frac{\hbar}{\tau}$, g_1, g_2 are respectively the degeneracies of states 1 and 2, use having been made of the relation $A_{12} = 1/\tau$ and the uncertainty principle $\Gamma \tau \approx \hbar$. Then as $N_\omega = I_\omega / c$, c being the velocity of light in free space, we have

$$P_\omega = \frac{\pi^2 c^2}{\hbar \omega^2} \cdot \frac{g_1}{g_2} \Gamma I_\omega .$$

Introducing the form factor $g(\omega)$ and considering ω and I_ω as average values in the band of $g(\omega)$, we can write the above as

$$P_\omega = \frac{\pi^2 c^2}{\hbar \omega^2} \cdot \frac{g_1}{g_2} \Gamma I_\omega g(\omega).$$

Take for $g(\omega)$ the Lorentz profile

$$g(\omega) = \frac{\hbar}{2\pi} \frac{\Gamma}{(E_2 - E_1 - \hbar\omega)^2 + \dfrac{\Gamma^2}{4}}.$$

At resonance,

$$E_2 - E_1 = \hbar\omega,$$

and so

$$g\left(\omega = \frac{E_2 - E_1}{\hbar}\right) = \frac{2\hbar}{\pi\Gamma}.$$

Hence

$$\sigma_A = \frac{\pi^2 c^2}{\hbar \omega^2} \cdot \frac{g_1}{g_2} \cdot \frac{2\hbar}{\pi} = \frac{2\pi c^2}{\omega^2} \cdot \frac{g_1}{g_2}.$$

For the yellow light of Na (D line), $g_1 = 2$, $g_2 = 6$, $\lambda = 5890$ Å, and

$$\sigma_A = \frac{1}{3} \cdot \frac{\lambda^2}{2\pi} = 1.84 \times 10^{-10} \text{ cm}^2.$$

For the D line of sodium, $\tau \approx 10^{-8}$ s and the line width at half intensity is

$$\Gamma \approx \frac{\hbar}{\tau} = 6.6 \times 10^{-8} \text{ eV}.$$

As

$$\Gamma = \Delta E = \hbar \Delta \omega = \hbar \Delta \left(\frac{2\pi c}{\lambda}\right) = 2\pi \hbar c \Delta\tilde{\nu},$$

the line width in wave numbers is

$$\Delta\tilde{\nu} = \frac{\Gamma}{2\pi\hbar c} \approx \frac{1}{2\pi c\tau} = 5.3 \times 10^{-4} \text{ cm}^{-1}.$$

1011

The cross section for electron impact excitation of a certain atomic level A is $\sigma_A = 1.4 \times 10^{-20}$ cm^2. The level has a lifetime $\tau = 2 \times 10^{-8}$ sec, and decays 10 per cent of the time to level B and 90 per cent of the time to level C (Fig. 1.3).

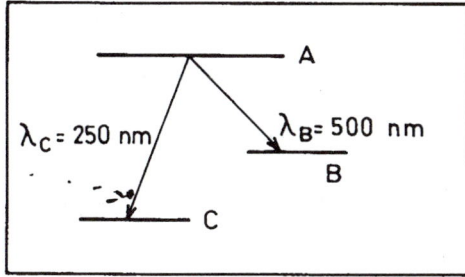

Fig. 1.3

(a) Calculate the equilibrium population per cm^3 in level A when an electron beam of 5 mA/cm^2 is passed through a vapor of these atoms at a pressure of 0.05 torr.

(b) Calculate the light intensity emitted per cm^3 in the transition $A \rightarrow B$, expressed in watts/steradian.

(Wisconsin)

Solution:

(a) According to Einstein's relation, the number of transitions $B, C \rightarrow A$ per unit time (rate of production of A) is

$$\frac{dN_{BC \rightarrow A}}{dt} = n_0 \sigma_A N_{BC} ,$$

and the number of decays $A \rightarrow B, C$ per unit time is

$$\frac{dN_{A \rightarrow BC}}{dt} = \left(\frac{1}{\tau} + n_0 \sigma_A \right) N_A ,$$

where N_{BC} and N_A are the numbers of atoms in the energy levels B, C and A respectively, n_0 is the number of electrons crossing unit area per unit time. At equilibrium,

$$\frac{dN_{BC \rightarrow A}}{dt} = \frac{dN_{A \rightarrow BC}}{dt} ,$$

giving

$$N_A = \frac{n_0 \sigma_A N}{\frac{1}{\tau} + 2n_0 \sigma_A} \approx n_0 \sigma_A N \tau , \qquad (N = N_A + N_{BC})$$

as $n_0 = 5 \times 10^{-3}/1.6 \times 10^{-19} = 3.1 \times 10^{16}$ cm^{-2} s^{-1} and so $\frac{1}{\tau} \gg 2n_0 \sigma_A$.

Hence the number of atoms per unit volume in energy level A at equilibrium is

$$n = \frac{N_A}{V} = \frac{\tau n_0 \sigma_A N}{V} = \frac{\tau n_0 \sigma_A p}{kT}$$

$$= 2 \times 10^{-8} \times 3.1 \times 10^{16} \times 1.4 \times 10^{-20} \times \frac{0.05 \times 1.333 \times 10^3}{1.38 \times 10^{-16} \times 300}$$

$$= 1.4 \times 10^4 \text{ cm}^{-3},$$

where we have taken the room temperature to be $T = 300$ K.

(b) The probability of atomic decay $A \to B$ is

$$\lambda_1 = \frac{0.1}{\tau}.$$

The wavelength of the radiation emitted in the transition $A \to B$ is given as $\lambda_B = 500$ nm. The corresponding light intensity I per unit volume per unit solid angle is then given by

$$4\pi I = n\lambda_1 hc/\lambda_B,$$

i.e.,

$$I = \frac{nhc}{40\pi\tau\lambda_B} = \frac{1.4 \times 10^4 \times 6.63 \times 10^{-27} \times 3 \times 10^{10}}{40\pi \times 2 \times 10^{-8} \times 500 \times 10^{-7}}$$

$$= 2.2 \times 10^{-2} \text{ erg} \cdot \text{s}^{-1} \text{ sr}^{-1} = 2.2 \times 10^{-9} \text{ W sr}^{-1}.$$

1012

The electric field that an atom experiences from its surroundings within a molecule or crystal can noticeably affect properties of the atomic ground state. An interesting example has to do with the phenomenon of angular momentum quenching in the iron atom of the hem group in the hemoglobin of your blood. Iron and hemoglobin are too complicated, however. So consider an atom containing one valence electron moving in a central atomic potential. It is in an $l = 1$ level. Ignore spin. We ask what happens to this level when the electron is acted on by the external potential arising from the atom's surroundings. Take this external potential to be

$$V_{\text{pert}} = Ax^2 + By^2 - (A+B)z^2$$

(the atomic nucleus is at the origin of coordinates) and treat it to lowest order.

(a) The $l = 1$ level now splits into three distinct levels. As you can confirm (and as we hint at) each has a wave function of the form

$$\Psi = (\alpha x + \beta y + \gamma z)f(r),$$

where $f(r)$ is a common central function and where each level has its own set of constants (α, β, γ), which you will need to determine. Sketch the energy level diagram, specifying the *relative* shifts ΔE in terms of the parameters A and B (i.e., compute the three shifts up to a common factor).

(b) More interesting: Compute the expectation value of L_z, the z component of angular momentum, for each of the three levels.

(Princeton)

Solution:

(a) The external potential field V can be written in the form

$$V = \frac{1}{2}(A+B)r^2 - \frac{3}{2}(A+B)z^2 + \frac{1}{2}(A-B)(x^2 - y^2).$$

The degeneracy of the state $n = 2$, $l = 1$ is 3 in the absence of perturbation, with wave functions

$$\Psi_{210} = \left(\frac{1}{32\pi a^3}\right)^{\frac{1}{2}} \frac{r}{a} \exp\left(-\frac{r}{2a}\right) \cos\theta,$$

$$\Psi_{21\pm1} = \mp \left(\frac{1}{64\pi a^3}\right)^{\frac{1}{2}} \frac{r}{a} \exp\left(-\frac{r}{2a}\right) \exp(\pm i\varphi) \sin\theta,$$

where $a = \hbar^2/\mu e^2$, μ being the reduced mass of the valence electron.

After interacting with the external potential field V, the wave functions change to

$$\Psi = a_1\Psi_{211} + a_2\Psi_{21-1} + a_3\Psi_{210}.$$

Perturbation theory for degenerate systems gives for the perturbation energy E' the following matrix equation:

$$\begin{pmatrix} C + A' - E' & B' & 0 \\ B' & C + A' - E' & 0 \\ 0 & 0 & C + 3A' - E' \end{pmatrix} \begin{pmatrix} a_1 \\ a_2 \\ a_2 \end{pmatrix} = 0\,,$$

where

$$C = \langle \Psi_{211} | \frac{1}{2}(A + B)r^2 | \Psi_{211} \rangle$$

$$= \langle \Psi_{21-1} | \frac{1}{2}(A + B)r^2 | \Psi_{21-1} \rangle$$

$$= \langle \Psi_{210} | \frac{1}{2}(A + B)r^2 | \Psi_{210} \rangle$$

$$= 15a^2(A + B)\,,$$

$$A' = -\langle \Psi_{211} | \frac{3}{2}(A + B)z^2 | \Psi_{211} \rangle$$

$$= -\langle \Psi_{21-1} | \frac{3}{2}(A + B)z^2 | \Psi_{21-1} \rangle$$

$$= -\frac{1}{3} \langle \Psi_{210} | \frac{3}{2}(A + B)z^2 | \Psi_{210} \rangle$$

$$= -9a^2(A + B)\,,$$

$$B' = \langle \Psi_{211} | \frac{1}{2}(A - B)(x^2 - y^2) | \Psi_{21-1} \rangle$$

$$= \langle \Psi_{21-1} | \frac{1}{2}(A - B)(x^2 - y^2) | \Psi_{211} \rangle$$

$$= -\frac{3}{2}a^2(A - B)\,.$$

Setting the determinant of the coefficients to zero, we find the energy corrections

$$E' = C + 3A'\,, C + A' \pm B'\,.$$

For $E' = C + 3A' = -12(A + B)a^2$, the wave function is

$$\Psi_1 = \Psi_{210} = \left(\frac{1}{32\pi a^3} \right)^{\frac{1}{2}} \frac{r}{a} \exp\left(-\frac{r}{2a} \right) \cos\theta = f(r)z\,,$$

where

$$f(r) = \left(\frac{1}{32\pi a^3}\right)^{\frac{1}{2}} \frac{1}{a} \cdot \exp\left(-\frac{r}{2a}\right),$$

corresponding to $\alpha = \beta = 0$, $\gamma = 1$.

For $E' = C + A' + B' = \frac{3}{2}(5A + 3B)a^2$, the wave function is

$$\Psi_2 = \frac{1}{\sqrt{2}}(\Psi_{211} + \Psi_{21-1}) = -i\left(\frac{1}{32\pi a^3}\right)^{\frac{1}{2}} \frac{r}{a} \exp\left(-\frac{r}{2a}\right) \sin\theta \sin\varphi$$

$$= -if(r)y,$$

corresponding to $\alpha = \gamma = 0$, $\beta = -i$.

For $E' = C + A' - B' = \frac{3}{2}(3A + 5B)a^2$, the wave function is

$$\Psi_3 = \frac{1}{\sqrt{2}}(\Psi_{211} - \Psi_{21-1}) = -f(r)x,$$

corresponding to $\alpha = -1$, $\beta = \gamma = 0$.

Thus the unperturbed energy level E_2 is, on the application of the perturbation V, split into three levels:

$$E_2 - 12(A + B)a^2, \quad E_2 + \frac{3}{2}(3A + 5B)a^2, \quad E_2 + \frac{3}{2}(5A + 3B)a^2,$$

as shown in Fig. 1.4.

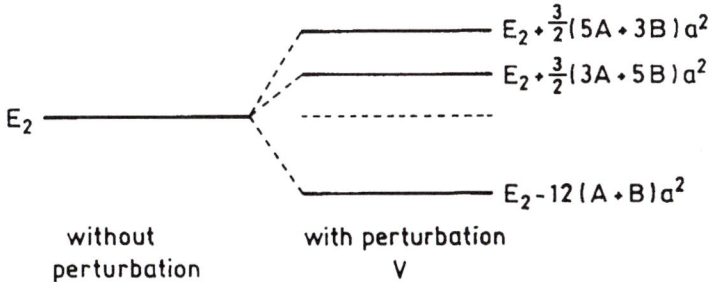

Fig. 1.4

(b) The corrected wave functions give

$$\langle\Psi_1|l_z|\Psi_1\rangle = \langle\Psi_2|l_z|\Psi_2\rangle = \langle\Psi_3|l_z|\Psi_3\rangle = 0.$$

Hence the expectation value of the z component of angular momentum is zero for all the three energy levels.

1013

The Thomas-Fermi model of atoms describes the electron cloud in an atom as a continuous distribution $\rho(x)$ of charge. An individual electron is assumed to move in the potential determined by the nucleus of charge Ze and of this cloud. Derive the equation for the electrostatic potential in the following stages.

(a) By assuming the charge cloud adjusts itself locally until the electrons at Fermi sphere have zero energy, find a relation between the potential ϕ and the Fermi momentum p_F.

(b) Use the relation derived in (a) to obtain an algebraic relation between the charge density $\rho(x)$ and the potential $\phi(x)$.

(c) Insert the result of (b) in Poisson's equation to obtain a nonlinear partial differential equation for ϕ.

(Princeton)

Solution:

(a) For a bound electron, its energy $E = \frac{p^2}{2m} - e\phi(\mathbf{x})$ must be lower than that of an electron on the Fermi surface. Thus

$$\frac{p_{\text{max}}^2}{2m} - e\phi(\mathbf{x}) = 0 ,$$

where $p_{\text{max}} = p_f$, the Fermi momentum.

Hence

$$p_f^2 = 2me\phi(\mathbf{x}) .$$

(b) Consider the electrons as a Fermi gas. The number of electrons filling all states with momenta 0 to p_f is

$$N = \frac{V p_f^3}{3\pi^2 \hbar^3} .$$

The charge density is then

$$\rho(\mathbf{x}) = \frac{eN}{V} = \frac{e p_f^3}{3\pi^2 \hbar^3} = \frac{e}{3\pi^2 \hbar^3} [2me\phi(\mathbf{x})]^{\frac{3}{2}} .$$

(c) Substituting $\rho(\mathbf{x})$ in Poisson's equation

$$\nabla^2 \phi = 4\pi \rho(\mathbf{x})$$

gives

$$\left(\frac{\partial^2}{\partial x^2} + \frac{\partial^2}{\partial y^2} + \frac{\partial^2}{\partial z^2} \right) \phi(\mathbf{x}) = \frac{4e}{3\pi\hbar^3} [2me\phi(\mathbf{x})]^{\frac{3}{2}} .$$

On the assumption that ϕ is spherically symmetric, the equation reduces to

$$\frac{1}{r} \frac{d^2}{dr^2} [r\phi(r)] = \frac{4e}{3\pi\hbar^3} [2me\phi(r)]^{\frac{3}{2}} .$$

1014

In a crude picture, a metal is viewed as a system of free electrons enclosed in a well of potential difference V_0. Due to thermal agitation, electrons with sufficiently high energies will escape from the well. Find and discuss the emission current density for this model.

(SUNY, Buffalo)

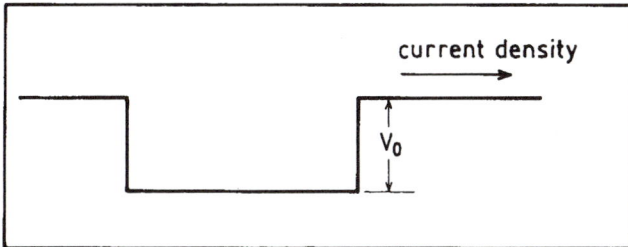

Fig. 1.5

Solution:

The number of states in volume element $dp_x dp_y dp_z$ in the momentum space is $dN = \frac{2}{\hbar^3} dp_x dp_y dp_z$. Each state ε has degeneracy $\exp(-\frac{\varepsilon - \mu}{kT})$, where ε is the energy of the electron and μ is the Fermi energy.

Only electrons with momentum component $p_z > (2mV_0)^{1/2}$ can escape from the potential well, the z-axis being selected parallel to the outward normal to the surface of the metal. Hence the number of electrons escaping from the volume element in time interval dt is

$$dN' = Av_z dt \frac{2}{h^3} dp_x dp_y dp_z \exp\left(-\frac{\varepsilon - \mu}{kT}\right) ,$$

where v_z is the velocity component of the electrons in the z direction which satisfies the condition $mv_z > (2mV_0)^{1/2}$, A is the area of the surface of the metal. Thus the number of electrons escaping from the metal surface per unit area per unit time is

$$R = \int_{-\infty}^{+\infty} \int_{-\infty}^{+\infty} \int_{(2mV_0)^{1/2}}^{+\infty} \frac{2v_z}{h^3} \exp\left(-\frac{\varepsilon - \mu}{kT}\right) dp_x dp_y dp_z$$

$$= \frac{2}{mh^3} \exp\left(\frac{\mu}{kT}\right) \int_{-\infty}^{+\infty} \exp\left(-\frac{p_x^2}{2mkT}\right) dp_x \int_{-\infty}^{+\infty} \exp\left(-\frac{p_y^2}{2mkT}\right) dp_y$$

$$\times \int_{(2mV_0)^{1/2}}^{+\infty} p_z \exp\left(-\frac{p_z^2}{2mkT}\right) dp_z$$

$$= \frac{4\pi mk^2 T^2}{h^3} \exp\left(\frac{\mu - V_0}{kT}\right) ,$$

and the emission current density is

$$J = -eR = -\frac{4\pi mek^2 T^2}{h^3} \exp\left(\frac{\mu - V_0}{kT}\right) ,$$

which is the Richardson–Dushman equation.

1015

A narrow beam of neutral particles with spin $1/2$ and magnetic moment μ is directed along the x-axis through a "Stern-Gerlach" apparatus, which splits the beam according to the values of μ_z in the beam. (The apparatus consists essentially of magnets which produce an inhomogeneous field $B_z(z)$ whose force on the particle moments gives rise to displacements Δz proportional to $\mu_z B_z$.)

(a) Describe the pattern of splitting for the cases:
(i) Beam polarized along $+z$ direction.

(ii) Beam polarized along $+x$ direction.

(iii) Beam polarized along $+y$ direction.

(iv) Beam unpolarized.

(b) For those cases, if any, with indistinguishable results, describe how one might distinguish among these cases by further experiments which use the above Stern-Gerlach apparatus and possibly some additional equipment.

(*Columbia*)

Solution:

(a) (i) The beam polarized along $+z$ direction is not split, but its direction is changed.

(ii) The beam polarized along $+x$ direction splits into two beams, one deflected to $+z$ direction, the other to $-z$ direction.

(iii) Same as for (ii).

(iv) The unpolarized beam is split into two beams, one deflected to $+z$ direction, the other to $-z$ direction.

(b) The beams of (ii) (iii) (iv) are indistinguishable. They can be distinguished by the following procedure.

(1) Turn the magnetic field to $+y$ direction. This distinguishes (iii) from (ii) and (iv), as the beam in (iii) is not split but deflected, while the beams of (ii) and (iv) each splits into two.

(2) Put a reflector in front of the apparatus, which changes the relative positions of the source and apparatus (Fig. 1.6). Then the beam of (ii) does not split, though deflected, while that of (iv) splits into two.

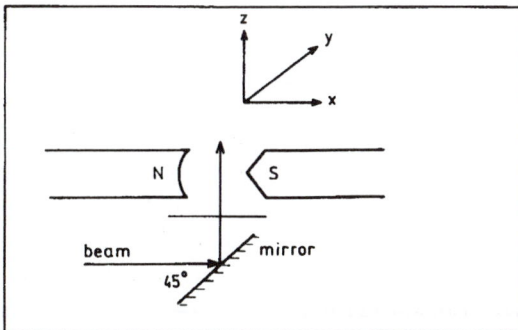

Fig. 1.6

1016

The range of the potential between two hydrogen atoms is approximately 4 Å. For a gas in thermal equilibrium, obtain a numerical estimate of the temperature below which the atom-atom scattering is essentially s-wave.

(*MIT*)

Solution:

The scattered wave is mainly s-wave when $ka \leq 1$, where a is the interaction length between hydrogen atoms, k the de Broglie wave number

$$k = \frac{p}{\hbar} = \frac{\sqrt{2mE_k}}{\hbar} = \frac{\sqrt{2m \cdot \frac{3}{2}k_B T}}{\hbar} = \frac{\sqrt{3mk_B T}}{\hbar},$$

where p is the momentum, E_k the kinetic energy, and m the mass of the hydrogen atom, and k_B is the Boltzmann constant. The condition

$$ka = \sqrt{3mk_B T} \cdot \frac{a}{\hbar} \leq 1$$

gives

$$T \leq \frac{\hbar^2}{3mk_B a^2} = \frac{(1.06 \times 10^{-34})^2}{3 \times 1.67 \times 10^{-27} \times 1.38 \times 10^{-23} \times (4 \times 10^{-10})^2}$$

$$\approx 1 \text{ K}$$

1017

(a) If you remember it, write down the differential cross section for Rutherford scattering in cm^2/sr. If you do not remember it, say so, and write down your best guess. Make sure that the Z dependence, energy dependence, angular dependence and dimensions are "reasonable". Use the expression you have just given, whether correct or your best guess, to evaluate parts (b–e) below.

An accelerator supplies a proton beam of 10^{12} particles per second and 200 MeV/c momentum. This beam passes through a 0.01-cm aluminum window. (Al density $\rho = 2.7$ gm/cm^3, Al radiation length $x_0 = 24$ gm/cm^2, $Z = 13$, $A = 27$).

(b) Compute the differential Rutherford scattering cross section in cm^2/sr at 30° for the above beam in Al.

(c) How many protons per second will enter a 1-cm radius circular counter at a distance of 2 meters and at an angle of 30° with the beam direction?

(d) Compute the integrated Rutherford scattering cross section for angles greater than 5°. (Hint: $\sin\theta d\theta = 4\sin\frac{\theta}{2}\cos\frac{\theta}{2}d\frac{\theta}{2}$)

(e) How many protons per second are scattered out of the beam into angles $> 5°$?

(f) Compute the projected rms multiple Coulomb scattering angle for the proton beam through the above window. Take the constant in the expression for multiple Coulomb scattering as 15 MeV/c.

(*UC, Berkeley*)

Solution:

(a) The differential cross section for Rutherford scattering is

$$\frac{d\sigma}{d\Omega} = \left(\frac{zZe^2}{2mv^2}\right)^2 \left(\sin\frac{\theta}{2}\right)^{-4}.$$

This can be obtained, to a dimensionless constant, if we remember

$$\frac{d\sigma}{d\Omega} \sim \left(\sin\frac{\theta}{2}\right)^{-4},$$

and assume that it depends also on ze, Ze and $E = \frac{1}{2}mv^2$.

Let

$$\frac{d\sigma}{d\Omega} = K(zZe^2)^x E^y \left(\sin\frac{\theta}{2}\right)^{-4},$$

where K is a dimensionless constant. Dimensional analysis then gives

$$[L]^2 = (e^2)^x E^y.$$

As

$$\left[\frac{e^2}{r}\right] = [E],$$

the above gives

$$x = 2, y = -x = -2 \,.$$

(b) For the protons,

$$\beta \equiv \frac{v}{c} = \frac{pc}{\sqrt{m^2 c^4 + p^2 c^2}} = \frac{200}{\sqrt{938^2 + 200^2}} = 0.2085 \,.$$

We also have

$$\frac{e^2}{mv^2} = r_0 \left(\frac{m_e}{m}\right) \left(\frac{v}{c}\right)^{-2} \,,$$

where $r_0 = \frac{e^2}{m_e c^2} = 2.82 \times 10^{-13}$ cm is the classical radius of electron. Hence at $\theta = 30°$,

$$\frac{d\sigma}{d\Omega} = \left(\frac{13}{2}\right)^2 r_0^2 \left(\frac{m_e}{m}\right)^2 \left(\frac{v}{c}\right)^{-4} \left(\sin \frac{\theta}{2}\right)^{-4}$$

$$= \left(\frac{6.5 \times 2.82 \times 10^{-13}}{1836 \times 0.2085^2}\right)^2 \times (\sin 15°)^{-4}$$

$$= 5.27 \times 10^{-28} \times (\sin 15°)^{-4} = 1.18 \times 10^{-25} \text{ cm}^2/\text{sr} \,.$$

(c) The counter subtends a solid angle

$$d\Omega = \frac{\pi (0.01)^2}{2^2} = 7.85 \times 10^{-5} \text{ sr} \,.$$

The number of protons scattered into it in unit time is

$$\delta n = n \left(\frac{\rho t}{27}\right) A_v \left(\frac{d\sigma}{d\Omega}\right) \delta\Omega$$

$$= 10^{12} \times \left(\frac{2.7 \times 0.01}{27}\right) \times 6.02 \times 10^{23} \times 1.18 \times 10^{-25} \times 7.85 \times 10^{-5}$$

$$= 5.58 \times 10^3 \text{ s}^{-1} \,.$$

(d)

$$\sigma_I = \int \frac{d\sigma}{d\Omega} d\Omega = 2\pi \int_{5°}^{180°} \left(\frac{Ze^2}{2mv^2} \right)^2 \frac{\sin\theta}{\sin^4 \frac{\theta}{2}} d\theta$$

$$= 8\pi \left(\frac{Ze^2}{2mv^2} \right)^2 \int_{5°}^{180°} \left(\sin \frac{\theta}{2} \right)^{-3} d\sin \frac{\theta}{2}$$

$$= 4\pi \left(\frac{Ze^2}{2mv^2} \right)^2 \left[\frac{1}{(\sin 2.5°)^2} - 1 \right]$$

$$= 4\pi \times 5.27 \times 10^{-28} \times \left[\frac{1}{(\sin 2.5°)^2} - 1 \right]$$

$$= 3.47 \times 10^{-24} \text{ cm}^2 .$$

(e) The number of protons scattered into $\theta \geq 5°$ is

$$\delta n = n \left(\frac{\rho t}{27} \right) A_v \sigma_I = 2.09 \times 10^9 \text{ s}^{-1} ,$$

where $A_v = 6.02 \times 10^{23}$ is Avogadro's number.

(f) The projected rms multiple Coulomb scattering angle for the proton beam through the Al window is given by

$$\theta_{\text{rms}} = \frac{kZ}{\sqrt{2}\beta p} \sqrt{\frac{t}{x_0}} \left[1 + \frac{1}{9} \ln \left(\frac{t}{x_0} \right) \right] ,$$

where k is a constant equal to 15 MeV/c. As $Z = 13$, $p = 200$ MeV/c, $\beta = 0.2085$, $t = 0.01 \times 2.7$ g cm^{-2}, $x_0 = 24$ g cm^{-2}, $t/x_0 = 1.125 \times 10^{-3}$, we have

$$\theta_{\text{rms}} = \frac{15 \times 13}{\sqrt{2} \times 0.2085 \times 200} \times \sqrt{1.125 \times 10^{-3}} \left[1 + \frac{1}{9} \ln(1.125 \times 10^{-3}) \right]$$

$$= 2.72 \times 10^{-2} \text{ rad} .$$

1018

Typical lifetime for an excited atom is $10^{-1}, 10^{-8}, 10^{-13}, 10^{-23}$ sec.

(*Columbia*)

Solution:

The answer is 10^{-8} s.

1019

An atom is capable of existing in two states: a ground state of mass M and an excited state of mass $M + \Delta$. If the transition from ground to excited state proceeds by the absorption of a photon, what must be the photon frequency in the laboratory where the atom is initially at rest?

(Wisconsin)

Solution:

Let the frequency of the photon be ν and the momentum of the atom in the excited state be p. The conservation laws of energy and momentum give

$$Mc^2 + h\nu = [(M + \Delta)^2 c^4 + p^2 c^2]^{1/2},$$

$$\frac{h\nu}{c} = p,$$

and hence

$$\nu = \frac{\Delta c^2}{h} \left(1 + \frac{\Delta}{2M}\right).$$

1020

If one interchanges the spatial coordinates of two electrons in a state of total spin 0:

(a) the wave function changes sign,
(b) the wave function is unchanged,
(c) the wave function changes to a completely different function.

(CCT)

Solution:

The state of total spin zero has even parity, i.e., spatial symmetry. Hence the wave function does not change when the space coordinates of the electrons are interchanged.

So the answer is (b).

1021

The Doppler width of an optical line from an atom in a flame is 10^6, 10^9, 10^{13}, 10^{16} Hz.

(Columbia)

Solution:

Recalling the principle of equipartition of energy $m\overline{v^2}/2 = 3kT/2$ we have for hydrogen at room temperature $mc^2 \approx 10^9$ eV, $T = 300$ K, and so

$$\beta = \frac{v}{c} \approx \frac{\sqrt{\overline{v^2}}}{c} = \sqrt{\frac{3kT}{mc^2}} \sim 10^{-5},$$

where $k = 8.6 \times 10^{-5}$ eV/K is Boltzmann's constant.

The Doppler width is of the order

$$\Delta\nu \approx \nu_0\beta.$$

For visible light, $\nu_0 \sim 10^{14}$ Hz. Hence $\Delta\nu \sim 10^9$ Hz.

1022

Estimate (order of magnitude) the Doppler width of an emission line of wavelength $\lambda = 5000$ Å emitted by argon $A = 40$, $Z = 18$, at $T = 300$ K.

(Columbia)

Solution:

The principle of equipartition of energy $\frac{1}{2}m\bar{v}^2 = \frac{3}{2}kT$ gives

$$v \approx \sqrt{\overline{v^2}} = c\sqrt{\frac{3kT}{mc^2}}$$

with $mc^2 = 40 \times 938$ MeV, $kT = 8.6 \times 10^{-5} \times 300 = 2.58 \times 10^{-2}$ eV. Thus

$$\beta = \frac{v}{c} = 1.44 \times 10^{-6}$$

and the (full) Doppler width is

$$\Delta\lambda \approx 2\beta\lambda = 1.44 \times 10^{-2} \text{ Å}.$$

1023

Typical cross section for low-energy electron-atom scattering is 10^{-16}, 10^{-24}, 10^{-32}, 10^{-40} cm².

(Columbia)

Solution:

The linear dimension of an atom is of the order 10^{-8} cm, so the cross section is of the order $(10^{-8})^2 = 10^{-16}$ cm².

1024

An electron is confined to the interior of a hollow spherical cavity of radius R with impenetrable walls. Find an expression for the pressure exerted on the walls of the cavity by the electron in its ground state.

(MIT)

Solution:

Suppose the radius of the cavity is to increase by dR. The work done by the electron in the process is $4\pi R^2 P dR$, causing a decrease of its energy by dE. Hence the pressure exerted by the electron on the walls is

$$P = -\frac{1}{4\pi R^2}\frac{dE}{dR}.$$

For the electron in ground state, the angular momentum is 0 and the wave function has the form

$$\Psi = \frac{1}{\sqrt{4\pi}}\frac{\chi(r)}{r},$$

where $\chi(r)$ is the solution of the radial part of Schrödinger's equation,

$$\chi''(r) + k^2\chi(r) = 0,$$

with $k^2 = 2mE/\hbar^2$ and $\chi(r) = 0$ at $r = 0$. Thus

$$\chi(r) = A\sin kr.$$

As the walls cannot be penetrated, $\chi(r) = 0$ at $r = R$, giving $k = \pi/R$. Hence the energy of the electron in ground state is

$$E = \frac{\pi^2\hbar^2}{2mR^2},$$

and the pressure is

$$P = -\frac{1}{4\pi R^2}\frac{dE}{dR} = \frac{\pi\hbar^2}{4mR^5}.$$

1025

A particle with magnetic moment $\boldsymbol{\mu} = \mu_0 \mathbf{s}$ and spin $\mathbf{s}$ of magnitude $1/2$ is placed in a constant magnetic field $\mathbf{B}$ pointing along the x-axis. At $t = 0$, the particle is found to have $s_z = +1/2$. Find the probabilities at any later time of finding the particle with $s_y = \pm 1/2$.

(*Columbia*)

Solution:

In the representation $(\mathbf{s}^2, s_x)$, the spin matrices are

$$\sigma_x = \begin{pmatrix} 1 & 0 \\ 0 & -1 \end{pmatrix}, \qquad \sigma_y = \begin{pmatrix} 0 & 1 \\ 1 & 0 \end{pmatrix}, \qquad \sigma_z = \begin{pmatrix} 0 & -i \\ i & 0 \end{pmatrix}$$

with eigenfunctions $\binom{1}{0}$, $\binom{1}{1}$, $\binom{1}{i}$ respectively. Thus the Hamiltonian of interaction between the magnetic moment of the particle and the magnetic field is

$$H = -\boldsymbol{\mu} \cdot \mathbf{B} = -\frac{\mu_0 B}{2}\begin{pmatrix} 1 & 0 \\ 0 & -1 \end{pmatrix},$$

and the Schrödinger equation is

$$i\hbar\frac{d}{dt}\begin{pmatrix} a(t) \\ b(t) \end{pmatrix} = -\frac{\mu_0 B}{2}\begin{pmatrix} 1 & 0 \\ 0 & -1 \end{pmatrix}\begin{pmatrix} a(t) \\ b(t) \end{pmatrix},$$

where $\begin{pmatrix} a(t) \\ b(t) \end{pmatrix}$ is the wave function of the particle at time t. Initially we have $\begin{pmatrix} a(0) \\ b(0) \end{pmatrix} = \frac{1}{\sqrt{2}}\binom{1}{i}$, and so the solution is

$$\begin{pmatrix} a(t) \\ b(t) \end{pmatrix} = \frac{1}{\sqrt{2}}\begin{pmatrix} \exp\left(i\dfrac{\mu_0 B t}{2\hbar}\right) \\ i\exp\left(-i\dfrac{\mu_0 B t}{2\hbar}\right) \end{pmatrix}.$$

Hence the probability of the particle being in the state $s_y = +1/2$ at time t is

$$\left| \frac{1}{\sqrt{2}}(1\ 1)\begin{pmatrix} a(t) \\ b(t) \end{pmatrix} \right|^2 = \frac{1}{4}\left| \exp\left(i\frac{\mu_0 Bt}{2\hbar}\right) + i\exp\left(-i\frac{\mu_0 Bt}{2\hbar}\right) \right|^2$$

$$= \frac{1}{2}\left(1 + \sin\frac{\mu_0 Bt}{\hbar}\right).$$

Similarly, the probability of the particle being in the state $s_y = -1/2$ at time t is $\frac{1}{2}(1 - \sin\frac{\mu_0 Bt}{\hbar})$.

1026

The ground state of the realistic helium atom is of course nondegenerate. However, consider a hypothetical helium atom in which the two electrons are replaced by two identical spin-one particles of negative charge. Neglect spin-dependent forces. For this hypothetical atom, what is the degeneracy of the ground state? Give your reasoning.

(CUSPEA)

Solution:

Spin-one particles are bosons. As such, the wave function must be symmetric with respect to interchange of particles. Since for the ground state the spatial wave function is symmetric, the spin part must also be symmetric. For two spin-1 particles the total spin S can be 2, 1 or 0. The spin wave functions for $S = 2$ and $S = 0$ are symmetric, while that for $S = 1$ is antisymmetric. Hence for ground state we have $S = 2$ or $S = 0$, the total degeneracy being

$$(2 \times 2 + 1) + (2 \times 0 + 1) = 6.$$

1027

A beam of neutrons (mass m) traveling with nonrelativistic speed v impinges on the system shown in Fig. 1.7, with beam-splitting mirrors at corners B and D, mirrors at A and C, and a neutron detector at E. The corners all make right angles, and neither the mirrors nor the beam-splitters affect the neutron spin. The beams separated at B rejoin coherently at D, and the detector E reports the neutron intensity I.

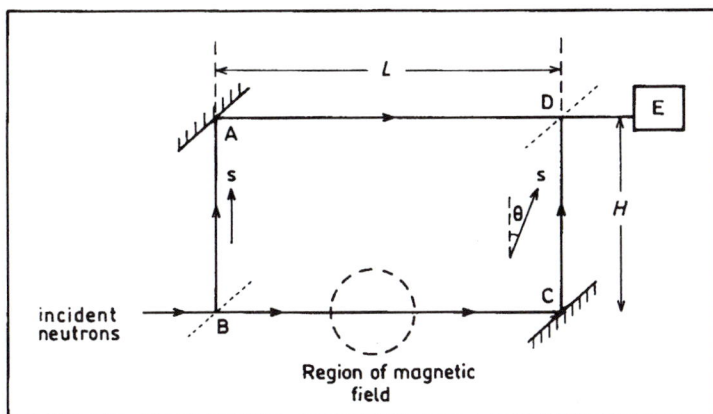

Fig. 1.7

(a) In this part of the problem, assume the system to be in a vertical plane (so gravity points down parallel to AB and DC). Given that detector intensity was I_0 with the system in a horizontal plane, derive an expression for the intensity I_g for the vertical configuration.

(b) For this part of the problem, suppose the system lies in a horizontal plane. A uniform magnetic field, pointing out of the plane, acts in the dotted region indicated which encompasses a portion of the leg BC. The incident neutrons are polarized with spin pointing along BA as shown. The neutrons which pass through the magnetic field region will have their spins pressed by an amount depending on the field strength. Suppose the spin expectation value presses through an angle θ as shown. Let $I(\theta)$ be the intensity at the detector E. Derive $I(\theta)$ as a function of θ, given that $I(\theta = 0) = I_0$.

(*Princeton*)

Solution:

(a) Assume that when the system is in a horizontal plane the two split beams of neutrons have the same intensity when they reach D, and so the wave functions will each have amplitude $\sqrt{I_0}/2$. Now consider the system in a vertical plane. As BA and CD are equivalent dynamically, they need not be considered. The velocities of neutrons v in BC and v_1 in AD are related through the energy equation

$$\frac{1}{2}mv^2 = \frac{1}{2}mv_1^2 + mgH \,,$$

giving

$$v_1 = \sqrt{v^2 - 2gH} \,.$$

When the two beams recombine at D, the wave function is

$$\Psi = \left[\frac{\sqrt{I_0}}{2} \exp\left(i\frac{mv_1}{\hbar}L \right) + \frac{\sqrt{I_0}}{2} \exp\left(i\frac{mv}{\hbar}L \right) \right] \exp\left(-i\frac{Et}{\hbar} \right) \exp(i\delta) \,,$$

and the intensity is

$$I_g = |\Psi|^2 = \frac{I_0}{2} + \frac{I_0}{2} \cos\left[\frac{mL(v - v_1)}{\hbar} \right] = I_0 \cos^2\left[\frac{mL(v - v_1)}{2\hbar} \right] \,.$$

If we can take $\frac{1}{2}mv^2 \gg mgH$, then $v_1 \approx v - \frac{gH}{v}$ and

$$I_g \approx I_0 \cos^2\left(\frac{mgHL}{2\hbar v} \right) \,.$$

(b) Take z-axis in the direction of BA and proceed in the representation of $(\mathbf{s}^2, s_z)$. At D the spin state is $\binom{1}{0}$ for neutrons proceeding along BAD and is $\begin{pmatrix} \cos\frac{\theta}{2} \\ \sin\frac{\theta}{2} \end{pmatrix}$ for those proceeding along BCD. Recombination gives

$$\Psi = \frac{\sqrt{I_0}}{2} \exp\left(-i\frac{Et}{\hbar} \right) \exp(i\delta) \left[\binom{1}{0} + \begin{pmatrix} \cos\frac{\theta}{2} \\ \sin\frac{\theta}{2} \end{pmatrix} \right]$$

$$= \frac{\sqrt{I_0}}{2} \exp\left(-i\frac{Et}{\hbar} \right) \exp(i\delta) \begin{pmatrix} 1 + \cos\frac{\theta}{2} \\ \sin\frac{\theta}{2} \end{pmatrix} \,,$$

and hence

$$I(\theta) = |\Psi|^2 = \frac{I_0}{4} \left[\left(1 + \cos\frac{\theta}{2} \right)^2 + \sin^2\frac{\theta}{2} \right] = I_0 \cos^2\frac{\theta}{4} \,.$$

1028

The fine structure of atomic spectral lines arises from

(a) electron spin-orbit coupling.
(b) interaction between electron and nucleus.
(c) nuclear spin.

(CCT)

Solution:

The answer is (a).

1029

Hyperfine splitting in hydrogen ground state is 10^{-7}, 10^{-5}, 10^{-3}, 10^{-1} eV.

(Columbia)

Solution:

For atomic hydrogen the experimental hyperfine line spacing is $\Delta\nu_{hf} = 1.42 \times 10^9$ s^{-1}. Thus $\Delta E = h\nu_{hf} = 4.14 \times 10^{-15} \times 1.42 \times 10^9 = 5.9 \times 10^{-6}$ eV. So the answer is 10^{-5} eV.

1030

The hyperfine structure of hydrogen

(a) is too small to be detected.
(b) arises from nuclear spin.
(c) arises from finite nuclear size.

(CCT)

Solution:

The answer is (b).

1031

Spin-orbit splitting of the hydrogen 2p state is 10^{-6}, 10^{-4}, 10^{-2}, 10^0 eV.

(Columbia)

Solution:

For the $2p$ state of hydrogen atom, $n = 2$, $l = 1$, $s = 1/2$, $j_1 = 3/2$, $j_2 = 1/2$. The energy splitting caused by spin-orbit coupling is given by

$$\Delta E_{ls} = \frac{hcR\alpha^2}{n^3 l \left(l + \frac{1}{2}\right)(l+1)} \left[\frac{j_1(j_1+1) - j_2(j_2+1)}{2}\right],$$

where R is Rydberg's constant and $hcR = 13.6$ eV is the ionization potential of hydrogen atom, $\alpha = \frac{1}{137}$ is the fine-structure constant. Thus

$$\Delta E_{ls} = \frac{13.6 \times (137)^{-2}}{2^3 \times \frac{3}{2} \times 2} \times \frac{1}{2}\left(\frac{15}{4} - \frac{3}{4}\right) = 4.5 \times 10^{-5} \text{ eV}.$$

So the answer is 10^{-4} eV.

1032

The Lamb shift is

(a) a splitting between the $1s$ and $2s$ energy levels in hydrogen.
(b) caused by vacuum fluctuations of the electromagnetic field.
(c) caused by Thomas precession.

(CCT)

Solution:

The answer is (b)

1033

The average speed of an electron in the first Bohr orbit of an atom of atomic number Z is, in units of the velocity of light,

(a) $Z^{1/2}$.
(b) Z.
(c) Z/137.

(CCT)

Solution:

Let the average speed of the electron be v, its mass be m, and the radius of the first Bohr orbit be a. As

$$\frac{mv^2}{a} = \frac{Ze^2}{a^2}, \qquad a = \frac{\hbar^2}{mZe^2},$$

We have

$$v = \frac{Ze^2}{\hbar} = Zc\alpha,$$

where $\alpha = \frac{e^2}{\hbar c} = \frac{1}{137}$ is the fine-structure constant. Hence the answer is (c).

1034

The following experiments were significant in the development of quantum theory. Choose TWO. In each case, briefly describe the experiment and state what it contributed to the development of the theory. Give an approximate date for the experiment.

(a) Stern-Gerlach experiment

(b) Compton Effect

(c) Franck-Hertz Experiment

(d) Lamb-Rutherford Experiment

(*Wisconsin*)

Solution:

(a) *Stern-Gerlach experiment.* The experiment was carried out in 1921 by Stern and Gerlach using apparatus as shown in Fig. 1.8. A highly collimated beam ($v \approx 500$ m/s) of silver atoms from an oven passes through the poles of a magnet which are so shaped as to produce an extremely non-uniform field (gradient of field $\sim 10^3$ T/m, longitudinal range ~ 4 cm) normal to the beam. The forces due to the interaction between the component μ_z of the magnetic moment in the field direction and the field gradient cause a deflection of the beam, whose magnitude depends on μ_z. Stern and Gerlach found that the beam split into two, rather than merely broadened, after crossing the field. This provided evidence for the space quantization of the angular momentum of an atom.

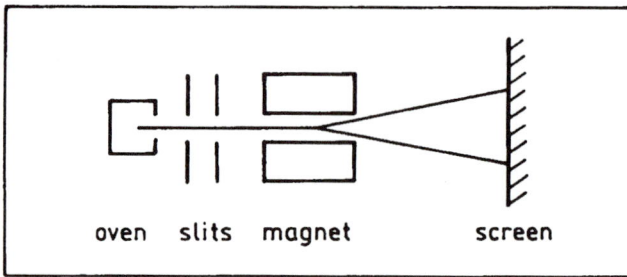

Fig. 1.8

(b) *Compton Effect.* A. H. Compton discovered that when monochromatic X-rays are scattered by a suitable target (Fig. 1.9), the scattered radiation consists of two components, one spectrally unchanged the other with increased wavelength. He further found that the change in wavelength of the latter is a function only of the scattering angle but is independent of the wavelength of the incident radiation and the scattering material. In 1923, using Einstein's hypothesis of light quanta and the conservation of momentum and energy, Compton found a relation between the change of wavelength and the scattering angle, $\Delta\lambda = \frac{h}{m_e c}(1-\cos\theta)$, which is in excellent agreement with the experimental results. Compton effect gives direct support to Einstein's theory of light quanta.

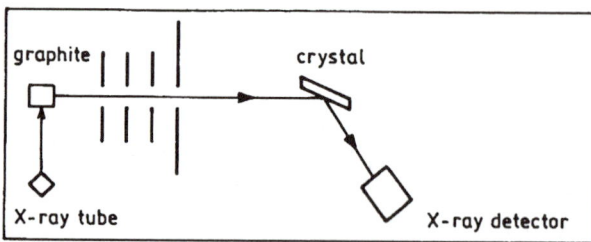

Fig. 1.9

(c) *Franck-Hertz experiment.* Carried out by Franck and Hertz in 1914, this experiment proved Bohr's theory of quantization of atomic energy states as well as provided a method to measure the energy spacing of quantum states. The experimental setup was as shown in Fig. 1.10. A glass

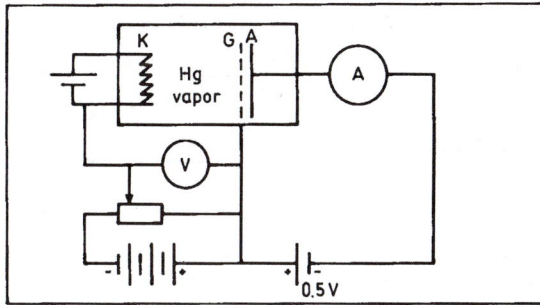

Fig. 1.10

vessel, filled with Hg vapor, contained cathode K, grid G and anode A. Thermoelectrons emitted from K were accelerated by an electric field to G, where a small retarding field prevented low energy electrons from reaching A. It was observed that the electric current detected by the ammeter A first increased with the accelerating voltage until it reached 4.1 V. Then the current dropped suddenly, only to increase again. At the voltages 9.0 V and 13.9 V, similar phenomena occurred. This indicated that the electron current dropped when the voltage increased by 4.9 V (the first drop at 4.1 V was due to the contact voltage of the instrument), showing that 4.9 eV was the first excited state of Hg above ground. With further improvements in the instrumentation Franck and Hertz were able to observe the higher excited states of the atom.

(d) *Lamb-Rutherford Experiment.* In 1947, when Lamb and Rutherford measured the spectrum of H atom accurately using an RF method, they found it different from the predictions of Dirac's theory, which required states with the same (n, j) but different l to be degenerate. Instead, they found a small splitting. The result, known as the Lamb shift, is satisfactorily explained by the interaction between the electron with its radiation field. The experiment has been interpreted as providing strong evidence in support of quantum electrodynamics.

The experimental setup was shown in Fig. 1.11. Of the hydrogen gas contained in a stove, heated to temperature 2500 K, about 64% was ionized (average velocity 8×10^3 m/s). The emitted atomic beam collided at B with a transverse electron beam of energy slightly higher than 10.2 eV and were excited to $2^2S_{1/2}$, $2^2P_{1/2}$, $2^2P_{3/2}$ states. The atoms in the P

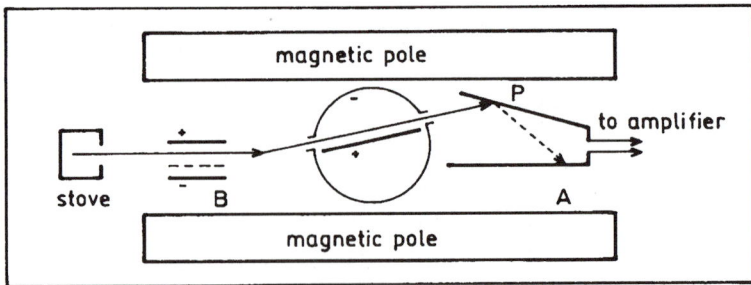

Fig. 1.11

states spontaneously underwent transition to the ground state $1^2S_{1/2}$ almost immediately whereas the $2^2S_{1/2}$ state, which is metastable, remained. Thus the atomic beam consisted of only $2^2S_{1/2}$ and $1^2S_{1/2}$ states when it impinged on the tungsten plate P. The work function of tungsten is less than 10.2 eV, so that the atoms in $2^2S_{1/2}$ state were able to eject electrons from the tungsten plate, which then flowed to A, resulting in an electric current between P and A, which was measured after amplification. The current intensity gave a measure of the numbers of atoms in the $2^2S_{1/2}$ state. A microwave radiation was then applied between the excitation and detection regions, causing transition of the $2^2S_{1/2}$ state to a P state, which almost immediately decayed to the ground state, resulting in a drop of the electric current. The microwave energy corresponding to the smallest electric current is the energy difference between the $2^2S_{1/2}$ and $2^2P_{1/2}$ states. Experimentally the frequency of Lamb shift was found to be 1057 MHz.

1035

(a) Derive from Coulomb's law and the simple quantization of angular momentum, the energy levels of the hydrogen atom.

(b) What gives rise to the doublet structure of the optical spectra from sodium?

(Wisconsin)

Solution:

(a) The Coulomb force between the electron and the hydrogen nucleus is

$$F = \frac{e^2}{4\pi\varepsilon_0 r^2} \cdot$$

In a simplest model, the electron moves around the nucleus in a circular orbit of radius r with speed v, and its orbital angular momentum $p_\phi = mvr$ is quantized according to the condition

$$p_\phi = n\hbar,$$

where $n = 1, 2, 3, \ldots$ and $\hbar = h/2\pi$, h being Planck's constant. For the electron circulating the nucleus, we have

$$m\frac{v^2}{r} = \frac{e^2}{4\pi\varepsilon_0 r^2},$$

and so

$$v = \frac{e^2}{4\pi\varepsilon_0 n\hbar} \cdot$$

Hence the quantized energies are

$$E_n = T + V = \frac{1}{2}mv^2 - \frac{e^2}{4\pi\varepsilon_0 r} = -\frac{1}{2}mv^2$$

$$= -\frac{1}{2}\frac{me^4}{(4\pi\varepsilon_0)^2\hbar^2 n^2},$$

with $n = 1, 2, 3, \ldots$.

(b) The doublet structure of the optical spectra from sodium is caused by the coupling between the orbital and spin angular momenta of the valence electron.

1036

We may generalize the semiclassical Bohr-Sommerfeld relation

$$\oint \mathbf{p} \cdot d\mathbf{r} = \left(n + \frac{1}{2}\right) 2\pi\hbar$$

(where the integral is along a closed orbit) to apply to the case where an electromagnetic field is present by replacing $\mathbf{p} \to \mathbf{p} - \frac{e\mathbf{A}}{c}$. Use this and the equation of motion for the linear momentum **p** to derive a quantized condition on the magnetic flux of a semiclassical electron which is in a magnetic field **B** in an arbitrary orbit. For electrons in a solid this condition can be restated in terms of the size S of the orbit in k-space. Obtain the quantization condition on S in terms of B. (Ignore spin effects)

(Chicago)

Solution:

Denote the closed orbit by C. Assume **B** is constant, then Newton's second law

$$\frac{d\mathbf{p}}{dt} = -\frac{e}{c}\frac{d\mathbf{r}}{dt} \times \mathbf{B}$$

gives

$$\oint_C \mathbf{p} \cdot d\mathbf{r} = -\frac{e}{c}\oint_C (\mathbf{r} \times \mathbf{B}) \cdot d\mathbf{r} = \frac{e}{c}\oint_C \mathbf{B} \cdot \mathbf{r} \times d\mathbf{r} = \frac{2e}{c}\int_S \mathbf{B} \cdot d\mathbf{S} = \frac{2e}{c}\Phi,$$

where Φ is the magnetic flux crossing a surface S bounded by the closed orbit. We also have, using Stokes' theorem,

$$-\frac{e}{c}\oint_C \mathbf{A} \cdot d\mathbf{r} = -\frac{e}{c}\int_S (\nabla \times \mathbf{A}) \cdot d\mathbf{S} = -\frac{e}{c}\int_S \mathbf{B} \cdot d\mathbf{S} = -\frac{e}{c}\Phi.$$

Hence

$$\oint \left(\mathbf{p} - \frac{e}{c}\mathbf{A}\right) \cdot d\mathbf{r} = \oint_C \mathbf{p} \cdot d\mathbf{r} - \frac{e}{c}\oint_C \mathbf{A} \cdot d\mathbf{r} = \frac{2e}{c}\Phi - \frac{e}{c}\Phi = \frac{e}{c}\Phi.$$

The generalized Bohr-Sommerfeld relation then gives

$$\Phi = \left(n + \frac{1}{2}\right)\frac{2\pi\hbar c}{e},$$

which is the quantization condition on the magnetic flux.

On a plane perpendicular to **B**,

$$\Delta p \equiv \hbar\Delta k = \frac{e}{c}B\Delta r,$$

i.e.,

$$\Delta r = \frac{\hbar c}{eB}\Delta k.$$

Hence the orbital area S in k-space and A in r-space are related by

$$A = \left(\frac{\hbar c}{eB}\right)^2 S.$$

Using the quantization condition on magnetic flux, we have

$$A = \frac{\Phi}{B} = \left(n + \frac{1}{2}\right)\frac{2\pi\hbar c}{eB},$$

or

$$\left(\frac{\hbar c}{eB}\right)^2 S = \left(n + \frac{1}{2}\right)\frac{2\pi\hbar c}{eB}.$$

Therefore the quantization condition on the orbital area S in k-space is

$$S = \left(n + \frac{1}{2}\right)\frac{2\pi e}{\hbar c}B.$$

1037

If a very small uniform-density sphere of charge is in an electrostatic potential $V(\mathbf{r})$, its potential energy is

$$U(\mathbf{r}) = V(\mathbf{r}) + \frac{r_0^2}{6}\nabla^2 V(\mathbf{r}) + \cdots$$

where $\mathbf{r}$ is the position of the center of the charge and r_0 is its very small radius. The "Lamb shift" can be thought of as the small correction to the energy levels of the hydrogen atom because the physical electron does have this property.

If the r_0^2 term of U is treated as a very small perturbation compared to the Coulomb interaction $V(\mathbf{r}) = -e^2/r$, what are the Lamb shifts for the $1s$ and $2p$ levels of the hydrogen atom? Express your result in terms of r_0 and fundamental constants. The unperturbed wave functions are

$$\psi_{1s}(\mathbf{r}) = 2a_B^{-3/2}\exp(-r/a_B)Y_0^0,$$

$$\psi_{2pm}(\mathbf{r}) = a_B^{-5/2}r\exp(-r/2a_B)Y_1^m/\sqrt{24},$$

where $a_B = \hbar^2/m_e e^2$.

(CUSPEA)

Solution:

As

$$\nabla^2 V(\mathbf{r}) = -e^2 \nabla^2 \frac{1}{r} = 4\pi e^2 \delta(\mathbf{r}),$$

where $\delta(\mathbf{r})$ is Dirac's delta function defined by

$$\nabla^2 \frac{1}{r} = -4\pi \delta(\mathbf{r}),$$

we have

$$\int \psi^* \nabla^2 V(\mathbf{r}) \psi d^3 \mathbf{r} = 4\pi e^2 \int \psi^*(\mathbf{r}) \psi(\mathbf{r}) \delta(\mathbf{r}) d^3 \mathbf{r} = 4\pi e^2 \psi^*(0) \psi(0).$$

Hence

$$\Delta E_{1s} = \frac{r_0^2}{6} \cdot 4\pi e^2 \psi_{1s}^*(0) \psi_{1s}(0)$$

$$= \frac{r_0^2}{6} \cdot 4\pi e^2 \cdot 4a_B^{-3} = \frac{8\pi e^2 r_0^2}{3} a_B^{-3},$$

$$\Delta E_{2p} = \frac{r_0^2}{6} \cdot 4\pi e^2 \psi_{2p}^*(0) \psi_{2p}(0) = 0.$$

1038

(a) Specify the dominant multipole (such as E1 (electric dipole), E2, E3 ..., M1, M2, M3...) for spontaneous photon emission by an excited atomic electron in each of the following transitions,

$$2p_{1/2} \to 1s_{1/2},$$

$$2s_{1/2} \to 1s_{1/2},$$

$$3d_{3/2} \to 2s_{1/2},$$

$$2p_{3/2} \to 2p_{1/2},$$

$$3d_{3/2} \to 2p_{1/2}.$$

(b) Estimate the transition rate for the first of the above transitions in terms of the photon frequency ω, the atomic radius a, and any other necessary physical constants. Give a rough numerical estimate of this rate for a typical atomic transition.

(c) Estimate the ratios of the other transition rates (for the other transitions in (a)) relative to the first one in terms of the same parameters as in (b).

(UC, Berkeley)

Solution:

(a) In multipole transitions for spontaneous photon emission, angular momentum conservation requires

$$|j_i - j_f| \leq L \leq j_i + j_f\,,$$

L being the order of transition, parity conservation requires

$$\Delta P = (-1)^L \text{ for electric multipole radiation}\,,$$

$$\Delta P = (-1)^{L+1} \text{ for magnetic multipole radiation}\,.$$

Transition with the smallest order L is the most probable. Hence for

$$2p_{1/2} \rightarrow 1s_{1/2} : L = 1, \Delta P = -, \text{ transition is E1}\,,$$

$$2s_{1/2} \rightarrow 1s_{1/2} : L = 0, \Delta P = +,$$

transition is a double-photon dipole transition,

$$3d_{3/2} \rightarrow 2s_{1/2} : L = 1, 2, \Delta P = +, \text{ transition is M1 or E2}\,,$$

$$2p_{3/2} \rightarrow 2p_{1/2} : L = 1, 2, \Delta P = +, \text{ transition is M1 or E2}\,,$$

$$3d_{3/2} \rightarrow 2p_{1/2} : L = 1, 2, \Delta P = -, \text{ transition is E1}\,.$$

(b) The probability of spontaneous transition from $2p_{1/2}$ to $1s_{1/2}$ per unit time is

$$A_{E1} = \frac{e^2 \omega^3}{3\pi \varepsilon_0 \hbar c^3} |\mathbf{r}_{12}|^2 = \frac{4}{3} \alpha \omega^3 \left(\frac{|\mathbf{r}_{12}|}{c} \right)^2,$$

where $\alpha = e^2/(4\pi\varepsilon_0 \hbar c) = 1/137$ is the fine-structure constant. As $|\mathbf{r}_{12}| \approx a$,

$$A_{E1} \approx \frac{4}{3} \alpha \omega^3 \left(\frac{a}{c} \right)^2.$$

With $a \sim 10^{-10}$ m, $\omega \sim 10^{16}$ s^{-1}, we have $A_{E1} \sim 10^9$ s^{-1}.

(c)

$$\frac{A(2^2s_{\frac{1}{2}} \to 1^2s_{\frac{1}{2}})}{A_{E1}} \approx 10 \left(\frac{\hbar}{mc\alpha}\right),$$

$$\frac{A(3d_{\frac{3}{2}} \to 2s_{\frac{1}{2}})}{A_{E1}} \approx (ka)^2,$$

$$\frac{A(2p_{\frac{3}{2}} - 2p_{\frac{1}{2}})}{A_{E1}} \approx (ka)^2,$$

where $k = \omega/c$ is the wave number of the photon,

$$\frac{A(3d_{3/2} \to 2p_{1/2})}{A_{E1}} \approx \left[\frac{\omega(3d_{3/2} \to 2p_{1/2})}{\omega(2p_{1/2} \to 1s_{1/2})}\right]^3.$$

1039

(a) What is the energy of the neutrino in a typical beta decay?

(b) What is the dependence on atomic number Z of the lifetime for spontaneous decay of the $2p$ state in the hydrogen-like atoms H, He$^+$, Li^{++}, etc.?

(c) What is the electron configuration, total spin S, total orbital angular momentum L, and total angular momentum J of the ground state of atomic oxygen?

(UC, Berkeley)

Solution:

(a) The energy of the neutrino emitted in a typical β-decay is $E_\nu \approx$ 1 MeV.

(b) The probability of spontaneous transition $2p \to 1s$ per unit time is **(Problem 1038(b))** $A \propto |\mathbf{r}_{12}|^2 \omega^3$, where

$$|\mathbf{r}_{12}|^2 = |\langle 1s(Zr)|\mathbf{r}|2p(Zr)\rangle|^2,$$

$|1s(Zr)\rangle$ and $|2p(Zr)\rangle$ being the radial wave functions of a hydrogen-like atom of nuclear charge Z, and

$$\omega = \frac{1}{\hbar}(E_2 - E_1).$$

As

$$1s(Zr)\rangle = \left(\frac{Z}{a_0}\right)^{\frac{3}{2}} 2e^{\frac{-Zr}{a_0}},$$

$$2p(Zr)\rangle = \left(\frac{Z}{2a_0}\right)^{\frac{3}{2}} \frac{Zr}{a_0\sqrt{3}} e^{-\frac{Zr}{2a_0}},$$

a_0 being a constant, we have for $Z > 1$,

$$|\mathbf{r}_{12}|^2 \propto Z^{-2}, \qquad \omega^3 \propto Z^6,$$

and so $A \propto Z^4$. Hence the lifetime τ is

$$\tau \propto \frac{1}{A} \propto Z^{-4}.$$

(c) The electron configuration of ground state atomic oxygen is $1s^2 2s^2 2p^4$. As the state has $S = 1$, $L = 1$, $J = 2$, it is designated $^3P_2$.

1040

Suppose that, because of small parity-violating forces, the $2^2 S_{1/2}$ level of the hydrogen atom has a small p-wave admixture:

$$\Psi(n = 2, j = 1/2) = \Psi_s(n = 2, j = 1/2, l = 0)$$

$$+ \varepsilon\Psi_p(n = 2, j = 1/2, l = 1).$$

What first-order radiation decay will de-excite this state? What is the form of the decay matrix element? What dose it become if $\varepsilon \to 0$ and why?

(*Wisconsin*)

Solution:

Electric dipole radiation will de-excite the p-wave part of this mixed state: $\Psi_p(n = 2, j = 1/2, l = 1) \to \Psi_s(n = 1, j = 1/2, l = 0)$. The $\Psi_s(n = 2, j = 1/2, l = 0)$ state will not decay as it is a metastable state. The decay matrix, i.e. the T matrix, is

$$\langle\Psi_f|T|\Psi_i\rangle = \varepsilon \int \Psi_f^* V(\mathbf{r})\Psi_i d^3r,$$

where, for electric dipole radiation, we have

$$V(\mathbf{r}) = -(-e\mathbf{r}) \cdot \mathbf{E} = erE\cos\theta\,,$$

taking the z-axis along the electric field. Thus

$$\langle \Psi_f | T | \Psi_i \rangle = \varepsilon eE \int R_{10} r R_{21} r^2 dr \int Y_{00} Y_{10} \cos\theta d\Omega$$

$$= \frac{\varepsilon eE}{\sqrt{2}a^3} \int_0^\infty r^3 \exp\left(-\frac{3r}{2a}\right) dr$$

$$= \frac{32}{27\sqrt{6}} \varepsilon eaE \int_\Omega Y_{00} Y_{10} \cos\theta d\Omega\,.$$

As

$$\cos\theta Y_{10} = \sqrt{\frac{4}{15}} Y_{20} + \sqrt{\frac{1}{3}} Y_{00}\,,$$

the last integral equals $\sqrt{\frac{1}{3}}$ and

$$\langle \Psi_f | T | \Psi_i \rangle = \left(\frac{2}{3}\right)^4 \sqrt{2}\varepsilon eaE\,.$$

If $\varepsilon \to 0$, the matrix element of the dipole transition $\langle \Psi_f | T | \Psi_i \rangle \to 0$ and no such de-excitation takes place. The excited state $\Psi_s(n = 2, j = 1/2, l = 0)$ is metastable. It cannot decay to the ground state via electric dipole transition (because $\Delta l \neq 1$). Nor can it do so via magnetic dipole or electric quadruple transition. It can only decay to the ground state by the double-photons transition $2^2 S_{1/2} \to 1^2 S_{1/2}$, which however has a very small probability.

1041

(a) The ground state of the hydrogen atom is split by the hyperfine interaction. Indicate the level diagram and show from first principles which state lies higher in energy.

(b) The ground state of the hydrogen molecule is split into total nuclear spin triplet and singlet states. Show from first principles which state lies higher in energy.

(Chicago)

Solution:

(a) The hyperfine interaction in hydrogen arises from the magnetic interaction between the intrinsic magnetic moments of the proton and the electron, the Hamiltonian being

$$H_{\text{int}} = -\boldsymbol{\mu}_p \cdot \mathbf{B}\,,$$

where $\mathbf{B}$ is the magnetic field produced by the magnetic moment of the electron and $\boldsymbol{\mu}_p$ is the intrinsic magnetic moment of the proton.

In the ground state, the electron charge density is spherically symmetric so that $\mathbf{B}$ has the same direction as the electron intrinsic magnetic moment $\boldsymbol{\mu}_e$. However as the electron is negatively charged, $\boldsymbol{\mu}_e$ is antiparallel to the electron spin angular momentum $\mathbf{s}_e$. For the lowest energy state of H_{int}, $\langle \boldsymbol{\mu}_p \cdot \boldsymbol{\mu}_e \rangle > 0$, and so $\langle \mathbf{s}_p \cdot \mathbf{s}_e \rangle < 0$. Thus the singlet state $F = 0$ is the ground state, while the triplet $F = 1$ is an excited state (see Fig. 1.12).

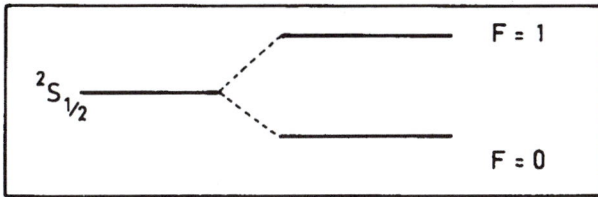

Fig. 1.12

(b) As hydrogen molecule consists of two like atoms, each having a proton (spin $\frac{1}{2}$) as nucleus, the nuclear system must have an antisymmetric state function. Then the nuclear spin singlet state ($S = 0$, antisymmetric) must be associated with a symmetric nuclear rotational state; thus $J = 0, 2, 4, \ldots$, with the ground state having $J = 0$. For the spin triplet state ($S = 1$, symmetric) the rotational state must have $J = 1, 3, \ldots$, with the ground state having $J = 1$. As the rotational energy is proportional to $J(J+1)$, the spin triplet ground state lies higher in energy.

1042

(a) In Bohr's original theory of the hydrogen atom (circular orbits) what postulate led to the choice of the allowed energy levels?

(b) Later de Broglie pointed out a most interesting relationship between the Bohr postulate and the de Broglie wavelength of the electron. State and derive this relationship.

(Wisconsin)

Solution:

(a) Bohr proposed the quantization condition

$$mvr = n\hbar \,,$$

where m and v are respectively the mass and velocity of the orbiting electron, r is the radius of the circular orbit, $n = 1, 2, 3, \ldots$. This condition gives descrete values of the electron momentum $p = mv$, which in turn leads to descrete energy levels.

(b) Later de Broglie found that Bohr's circular orbits could exactly hold integral numbers of de Broglie wavelength of the electron. As

$$pr = n\hbar = \frac{nh}{2\pi} \,,$$

$$2\pi r = n\frac{h}{p} = n\lambda \,,$$

where λ is the de Broglie wavelength, which is associated with the group velocity of matter wave.

1043

In radio astronomy, hydrogen atoms are observed in which, for example, radiative transitions from $n = 109$ to $n = 108$ occur.

(a) What are the frequency and wavelength of the radiation emitted in this transition?

(b) The same transition has also been observed in excited helium atoms. What is the ratio of the wavelengths of the He and H radiation?

(c) Why is it difficult to observe this transition in laboratory experiment?

(Wisconsin)

Solution:

(a) The energy levels of hydrogen, in eV, are

$$E_n = -\frac{13.6}{n^2}.$$

For transitions between excited states $n = 109$ and $n = 108$ we have

$$h\nu = \frac{13.6}{108^2} - \frac{13.6}{109^2},$$

giving

$$\nu = 5.15 \times 10^9 \text{ Hz},$$

or

$$\lambda = c/\nu = 5.83 \text{ cm}.$$

(b) For such highly excited states the effective nuclear charge of the helium atom experienced by an orbital electron is approximately equal to that of a proton. Hence for such transitions the wavelength from He approximately equals that from H.

(c) In such highly excited states, atoms are easily ionized by colliding with other atoms. At the same time, the probability of a transition between these highly excited states is very small. It is very difficult to produce such environment in laboratory in which the probability of a collision is very small and yet there are sufficiently many such highly excited atoms available. (However the availability of strong lasers may make it possible to stimulate an atom to such highly excited states by multiphoton excitation.)

1044

Sketch the energy levels of atomic Li for the states with $n = 2, 3, 4$. Indicate on the energy diagram several lines that might be seen in emission and several lines that might be seen in absorption. Show on the same diagram the energy levels of atomic hydrogen for $n = 2, 3, 4$.

(Wisconsin)

Solution:

As most atoms remain in the ground state, the absorption spectrum arises from transitions from $2s$ to np states ($n = 2, 3, 4$). In Fig. 1.13,

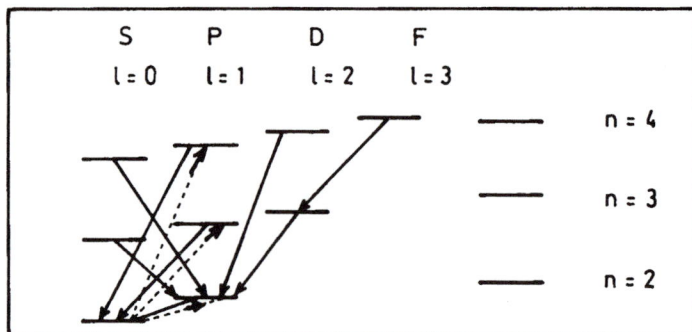

Fig. 1.13

the dashed lines represent absorption transitions, the solid lines, emission transitions.

1045

The "plum pudding" model of the atom proposed by J. J. Thomson in the early days of atomic theory consisted of a sphere of radius a of positive charge of total value Ze. Z is an integer and e is the fundamental unit of charge. The electrons, of charge $-e$, were considered to be point charges embedded in the positive charge.

(a) Find the force acting on an electron as a function of its distance r from the center of the sphere for the element hydrogen.

(b) What type of motion does the electron execute?

(c) Find an expression for the frequency for this motion.

(Wisconsin)

Solution:

(a) For the hydrogen atom having $Z = 1$, radius a, the density of positive charge is

$$\rho = \frac{e}{\frac{4}{3}\pi a^3} = \frac{3e}{4\pi a^3}.$$

When an electron is at a distance r from the center of the sphere, only the positive charge inside the sphere of radius r can affect the electron and so the electrostatic force acting on the electron is

$$F(r) = -\frac{e}{4\pi\varepsilon_0 r^2} \cdot \frac{4}{3}\pi r^3 \rho = -\frac{e^2 r}{4\pi\varepsilon_0 a^3} \,,$$

pointing toward the center of the sphere.

(b) The form of $F(r)$ indicates the motion of the electron is simple harmonic.

(c) $F(r)$ can be written in the form

$$F(r) = -kr \,,$$

where $k = \frac{e^2}{4\pi\varepsilon_0 a^3}$. The angular frequency of the harmonic motion is thus

$$\omega = \sqrt{\frac{k}{m}} = \sqrt{\frac{e^2}{4\pi\varepsilon_0 a^3 m}} \,,$$

where m is the mass of electron.

1046

Lyman alpha, the $n = 1$ to $n = 2$ transition in atomic hydrogen, is at 1215 Å.

(a) Define the wavelength region capable of photoionizing a H atom in the ground level $(n = 1)$.

(b) Define the wavelength region capable of photoionizing a H atom in the first excited level $(n = 2)$.

(c) Define the wavelength region capable of photoionizing a He$^+$ ion in the ground level $(n = 1)$.

(d) Define the wavelength region capable of photoionizing a He$^+$ ion in the first excited level $(n = 2)$.

(Wisconsin)

Solution:

(a) A spectral series of a hydrogen-like atom has wave numbers

$$\tilde{\nu} = Z^2 R\left(\frac{1}{n^2} - \frac{1}{m^2}\right) \,,$$

where Z is the nuclear charge, R is the Rydberg constant, and n, m are positive integers with $m > n$. The ionization energy of the ground state of H atom is the limit of the Lyman series $(n = 1)$, the wave number being

$$\tilde{\nu}_0 = \frac{1}{\lambda_0} = R.$$

For the alpha line of the Lyman series,

$$\tilde{\nu}_\alpha = \frac{1}{\lambda_\alpha} = R\left(1 - \frac{1}{2^2}\right) = \frac{3}{4}R = \frac{3}{4\lambda_0}.$$

As $\lambda_\alpha = 1215$ Å, $\lambda_0 = 3\lambda_\alpha/4 = 911$ Å. Hence the wavelength of light that can photoionize H atom in the ground state must be shorter than 911 Å.

(b) The wavelength should be shorter than the limit of the Balmer series $(n = 2)$, whose wave number is

$$\tilde{\nu} = \frac{1}{\lambda} = \frac{R}{2^2} = \frac{1}{4\lambda_0}.$$

Hence the wavelength should be shorter than $4\lambda_0 = 3645$ Å.

(c) The limiting wave number of the Lyman series of He$^+$ $(Z = 2)$ is

$$\tilde{\nu} = \frac{1}{\lambda} = \frac{Z^2 R}{1^2} = 4R = \frac{4}{\lambda_0}.$$

The wavelength that can photoionize the He$^+$ in the ground state must be shorter than $\lambda_0/4 = 228$ Å.

(d) The wavelength should be shorter than $1/R = \lambda_0 = 1215$ Å.

1047

A tritium atom in its ground state beta-decays to He$^+$.

(a) Immediately after the decay, what is the probability that the helium ion is in its ground state?
(b) In the 2s state?
(c) In the 2p state?
(Ignore spin in this problem.)

(UC, Berkeley)

Solution:

At the instant of β-decay continuity of the wave function requires

$$|1s\rangle_H = a_1|1s\rangle_{He^+} + a_2|2s\rangle_{He^+} + a_3|2p\rangle_{He^+} + \cdots,$$

where

$$|1s\rangle = R_{10}(r)Y_{00}, \qquad |2s\rangle = R_{20}(r)Y_{00}, \qquad |2p\rangle = R_{21}(r)Y_{10},$$

with

$$R_{10} = \left(\frac{Z}{a}\right)^{\frac{3}{2}} 2\exp\left(-\frac{Zr}{a}\right), \qquad R_{20} = \left(\frac{Z}{2a}\right)^{\frac{3}{2}} \left(2 - \frac{Zr}{a}\right)\exp\left(-\frac{Zr}{2a}\right),$$

$$R_{21} = \left(\frac{Z}{2a}\right)^{\frac{3}{2}} \frac{Zr}{a\sqrt{3}}\exp\left(-\frac{Zr}{2a}\right), \qquad a = \frac{\hbar^2}{me^2}.$$

(a)

$$a_1 = _{He^+}\langle 1s|1s\rangle_H = \int_0^\infty \frac{2}{a^{3/2}}\exp\left(-\frac{r}{a}\right)\cdot 2\left(\frac{2}{a}\right)^{3/2}$$

$$\times \exp\left(-\frac{2r}{a}\right)\cdot r^2 dr \int Y_{00}^2 d\Omega = \frac{16\sqrt{2}}{27}.$$

Accordingly the probability of finding the He$^+$ in the ground state is

$$W\langle 1s\rangle = |a_1|^2 = \frac{512}{729}.$$

(b)

$$a_2 = _{He^+}\langle 2s|1s\rangle_H = \int_0^\infty \frac{2}{a^{3/2}}\exp\left(-\frac{r}{a}\right)\cdot \frac{1}{\sqrt{2}}\left(\frac{2}{a}\right)^{3/2}\left(1 - \frac{r}{a}\right)$$

$$\times \exp\left(-\frac{r}{a}\right)\cdot r^2 dr \int Y_{00}^2 d\Omega = -\frac{1}{2}.$$

Hence the probability of finding the He$^+$ in the 2s state is

$$W\langle 2s\rangle = |a_2|^2 = \frac{1}{4}.$$

(c)

$$a_3 = {}_{He^+}\langle 2p|1s\rangle_H = \int_0^\infty \frac{2}{a^{3/2}} \exp\left(-\frac{r}{a}\right) \cdot \frac{1}{2\sqrt{6}}\left(\frac{2}{a}\right)^{3/2} \cdot \frac{2r}{a}$$

$$\times \exp\left(-\frac{r}{a}\right) \cdot r^2 dr \int Y_{10}^* Y_{00} d\Omega = 0.$$

Hence the probability of finding the He^+ in the $2p$ state is

$$W\langle 2p\rangle = |a_3|^2 = 0.$$

1048

Consider the ground state and $n = 2$ states of hydrogen atom.

Indicate in the diagram (Fig. 1.14) the complete spectroscopic notation for all four states. There are four corrections to the indicated level structure that must be considered to explain the various observed splitting of the levels. These corrections are:

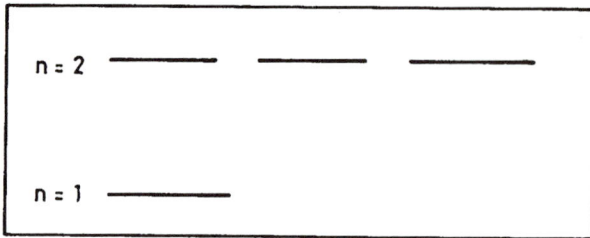

Fig. 1.14

(a) Lamb shift,
(b) fine structure,
(c) hyperfine structure,
(d) relativistic effects.

(1) Which of the above apply to the $n = 1$ state?
(2) Which of the above apply to the $n = 2$, $l = 0$ state? The $n = 2$, $l = 1$ state?

(3) List in order of decreasing importance these four corrections. (i.e. biggest one first, smallest last). Indicate if some of the corrections are of the same order of magnitude.

(4) Discuss briefly the physical origins of the hyperfine structure. Your discussion should include an appropriate mention of the Fermi contact potential.

(Wisconsin)

Solution:

The spectroscopic notation for the ground and first excited states of hydrogen atom is shown in Fig. 1.15.

Three corrections give rise to the fine structure for hydrogen atom:

$$E_f = E_m + E_D + E_{so} ,$$

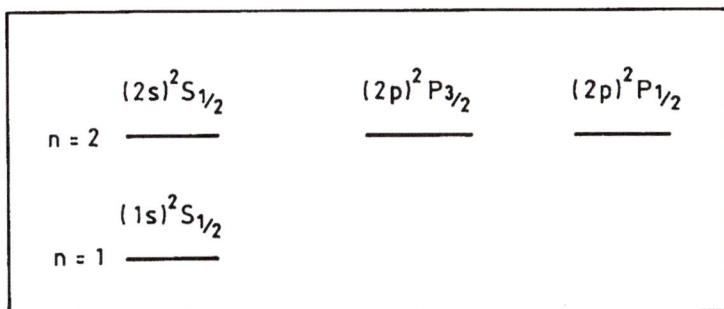

Fig. 1.15

where E_m is caused by the relativistic effect of mass changing with velocity, E_D, the Darwin term, arises from the relativistic non-locality of the electron, E_{so} is due to the spin-orbit coupling of the electron. They are given by

$$E_m = -\frac{\alpha^2 Z^4}{4n^4} \left(\frac{4n}{l + \dfrac{1}{2} - 3} \right) \times 13.6 \text{ eV} ,$$

$$E_D = \frac{\alpha^2 Z^4}{n^3} \delta_{l0} \times 13.6 \text{ eV},$$

$$E_{so} = \begin{cases} (1 - \delta_{l0}) \dfrac{\alpha^2 Z^4 l}{n^3 l(l+1)(2l+1)} \times 13.6 \text{ eV}, & \left(j = l + \dfrac{1}{2}\right) \\[4mm] -(1 - \delta_{l0}) \dfrac{\alpha^2 Z^4 (l+1)}{n^3 l(l+1)(2l+1)} \times 13.6 \text{ eV}. & \left(j = l - \dfrac{1}{2}\right) \end{cases}$$

where α is the fine-structure constant, and δ_{l0} is the usual Kronecker delta.

Lamb shift arises from the interaction between the electron and its radiation field, giving rise to a correction which, when expanded with respect to $Z\alpha$, has the first term

$$E_L = k(l) \frac{\alpha(Z\alpha)^4 mc^2}{2\pi n^3}$$

$$= k(l) \frac{\alpha^3 Z^4}{\pi n^3} \times 13.6 \text{ eV},$$

where $k(l)$ is a parameter related to l.

Hyperfine structure arises from the coupling of the total angular momentum of the electron with the nuclear spin.

(1) For the $n = 1$ state ($l = 0$), E_m, E_D, E_L can only cause the energy level to shift as a whole. As $E_{so} = 0$ also, the fine-structure correction does not split the energy level. On the other hand, the hyperfine structure correction can cause a splitting as shown in Fig. 1.16.

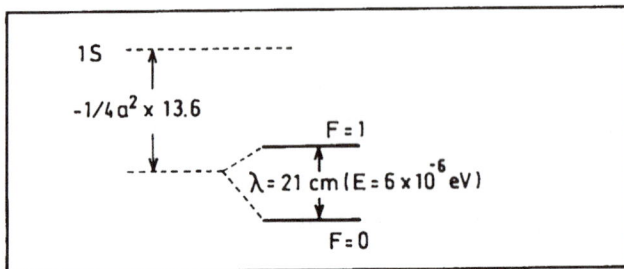

Fig. 1.16

(2) For the $n = 2$ state ($l = 0$ and $l = 1$), the fine-structure correction causes the most splitting in the $l = 1$ level, to which the hyperfine structure correction also contributes (see Fig. 1.17).

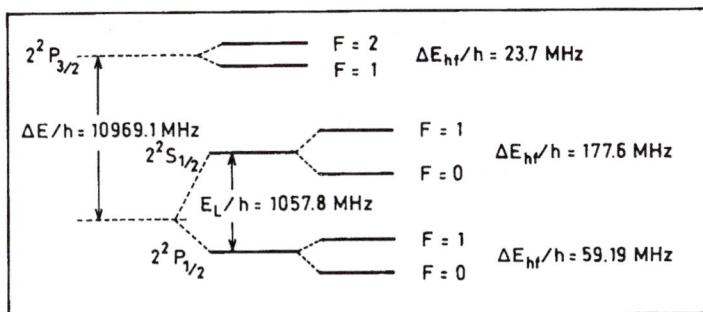

Fig. 1.17

(3) E_m, E_D, E_{so} are of the same order of magnitude > Lamb shift $\gtrsim$ hyperfine structure.

(4) The hyperfine structure can be separated into three terms:

(a) Interaction between the nuclear magnetic moment and the magnetic field at the proton due to the electron's orbital motion,

(b) dipole-dipole interaction between the electron and the nuclear magnetic moment,

(c) the Fermi contact potential due to the interaction between the spin magnetic moment of the electron and the internal magnetic field of the proton.

1049

Using the Bohr model of the atom,

(a) derive an expression for the energy levels of the He^+ ion.

(b) calculate the energies of the $l = 1$ state in a magnetic field, neglecting the electron spin.

(Wisconsin)

Solution:

(a) Let the radius of the orbit of the electron be r, and its velocity be v. Bohr assumed that the angular momentum L_ϕ is quantized:

$$L_\phi = mvr = n\hbar. \qquad (n = 1, 2, 3 \ldots)$$

The centripetal force is provided by the Coulomb attraction and so

$$m\frac{v^2}{r} = \frac{2e^2}{4\pi\varepsilon_0 r^2}.$$

Hence the energy of He^+ is

$$E_n = \frac{1}{2}mv^2 - \frac{2e^2}{4\pi\varepsilon_0 r} = -\frac{1}{2}mv^2 = -\frac{2me^4}{(4\pi\varepsilon_0)^2 n^2 \hbar^2}.$$

(b) The area of the electron orbit is

$$A = \int_0^{2\pi} \frac{r}{2} \cdot r d\phi = \frac{1}{2}\int_0^T r^2 \omega dt = \frac{L_\phi}{2m}T,$$

where $\omega = \frac{d\phi}{dt}$, the angular velocity, is given by $L_\phi = mr^2\omega$, and T is the period of circular motion. For $l = 1$, $L_\phi = \hbar$ and the magnetic moment of the electron due to its orbital motion is

$$\mu = IA = -\frac{e}{T}A = -\frac{e\hbar}{2m},$$

where I is the electric current due to the orbital motion of the electron. The energy arising from interaction between the $l = 1$ state and a magnetic field $\mathbf{B}$ is

$$\Delta E = -\mu \cdot \mathbf{B} = \begin{cases} \dfrac{e\hbar}{2m}B, & (\mu//\mathbf{B}) \\[2mm] 0, & (\mu \perp \mathbf{B}) \\[2mm] -\dfrac{e\hbar}{2m}B. & (\mu// - \mathbf{B}) \end{cases}$$

1050

An atom has a nucleus of charge Z and one electron. The nucleus has a radius R, inside which the charge (protons) is uniformly distributed. We want to study the effect of the finite size of the nucleus on the electron levels:

(a) Calculate the potential taking into account the finite size of the nucleus.

(b) Calculate the level shift due to the finite size of the nucleus for the $1s$ state of ^{208}Pb using perturbation theory, assuming that R is much smaller than the Bohr radius and approximating the wave function accordingly.

(c) Give a numerical answer to (b) in cm^{-1} assuming $R = r_0 A^{1/3}$, $r_0 = 1.2$ fermi.

(Wisconsin)

Solution:

(a) For $r \geq R$.

$$V(r) = -\frac{Ze^2}{4\pi\varepsilon_0 r}.$$

For $r < R$,

$$V(r) = -\frac{Ze^2}{4\pi\varepsilon_0 r} \cdot \left(\frac{r}{R}\right)^3 - \int_r^R \frac{e\rho 4\pi r'^2}{r'} dr' = -\frac{Ze^2}{8\pi\varepsilon_0 R^3}(3R^2 - r^2),$$

where

$$\rho = \frac{Ze}{\frac{4}{3}\pi r^3}.$$

(b) Taking the departure of the Hamiltonian from that of a point nucleus as perturbation, we have

$$H' = \begin{cases} \dfrac{Ze^2}{4\pi\varepsilon_0 r} - \dfrac{Ze^2}{4\pi\varepsilon_0 R}\left(\dfrac{3}{2} - \dfrac{r^2}{2R^2}\right) & \text{for } r < R, \\ 0 & \text{for } r \geq R. \end{cases}$$

The $1s$ wave function of ^{208}Pb is

$$|1s\rangle = 2\left(\frac{Z}{a_0}\right)^{3/2} \exp\left(-\frac{2r}{a_0}\right) \cdot \frac{1}{\sqrt{4\pi}},$$

where $Z = 82$, a_0 is the Bohr radius. Taking the approximation $r \ll a_0$, i.e., $\exp(-\frac{2r}{a_0}) \approx 1$, the energy shift is

$$\Delta E = \langle 1s|H'|1s \rangle$$

$$= -\frac{4Z^4 e^2}{4\pi\varepsilon_0 a_0^3} \int_0^R \left(\frac{3}{2R} - \frac{r^2}{2R^3} - \frac{1}{r} \right) r^2 dr$$

$$= \frac{4}{5} Z^2 |E_0| \left(\frac{R}{a_0} \right)^2 ,$$

where $E_0 = -\frac{Z^2 e^2}{(4\pi\varepsilon_0)2a_0}$ is the ground state energy of a hydrogen-like atom.

(c)

$$\Delta E = \frac{4}{5} \times 82^2 \times (82^2 \times 13.6) \times \left(\frac{1.2 \times 10^{-19} \times 208^{\frac{1}{3}}}{5.29 \times 10^{-9}} \right)^2 = 8.89 \text{ eV} ,$$

$$\Delta \tilde{\nu} = \frac{\Delta E}{hc} \approx 7.2 \times 10^4 \text{ cm}^{-1} .$$

1051

If the proton is approximated as a uniform charge distribution in a sphere of radius R, show that the shift of an s-wave atomic energy level in the hydrogen atom, from the value it would have for a point proton, is approximately

$$\Delta E_{ns} \approx \frac{2\pi}{5} e^2 |\Psi_{ns}(0)|^2 R^2 ,$$

using the fact that the proton radius is much smaller than the Bohr radius. Why is the shift much smaller for non-s states?

The $2s$ hydrogenic wave function is

$$(2a_0)^{-3/2} \pi^{-1/2} \left(1 - \frac{r}{2a_0} \right) \exp \left(-\frac{r}{2a_0} \right) .$$

What is the approximate splitting (in eV) between the $2s$ and $2p$ levels induced by this effect? [$a_0 \approx 5 \times 10^{-9}$ cm for H, $R \approx 10^{-13}$ cm.]

(*Wisconsin*)

Solution:

The perturbation caused by the finite volume of proton is (**Problem 1050**)

$$H' = \begin{cases} 0, & (r \geq R) \\ \dfrac{e^2}{r} - \dfrac{e^2}{R}\left(\dfrac{3}{2} - \dfrac{r^2}{2R^2}\right). & (r < R) \end{cases}$$

The unperturbed wave function is

$$\Psi_{ns} = N_{n0} \exp\left(-\frac{r}{na_0}\right) F\left(-n+1, 2, \frac{2r}{na_0}\right) Y_{00},$$

where

$$N_{n0} = \frac{2}{(na_0)^{3/2}} \sqrt{\frac{n!}{(n-1)!}} \approx \frac{2}{(na_0)^{3/2}},$$

$$F\left(-n+1, 2, \frac{2r}{na_0}\right) = 1 - \frac{n-1}{2} \cdot \frac{2r}{na_0} + \frac{(n-1)(n-2)}{2\cdot 3}$$

$$\times \frac{1}{2!} \left(\frac{2r}{na_0}\right)^2 + \cdots.$$

Taking the approximation $r \ll a_0$, we have

$$F\left(-n+1, 2, \frac{2r}{na_0}\right) \approx 1, \qquad \exp\left(-\frac{r}{na_0}\right) \approx 1,$$

and so

$$\Psi_{ns} = N_{n0}Y_{00} = \frac{2}{(na_0)^{3/2}}Y_{00},$$

$$\Delta E_{ns} = \langle \Psi_{ns}^* | H' | \Psi_{ns} \rangle = \int_0^R \left[\frac{e^2}{r} - \frac{e^2}{R}\left(\frac{3}{2} - \frac{r^2}{2R^2}\right)\right] \Psi_{ns}^* \Psi_{ns} r^2 dr d\Omega$$

$$= \frac{2\pi}{5} \frac{e^2 R^2}{\pi(na_0)^3}.$$

Using

$$\Psi_{ns}(0) = \frac{2}{(na_0)^{3/2}} \cdot \frac{1}{\sqrt{4\pi}} = \frac{1}{\sqrt{\pi}(na_0)^{3/2}},$$

we have

$$\Delta E_{ns} = \frac{2\pi}{5} e^2 |\Psi_{ns}(0)|^2 R^2 .$$

As the non-*s* wave functions have a much smaller fraction inside the nucleus and so cause smaller perturbation, the energy shift is much smaller. For hydrogen atom, since $\Delta E_{2p} \ll \Delta E_{2s}$,

$$\Delta E_{ps} = \Delta E_{2s} - \Delta E_{2p} \approx \Delta E_{2s}$$

$$= \frac{2\pi}{5} e^2 |\Psi_{2s}(0)|^2 R ,$$

where

$$\Psi_{2s}(0) = (2a_0)^{-3/2} \pi^{-1/2} .$$

Hence

$$\Delta E_{ps} \approx \frac{2\pi}{5} e^2 [(2a_0)^{-3/2} \pi^{-1/2}]^2 R^2$$

$$= \frac{e^2 R^2}{20 a_0^3} = \left(\frac{e^2}{\hbar c}\right)^2 \cdot \frac{R^2 m c^2}{20 a_0^2}$$

$$= \left(\frac{1}{137}\right)^2 \times \frac{10^{-26} \times 0.511 \times 10^6}{20 \times (5 \times 10^{-9})^2} \approx 5.4 \times 10^{-10} \text{ eV} .$$

1052

The ground state of hydrogen atom is $1s$. When examined very closely, it is found that the level is split into two levels.

(a) Explain why this splitting takes place.

(b) Estimate numerically the energy difference between these two levels.

(Columbia)

Solution:

(a) In the fine-structure spectrum of hydrogen atom, the ground state $1s$ is not split. The splitting is caused by the coupling between the magnetic moments of the nuclear spin and the electron spin: $\hat{\mathbf{F}} = \hat{\mathbf{I}} + \hat{\mathbf{J}}$. As $I = 1/2$, $J = 1/2$, the total angular momentum is $F = 1$ or $F = 0$, corresponding to the two split energy levels.

(b) The magnetic moment of the nucleus (proton) is $\boldsymbol{\mu} = \mu_N \boldsymbol{\sigma}_N$, where $\boldsymbol{\sigma}_N$ is the Pauli matrix operating on the nuclear wave function, inducing a magnetic field $\mathbf{H}_m = \nabla \times \nabla \times (\frac{\mu_N \boldsymbol{\sigma}_N}{r})$. The Hamiltonian of the interaction between $\mathbf{H}_m$ and the electron magnetic moment $\boldsymbol{\mu} = -\mu_e \boldsymbol{\sigma}_e$ is

$$\hat{H} = -\boldsymbol{\mu} \cdot \hat{\mathbf{H}}_m = \mu_e \mu_N \boldsymbol{\sigma}_e \cdot \nabla \times \nabla \times \left(\frac{\boldsymbol{\sigma}_N}{r}\right) .$$

Calculation gives the hyperfine structure splitting as (**Problem 1053**)

$$\Delta E = A' \mathbf{I} \cdot \mathbf{J} ,$$

where

$$A' \sim \frac{\mu_e \mu_N}{e^2 a_0^3} \approx \left(\frac{m_e}{m_N}\right) \frac{m_e c^2}{4} \cdot \left(\frac{e^2}{\hbar c}\right)^4$$

$$\approx \frac{1}{2000} \cdot \frac{0.51 \times 10^6}{4} \times \left(\frac{1}{137}\right)^4$$

$$\approx 2 \times 10^{-7} \text{ eV} ,$$

m_e, m_N, c, a_0 being the electron mass, nucleon mass, velocity of light, Bohr radius respectively.

1053

Derive an expression for the splitting in energy of an atomic energy level produced by the hyperfine interaction. Express your result in terms of the relevant angular momentum quantum numbers.

(SUNY, Buffalo)

Solution:

The hyperfine structure is caused by the interaction between the magnetic field produced by the orbital motion and spin of the electron and the nuclear magnetic moment $\mathbf{m}_N$. Taking the site of the nucleus as origin, the magnetic field caused by the orbital motion of the electron at the origin is

$$\mathbf{B}_e(0) = \frac{\mu_0 e}{4\pi} \frac{\mathbf{v} \times \mathbf{r}}{r^3} = -\frac{2\mu_0 \mu_B}{4\pi \hbar} \frac{1}{r^3} ,$$

where $\mathbf{v}$ is the velocity of the electron in its orbit, $\mathbf{l} = m\mathbf{r} \times \mathbf{v}$ is its orbital angular momentum, $\mu_B = \frac{e\hbar}{2m}$, m being the electron mass, is the Bohr magneton.

The Hamiltonian of the interaction between the nuclear magnetic moment $\mathbf{m}_N$ and $\mathbf{B}_e(0)$ is

$$H_{lI} = -\mathbf{m}_N \cdot \mathbf{B}_e(0) = \frac{2\mu_0 g_N \mu_N \mu_B}{4\pi\hbar^2 r^3} \mathbf{l} \cdot \mathbf{I},$$

where $\mathbf{I}$ is the nuclear spin, μ_N the nuclear magneton, g_N the Landé g-factor of the nucleon.

At $\mathbf{r} + \mathbf{r}'$, the vector potential caused by the electron magnetic moment $\mathbf{m}_s = -\frac{2\mu_B \mathbf{s}}{\hbar}$ is $\mathbf{A} = \frac{\mu_0}{4\pi} \mathbf{m}_s \times \frac{\mathbf{r}'}{r'^3}$, $\mathbf{r}'$ being the radius vector from $\mathbf{r}$ to the field point. So the magnetic field is

$$\mathbf{B}_s = \nabla \times \mathbf{A} = \frac{\mu_0}{4\pi} \nabla \times \left(\mathbf{m}_s \times \frac{\mathbf{r}'}{r'^3} \right)$$

$$= \frac{2\mu_0 \mu_B}{4\pi\hbar} \nabla' \times \left(\mathbf{s} \times \nabla' \frac{1}{r'} \right) = \frac{2\mu_0 \mu_B}{4\pi\hbar} \left[\mathbf{s} \nabla'^2 \frac{1}{r'} - (\mathbf{s} \cdot \nabla') \nabla' \frac{1}{r'} \right]$$

$$= -\frac{2\mu_0 \mu_B}{4\pi\hbar} \left[4\pi \mathbf{s} \delta(\mathbf{r}') + (\mathbf{s} \cdot \nabla') \nabla' \frac{1}{r'} \right].$$

Letting $\mathbf{r}' = -\mathbf{r}$, we get the magnetic field caused by $\mathbf{m}_s$ at the origin:

$$\mathbf{B}_s(0) = -\frac{2\mu_0 \mu_B}{4\pi\hbar} \left[4\pi \mathbf{s} \delta(\mathbf{r}) + (\mathbf{s} \cdot \nabla) \nabla \frac{1}{r} \right].$$

Hence the Hamiltonian of the interaction between $\mathbf{m}_N = \frac{g_N \mu_N \mathbf{I}}{\hbar}$ and $\mathbf{B}_s(0)$ is

$$H_{sI} = -\mathbf{m}_N \cdot \mathbf{B}_s(0)$$

$$= \frac{2\mu_0 g_N \mu_N \mu_B}{4\pi\hbar^2} \left[4\pi \mathbf{I} \cdot \mathbf{s} \delta(\mathbf{r}) + (\mathbf{s} \cdot \nabla) \left(\mathbf{I} \cdot \nabla \frac{1}{r} \right) \right].$$

The total Hamiltonian is then

$$H_{hf} = H_{lI} + H_{sI}$$

$$= \frac{2\mu_0 g_N \mu_N \mu_B}{4\pi\hbar^2} \left[\frac{\mathbf{l} \cdot \mathbf{I}}{r^3} + 4\pi \mathbf{s} \cdot \mathbf{I} \delta(\mathbf{r}) + (\mathbf{s} \cdot \nabla) \left(\mathbf{I} \cdot \nabla \frac{1}{r} \right) \right].$$

In zeroth order approximation, the wave function is $|lsjIFM_F\rangle$, where l, s and j are respectively the quantum numbers of orbital angular momentum, spin and total angular momentum of the electron, I is the quantum number of the nuclear spin, F is the quantum number of the total angular momentum of the atom and M_F is of its z-component quantum number. Hence in first order perturbation the energy correction due to H_{hf} is

$$\Delta E = \langle lsjIFM_F|H_{hf}|lsjIFM_F\rangle \,.$$

If $l \neq 0$, the wave function is zero at the origin and we only need to consider H_{hf} for $\mathbf{r} \neq 0$. Thus

$$H_{hf} = \frac{2\mu_0 g_N \mu_N \mu_B}{4\pi\hbar^2} \left[\frac{\mathbf{I}\cdot\mathbf{l}}{r^3} + (\mathbf{s}\cdot\nabla)\left(\mathbf{I}\cdot\nabla\frac{1}{r}\right) \right]$$

$$= \frac{2\mu_0 g_N \mu_N \mu_B}{4\pi\hbar^2 r^3} \mathbf{G}\cdot\mathbf{I} \,,$$

where

$$\mathbf{G} = 1 + 3\frac{(\mathbf{s}\cdot\mathbf{r})\mathbf{r}}{r^2} \,.$$

Hence

$$\Delta E = \frac{2\mu_0 g_N \mu_N \mu_B}{4\pi\hbar^2} \left\langle \frac{1}{r^3}\mathbf{G}\cdot\mathbf{I} \right\rangle$$

$$= \frac{\mu_0 g_N \mu_N \mu_B}{4\pi} \cdot \frac{l(l+1)}{j(j+1)} \cdot [F(F+1) - I(I+1) - j(j+1)] \left\langle \frac{1}{r^3} \right\rangle$$

$$= \frac{\mu_0 g_N \mu_N \mu_B}{4\pi} \cdot \frac{Z^3}{a_0^3 n^3 \left(l+\frac{1}{2}\right) j(j+1)} \cdot [F(F+1)$$

$$- I(I+1) - j(j+1)] \,,$$

where a_0 is the Bohr radius and Z is the atomic number of the atom.

For $l = 0$, the wave function is spherically symmetric and

$$\Delta E = \frac{2\mu_0 g_N \mu_N \mu_B}{4\pi\hbar^2} \left[4\pi\langle\mathbf{s}\cdot\mathbf{I}\delta(\mathbf{r})\rangle + \left\langle (\mathbf{s}\cdot\nabla)\left(\mathbf{I}\cdot\nabla\frac{1}{r}\right) \right\rangle \right] \,.$$

As

$$\left\langle (\mathbf{s} \cdot \nabla)\left(\mathbf{I} \cdot \nabla \frac{1}{r}\right)\right\rangle = \left\langle \sum_{i,j=1}^{3} s_i I_j \frac{\partial^2}{\partial x_i \partial x_j}\left(\frac{1}{r}\right)\right\rangle$$

$$= \left\langle \sum_{i,j=1}^{3} s_i I_j \frac{\partial^2}{\partial x_i^2}\left(\frac{1}{r}\right)\right\rangle + \left\langle \sum_{\substack{i,j=1 \\ i \neq j}}^{3} s_i I_j \frac{\partial^2}{\partial x_i \partial x_j}\left(\frac{1}{r}\right)\right\rangle$$

$$= \frac{1}{3}\left\langle \mathbf{s} \cdot \mathbf{I} \nabla^2 \left(\frac{1}{r}\right)\right\rangle = -\frac{4\pi}{3}\langle \mathbf{s} \cdot \mathbf{I}\delta(\mathbf{r})\rangle \,,$$

we have

$$\Delta E = \frac{2\mu_0 g_N \mu_N \mu_B}{4\pi \hbar^2} \cdot \frac{8\pi}{3} \langle \mathbf{s} \cdot \mathbf{I}\delta(\mathbf{r})\rangle$$

$$= \frac{\mu_0 g_N \mu_N \mu_B}{4\pi}[F(F+1) - I(I+1) - s(s+1)] \cdot \frac{8\pi}{3}\langle \delta(\mathbf{r})\rangle$$

$$= \frac{2\mu_0 g_N \mu_N \mu_B}{3\pi} \cdot \frac{Z^3}{a_0^3 n^3} \cdot [F(F+1) - I(I+1) - s(s+1)] \,.$$

1054

What is meant by the fine structure and hyperfine structure of spectral lines? Discuss their physical origins. Give an example of each, including an estimate of the magnitude of the effect. Sketch the theory of one of the effects.

(Princeton)

Solution:

(a) *Fine structure:* The spectral terms as determined by the principal quantum number n and the orbital angular momentum quantum number l are split due to a coupling between the electron spin $\mathbf{s}$ and orbital angular momentum $\mathbf{l}$. Consequently the spectral lines arising from transitions between the energy levels are each split into several lines. For example, the spectral line arising from the transition $3p \rightarrow 3s$ of Na atom shows a doublet structure, the two yellow lines D_1 (5896 Å), D_2 (5890 Å) which are close to each other.

As an example of numerical estimation, consider the fine structure in hydrogen.

The magnetic field caused by the orbital motion of the electron is $B = \frac{\mu_0 ev}{4\pi r^2}$. The dynamic equation $\frac{mv^2}{r} = \frac{e^2}{4\pi\varepsilon_0 r^2}$ and the quantization condition $mvr = n\hbar$ give $v = \alpha c/n$, where $\alpha = \frac{e^2}{\hbar c}$ is the fine-structure constant, $n = 1, 2, 3, \ldots$. For the ground state $n = 1$. Then the interaction energy between the spin magnetic moment μ_s of the electron and the magnetic field B is

$$\Delta E \approx -\mu_s B \approx \frac{\mu_0 \mu_B \alpha ec}{4\pi r^2} \, ,$$

where $\mu_s = -\frac{e\hbar}{2m} = -\mu_B$, the Bohr magnetron. Take $r \approx 10^{-10}$ m, we find

$$\Delta E \approx 10^{-7} \times 10^{-23} \times 10^{-2} \times 10^{-19} \times 10^8 / 10^{-20} \approx 10^{-23} \text{ J} \approx 10^{-4} \text{ eV} \, .$$

Considering an electron moving in a central potential $V(\mathbf{r}) = -\frac{Ze^2}{4\pi\varepsilon_0 r}$, the interaction Hamiltonian between its orbital angular momentum about the center, l, and spin s can be obtained quantum mechanically following the same procedure as

$$H' = \frac{1}{2m^2c^2} \frac{1}{r} \frac{dV}{dr} (\mathbf{s} \cdot \mathbf{l}) \, .$$

Taking H' as perturbation we then obtain the first order energy correction

$$\Delta E_{nlj} = \langle H' \rangle = \frac{Rhc\alpha^2 Z^4 \left[j(j+1) - l(l+1) - \frac{3}{4} \right]}{2n^3 l \left(l + \frac{1}{2} \right)(l+1)} \, ,$$

where R is the Rydberg constant, j is the total angular momentum of the electron.

As states with different j have different ΔE_{nlj}, an energy level (n, l) is split into two levels with $j = l + 1/2$ and $j = l - 1/2$.

(b) *Hyperfine structure:* Taking into account the coupling between the nuclear spin I and the total angular momentum j of the orbiting electron, an energy level determined by j will be split further, forming a hyperfine structure. Using an instrument of high resolution, we can see that the D_1 spectral line of Na atom is actually composed of two lines with a separation of 0.023 Å, and the D_2 line is composed of two lines separated by 0.021 Å.

For ground state hydrogen atom, the magnetic field caused by the electron at the nucleus is $B = \frac{\mu_0}{4\pi} \frac{ev}{a^2}$, where a is the Bohr radius. The hyperfine structure splitting is

$$\Delta E \approx \mu_N B \approx \frac{\mu_0}{4\pi} \frac{\mu_N e a c}{a^2}$$

$$\approx 10^{-7} \times \frac{5 \times 10^{-27} \times 1.6 \times 10^{-19} \times 3 \times 10^8}{137 \times (0.53 \times 10^{-10})^2} \text{ J}$$

$$\approx 10^{-7} \text{ eV}.$$

A theory of hyperfine structure is outlined in **Problem 1053**.

1055

Calculate, to an order of magnitude, the following properties of the $2p$-$1s$ electromagnetic transition in an atom formed by a muon and a strontium nucleus ($Z = 38$):

(a) the fine-structure splitting,
(b) the natural line width. (Hint: the lifetime of the $2p$ state of hydrogen is 10^{-9} sec)

(*Princeton*)

Solution:

Taking into account the hyperfine structure corrections, the energy levels of a hydrogen-like atom are given by

$$E = E_0 + \Delta E_r + \Delta E_{ls}$$

$$= \begin{cases} -\dfrac{RhcZ^2}{n^2} - \dfrac{Rhc\alpha^2 Z^4}{n^3}\left(\dfrac{1}{l} - \dfrac{3}{4n}\right), & \left(j = l - \dfrac{1}{2}\right) \\[3mm] -\dfrac{RhcZ^2}{n^2} - \dfrac{Rhc\alpha^2 Z^4}{n^3}\left(\dfrac{1}{l+1} - \dfrac{3}{4n}\right). & \left(j = l + \dfrac{1}{2}\right) \end{cases}$$

The $1s$ state is not split, but the $2p$ state is split into two substates corresponding to $j = 1/2$ and $j = 3/2$. The energy difference between the two lines of $2p \rightarrow 1s$ is

$$\Delta E = \frac{Rhc\alpha^2 Z^4}{n^3}\left(\frac{1}{l} - \frac{1}{l+1}\right),$$

where $Z = 38$, $n = 2$, $l = 1$, $R = m_\mu R_H / m_e \approx 200 R_H = 2.2 \times 10^9$ m^{-1}, $\alpha = \frac{1}{137}$. Hence

$$\Delta E = \frac{2.2 \times 10^9 \times 4.14 \times 10^{-15} \times 3 \times 10^8 \times 38^4}{2^3 \times 137^2 \times 2} = 1.9 \times 10^4 \text{ eV}.$$

(b) The lifetime of the $2p$ state of μ-mesic atom is

$$\tau_\mu = \frac{1}{Z^4} \cdot \frac{m_e}{m_\mu} \tau_H = 2.4 \times 10^{-18} \text{ s}.$$

The uncertainty principle gives the natural width of the level as

$$\Gamma \approx \hbar / \tau_\mu = 2.7 \times 10^2 \text{ eV}.$$

1056

The lowest-energy optical absorption of neutral alkali atoms corresponds to a transition $ns \to (n+1)p$ and gives rise to a characteristic doublet structure. The intensity ratio of these two lines for light alkalis is 2; but as Z increases, so does the ratio, becoming 3.85 for Cs $(6s \to 7p)$.

(a) Write an expression for the spin-orbit operator $N(r)$.

(b) In a hydrogenic atom, is this operator diagonal in the principal quantum number n? Is it diagonal in J?

(c) Using the following data, evaluate approximately the lowest order correction to the intensity ratio for the Cs doublet:

E_n = energy of the np state in cm^{-1},

I_n = transition intensity for the unperturbed states from the $6s$ state to the np state,

$$I_6 / I_7 = 1.25, \qquad I_8 / I_7 = 0.5,$$

Δn = spin-orbit splitting of the np state in cm^{-1},

$$\Delta_6 = 554 \qquad E_6 = -19950,$$

$$\Delta_7 = 181 \qquad E_7 = -9550,$$

$$\Delta_8 = 80 \qquad E8 = -5660.$$

In evaluating the terms in the correction, you may assume that the states can be treated as hydrogenic.

HINT: For small r, the different hydrogenic radial wave functions are proportional: $f_m(r) = k_{mn} f_n(r)$, so that, to a good approximation, $\langle 6p|N(r)|6p \rangle \approx k_{67} \langle 7p|N(r)|6p \rangle \approx k_{67}^2 \langle 7p|N(r)|7p \rangle$.

(Princeton)

Solution:

(a) The spin-orbit interaction Hamiltonian is

$$N(r) = \frac{1}{2\mu^2 c^2 r} \frac{dV}{dr} \hat{\mathbf{s}} \cdot \hat{\mathbf{l}}$$

$$= \frac{1}{4\mu^2 c^2 r} \frac{dV}{dr} (\hat{\mathbf{j}}^2 - \hat{\mathbf{l}}^2 - \hat{\mathbf{s}}^2),$$

where μ is the reduced mass, and $V = -\frac{Ze^2}{4\pi\varepsilon_0 r}$.

(b) The Hamiltonian is $H = H_0 + N(r)$. For hydrogen atom, $[H_0, N(r)] \neq 0$, so in the principal quantum number n, $N(r)$ is not diagonal. Generally,

$$\langle nlm|N(r)|klm \rangle \neq 0 .$$

In the total angular momentum j (with fixed n), since $[N(r), \hat{\mathbf{j}}^2] = 0$, $N(r)$ is diagonal.

(c) The rate of induced transition is

$$W_{k'k} = \frac{4\pi^2 e^2}{3\hbar^2} |\mathbf{r}_{k'k}|^2 \rho(\omega_{k'k})$$

and the intensity of the spectral line is $I(\omega_{k'k}) \propto \hbar\omega_{k'k} W_{k'k}$.

With coupling between spin and orbital angular momentum, each np energy level of alkali atom is split into two sub-levels, corresponding to $j = 3/2$ and $j = 1/2$. However as the s state is not split, the transition $ns \rightarrow (n+1)p$ will give rise to a doublet. As the splitting of the energy level is very small, the frequencies of the $ns \rightarrow (n+1)p$ double lines can be taken to be approximately equal and so $I \propto |\mathbf{r}_{k'k}|^2$.

The degeneracy of the $j = 3/2$ state is 4, with $j_z = 3/2, 1/2, -1/2, -3/2$; the degeneracy of the $j = 1/2$ state is 2, with $j_z = 1/2, -1/2$. In the zeroth order approximation, the intensity ratio of these two lines is

$$\frac{I\left(j=\frac{3}{2}\right)}{I\left(j=\frac{1}{2}\right)} = \frac{\displaystyle\sum_{j_z}\left|\left\langle (n+1)p\frac{3}{2}\middle|\mathbf{r}\middle|ns\right\rangle\right|^2}{\displaystyle\sum_{j_z}\left|\left\langle (n+1)p\frac{1}{2}\middle|\mathbf{r}\middle|ns\right\rangle\right|^2} \approx 2\,,$$

as given. In the above $|(n+1)p, 1/2\rangle$, $|(n+1)p, 3/2\rangle$ are respectively the zeroth order approximate wave functions of the $j = 1/2$ and $j = 3/2$ states of the energy level $(n+1)p$.

To find the intensity ratio of the two lines of $6s \to 7p$ transition of Cs atom, take $N(r)$ as perturbation. First calculate the approximate wave functions:

$$\Psi_{3/2} = \left|7p\frac{3}{2}\right\rangle + \sideset{}{'}\sum_{n=6}^{\infty} \frac{\left\langle np\frac{3}{2}\middle|N(r)\middle|7p\frac{3}{2}\right\rangle}{E_7 - E_n}\left|np\frac{3}{2}\right\rangle\,,$$

$$\Psi_{1/2} = \left|7p\frac{1}{2}\right\rangle + \sideset{}{'}\sum_{n=6}^{\infty} \frac{\left\langle np\frac{1}{2}\middle|N(r)\middle|7p\frac{1}{2}\right\rangle}{E_7 - E_n}\left|np\frac{1}{2}\right\rangle\,,$$

and then the matrix elements:

$$|\langle\Psi_{3/2}|\mathbf{r}|6s\rangle|^2 = \left|\left\langle 7p\frac{3}{2}\middle|\mathbf{r}\middle|6s\right\rangle + \sideset{}{'}\sum_{n=6}^{\infty}\frac{\left\langle np\frac{3}{2}\middle|N(\mathbf{r})\middle|7p\frac{3}{2}\right\rangle}{E_7 - E_n}\left\langle np\frac{3}{2}\middle|\mathbf{r}\middle|6s\right\rangle\right|^2$$

$$\approx \left|\left\langle 7p\frac{3}{2}\middle|\mathbf{r}\middle|6s\right\rangle\right|^2\left|1 + \sideset{}{'}\sum_{n=6}^{\infty}\frac{\left\langle np\frac{3}{2}\middle|N(\mathbf{r})\middle|7p\frac{3}{2}\right\rangle}{E_7 - E_n}\sqrt{\frac{I_n}{I_7}}\right|^2\,,$$

$$|\langle\Psi_{1/2}|\mathbf{r}|6s\rangle|^2 = \left|\left\langle 7p\frac{1}{2}\middle|\mathbf{r}\middle|6s\right\rangle + \sideset{}{'}\sum_{n=6}^{\infty}\frac{\left\langle np\frac{1}{2}\middle|N(\mathbf{r})\middle|7p\frac{1}{2}\right\rangle}{E_7 - E_n}\left\langle np\frac{1}{2}\middle|\mathbf{r}\middle|6s\right\rangle\right|^2$$

$$\approx \left|\left\langle 7p\frac{1}{2}\middle|\mathbf{r}\middle|6s\right\rangle\right|^2\left|1 + \sideset{}{'}\sum_{n=6}^{\infty}\frac{\left\langle np\frac{1}{2}\middle|N(\mathbf{r})\middle|7p\frac{1}{2}\right\rangle}{E_7 - E_n}\sqrt{\frac{I_n}{I_7}}\right|^2\,,$$

where

$$\frac{\left\langle np\frac{3}{2}\middle|\mathbf{r}\middle|6s\right\rangle}{\left\langle 7p\frac{3}{2}\middle|\mathbf{r}\middle|6s\right\rangle} \approx \frac{\left\langle np\frac{1}{2}\middle|\mathbf{r}\middle|6s\right\rangle}{\left\langle 7p\frac{1}{2}\middle|\mathbf{r}\middle|6s\right\rangle} \approx \sqrt{\frac{I_n}{I_7}}\,.$$

As

$$N(r) = \frac{1}{4\mu^2 c^2 r}\frac{dV}{dr}(\hat{\mathbf{j}}^2 - \hat{\mathbf{l}}^2 - \hat{\mathbf{s}}^2)$$

$$= F(r)(\hat{\mathbf{j}}^2 - \hat{\mathbf{l}}^2 - \hat{\mathbf{s}}^2)\,,$$

where

$$F(r) \equiv \frac{1}{4\mu^2 c^2 r}\frac{dV}{dr}\,,$$

we have

$$\left\langle np\frac{3}{2}\middle|N(\mathbf{r})\middle|7p\frac{3}{2}\right\rangle = \left\langle np\frac{3}{2}\middle|F(r)(\hat{\mathbf{j}}^2 - \hat{\mathbf{l}}^2 - \hat{\mathbf{s}}^2)\middle|7p\frac{3}{2}\right\rangle$$

$$= \left[\frac{3}{2}\times\left(\frac{3}{2}+1\right) - 1\times(1+1) - \frac{1}{2}\times\left(\frac{1}{2}+1\right)\right]\hbar^2$$

$$\times\langle np|F(r)|7p\rangle = \hbar^2\langle np|F(r)|7p\rangle\,,$$

$$\left\langle np\frac{1}{2}\middle|N(\mathbf{r})\middle|7p\frac{1}{2}\right\rangle = -2\hbar^2\langle np|F(r)|7p\rangle\,.$$

For $n = 7$, as

$$\Delta_7 = \left\langle 7p\frac{3}{2}\middle|N(r)\middle|7p\frac{3}{2}\right\rangle - \left\langle 7p\frac{1}{2}\middle|N(r)\middle|7p\frac{1}{2}\right\rangle = 3\hbar^2\langle 7p|F(r)|7p\rangle\,,$$

we have

$$\langle 7p|F(r)|7p\rangle = \frac{\Delta_7}{3\hbar^2}\,.$$

For $n = 6$, we have

$$\left\langle 6p\frac{3}{2}\middle|N(r)\middle|7p\frac{3}{2}\right\rangle = \hbar^2\langle 6p|F(r)|7p\rangle = \hbar^2 k_{67}\langle 7p|F(r)|7p\rangle = \frac{k_{67}}{3}\Delta_7\,,$$

$$\left\langle 6p\frac{1}{2}\middle|N(r)\middle|7p\frac{1}{2}\right\rangle = -2\hbar^2\langle 6p|F(r)|7p\rangle$$

$$= -2\hbar^2 k_{67}\langle 7p|F(r)|7p\rangle = -\frac{2k_{67}}{3}\Delta_7\,.$$

For $n = 8$, we have

$$\left\langle 8p\frac{3}{2} \middle| N(r) \middle| 7p\frac{3}{2} \right\rangle = \frac{k_{87}}{3}\Delta_7 ,$$

$$\left\langle 8p\frac{1}{2} \middle| N(r) \middle| 7p\frac{1}{2} \right\rangle = -\frac{2k_{87}}{3}\Delta_7 .$$

In the above

$$k_{67} = \frac{\langle 6p|F(r)|7p\rangle}{\langle 7p|F(r)|7p\rangle} ,$$

$$k_{87} = \frac{\langle 8p|F(r)|7p\rangle}{\langle 7p|F(r)|7p\rangle} .$$

Hence

$$|\langle \Psi_{3/2}|\mathbf{r}|6s\rangle|^2 = \left|\left\langle 7p\frac{3}{2} \middle| \mathbf{r} \middle| 6s \right\rangle\right|^2$$

$$\times \left|1 + \frac{k_{67}\Delta_7}{3(E_7 - E_6)}\sqrt{\frac{I_6}{I_7}} + \frac{k_{87}\Delta_7}{3(E_7 - E_8)}\sqrt{\frac{I_8}{I_7}}\right|^2 ,$$

$$|\langle \Psi_{1/2}|\mathbf{r}|6s\rangle|^2 = \left|\left\langle 7p\frac{1}{2} \middle| \mathbf{r} \middle| 6s \right\rangle\right|^2$$

$$\times \left|1 - \frac{2k_{67}\Delta_7}{3(E_7 - E_6)}\sqrt{\frac{I_6}{I_7}} - \frac{2k_{87}\Delta_7}{3(E_7 - E_8)}\sqrt{\frac{I_8}{I_7}}\right|^2 .$$

As

$$\Delta_6 = \left\langle 6p\frac{3}{2} \middle| N(r) \middle| 6p\frac{3}{2} \right\rangle - \left\langle 6p\frac{1}{2} \middle| N(r) \middle| 6p\frac{1}{2} \right\rangle = 3\hbar^2 \langle 6p|F(r)|6p\rangle$$

$$= 3\hbar^2 k_{67}^2 \langle 7p|F(r)|7p\rangle = k_{67}^2 \Delta_7 ,$$

we have

$$k_{67} = \sqrt{\frac{\Delta_6}{\Delta_7}} ,$$

and similarly

$$k_{87} = \sqrt{\frac{\Delta_8}{\Delta_7}} .$$

Thus

$$\frac{I\left(j = \frac{3}{2}\right)}{I\left(j = \frac{1}{2}\right)} = \frac{\sum_{j_z} |\langle \Psi_{3/2}|\mathbf{r}|6s\rangle|^2}{\sum_{j_z} |\langle \Psi_{1/2}|\mathbf{r}|6s\rangle|^2}$$

$$\approx 2 \left|\frac{1 + \dfrac{k_{67}\Delta_7}{3(E_7 - E_6)}\sqrt{\dfrac{I_6}{I_7}} + \dfrac{k_{87}\Delta_7}{3(E_7 - E_8)}\sqrt{\dfrac{I_8}{I_7}}}{1 - \dfrac{2k_{67}\Delta_7}{3(E_7 - E_6)}\sqrt{\dfrac{I_6}{I_7}} - \dfrac{2k_{87}\Delta_7}{3(E_7 - E_8)}\sqrt{\dfrac{I_8}{I_7}}}\right|^2$$

$$= 2 \left|\frac{1 + \dfrac{\sqrt{\Delta_6 \Delta_7}}{3(E_7 - E_6)}\sqrt{\dfrac{I_6}{I_7}} + \dfrac{\sqrt{\Delta_8 \Delta_7}}{3(E_7 - E_8)}\sqrt{\dfrac{I_8}{I_7}}}{1 - \dfrac{2\sqrt{\Delta_6 \Delta_7}}{3(E_7 - E_6)}\sqrt{\dfrac{I_6}{I_7}} - \dfrac{2\sqrt{\Delta_8 \Delta_7}}{3(E_7 - E_8)}\sqrt{\dfrac{I_8}{I_7}}}\right|^2 = 3.94,$$

using the data supplied.

1057

An atomic clock can be based on the (21-cm) ground-state hyperfine transition in atomic hydrogen. Atomic hydrogen at low pressure is confined to a small spherical bottle ($r \ll \lambda = 21$ cm) with walls coated by Teflon. The magnetically neutral character of the wall coating and the very short "dwell-times" of the hydrogen on Teflon enable the hydrogen atom to collide with the wall with little disturbance of the spin state. The bottle is shielded from external magnetic fields and subjected to a controlled weak and uniform field of prescribed orientation. The resonant frequency of the gas can be detected in the absorption of 21-cm radiation, or alternatively by subjecting the gas cell to a short radiation pulse and observing the coherently radiated energy.

(a) The Zeeman effect of these hyperfine states is important. Draw an energy level diagram and give quantum numbers for the hyperfine substates of the ground state as functions of field strength. Include both the weak and strong field regions of the Zeeman pattern.

(b) How can the energy level splitting of the strong field region be used to obtain a measure of the g-factor for the proton?

(c) In the weak field case one energy-level transition is affected little by the magnetic field. Which one is this? Make a rough estimate of the maximum magnetic field strength which can be tolerated with the resonance frequency shifted by $\Delta\nu < 10^{-10}\,\nu$.

(d) There is no Doppler broadening of the resonance line. Why is this?

(*Princeton*)

Solution:

(a) Taking account of the hyperfine structure and the Zeeman effect, two terms are to be added to the Hamiltonian of hydrogen atom:

$$H_{hf} = A\mathbf{I}\cdot\mathbf{J}\,, \qquad (A > 0)$$

$$H_B = -\boldsymbol{\mu}\cdot\mathbf{B}\,.$$

For the ground state of hydrogen,

$$I = \frac{1}{2}, \qquad J = \frac{1}{2}$$

$$\boldsymbol{\mu} = -g_e\frac{e\hbar}{2m_ec}\frac{\mathbf{J}}{\hbar} + g_p\frac{e\hbar}{2m_pc}\frac{\mathbf{I}}{\hbar}\,.$$

Letting

$$\frac{e\hbar}{2m_ec} = \mu_B\,, \qquad \frac{e\hbar}{2m_pc} = \mu_N$$

and using units in which $\hbar = 1$ we have

$$\boldsymbol{\mu} = -g_e\mu_B\mathbf{J} + g_p\mu_N\mathbf{I}\,.$$

(1) *Weak magnetic field case.* $\langle H_{nf}\rangle \gg \langle H_B\rangle$, we couple $\mathbf{I}$, $\mathbf{J}$ as $\mathbf{F} = \mathbf{I} + \mathbf{J}$. Then taking H_{hf} as the main Hamiltonian and H_B as perturbation we solve the problem in the representation of $\{\hat{\mathbf{F}}^2, \hat{\mathbf{I}}^2, \hat{\mathbf{J}}^2, \hat{\mathbf{F}}_z\}$. As

$$H_{hf} \doteq \frac{A}{2}(\hat{\mathbf{F}}^2 - \hat{\mathbf{I}}^2 - \hat{\mathbf{J}}^2) = \frac{A}{2}\left(\hat{\mathbf{F}}^2 - \frac{1}{2}\cdot\frac{3}{2} - \frac{1}{2}\cdot\frac{3}{2}\right) = \frac{A}{2}\left(\hat{\mathbf{F}}^2 - \frac{3}{2}\right),$$

we have

$$\Delta E_{hf} = \begin{cases} -\dfrac{3}{4}A & \text{for } F = 0 \\[2mm] \dfrac{1}{4}A & \text{for } F = 1\,. \end{cases}$$

In the subspace of $\{\hat{\mathbf{F}}^2, \hat{\mathbf{F}}_z\}$, the Wigner-Ecart theory gives

$$\langle \mu \rangle = \frac{(-g_e\mu_B \mathbf{J} + g_p\mu_N \mathbf{I}) \cdot \mathbf{F}}{F^2} \mathbf{F} \, .$$

As for $I = J = \frac{1}{2}$,

$$\mathbf{J} \cdot \mathbf{F} = \frac{1}{2}(\hat{\mathbf{F}}^2 + \hat{\mathbf{J}}^2 - \hat{\mathbf{I}}^2) = \frac{1}{2}\hat{\mathbf{F}}^2 \, ,$$

$$\mathbf{I} \cdot \mathbf{F} = \frac{1}{2}(\hat{\mathbf{F}}^2 + \hat{\mathbf{I}}^2 - \hat{\mathbf{J}}^2) = \frac{1}{2}\hat{\mathbf{F}}^2 \, ,$$

we have

$$\langle \mu \rangle = -\frac{g_e\mu_B - g_p\mu_N}{2}\hat{\mathbf{F}} \, .$$

Then as

$$H_B = -\boldsymbol{\mu} \cdot \mathbf{B} = \frac{g_e\mu_B - g_p\mu_N}{2} B\hat{F}_z \, ,$$

we have

$$\Delta E_B = \begin{cases} E_1, & (F_z = 1) \\ 0, & (F_z = 0) \\ -E_1, & (F_z = -1) \end{cases}$$

where

$$E_1 = \frac{g_e\mu_B - g_p\mu_N}{2} B \, .$$

(2) *Strong magnetic field case.* As $\langle H_B \rangle \gg \langle H_{hf} \rangle$, we can treat H_B as the main Hamiltonian and H_{hf} as perturbation. With $\{\hat{\mathbf{J}}^2, \hat{\mathbf{I}}^2, \hat{\mathbf{J}}_z, \hat{\mathbf{I}}_z\}$ as a complete set of mechanical quantities, the base of the subspace is $|++\rangle$, $|+-\rangle$, $|-+\rangle$, $|--\rangle$ (where $|++\rangle$ means $J_z = +1/2$, $I_z = +1/2$, etc.). The energy correction is

$$\Delta E = \langle H_{hf} + H_B \rangle = \langle A I_z J_z \rangle + g_e\mu_B B\langle J_z \rangle - g_p\mu_N B\langle I_z \rangle$$

$$= \begin{cases} E_1 + \dfrac{A}{4} & \text{for } |++\rangle, \\ E_2 - \dfrac{A}{4} & \text{for } |+-\rangle, \\ -E_2 - \dfrac{A}{4} & \text{for } |-+\rangle, \\ -E_1 + \dfrac{A}{4} & \text{for } |--\rangle, \end{cases}$$

where

$$E_1 = \frac{g_e\mu_B - g_p\mu_N}{2}B,$$

$$E_2 = \frac{g_e\mu_B + g_p\mu_N}{2}B.$$

The quantum numbers of the energy sublevels are given below and the energy level scheme is shown in Fig. 1.18.

quantum numbers	(F, J, I, F_z),	(J, I, J_z, I_z)
sublevel	$(1, 1/2, 1/2, 1)$	$(1/2, 1/2, 1/2, -1/2)$
	$(1, 1/2, 1/2, 0)$	$(1/2, 1/2, 1/2, 1/2)$
	$(1, 1/2, 1/2, -1)$	$(1/2, 1/2, -1/2, -1/2)$
	$(0, 1/2, 1/2, 0)$	$(1/2, 1/2, -1/2, 1/2)$

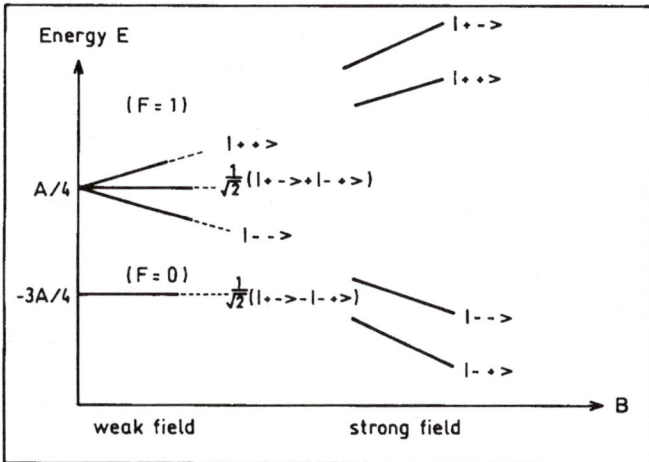

Fig. 1.18

(b) In a strong magnetic field, the gradients of the energy levels with respect to B satisfy the relation

$$\frac{\dfrac{\Delta E_{|+-\rangle}}{\Delta B} + \dfrac{\Delta E_{|--\rangle}}{\Delta B}}{\dfrac{\Delta E_{|+-\rangle}}{\Delta B} + \dfrac{\Delta E_{|++\rangle}}{\Delta B}} = \frac{g_p \mu_N}{g_e \mu_B} \,,$$

which may be used to determine g_p if the other quantities are known.

(c) In a weak magnetic field, the states $|F = 1, F_z = 0\rangle$, $|F = 0, F_z = 0\rangle$ are not appreciably affected by the magnetic field, so is the transition energy between these two states. This conclusion has been reached for the case of weak magnetic field $(A \gg E_1)$ considering only the first order effect. It may be expected that the effect of magnetic field on these two states would appear at most as second order of E_1/A. Thus the dependence on B of the energy of the two states is

$$\left(\frac{E_1}{A}\right)^2 \cdot A = \frac{E_1^2}{A} \,,$$

and so

$$\frac{\Delta \nu}{\nu} = \frac{\Delta E}{E} \approx \frac{\dfrac{E_1^2}{A}}{\dfrac{A}{4} - \left(-\dfrac{3A}{4}\right)} = \frac{E_1^2}{A^2} \approx \left(\frac{g_e \mu_B B}{2A}\right)^2 \,,$$

neglecting $g_p \mu_N$. For $\Delta \nu / \nu < 10^{-10}$ and the 21-cm line we have

$$A = \frac{1}{4}A - \left(-\frac{3}{4}A\right) = h\nu = \frac{2\pi \hbar c}{\lambda} \approx \frac{2\pi \times 2 \times 10^{-5}}{21} = 6 \times 10^{-6} \text{ eV} \,,$$

and so

$$B \le \left(\frac{2A}{g_e \mu_B}\right)^2 \times 10^{-5} = \left(\frac{2 \times 6 \times 10^{-6}}{2 \times 6 \times 10^{-9}}\right) \times 10^{-5} = 10^{-2} \text{ Gs} \,.$$

(d) The resonance energy is very small. When photon is emitted, the ratio of the recoil energy of the nucleon to that of the photon E, $\Delta E/E \ll 1$. Hence the Doppler broadening caused by recoiling can be neglected.

1058

Consider an atom formed by the binding of an Ω^- particle to a bare Pb nucleus $(Z = 82)$.

(a) Calculate the energy splitting of the $n = 10$, $l = 9$ level of this atom due to the spin-orbit interaction. The spin of the Ω^- particle is $3/2$. Assume a magnetic moment of $\mu = \frac{e\hbar}{2mc}g\mathbf{p}_s$ with $g = 2$ and $m = 1672$ MeV/c^2.

Note:

$$\left\langle \frac{1}{r^3} \right\rangle = \left(\frac{mc^2}{\hbar c} \right)^3 (\alpha Z)^3 \frac{1}{n^3 l \left(l + \dfrac{1}{2} \right) (l + 1)}$$

for a particle of mass m bound to a charge Z in a hydrogen-like state of quantum numbers (n, l).

(b) If the Ω^- has an electric quadrupole moment $Q \sim 10^{-26}$ cm² there will be an additional energy shift due to the interaction of this moment with the Coulomb field gradient $\partial E_z / \partial z$. Estimate the magnitude of this shift; compare it with the results found in (a) and also with the total transition energy of the $n = 11$ to $n = 10$ transition in this atom.

(Columbia)

Solution:

(a) The energy of interaction between the spin and orbital magnetic moments of the Ω^- particle is

$$\Delta E_{ls} = Z \boldsymbol{\mu}_l \cdot \boldsymbol{\mu}_s \left\langle \frac{1}{r^3} \right\rangle \,,$$

where

$$\boldsymbol{\mu}_l = \frac{e}{2mc} \mathbf{p}_l, = \frac{e\hbar}{2mc} \mathbf{l} \,,$$

$$\boldsymbol{\mu}_s = \frac{e}{mc} \mathbf{p}_s, = \frac{e\hbar}{mc} \mathbf{s} \,,$$

$\mathbf{p}_l$, $\mathbf{p}_s$ being the orbital and spin angular momenta. Thus

$$\Delta E_{ls} = \frac{Ze^2 \hbar^2}{2m^2 c^2} \left\langle \frac{1}{r^3} \right\rangle \mathbf{l} \cdot \mathbf{s} \,.$$

As

$$\mathbf{l} \cdot \mathbf{s} = \frac{1}{2} [(\mathbf{l} + \mathbf{s})^2 - l^2 - s^2] \,,$$

we have

$$\Delta E_{ls} = \frac{Ze^2 \hbar^2}{2m^2 c^2} \left\langle \frac{1}{r^3} \right\rangle \frac{(\mathbf{j}^2 - \mathbf{l}^2 - \mathbf{s}^2)}{2}$$

$$= \frac{(Z\alpha)^4 mc^2}{4} \left[\frac{j(j+1) - l(l+1) - s(s+1)}{n^3 l \left(l + \dfrac{1}{2} \right) (l + 1)} \right] \,.$$

With $Z = 82$, $m = 1672$ MeV/c^2, $s = 3/2$, $n = 10$, $l = 9$, $\alpha = \frac{1}{137}$, and $\langle 1/r^3 \rangle$ as given, we find $\Delta E_{ls} = 62.75 \times [j(j+1) - 93.75]$ eV. The results are given in the table below.

j	ΔE_{ls} (eV)	Level splitting (eV)
19/2	377	1193
17/2	−816	1067
15/2	−1883	941
13/2	−2824	

(b) The energy shift due to the interaction between the electric quadrupole moment Q and the Coulomb field gradient $\frac{\partial E_z}{\partial z}$ is

$$\Delta E_Q \approx Q \left\langle \frac{\partial E_z}{\partial z} \right\rangle,$$

where $\frac{\partial E_z}{\partial z}$ is the average value of the gradient of the nuclear Coulomb field at the site of Ω^-. As

$$\left\langle \frac{\partial E_z}{\partial z} \right\rangle \approx -\left\langle \frac{1}{r^3} \right\rangle,$$

we have

$$\Delta E_Q \approx -Q \left\langle \frac{1}{r^3} \right\rangle$$

in the atomic units of the hyperon atom which have units of length and energy, respectively,

$$a = \frac{\hbar^2}{me^2} = \frac{\hbar c}{mc^2} \left(\frac{\hbar c}{e^2} \right) = \frac{1.97 \times 10^{11}}{1672} \times 137 = 1.61 \times 10^{-12} \text{ cm},$$

$$\varepsilon = \frac{me^4}{\hbar^2} = mc^2 \left(\frac{e^2}{\hbar c} \right)^2 = \frac{1672 \times 10^6}{137^2} = 8.91 \times 10^4 \text{ eV}.$$

For $n = 10$, $l = 9$, $\langle \frac{1}{r^3} \rangle = 1.53 \times 10^{35}$ cm^{-3} ≈ 0.6 a.u. With $Q \approx 10^{-26}$ cm^2 $\approx 4 \times 10^{-3}$ a.u., we have

$$\Delta E_Q \approx 2.4 \times 10^{-3} \text{ a.u.} \approx 2 \times 10^2 \text{ eV}.$$

The total energy resulting from a transition from $n = 11$ to $n = 10$ is

$$\Delta E = \frac{Z^2 mc^2}{2} \left(\frac{e^2}{\hbar c}\right)^2 \left(\frac{1}{10^2} - \frac{1}{11^2}\right)$$

$$= \frac{82^2 \times 1672 \times 10^6}{2 \times 137^2} \left(\frac{1}{10^2} - \frac{1}{11^2}\right)$$

$$\approx 5 \times 10^5 \text{ eV}.$$

1059

What is the energy of the photon emitted in the transition from the $n = 3$ to $n = 2$ level of the μ^- mesic atom of carbon? Express it in terms of the γ energy for the electronic transition from $n = 2$ to $n = 1$ of hydrogen, given that $m_\mu/m_e = 210$.

(*Wisconsin*)

Solution:

The energy of the μ^- atom of carbon is

$$E_n(\mu) = \frac{Z^2 m_\mu}{m_e} E_n(H),$$

where $E_n(H)$ is the energy of the corresponding hydrogen atom, and $Z = 6$.

The energy of the photon emitted in the transition from $n = 3$ to $n = 2$ level of the mesic atom is

$$\Delta E = \frac{Z^2 m_\mu}{m_e}[E_3(H) - E_2(H)].$$

As

$$-E_n(H) \propto \frac{1}{n^2},$$

we have

$$\frac{36}{5}[E_3(H) - E_2(H)] = \frac{4}{3}[E_2(H) - E_1(H)],$$

and hence

$$\Delta E = \frac{5Z^2 m_\mu}{27 m_e}[E_2(H) - E_1(H)]$$

$$= 1400[E_2(H) - E_1(H)],$$

where $E_2(H) - E_1(H)$ is the energy of the photon emitted in the transition from $n = 2$ to $n = 1$ level of hydrogen atom.

1060

The muon is a relatively long-lived elementary particle with mass 207 times the mass of electron. The electric charge and all known interactions of the muon are identical to those of the electron. A "muonic atom" consists of a neutral atom in which one electron is replaced by a muon.

(a) What is the binding energy of the ground state of muonic hydrogen?

(b) What ordinary chemical element does muonic lithium ($Z = 3$) resemble most? Explain your answer.

(*MIT*)

Solution:

(a) By analogy with the hydrogen atom, the binding energy of the ground state of the muonic atom is

$$E_\mu = \frac{m_\mu e^4}{2\hbar^2} = 207 E_H = 2.82 \times 10^3 \text{ eV} \,.$$

(b) A muonic lithium atom behaves chemically most like a He atom. As μ and electron are different fermions, they fill their own orbits. The two electrons stay in the ground state, just like those in the He atom, while the μ stays in its own ground state, whose orbital radius is $1/207$ of that of the electrons. The chemical properties of an atom is determined by the number of its outer most shell electrons. Hence the mesic atom behaves like He, rather than like Li.

1061

The Hamiltonian for a $(\mu^+ e^-)$ atom in the $n = 1$, $l = 0$ state in an external magnetic field is

$$H = a\mathbf{S}_\mu \cdot \mathbf{S}_e + \frac{|e|}{m_e c} \mathbf{S}_e \cdot \mathbf{B} - \frac{|e|}{m_\mu c} \mathbf{S}_\mu \cdot \mathbf{B} \,.$$

(a) What is the physical significance of each term? Which term dominates in the interaction with the external field?

(b) Choosing the z-axis along $\mathbf{B}$ and using the notation (F, M_F), where $\mathbf{F} = \mathbf{S}_\mu + \mathbf{S}_e$, show that $(1, +1)$ is an eigenstate of H and give its eigenvalue.

(c) An RF field can be applied to cause transition to the state $(0,0)$. Describe quantitatively how an observation of the decay $\mu^+ \to e^+ \nu_e \bar{\nu}_\mu$ could be used to detect the occurrence of this transition.

(*Wisconsin*)

Solution:

(a) In the Hamiltonian, the first term, $a\mathbf{S}_\mu \cdot \mathbf{S}_e$, describes the electromagnetic interaction between μ^+ and e^-, the second and third terms respectively describe the interactions between the electron and μ^+ with the external magnetic field.

(b) Denote the state of $F = 1$, $M_F = +1$ with Ψ. As $\mathbf{F} = \mathbf{S}_\mu + \mathbf{S}_e$, we have

$$\mathbf{S}_\mu \cdot \mathbf{S}_e = \frac{1}{2}(\mathbf{F}^2 - \mathbf{S}_\mu^2 - \mathbf{S}_e^2),$$

and hence

$$\mathbf{S}_\mu \cdot \mathbf{S}_e \Psi = \frac{1}{2}(\mathbf{F}^2\Psi - \mathbf{S}_\mu^2\Psi - \mathbf{S}_e^2\Psi) = \frac{\hbar^2}{2}\left(2\Psi - \frac{3}{4}\Psi - \frac{3}{4}\Psi\right) = \frac{\hbar^2}{4}\Psi.$$

In the common eigenvector representation of $\mathbf{S}_e^z$, $\mathbf{S}_\mu^z$, the Ψ state is represented by the spinor

$$\Psi = \begin{pmatrix} 1 \\ 0 \end{pmatrix}_e \otimes \begin{pmatrix} 1 \\ 0 \end{pmatrix}_\mu.$$

Then

$$\mathbf{S}_e^z \Psi = \frac{\hbar}{2}\sigma_e^z \Psi = \frac{\hbar}{2}\Psi,$$

$$\mathbf{S}_\mu^z \Psi = \frac{\hbar}{2}\sigma_\mu^z \Psi = \frac{\hbar}{2}\Psi,$$

and so

$$H = a\mathbf{S}_\mu \cdot \mathbf{S}_e \Psi + \frac{e}{m_e c}B\mathbf{S}_e^z\Psi - \frac{e}{m_\mu c}B\mathbf{S}_\mu^z\Psi$$

$$= a\frac{\hbar^2}{4}\Psi + \frac{eB}{m_e c}\cdot\frac{\hbar}{2}\Psi - \frac{eB}{m_\mu c}\cdot\frac{\hbar}{2}\Psi$$

$$= \left(\frac{1}{4}a\hbar^2 + \frac{eB}{2m_e c}\hbar - \frac{eB}{2m_\mu c}\hbar\right)\Psi.$$

Hence the $(1, +1)$ state is an eigenstate of H with eigenvalue

$$\left(\frac{1}{4} a \hbar^2 + \frac{eB}{2m_e c} \hbar - \frac{eB}{2m_\mu c} \hbar \right).$$

(c) The two particles in the state $(1, +1)$ have parallel spins, while those in the state $(0,0)$ have anti-parallel spins. So relative to the direction of spin of the electron, the polarization directions of μ^+ in the two states are opposite. It follows that the spin of the positrons arising from the decay of μ^+ is opposite in direction to the spin of the electron. An (e^+e^-) pair annihilate to give rise to 3γ or 2γ in accordance with whether their spins are parallel or antiparallel. Therefore if it is observed that the (e^+e^-) pair arising from the decay $\mu^+ \to e^+ \nu_e \tilde{\mu}_\mu$ annihilate to give 2γ, then it can be concluded that the transition is between the states $(1, +1)$ and $(0,0)$.

1062

Muonic atoms consist of mu-mesons (mass $m_\mu = 206 m_e$) bound to atomic nuclei in hydrogenic orbits. The energies of the mu mesic levels are shifted relative to their values for a point nucleus because the nuclear charge is distributed over a region with radius R. The effective Coulomb potential can be approximated as

$$V(r) = \begin{cases} -\dfrac{Ze^2}{r}, & (r \geq R) \\[2mm] -\dfrac{Ze^2}{R}\left(\dfrac{3}{2} - \dfrac{r^2}{2R^2}\right). & (r < R) \end{cases}$$

(a) State qualitatively how the energies of the $1s$, $2s$, $2p$, $3s$, $3p$, $3d$ muonic levels will be shifted absolutely and relative to each other, and explain physically any differences in the shifts. Sketch the unperturbed and perturbed energy level diagrams for these states.

(b) Give an expression for the first order change in energy of the $1s$ state associated with the fact that the nucleus is not point-like.

(c) *Estimate* the $2s$–$2p$ energy shift under the assumption that $R/a_\mu \ll$ 1, where a_μ is the "Bohr radius" for the muon and show that this shift gives a measure of R.

(d) When is the method of part (b) likely to fail? Does this method underestimate or overestimate the energy shift. Explain your answer in physical terms.

Useful information:

$$\Psi_{1s} = 2N_0 \exp\left(-\frac{r}{a_\mu}\right) Y_{00}(\theta, \phi),$$

$$\Psi_{2s} = \frac{1}{\sqrt{8}} N_0 \left(2 - \frac{r}{a_\mu}\right) \exp\left(-\frac{r}{2a_\mu}\right) Y_{00}(\theta, \phi),$$

$$\Psi_{2p} = \frac{1}{\sqrt{24}} N_0 \frac{r}{a_\mu} \exp\left(-\frac{r}{2a_\mu}\right) Y_{1m}(\theta, \phi),$$

$$N_0 = \frac{1}{a_\mu^{3/2}}.$$

(Wisconsin)

Solution:

(a) If nuclear charge is distributed over a finite volume, the intensity of the electric field at a point inside the nucleus is smaller than that at the same point if the nucleus is a point. Consequently the energy of the same state is higher in the former case. The probability of a 1s state electron staying in the nucleus is larger than that in any other state, so the effect of a finite volume of the nucleus on its energy level, i.e. the energy shift, is largest. Next come 2s, 3s, 2p, 3p, 3d, etc. The energy levels are shown in Fig. 1.19.

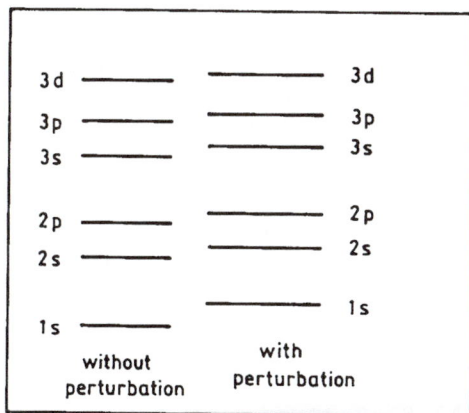

Fig. 1.19

(b) The perturbation potential due to the limited volume of nucleus has the form

$$\Delta V = \begin{cases} 0, & (r \geq R) \\ \dfrac{Ze^2}{R}\left(\dfrac{r^2}{2R^2} - \dfrac{3}{2} + \dfrac{R}{r}\right). & (r < R) \end{cases}$$

The first order energy correction of the 1s state with the approximation $\dfrac{R}{a_\mu} \ll 1$ is

$$\Delta E_{1s} = \int \Psi_{1s}^* \Delta V \Psi_{1s} d\tau$$

$$= \frac{Ze^2}{R} 4N_0^2 \int_0^R \exp\left(-\frac{2r}{a_\mu}\right) \cdot \left(\frac{r^2}{2R^2} - \frac{3}{2} + \frac{R}{r}\right) r^2 dr$$

$$\approx \frac{Ze^2}{R} 4N_0^2 \int_0^R \left(\frac{r^2}{2R^2} - \frac{3}{2} + \frac{R}{r}\right) r^2 dr$$

$$= \frac{2Ze^2R^2}{5a_\mu^3} \cdot$$

(c) The energy shifts for the 2s and 2p states are

$$\Delta E_{2s} = \int \Psi_{2s}^* \Delta V \Psi_{2s} d\tau$$

$$= \frac{Ze^2N_0^2}{8R} \int_0^R \left(2 - \frac{r}{a_\mu}\right)^2 \exp\left(-\frac{r}{a_\mu}\right) \cdot \left(\frac{r^2}{2R^2} - \frac{3}{2} + \frac{R}{r}\right) r^2 dr$$

$$\approx \frac{Ze^2N_0^2}{8R} \int_0^R 4\left(\frac{r^2}{2R^2} - \frac{3}{2} + \frac{R}{r}\right) r^2 dr$$

$$= \frac{Ze^2R^2}{20a_\mu^3} \,,$$

$$\Delta E_{2p} = \int \Psi_{2p}^* \Delta V \Psi_{2p} d\tau$$

$$= \frac{Ze^2N_0^2}{24a_\mu^2 R} \int_0^R r^2 \exp\left(-\frac{r}{a_\mu}\right) \cdot \left(\frac{r^2}{2R^2} - \frac{3}{2} + \frac{R}{r}\right) r^2 dr$$

$$\approx \frac{Ze^2 N_0^2}{24a_\mu^2 R} \int_0^R r^2 \left(\frac{r^2}{2R^2} - \frac{3}{2} + \frac{R}{r} \right) r^2 dr$$

$$= \frac{3Ze^2 R^4}{3360a_\mu^5} \ll \Delta E_{2s} \, .$$

Hence the relative shift of $2s-2p$ is

$$\Delta E_{sp} \approx \Delta E_{2s} = \frac{Ze^2 R^2}{20a_\mu^3} \, .$$

Thus R can be estimated from the relative shift of the energy levels.

(d) For large Z, $a_\mu = \frac{\hbar^2}{Zm_\mu e^2}$ becomes so small that $\frac{R}{a_\mu} \geq 1$. When $\frac{R}{a_\mu} \geq \frac{\sqrt{5}}{2}$, we have, using the result of (b),

$$\Delta E_{1s} = \frac{2Ze^2 R^2}{5a_\mu^3} = \frac{4}{5}|E_{1s}^0| \left(\frac{R}{a_\mu} \right)^2 > |E_{1s}^0| \, ,$$

where

$$E_{1s}^0 = -\frac{m_\mu Z^2 e^4}{2\hbar^2} \, .$$

This means that $E_{1s} = E_{1s}^0 + \Delta E_{1s} > 0$, which is contradictory to the fact that E_{1s}, a bound state, is negative. Hence ΔE_{1s} as given by (b) is higher than the actual value. This is because we only included the zeroth order term in the expansion of $\exp(-\frac{2r}{a_\mu})$. Inclusion of higher order terms would result in more realistic values.

1063

Consider the situation which arises when a negative muon is captured by an aluminum atom (atomic number $Z = 13$). After the muon gets inside the "electron cloud" it forms a hydrogen-like muonic atom with the aluminum nucleus. The mass of the muon is 105.7 MeV.

(a) Compute the wavelength (in Å) of the photon emitted when this muonic atom decays from the $3d$ state. (Sliderule accuracy; neglect nuclear motion).

(b) Compute the mean life of the above muonic atom in the 3d state, taking into account the fact that the mean life of a hydrogen atom in the 3d state is 1.6×10^{-8} sec.

(UC, Berkeley)

Solution:

There are two energy levels in each of the 3d, 3p, 2p states, namely $3^2 D_{5/2}$ and $3^2 D_{3/2}$, $3^2 P_{3/2}$ and $3^2 P_{1/2}$, $2^2 P_{3/2}$ and $2^2 P_{1/2}$, respectively. There is one energy level each, $3^2 S_{1/2}$, $2^2 S_{1/2}$ and $1^2 S_{1/2}$, in the 3s, 2s and 1s states respectively.

The possible transitions are:

$$3^2 D_{5/2} \to 3^2 P_{3/2}, 3^2 D_{5/2} \to 2^2 P_{3/2}, 3^2 D_{3/2} \to 3^2 P_{1/2},$$

$$3^2 D_{3/2} \to 2^2 P_{3/2}, 3^2 D_{3/2} \to 2^2 P_{1/2},$$

$$3^2 P_{3/2} \to 3^2 S_{1/2}, 3^2 P_{3/2} \to 2^2 S_{1/2}, 3^2 P_{3/2} \to 1^2 S_{1/2},$$

$$3^2 P_{1/2} \to 2^2 S_{1/2}, 3^2 P_{1/2} \to 1^2 S_{1/2},$$

$$3^2 S_{1/2} \to 2^2 P_{3/2}, 3^2 S_{1/2} \to 2^2 P_{1/2}, 2^2 P_{3/2} \to 2^2 S_{1/2},$$

$$2^2 P_{3/2} \to 1^2 S_{1/2}, 2^2 P_{1/2} \to 1^2 S_{1/2}.$$

(a) The hydrogen-like mesic atom has energy

$$E = E_0 \left[\frac{1}{n^2} + \frac{\alpha^2 Z^2}{n^3} \left(\frac{1}{j + \dfrac{1}{2}} - \frac{3}{4n} \right) \right],$$

where

$$E_0 = -\frac{2\pi^2 m_\mu e^4 Z^2}{(4\pi\varepsilon_0)^2 h^2} = -13.6 \times \frac{105.7}{0.511} \times 13^2 = -4.754 \times 10^5 \text{ eV},$$

$\alpha = \frac{1}{137}$. Thus

$$\Delta E(3^2 D_{5/2} \to 3^2 P_{3/2}) = 26.42 \text{ eV},$$

$$\Delta E(3^2 D_{5/2} \to 2^2 P_{3/2}) = 6.608 \times 10^4 \text{ eV},$$

$$\Delta E(3^2 D_{3/2} \to 3^2 P_{1/2}) = 79.27 \text{ eV},$$

$$\Delta E(3^2 D_{3/2} \to 2^2 P_{3/2}) = 6.596 \times 10^4 \text{ eV},$$

$$\Delta E(3^2 D_{3/2} \to 2^2 P_{1/2}) = 6.632 \times 10^4 \text{ eV},$$

$$\Delta E(3^2 P_{3/2} \to 3^2 S_{1/2}) = 79.27 \text{ eV},$$

$$\Delta E(3^2 P_{3/2} \to 2^2 S_{1/2}) = 6.632 \times 10^4 \text{ eV},$$

$$\Delta E(3^2 P_{3/2} \to 1^2 S_{1/2}) = 4.236 \times 10^5 \text{ eV},$$

$$\Delta E(3^2 P_{1/2} \to 2^2 S_{1/2}) = 6.624 \times 10^4 \text{ eV},$$

$$\Delta E(3^2 P_{1/2} \to 1^2 S_{1/2}) = 4.235 \times 10^5 \text{ eV},$$

$$\Delta E(3^2 S_{1/2} \to 2^2 P_{3/2}) = 6.598 \times 10^5 \text{ eV},$$

$$\Delta E(3^2 S_{1/2} \to 2^2 P_{1/2}) = 6.624 \times 10^4 \text{ eV},$$

$$\Delta E(2^2 P_{3/2} \to 2^2 S_{1/2}) = 267.5 \text{ eV},$$

$$\Delta E(2^2 P_{3/2} \to 1^2 S_{1/2}) = 3.576 \times 10^5 \text{ eV},$$

$$\Delta E(2^2 P_{1/2} \to 1^2 S_{1/2}) = 3.573 \times 10^5 \text{ eV}.$$

Using the relation $\lambda = \frac{hc}{\Delta E} = \frac{12430}{\Delta E(\text{eV})}$ Å, we obtain the wavelengths of the photons emitted in the decays of the $3d$ state: $\lambda = 470$ Å, 0.188 Å, 0.157 Å, 0.188 Å, 0.187 Å in the above order.

(b) The probability of a spontaneous transition is

$$P \propto \frac{e^2 \omega^3}{\hbar c^3} R^2$$

with

$$\omega \propto \frac{m_\mu (Z e^2)^2}{\hbar^3}, \qquad R \propto \frac{\hbar^2}{m_\mu Z e^2}.$$

Thus

$$P \propto m_\mu (Z e^2)^4.$$

As the mean life of the initial state is

$$\tau = \frac{1}{P},$$

the mean life of the $3d$ state of the μ mesic atom is

$$\tau = \frac{m_e \tau_0}{m_\mu Z^4} = 2.7 \times 10^{-15} \text{ s}.$$

where $\tau_0 = 1.6 \times 10^{-8}$ s is the mean life of a $3d$ state hydrogen atom.

1064

One method of measuring the charge radii of nuclei is to study the characteristic X-rays from exotic atoms.

(a) Calculate the energy levels of a μ^- in the field of a nucleus of charge Ze assuming a point nucleus.

(b) Now assume the μ^- is completely inside a nucleus. Calculate the energy levels assuming the nucleus is a uniform charge sphere of charge Ze and radius ρ.

(c) Estimate the energy of the K X-ray from muonic $^{208}Pb_{82}$ using the approximations in (a) or (b). Discuss the validity of these approximations.

NOTE: $m_\mu = 200 m_e$.

(Princeton)

Solution:

(a) The energy levels of μ^- in the field of a point nucleus with charge Ze are given by (**Problem 1035**)

$$E_n = Z^2 \frac{m_\mu}{m_e} E_n(H) = -Z^2 \times 200 \times \frac{13.6}{n^2}$$

$$= -\frac{2.72 \times 10^3}{n^2} Z^2 \text{ eV},$$

where $E_n(H)$ is the corresponding energy level of a hydrogen atom.

(b) The potential for μ^- moving in a uniform electric charge sphere of radius ρ is (**Problem 1050(a)**)

$$V(r) = -\frac{Ze^2}{\rho}\left(\frac{3}{2} - \frac{r^2}{2\rho^2}\right) = -\frac{3Ze^2}{2\rho} + \frac{1}{2}\left(\frac{Ze^2}{\rho^3}\right)r^2.$$

The dependence of the potential on r suggests that the μ^- may be treated as an isotropic harmonic oscillator of eigenfrequency $\omega = \sqrt{\frac{Ze^2}{m_\mu \rho^3}}$. The energy levels are therefore

$$E_n = \hbar\omega\left(n + \frac{3}{2}\right) - \frac{3Ze^2}{2\rho},$$

where $n = 0, 1, 2, \ldots$, $\rho \approx 1.2 \times 10^{-13} A^{1/3}$ cm.

(c) K X-rays are emitted in the transitions of electron energy levels $n \geq 2$ to the $n = 1$ level.

The point-nucleus model (a) gives the energy of the X-rays as

$$\Delta E = E_2 - E_1 = -2.72 \times 10^3 \times 82^2 \left(\frac{1}{2^2} - 1 \right) = 1.37 \times 10^7 \text{ eV}.$$

The harmonic oscillator model (b) gives the energy of the X-rays as

$$\Delta E = E_2 - E_1 = \hbar \omega = \hbar \left(\frac{c}{\rho} \right) \sqrt{Z \frac{r_0}{\rho} \frac{m_e}{m_\mu}} = \frac{6.58 \times 10^{-16} \times 3 \times 10^{10}}{1.2 \times 10^{-13}}$$

$$\times \sqrt{\frac{82 \times 2.82 \times 10^{-13}}{208 \times 200 \times 1.2 \times 10^{-13}}} = 1.12 \times 10^7 \text{ eV},$$

where $r_0 = \frac{e^2}{m_e c^2} = 2.82 \times 10^{-13}$ cm is the classical radius of electron.

Discussion: As μ^- is much heavier than electron, it has a larger probability of staying inside the nucleus (first Bohr radius $a_0 \propto \frac{1}{m}$), which makes the effective nuclear charge $Z^* < Z$. Thus we may conclude that the energy of K X-rays as given by the point-nucleus model is too high. On the other hand, as the μ^- does have a finite probability of being outside the nucleus, the energy of the K X-rays as given by the harmonic oscillator model would be lower than the true value. As the probability of the μ^- being outside the nucleus decreases faster than any increase of Z, the harmonic oscillator model is closer to reality as compared to the point-nuclear model.

1065

A proposal has been made to study the properties of an atom composed of a π^+ ($m_{\pi^+} = 273.2\, m_e$) and a μ^- ($m_{\mu^-} = 206.77\, m_e$) in order to measure the charge radius of π^+ assuming that its charge is spread uniformly on a spherical shell of radius $r_0 = 10^{-13}$ cm and that the μ^- is a point charge. Express the potential as a Coulomb potential for a point charge plus a perturbation and use perturbation theory to calculate a numerical value for the percentage shift in the 1s–2p energy difference Δ (neglect spin orbit effects and Lamb shift). Given

$$a_0 = \frac{\hbar^2}{m e^2},$$

$$R_{10}(r) = \left(\frac{1}{a_0}\right)^{3/2} 2 \exp\left(-\frac{r}{a_0}\right),$$

$$R_{21}(r) = \left(\frac{1}{2a_0}\right)^{3/2} \frac{r}{a_0} \exp\left(-\frac{r}{a_0}\right) \cdot \frac{1}{\sqrt{3}}.$$

(Wisconsin)

Solution:

The potential function is

$$V(r) = \begin{cases} -e^2/r, & (r > r_0) \\ -e^2/r_0. & (r < r_0) \end{cases}$$

The Hamiltonian can be written as $H = H_0 + H'$, where H_0 is the Hamiltonian if π^+ is treated as a point charge, H' is taken as perturbation, being

$$H' = \begin{cases} 0, & (r > r_0) \\ e^2\left(\frac{1}{r} - \frac{1}{r_0}\right). & (r < r_0) \end{cases}$$

The shift of $1s$ level caused by H', to first order approximation, is

$$\Delta E_{1s} = \int \Psi_{1s}^* H' \Psi_{1s} d\tau = \int_0^{r_0} R_{10}^2(r) e^2 \left(\frac{1}{r} - \frac{1}{r_0}\right) r^2 dr \approx \frac{2e^2 r_0^2}{3a_0^3},$$

assuming $r_0 \ll a_0$. The shift of $2p$ level is

$$\Delta E_{2p} = \int \Psi_{2p}^* H' \Psi_{2p} d\tau = \int_0^{r_0} R_{21}^2(r) e^2 \left(\frac{1}{r} - \frac{1}{r_0}\right) r^2 dr$$

$$\approx \frac{e^2 r_0^4}{480 a_0^5} \ll \Delta E_{1s},$$

using the same approximation. Thus

$$\Delta E_{1s} - \Delta E_{2p} \approx \Delta E_{1s} = \frac{2e^2 r_0^2}{3a_0^3}.$$

Without considering the perturbation, the energy difference of $1s$–$2p$ is

$$\Delta = -\frac{me^4}{2\hbar^2}\left(\frac{1}{2^2} - 1\right) = \frac{3me^4}{8\hbar^2} = \frac{3e^2}{8a_0}.$$

Hence

$$\frac{\Delta E_{1s} - \Delta E_{2p}}{\Delta} \approx \frac{16}{9} \left(\frac{r_0}{a_0} \right)^2 .$$

As

$$m = \frac{m_{\mu^-} m_{\pi^+}}{m_{\mu^-} + m_{\pi^+}} = 117.7 m_e ,$$

we have

$$a_0 = \frac{\hbar^2}{me^2} = \left(\frac{\hbar^2}{m_e e^2} \right) \frac{m_e}{m} = \frac{0.53 \times 10^{-8}}{117.7} = 4.5 \times 10^{-11} \text{ cm} ,$$

and hence

$$\frac{\Delta E_{1s} - \Delta E_{2p}}{\Delta} = \frac{16}{9} \times \left(\frac{10^{-3}}{4.5 \times 10^{-11}} \right)^2 = 8.8 \times 10^{-6} .$$

1066

A μ^- meson (a heavy electron of mass $M = 210 m_e$ with m_e the electron mass) is captured into a circular orbit around a proton. Its initial radius $R \approx$ the Bohr radius of an electron around a proton. Estimate how long (in terms of R, M and m_e) it will take the μ^- meson to radiate away enough energy to reach its ground state. Use classical arguments, including the expression for the power radiated by a nonrelativistic accelerating charged particle.

(*CUSPEA*)

Solution:

The energy of the μ^- is

$$E(r) = K(r) - \frac{e^2}{r} = -\frac{e^2}{2r} ,$$

where $K(r)$ is the kinetic energy.

The radiated power is $P = \frac{2e^2 a^2}{3c^3}$, where

$$a = \frac{F_{Coul}}{M} = \frac{e^2}{r^2 M}$$

is the centripetal acceleration. Energy conservation requires

$$\frac{dE}{dt} = -P,$$

i.e.,

$$\frac{e^2}{2r^2}\frac{dr}{dt} = -\frac{2e^2}{3c^3} \cdot \frac{e^4}{r^4 M^2}.$$

Integration gives

$$R^3 - r^3 = \frac{4}{c^3} \cdot \frac{e^4}{M^2}t,$$

where R is the radius of the initial orbit of the μ mesion, being

$$R \approx \frac{\hbar^2}{me^2}.$$

At the μ ground state the radius of its orbit is the Bohr radius of the mesic atom

$$r_0 = \frac{\hbar^2}{Me^2},$$

and the time t taken for the μ meson to spiral down to this state is given by

$$\left(\frac{\hbar^2}{e^2}\right)^3 \left(\frac{1}{m^3} - \frac{1}{M^3}\right) = \frac{4e^4}{c^3 M^2}t.$$

Since $M \gg m$, we have

$$t \approx \frac{M^2 c^3 R^3}{4e^4} = \left(\frac{M}{m}\right)^2 \left(\frac{mc^2}{e^2}\right)^2 \frac{R^3}{4c}$$

$$= 210^2 \times \left(\frac{5.3 \times 10^{-9}}{2.82 \times 10^{-13}}\right)^2 \times \frac{5.3 \times 10^{-9}}{4 \times 3 \times 10^{10}} = 6.9 \times 10^{-7} \text{ s}.$$

1067

Consider a hypothetical universe in which the electron has spin 3/2 rather than spin 1/2.

(a) Draw an energy level diagram for the $n = 3$ states of hydrogen in the absence of an external magnetic field. Label each state in spectroscopic notation and indicate which states have the same energy. Ignore hyperfine structure (interaction with the nuclear spin).

(b) Discuss qualitatively the energy levels of the two-electron helium atom, emphasizing the differences from helium containing spin 1/2 electrons.

(c) At what values of the atomic number would the first two inert gases occur in this universe?

(Columbia)

Solution:

(a) Consider a hydrogen atom having electron of spin 3/2. For $n = 3$, the possible quantum numbers are given in Table 1.1.

Table 1.1

n	l	j
	0	3/2
3	1	5/2, 3/2, 1/2
	2	7/2, 5/2, 3/2, 1/2

If fine structure is ignored, these states are degenerate with energy

$$E_n = -\frac{RhcZ^2}{n^2}$$

where $Z = 1, n = 3, R$ is the Rydberg constant, c is the speed of light.

If the relativistic effect and spin-orbit interactions are taken into account, the energy changes into $E = E_0 + \Delta E$ and degeneracy disappears, i.e., different states have different energies.

(1) For $l = 0$ and $j = 3/2$, there is only the correction ΔE_r arising from the relativistic effect, i.e.,

$$\Delta E = \Delta E_r = -A \left(\frac{1}{l + \frac{1}{2}} - \frac{3}{4n} \right) = -\frac{7}{4}A,$$

where $A = Rhc\alpha^2 Z^4/n^3$, α being the fine structure constant.

(2) For $l \neq 0$, in addition to ΔE_r there is also the spin-orbital coupling correction ΔE_{ls}, so that

$$\Delta E = \Delta E_r + \Delta E_{ls} = -A \left(\frac{1}{l + \frac{1}{2}} - \frac{3}{4n} \right)$$

$$+A\frac{1}{l\left(l+\dfrac{1}{2}\right)(l+1)}\cdot\frac{j(j+1)-l(l+1)-s(s+1)}{2}.$$

(i) For $l = 1$,

$$\Delta E = \left[\frac{1}{6}j(j+1)-\frac{11}{8}\right]A\,,$$

Thus for

$$j=\frac{5}{2}\,,\qquad \Delta E = \frac{1}{12}A\,,$$

$$j=\frac{3}{2}\,,\qquad \Delta E = -\frac{3}{4}A\,,$$

$$j=\frac{1}{2}\,,\qquad \Delta E = -\frac{5}{4}A\,.$$

(ii) For $l = 2$,

$$\Delta E = \left[\frac{1}{30}j(j+1)-\frac{19}{40}\right]A\,,$$

Thus for

$$j=\frac{7}{2}\,,\qquad \Delta E = \frac{1}{20}A\,,$$

$$j=\frac{5}{2}\,,\qquad \Delta E = -\frac{11}{60}A\,,$$

$$j=\frac{3}{2}\,,\qquad \Delta E = -\frac{7}{20}A\,.$$

$$j=\frac{1}{2}\,,\qquad \Delta E = -\frac{9}{20}A\,.$$

The energy level scheme for $n = 3$ of the hydrogen atom is shown in Fig. 1.20.

(b) Table 1.2 shows the single-electron energy levels of the helium atoms (electron spins 1/2 and 3/2).

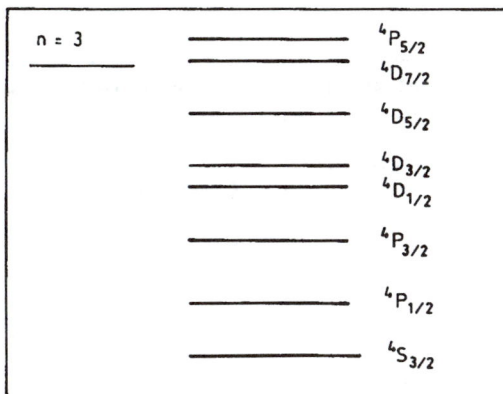

Fig. 1.20

Table 1.2

		He $(s = 3/2)$	He $(s = 1/2)$
$n_1 = 1$ $n_2 = 1$	Total electron spin	$S = 0, 2$	$S = 0$
$l = 0$	energy level	$^1S_0, \, ^5S_2$	$^1S_0$
$n_1 = 1$ $n_2 = 2$	Total electron spin	$S = 0, 1, 2, 3$	$S = 0, 1$
$l_2 = 0, 1$	energy level	$l_2 = 0 : {}^1S_0, \, ^3S_1, \, ^5S_2, \, ^7S_3$ $l_2 = 1: \, ^1P_1, \, ^3P_{2,1,0}, \, ^5P_{3,2,1}$ $^7P_{4,3,2}$	$l_2 = 0: \, ^1S_0, \, ^3S_1$ $l_2 = 1: \, ^1P_1, \, ^3P_{2,1,0}$

(c) If the electron spin were 3/2, the atomic numbers Z of the first two inert elements would be 4 and 20.

1068

Figure 1.21 shows the ground state and first four excited states of the helium atom.

(a) Indicate on the figure the complete spectroscopic notation of each level.

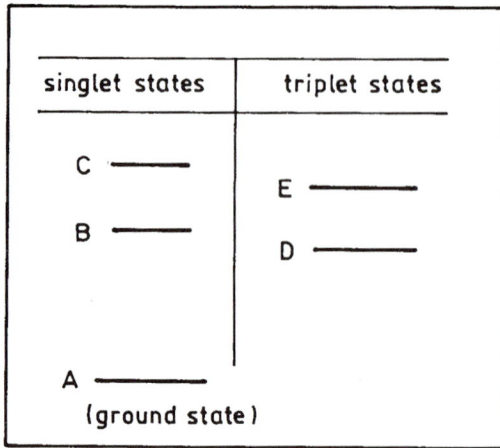

Fig. 1.21

(b) Indicate, with arrows on the figure, the allowed radiative dipole transitions.

(c) Give a qualitative reason why level B is lower in energy than level C.

(*Wisconsin*)

Solution:

(a) The levels in Fig. 1.21 are as follows:

A: 1^1S_0, constituted by $1s^2$,
B: 2^1S_0, constituted by $1s2s$,
C: 2^1P_1, constituted by $1s2p$,
D: 2^3S_1, constituted by $1s2s$,
E: $2^3P_{2,1,0}$, constituted by $1s2p$.

(b) The allowed radiative dipole transitions are as shown in Fig. 1.22. (Selection rules $\Delta L = \pm 1$, $\Delta S = 0$)

(c) In the C state constituted by $1s2p$, one of the electrons is excited to the $2p$ orbit, which has a higher energy than that of $2s$. The main reason is that the effect of the screening of the nuclear charge is larger for the p orbit.

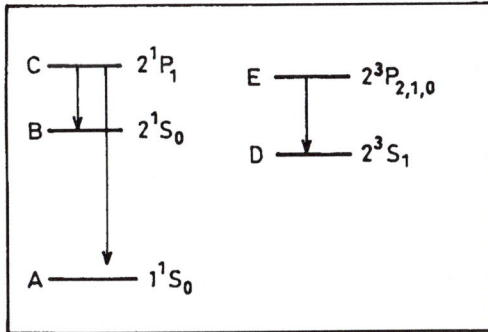

Fig. 1.22

1069

Figure 1.23 shows the ground state and the set of $n = 2$ excited states of the helium atom. Reproduce the diagram in your answer giving

(a) the spectroscopic notation for all 5 levels,

(b) an explanation of the source of ΔE_1,

(c) an explanation of the source of ΔE_2,

(d) indicate the allowed optical transitions among these five levels.

(*Wisconsin*)

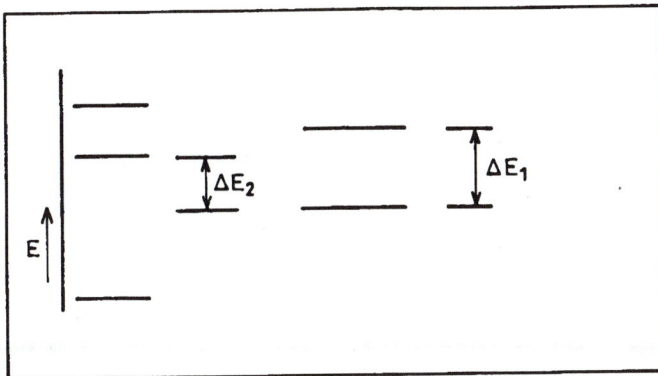

Fig. 1.23

Solution:

(a) See **Problem 1068(a)**.

(b) ΔE_1 is the difference in energy between different electronic configurations with the same S. The ^3P states belong to the configuration of $1s2p$, which has one electron in the $1s$ orbit and the other in the $2p$ orbit. The latter has a higher energy because the screening of the nuclear charge is greater for the p electron.

(c) ΔE_2 is the energy difference between levels of the same L in the same electronic configuration but with different S. Its origin lies in the Coulomb exchange energy.

(d) See **Problem 1068(b)**.

1070

Figure 1.24 is an energy level diagram for the ground state and first four excited states of a helium atom.

(a) On a copy of the figure, give the complete spectroscopic notation for each level.

(b) List the possible electric-dipole allowed transitions.

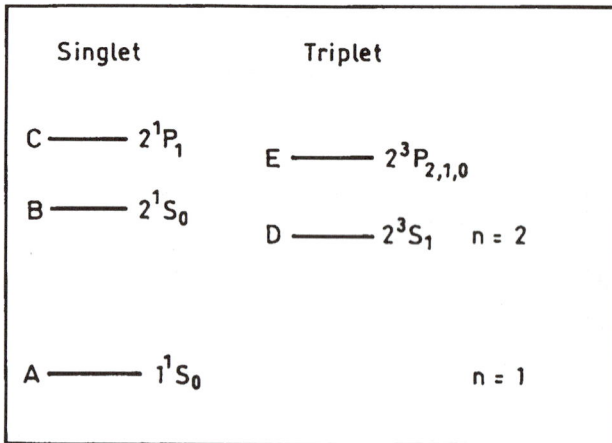

Fig. 1.24

(c) List the transitions between those levels that would be possible for an allowed 2-photon process (both photons electric dipole).

(d) Given electrons of sufficient energy, which levels could be populated as the result of electrons colliding with ground state atoms?

(Wisconsin)

Solution:

(a) (b) See **problem 1068**.

(c) The selection rule for a 2-photon process are

(1) conservation of parity,

(2) $\Delta J = 0, \pm 2$.

Accordingly the possible 2-photon process is

$$(1s2s)^1 S_0 \rightarrow (1s^2)^1 S_0 \,.$$

The transition $(1s2s)^3 S$ to $(1s^2)^1 S_0$ is also possible via the 2-photon process with a rate $10^{-8} \sim 10^{-9}$ s^{-1}. It has however been pointed out that the transition $2^3 S_1 \rightarrow 1^1 S_0$ could proceed with a rate $\sim 10^{-4}$ s via magnetic dipole radiation, attributable to some relativistic correction of the magnetic dipole operator relating to spin, which need not satisfy the condition $\Delta S = 0$.

(d) The $(1s2s)^1 S_0$ and $(1s2s)^3 S_1$ states are metastable. So, besides the ground state, these two levels could be populated by many electrons due to electrons colliding with ground state atoms.

1071

Sketch the low-lying energy levels of atomic He. Indicate the atomic configuration and give the spectroscopic notation for these levels. Indicate several transitions that are allowed in emission, several transitions that are allowed in absorption, and several forbidden transitions.

(Wisconsin)

Solution:

The energy levels of He are shown in Fig. 1.25.

According to the selection rules $\Delta S = 0$, $\Delta L = \pm 1$, $\Delta J = 0, \pm 1$ (except $0 \rightarrow 0$), the allowed transitions are: $3^1 S_0 \rightarrow 2^1 P_1$, $3^3 S_1 \rightarrow 2^3 P_{2,1,0}$, $2^1 P_1 \rightarrow 1^1 S_0$, $2^1 P_1 \rightarrow 2^1 S_0$, $3^3 D_1 \rightarrow 3^3 P_0$, $3^3 D_{2,1} \rightarrow 3^3 P_1$, $3^3 D_{3,2,1} \rightarrow 3^3 P_2$,

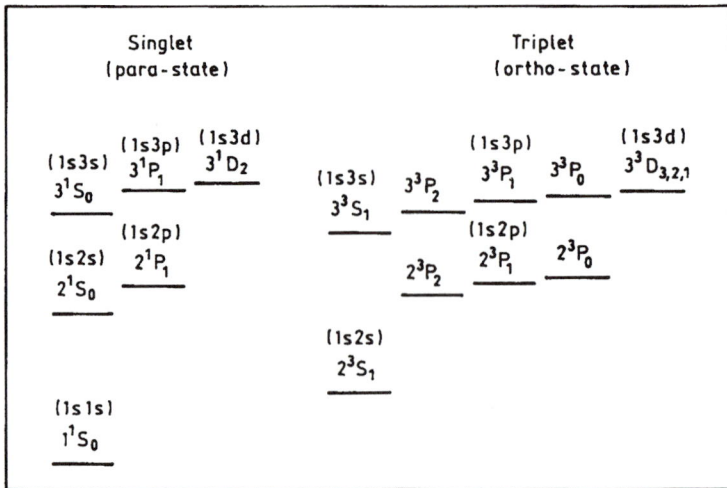

Fig. 1.25

$3^1D_2 \to 3^1P_1$, $3^1D_2 \to 2^1P_1$, $3^3D_1 \to 2^3P_{1,0}$, $3^3D_{3,2,1} \to 2^3P_2$, $3^3P_{2,1,0} \to 2^3S_1$. The reverse of the above are the allowed absorption transitions. Transitions between singlet and triplet states $(\Delta S \neq 0)$ are forbidden, e.g. $2^3S_1 \to 1^1S_0$, $2^1P_1 \to 2^3S_1$.

1072

Sketch the energy level diagram for a helium atom in the $1s3d$ configuration, taking into account Coulomb interaction and spin-orbit coupling.

(UC, Berkeley)

Solution:

See **Problem 1100**.

1073

For helium atom the only states of spectroscopic interest are those for which at least one electron is in the ground state. It can be constructed from orthonormal orbits of the form

$$\Psi_\pm(1,2) = \frac{1}{\sqrt{2}}[\Phi_{1s}(1)\Phi_{nlm}(2) \pm \Phi_{nlm}(1)\Phi_{1s}(2)] \times \text{spin wave function}.$$

The para-states correspond to the + sign and the ortho-states to the − sign.

(a) Determine for which state the ortho- or the corresponding para-state has the lowest energy. (i.e. most negative).

(b) Present an argument showing for large n that the energy difference between corresponding ortho- and para-states should become small.

(SUNY, Buffalo)

Solution:

(a) For fermions like electrons the total wave function of a system must be antisymmetric.

If both electrons of a helium atom are in $1s$ orbit, Pauli's principle requires that their spins be antiparallel, i.e. the total spin function be antisymmetric. Then the spatial wave function must be symmetric and the state is the para-state 1^1S_0.

If only one electron is in $1s$ orbit, and the other is in the nlm-state, where $n \neq 1$, their spins may be either parallel or antiparallel and the spatial wave functions are, respectively,

$$\Psi_\mp = \frac{1}{\sqrt{2}}[\Phi_{1s}(1)\Phi_{nlm}(2) \mp \Phi_{nlm}(1)\Phi_{1s}(2)].$$

Ignoring magnetic interactions, consider only the Coulomb repulsion between the electrons and take as perturbation $H' = e^2/r_{12}$, r_{12} being the distance between the electrons. The energy correction is then

$$W'_\mp = \frac{1}{2} \iint [\Phi^*_{1s}(1)\Phi^*_{nlm}(2) \mp \Phi^*_{nlm}(1)\Phi^*_{1s}(2)]$$

$$\times \frac{e^2}{r_{12}}[\Phi_{1s}(1)\Phi_{nlm}(2) \mp \Phi_{nlm}(1)\Phi_{1s}(2)]d\tau_1 d\tau_2$$

$$= J \mp K$$

with

$$J = \iint \frac{e^2}{r_{12}}|\Phi_{1s}(1)\Phi_{nlm}(2)|^2 d\tau_1 d\tau_2,$$

$$K = \iint \frac{e^2}{r_{12}}\Phi^*_{1s}(1)\Phi_{nlm}(1)\Phi^*_{nlm}(2)\Phi_{1s}(2)d\tau_1 d\tau_2.$$

Hence the ortho-state (− sign above) has lower corrected energy. Thus para-helium has ground state 1^1S_0 and ortho-helium has ground state 2^3S_1, which is lower in energy than the 2^1S_0 state of para-helium (see Fig. 1.25).

(b) As n increases the mean distance r_{12} between the electrons increases also. This means that the energy difference 2K between the para- and ortho-states of the same electron configuration decreases as n increases.

1074

(a) Draw and qualitatively explain the energy level diagram for the $n = 1$ and $n = 2$ levels of helium in the nonrelativistic approximation.

(b) Draw and discuss a similar diagram for hydrogen, including all the energy splitting that are actually present.

(CUSPEA)

Solution:

(a) In the lowest energy level ($n = 1$) of helium, both electrons are in the lowest state $1s$. Pauli's principle requires the electrons to have antiparallel spins, so that the $n = 1$ level is a singlet. On account of the repulsion energy between the electrons, e^2/r_{12}, the ground state energy is higher than $2Z^2E_0 = 8E_0$, where $E_0 = -\frac{me^4}{2\hbar^2} = -13.6$ eV is the ground state energy of hydrogen atom.

In the $n = 2$ level, one electron is in $1s$ state while the other is in a higher state. The two electrons can have antiparallel or parallel spins (singlet or triplet states). As the probability for the electrons to come near each other is larger in the former case, its Coulomb repulsion energy between the electrons, e^2/r_{12}, is also larger. Hence in general a singlet state has higher energy than the corresponding triplet state (Fig. 1.26).

(b) The energy levels of hydrogen atom for $n = 1$ and $n = 2$ are shown in Fig. 1.27. If one considers only the Coulomb interaction between the nucleus and electron, the (Bohr) energy levels are given by

$$E_n = -\frac{m_e e^4}{2\hbar^2 n^2},$$

which is a function of n only. If the relativistic effect and the spin-orbit interaction of the electron are taken into account, the $n = 2$ level splits into two levels with a spacing $\approx \alpha^2 E_2$, where α is the fine structure constant.

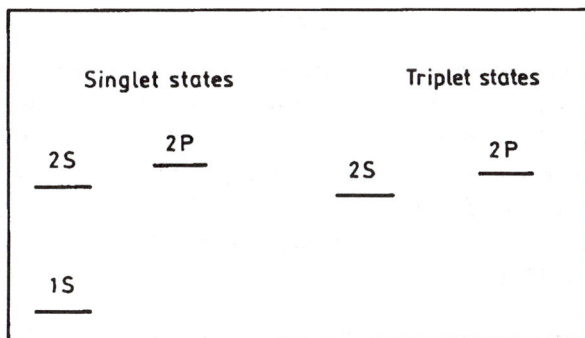

Fig. 1.26

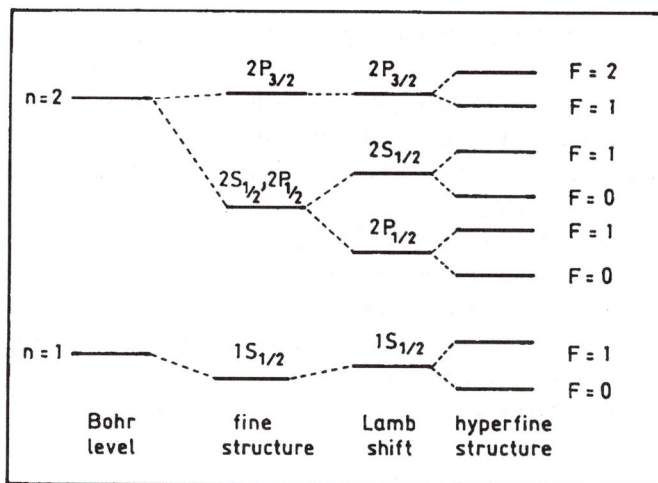

Fig. 1.27

If one considers, further, the interaction between the electron and its own magnetic field and vacuum polarization, Lamb shift results splitting the degenerate $2S_{1/2}$ and $2P_{1/2}$ states, the splitting being of the order $m_e c^2 \alpha^5$.

In addition, the levels split further on account of the interactions between the spin and orbital motions of the electron and the nuclear magnetic moment, giving rise to a hyperfine structure with spacing about $1/10$ of the Lamb shift for the same n.

1075

(a) The $1s2s$ configuration of the helium atom has two terms $^3S_1$ and $^1S_0$ which lie about 20 eV above the ground state. Explain the meaning of the spectroscopic notation. Also give the reason for the energy splitting of the two terms and estimate the order of magnitude of the splitting.

(b) List the ground-state configurations and the lowest-energy terms of the following atoms: He, Li, Be, B, C, N, O, F and A.

Possible useful numbers:

$$a_B = 0.529 \times 10^{-8} \text{ cm}, \quad \mu_B = 9.27 \times 10^{-21} \text{ erg/gauss}, \quad e = 4.8 \times 10^{-10} \text{ esu}.$$

(Princeton)

Solution:

(a) The spectroscopic notation indicates the state of an atom. For example in $^3S_1$, the superscript 3 indicates the state is a triplet ($3 = 2S+1$), the subscript 1 is the total angular momentum quantum number of the atom, $J = S + L = 1$, S labels the quantum state corresponding to the orbital angular momentum quantum number $L = 0$ (S for $L = 0$, P for $L = 1$, D for $L = 2$, etc.).

The split in energy of the states $^1S_0$ and $^3S_1$ arises from the difference in the Coulomb interaction energy between the electrons due to their different spin states. In the $1s2s$ configuration, the electrons can have antiparallel or parallel spins, giving rise to singlet and triplet states of helium, the approximate energy of which can be obtained by perturbation calculations to be (**Problem 1073**)

$$E(\text{singlet}) = -\frac{Z^2 e^2}{2a_0}\left(1 + \frac{1}{2^2}\right) + J + K,$$

$$E(\text{triplet}) = -\frac{Z^2 e^2}{2a_0}\left(1 + \frac{1}{2^2}\right) + J - K,$$

where J is the average Coulomb energy between the electron clouds, K is the exchange energy. The splitting is

$$\Delta E = 2K$$

with

$$K = e^2 \iint d^3x_1 d^3x_2 \frac{1}{r_{12}} \Psi^*_{100}(r_1)\Psi_{200}(r_1)\Psi_{100}(r_2)\Psi^*_{200}(r_2)$$

$$= \frac{4Z^6 e^2}{a_0^6} \left[\int_0^\infty r_1^2 \left(1 - \frac{Zr_1}{2a_0}\right) \exp\left(-\frac{3Zr_1}{2a_0}\right) dr_1 \right]^2$$

$$\approx \frac{2^4 Z e^2}{3^6 a_0}.$$

Thus

$$K = \frac{2^5 e^2}{3^6 a_0} = \frac{2^5}{3^6} \frac{me^4}{\hbar^2} = \frac{2^5}{3^6} \left(\frac{e^2}{\hbar c}\right)^2 mc^2$$

$$= \frac{2^5}{3^6} \left(\frac{1}{137}\right)^2 \times 0.511 \times 10^6 = 1.2 \text{ eV},$$

and $\Delta E \approx 2$ eV.

(b)

Atom	Ground state configuration	Lowest-energy spectral term
He	$1s^2$	$^1S_0$
Li	$1s^2 2s^1$	$^2S_{1/2}$
Be	$1s^2 2s^2$	$^1S_0$
B	$1s^2 2s^2 2p^1$	$^2P_{1/2}$
C	$1s^2 2s^2 2p^2$	$^3P_0$
N	$1s^2 2s^2 2p^3$	$^4S_{3/2}$
O	$1s^2 2s^2 2p^4$	$^3P_2$
F	$1s^2 2s^2 2p^5$	$^2P_{3/2}$
A	$1s^2 2s^2 2p^6 3s^2 3p^6$	$^1S_0$

1076

Use a variational method, a perturbation method, sum rules, and/or other method to obtain crude estimates of the following properties of the helium atom:

(a) the minimum energy required to remove both electrons from the atom in its ground state,

(b) the minimum energy required to remove one electron from the atom in its lowest F state ($L = 3$), and

(c) the electric polarizability of the atom in its ground state. (The lowest singlet P state lies ~ 21 eV above the ground state.)

(Princeton)

Solution:

(a) In the *perturbation method*, the Hamiltonian of helium atom is written as

$$H = \frac{p_1^2}{2m_e} + \frac{p_2^2}{2m_e} - \frac{2e^2}{r_1} - \frac{2e^2}{r_2} + \frac{e^2}{r_{12}} = H_0 + \frac{e^2}{r_{12}},$$

where

$$H_0 = \frac{p_1^2}{2m_e} + \frac{p_2^2}{2m_e} - \frac{2e^2}{r_1} - \frac{2e^2}{r_2}$$

is considered the unperturbed Hamiltonian, and the potential due to the Coulomb repulsion between the electrons as perturbation. The zero-order approximate wave function is then

$$\psi = \psi_{100}(r_1)\psi_{100}(r_2),$$

where

$$\psi_{100}(r) = \frac{1}{\sqrt{\pi}} \left(\frac{2}{a}\right)^{3/2} e^{-2r/a},$$

a being the Bohr radius. The zero-order (unperturbed) ground state energy is

$$E^{(0)} = 2\left(-\frac{2^2 e^2}{2a}\right) = -\frac{4e^2}{a},$$

where the factor 2 is for the two $1s$ electrons. The energy correction in first order perturbation is

$$E^{(1)} = \int |\psi_{100}|^2 \frac{e^2}{r_{12}} d\mathbf{r}_1 d\mathbf{r}_2 = \frac{5e^2}{4a}.$$

Hence the corrected ground state energy is

$$E = -\frac{4e^2}{a} + \frac{5e^2}{4a} = -\frac{11}{2} \cdot \frac{e^2}{2a} = -\frac{11}{2} \times 13.6 = -74.8 \text{ eV},$$

and the ionization energy of ground state helium atom, i.e. the energy required to remove both electrons from the atom, is

$$E_I = -E = 74.8 \text{ eV}.$$

In the *variational method*, take as the trial wave function

$$\psi = \frac{\lambda^3}{\pi a^3} e^{-\lambda(r_1+r_2)/a}.$$

We then calculate

$$\langle H \rangle = \iint \psi^* \left(-\frac{\hbar^2}{2m_e} \nabla_1^2 - \frac{\hbar^2}{2m_e} \nabla_2^2 - \frac{2e^2}{r_1} - \frac{2e^2}{r_2} + \frac{e^2}{r_{12}} \right) \psi d\mathbf{r}_1 d\mathbf{r}_2$$

$$= \left(2\lambda^2 - \frac{27}{4}\lambda \right) E_H,$$

where

$$E_H = \frac{e^2}{2a} = 13.6 \text{ eV}.$$

Minimizing $\langle H \rangle$ by taking

$$\frac{\partial \langle H \rangle}{\partial \lambda} = 0,$$

we find $\lambda = \frac{27}{16}$ and so

$$\langle H \rangle = \frac{27}{16} \left(\frac{27}{8} - \frac{27}{4} \right) E_H = -77.5 \text{ eV}.$$

The ionization energy is therefore $E_I = -\langle H \rangle = 77.5$ eV, in fairly good agreement with the perturbation calculation.

(b) In the lowest F state the electron in the $l = 3$ orbit is so far from the nucleus that the latter together with the $1s$ electron can be treated as a core of charge $+e$. Thus the excited atom can be considered as a hydrogen atom in the state $n = 4$. The ionizaion energy E_I, i.e. the energy required to remove one electron from the atom, is

$$E_I = -E = \frac{Ze^2}{2a4^2} = \frac{1}{16} \left(\frac{e^2}{2a} \right) = \frac{13.6}{16} = 0.85 \text{ eV}.$$

(c) Consider a perturbation u. The wave function and energy for the ground state, correct to first order, are

$$\Psi = \Psi_0 + \sum_{n\neq 0} \frac{u_{n0}}{E_0 - E_n}\Psi_n, \qquad E = E_0 + u_{00} + \sum_{n\neq 0} \frac{(u_{n0})^2}{E_0 - E_n},$$

where Ψ_0, E_n are the unperturbed wave function and energy, and $u_{n0} \equiv \langle 0|u|n\rangle$. Write

$$\sum_{n\neq 0} u_{n0}\Psi_n = \sum_{n=0} u_{n0}\psi_n - u_{00}\psi_0 = u\psi_0 - u_{00}\psi_0,$$

with $u\psi_0 = \sum_{n=0} u_{n0}\psi_n$. Then

$$\Psi \approx \Psi_0\left(1 + \frac{u - u_{00}}{E'}\right),$$

E' being the average of $E_0 - E_n$.

The average total kinetic energy of the electrons is calculated using a variational method with $\psi = (1 + \lambda u)\psi_0$ as trial function:

$$\langle T \rangle = \frac{\int \Psi_0^*(1 + \lambda u)\hat{T}\Psi_0(1 + \lambda u)d\mathbf{r}}{\int \Psi_0^*\Psi_0(1 + \lambda u)^2 d\mathbf{r}},$$

where

$$\hat{T} = \frac{1}{2m_e}(p_1^2 + p_2^2) = -\frac{\hbar^2}{2m_e}(\nabla_1^2 + \nabla_2^2),$$

or, in atomic units ($a_0 = \hbar = e = 1$),

$$\hat{T} = -\frac{1}{2}\sum_{i=1}^{2}\nabla_i^2.$$

Thus

$$\hat{T} \propto -\frac{1}{2}\sum_{i=1}^{2}\frac{1}{2}\int\{\Psi_0^*(1 + \lambda u)\nabla_i^2(1 + \lambda u)\Psi_0 + \Psi_0(1 + \lambda u)$$

$$\times \nabla_i^2(1 + \lambda u)\Psi_0^*\}d\mathbf{r}$$

$$= -\frac{1}{2}\sum_{i=1}^{2}\frac{1}{2}\int\{\Psi_0^*(1 + \lambda u)^2\nabla_i^2\Psi_0 + \Psi_0(1 + \lambda u)^2\nabla_i^2\Psi_0^*$$

$$+ 2\lambda\Psi_0\Psi_0^*(1 + \lambda u)\nabla_i^2 u + 2\lambda(1 + \lambda u)\nabla_i(\Psi_0\Psi_0^*)\cdot\nabla_i u\}d\mathbf{r}.$$

Consider

$$\sum_i \int \nabla_i \cdot [\psi_0 \psi_0^* (1 + \lambda u) \nabla_i u] d\mathbf{r} = \oint_S \psi_0 \psi_0^* (1 + \lambda u) \sum_i \nabla_i u \cdot d\mathbf{S} = 0$$

by virtue of Gauss' divergence theorem and the fact that $-\nabla_i u$ represents the mutual repulsion force between the electrons. As

$$\nabla_i \cdot [\Psi_0 \Psi_0^* (1 + \lambda u) \nabla_i u] = \Psi_0 \Psi_0^* (1 + \lambda u) \nabla_i^2 u + (1 + \lambda u) \nabla_i (\Psi_0 \Psi_0^*) \cdot \nabla_i u$$
$$+ \lambda \Psi_0 \Psi_0^* \nabla_i u \cdot \nabla_i u,$$

we can write

$$\int \{ \Psi_0 \Psi_0^* (1 + \lambda u) \nabla_i^2 u + (1 + \lambda u) \nabla_i (\Psi_0 \Psi_0^*) \cdot \nabla_i u \} d\mathbf{r}$$

$$= -\lambda \int \Psi_0 \Psi_0^* \nabla_i u \cdot \nabla_i u d\mathbf{r} .$$

Hence

$$\langle T \rangle \propto - \frac{1}{2} \sum_{i=1}^{2} \frac{1}{2} \int [\Psi_0^* (1 + \lambda u)^2 \nabla_i^2 \Psi_0 + \Psi_0 (1 + \lambda u)^2 \nabla_i^2 \Psi_0^*] d\mathbf{r}$$

$$+ \frac{\lambda^2}{2} \sum_{i=1}^{2} \int \Psi_0 \Psi_0^* \nabla_i u \cdot \nabla_i u d\mathbf{r} .$$

The total energy E can be similarly obtained by considering the total Hamiltonian

$$\hat{H} = \hat{H}_0 + \hat{T} + u .$$

As $\hat{H}$ and $(1 + \lambda u)$ commute, we have

$$\langle H \rangle = \frac{\dfrac{1}{2} \int (1 + \lambda u)^2 (\Psi_0^* \hat{H} \Psi_0 + \Psi_0 \hat{H} \Psi_0^*) d\mathbf{r} + \dfrac{\lambda^2}{2} \sum_{i=1}^{2} \int \Psi_0^* \Psi_0 \nabla_i u \cdot \nabla_i u d\mathbf{r}}{\displaystyle\int \Psi_0^* \Psi_0 (1 + \lambda u)^2 d\mathbf{r}}$$

$$= E_0 + \frac{\dfrac{1}{2} \int \Psi_0^* u (1 + \lambda u)^2 \Psi_0 d\mathbf{r} + \dfrac{\lambda^2}{2} \sum_{i=1}^{2} \int \Psi_0^* \Psi_0 \nabla_i u \cdot \nabla_i u d\mathbf{r}}{\displaystyle\int \Psi_0^* \Psi_0 (1 + \lambda u)^2 d\mathbf{r}}$$

$$= E_0 + \frac{(u)_{00} + 2\lambda(u^2)_{00} + \lambda^2(u^3)_{00} + \frac{1}{2}\lambda^2 \sum_{i=1}^{2} \int [\nabla_i u \cdot \nabla_i u]_{00} d\mathbf{r}}{1 + 2\lambda(u)_{00} + \lambda^2(u^2)_{00}},$$

where E_0 is given by $\hat{H}\psi_0 = E_0\psi_0$, $(u)_{00} = \int \Psi_0^* u \Psi_0 d\mathbf{r}$, $(u^2)_{00} = \int \Psi_0^* u^2 \Psi_0 d\mathbf{r}$, etc. Neglecting the third and higher order terms, we have the energy correction

$$\Delta E \approx (u)_{00} + 2\lambda(u^2)_{00} - 2\lambda(u)_{00}^2 + \frac{1}{2}\lambda^2 \sum_{i=1}^{2} [(\nabla_i u) \cdot (\nabla_i u)]_{00}.$$

Minimizing ΔE by putting

$$\frac{d\Delta E}{d\lambda} = 0,$$

we obtain

$$2(u^2)_{00} - 2(u)_{00}^2 + \lambda \sum_{i=1}^{2} [(\nabla_i u) \cdot (\nabla_i u)]_{00} = 0,$$

or

$$\lambda = \frac{2[(u)_{00}^2 - (u^2)_{00}]}{\sum\limits_{i=1}^{2} [\nabla_i u \cdot \nabla_i u]_{00}}.$$

This gives

$$\Delta E = (u)_{00} - \frac{2[(u)_{00}^2 - (u^2)_{00}]^2}{\sum\limits_{i=1}^{2} [\nabla_i u \cdot \nabla_i u]_{00}}.$$

Consider a He atom in an electric field of strength ε whose direction is taken to be that of the z-axis. Then

$$u = -\varepsilon(z_1 + z_2) \equiv -\varepsilon z.$$

As the matrix element $(u)_{00}$ is zero for a spherically symmetric atom, we have

$$\Delta E \approx -\frac{2[(z^2)_{00}]^2 \varepsilon^4}{2\varepsilon^2} = -[(z^2)_{00}]^2 \varepsilon^2.$$

The energy correction is related to the electric field by

$$\Delta E = -\frac{1}{2}\alpha\varepsilon^2,$$

where α is the polarizability. Hence

$$\alpha = 2[(z^2)_{00}]^2 = 2\langle(z_1 + z_2)^2\rangle^2 \,.$$

As $\langle z_1^2 \rangle = \langle z_2^2 \rangle \approx a'^2 = \frac{a_0^2}{Z^2}$, $\langle z_1 z_2 \rangle = 0$, where a_0 is the Bohr radius, using $Z = 2$ for He we have

$$\alpha = \frac{8\hbar^2}{e^2 m_e} \frac{a_0^4}{2^4} \approx \frac{1}{2} a_0^3$$

in usual units. If the optimized $Z = \frac{27}{16}$ from (a) is used,

$$\alpha = 8 \left(\frac{16}{27}\right)^4 a_0^3 = 0.98 a_0^3 \,.$$

1077

Answer each of the following questions with a brief, and, where possible, quantitative statement. Give your reasoning.

(a) A beam of neutral atoms passes through a Stern-Gerlach apparatus. Five equally spaced lines are observed. What is the total angular momentum of the atom?

(b) What is the magnetic moment of an atom in the state $^3P_0$? (Disregard nuclear effects)

(c) Why are noble gases chemically inert?

(d) Estimate the energy density of black body radiation in this room in erg/cm^3. Assume the walls are black.

(e) In a hydrogen gas discharge both the spectral lines corresponding to the transitions $2\,^2P_{1/2} \rightarrow 1\,^2S_{1/2}$ and $2\,^2P_{3/2} \rightarrow 1\,^2S_{1/2}$ are observed. Estimate the ratio of their intensities.

(f) What is the cause for the existence of two independent term-level schemes, the singlet and the triplet systems, in atomic helium?

(Chicago)

Solution:

(a) The total angular momentum of an atom is

$$P_J = \sqrt{J(J+1)}\hbar \,.$$

As the neutral-atom beam splits into five lines, we have $2J + 1 = 5$, or $J = 2$. Hence

$$P_J = \sqrt{6}\hbar.$$

(b) The state has total angular momentum quantum number $J = 0$. Hence its magnetic moment is $M = g\mu_B\sqrt{J(J+1)} = 0$.

(c) The electrons of a noble gas all lie in completed shells, which cannot accept electrons from other atoms to form chemical bonds. Hence noble gases are chemically inert.

(d) The energy density of black body radiation is $u = 4J_u/c$, where J_u is the radiation flux density given by the Stefan-Boltzmann's law

$$J_u = \sigma T^4,$$

$$\sigma = 5.669 \times 10^{-5} \ \mathrm{erg\,cm^{-2}\,K^{-4}\,s^{-1}}.$$

At room temperature, $T = 300$ K, and

$$u = \frac{4}{3 \times 10^{10}} \times 5.669 \times 10^{-5} \times 300^4$$

$$= 6.12 \times 10^{-5} \ \mathrm{erg \cdot cm^{-3}}.$$

(e) The degeneracies of $2^2P_{1/2}$ and $2^2P_{3/2}$ are 2 and 4 respectively, while the energy differences between each of them and $1^2S_{1/2}$ are approximately equal. Hence the ratio of the intensities of the spectral lines $(2^2P_{1/2} \to 1^2S_{1/2})$ to $(2^2P_{3/2} \to 1^2S_{1/2})$ is 1:2.

(f) The LS coupling between the two electrons of helium produces $S = 0$ (singlet) and $S = 1$ (triplet) states. As the transition between them is forbidden, the spectrum of atomic helium consists of two independent systems (singlet and triplet).

1078

(a) Make a table of the atomic ground states for the following elements: H, He, Be, B, C, N, indicating the states in spectroscopic notation. Give J only for S states.

(b) State Hund's rule and give a physical basis for it.

(Wisconsin)

Solution:

(a) The atomic ground states of the elements are as follows:

element:	H	He	Li	Be	B	C	N
ground state:	$^2S_{1/2}$	$^1S_0$	$^2S_{1/2}$	$^1S_0$	$^2P_{1/2}$	$^3P_0$	$^4S_{3/2}$

(b) For a statement of Hund's rules see **Problem 1008.** Hund's rules are empirical rules based on many experimental results and their application is consequently restricted. First, they are reliable only for determining the lowest energy states of atoms, except those of very heavy elements. They fail in many cases when used to determine the order of energy levels. For example, for the electron configuration $1s^2 2s 2p^3$ of Carbon, the order of energy levels is obtained experimentally as $^5S <^3 D <^1 D <^3 S <^1 P$. It is seen that although ^3S is a higher multiplet, its energy is higher than that of ^1D. For higher excited states, the rules may also fail. For instance, when one of the electrons of Mg atom is excited to d-orbital, the energy of ^1D state is lower than that of ^3D state.

Hund's rules can be somewhat understood as follows. On account of Pauli's exclusion principle, equivalent electrons of parallel spins tend to avoid each other, with the result that their Coulomb repulsion energy, which is positive, tends to be smaller. Hence energies of states with most parallel spins (with largest S) will be the smallest. However the statement regarding states of maximum angular momentum cannot be so readily explained.

1079

(a) What are the terms arising from the electronic configuration $2p3p$ in an (LS) Russell-Saunders coupled atom? Sketch the level structure, roughly show the splitting, and label the effect causing the splitting.

(b) What are the electric-dipole transition selection rules for these terms?

(c) To which of your forbidden terms could electric dipole transitions from a $^3P_1$ term be made?

(Wisconsin)

Solution:

(a) The spectroscopic terms arising from the electronic configuration $2p3p$ in LS coupling are obtained as follows.

As $l_1 = l_2 = 1$, $s_1 = s_2 = \frac{1}{2}$, $\mathbf{L} = \mathbf{l}_1 + \mathbf{l}_2$, $\mathbf{S} = \mathbf{s}_1 + \mathbf{s}_2$, $\mathbf{J} = \mathbf{L} + \mathbf{S}$, we can have $S = 1, 0$, $L = 2, 1, 0$, $J = 3, 2, 1, 0$.

For $S = 0$, $L = 2$, $J = 2$: $^1D_2$, $L = 1$, $J = 1$: $^1P_1$, $L = 0$, $J = 0$: $^1S_0$. For $S = 1$, $L = 2$, $J = 3, 2, 1, 0$: $^3D_{3,2,1}$, $L = 1$, $J = 2, 1, 0$: $^3P_{2,1,0}$, $L = 0$, $J = 1$: $^3S_1$. Hence the terms are

$$\text{singlet}: \quad ^1S_0, \quad ^1P_1, \quad ^1D_2$$
$$\text{triplet}: \quad ^3S_1, \quad ^3P_{2,1,0}, \quad ^3D_{3,2,1}$$

The corresponding energy levels are shown in Fig. 1.28.

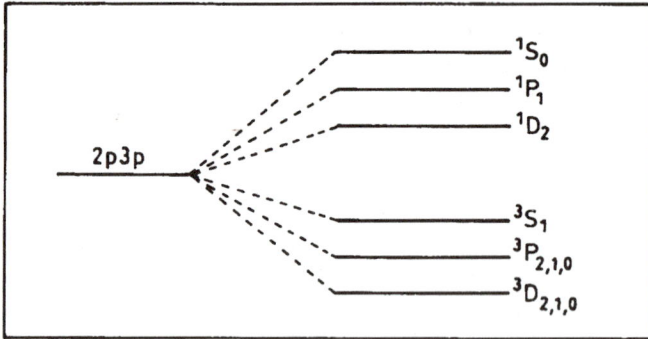

Fig. 1.28

Splitting of spectroscopic terms of different S is caused by the Coulomb exchange energy. Splitting of terms of the same S but different L is caused by the Coulomb repulsion energy. Splitting of terms of the same L, S but different J is caused by the coupling between orbital angular momentum and spin, i.e., by magnetic interaction.

(b) Selection rules for electric-dipole transitions are

(i) Parity must be reversed: even $\leftrightarrow$ odd.

(ii) Change in quantum numbers must satisfy

$$\Delta S = 0, \quad \Delta L = 0, \pm 1, \quad \Delta J = 0, \pm 1 \quad (\text{excepting } 0 \to 0).$$

Electric-dipole transition does not take place between these spectral terms because they have the same parity.

(c) If the $^3P_1$ state considered has odd parity, it can undergo transition to the forbidden spectral terms $^3S_0$, $^3P_{2,1,0}$, $^3D_{2,1}$.

1080

The atoms of lead vapor have the ground state configuration $6s^2 6p^2$.

(a) List the quantum numbers of the various levels of this configuration assuming LS coupling.

(b) State whether transitions between these levels are optically allowed, i.e., are of electric-dipole type. Explain why or why not.

(c) Determine the total number of levels in the presence of a magnetic field **B**.

(d) Determine the total number of levels when a weak electric field **E** is applied together with **B**.

(*Chicago*)

Solution:

(a) The two $6s$ electrons fill the first subshell. They must have anti-parallel spins, forming state $^1S_0$. Of the two $6p$ electrons, their orbital momenta can add up to a total $L = 0, 1, 2$. Their total spin quantum number S is determined by Pauli's exclusion principle for electrons in the same subshell, which requires $L + S =$ even (**Problem 2054(a)**). Hence $S = 0$ for $L = 0, 2$ and $S = 1$ for $L = 1$. The configuration thus has three "terms" with different L and S, and five levels including the fine structure levels with equal L and S but different J. The spectroscopic terms for configuration are therefore

$$^1S_0, {}^3P_{0,1,2}, {}^1D_2 .$$

(b) Electric-dipole transitions among these levels which have the same configuration are forbidden because the levels have the same parity.

(c) In a magnetic field each level with quantum number J splits into $2J + 1$ components with different M_J. For the $6p^2$ levels listed above the total number of sublevels is $1 + 1 + 3 + 5 + 5 = 15$.

(d) The electric field **E** perturbs the sublevels but causes no further splitting because the sublevels have no residual degeneracy. In other words, the applied electric field does not cause new splitting of the energy levels, whose total number is still 15.

1081

Consider a multi-electron atom whose electronic configuration is $1s^2 2s^2 2p^6 3s^2 3p^6 3d^{10} 4s^2 4p4d$.

(a) Is this element in the ground state? If not, what is the ground state?

(b) Suppose a Russell-Saunders coupling scheme applies to this atom. Draw an energy level diagram roughly to scale beginning with a single unperturbed configuration and then taking into account the various interactions, giving the perturbation term involved and estimating the energy split. Label the levels at each stage of the diagram with the appropriate term designation.

(c) What are the allowed transitions of this state to the ground state, if any?

(Columbia)

Solution:

(a) The atom is not in the ground state, which has the outermost-shell electronic configuration $4p^2$, corresponding to atomic states $^1D_2$, $^3P_{2,1,0}$ and $^1S_0$ (**Problem 1080**), among which $^3P_0$ has the lowest energy.

(b) The energy correction arising from LS coupling is

$$\Delta E = a_1 \mathbf{s}_1 \cdot \mathbf{s}_1 + a_2 \mathbf{l}_1 \cdot \mathbf{l}_2 + A\mathbf{L} \cdot \mathbf{S}$$

$$= \frac{a_1}{2}[S(S+1) - s_1(s_1+1) - s_2(s_2+1)] + \frac{a_2}{2}[L(L+1)$$

$$- l_1(l_1+1) - l_2(l_2+1)] + \frac{A}{2}[J(J+1) - L(L+1) - S(S+1)],$$

where a_1, a_2, A can be positive or negative. The energy levels can be obtained in three steps, namely, by plotting the splittings caused by S, L and J successively. The energy levels are given in Fig. 1.29.

(c) The selection rules for electric-dipole transitions are:

$$\Delta S = 0, \Delta L = 0, \pm 1, \Delta J = 0, \pm 1$$

(except $0 \to 0$).

The following transitions are allowed:

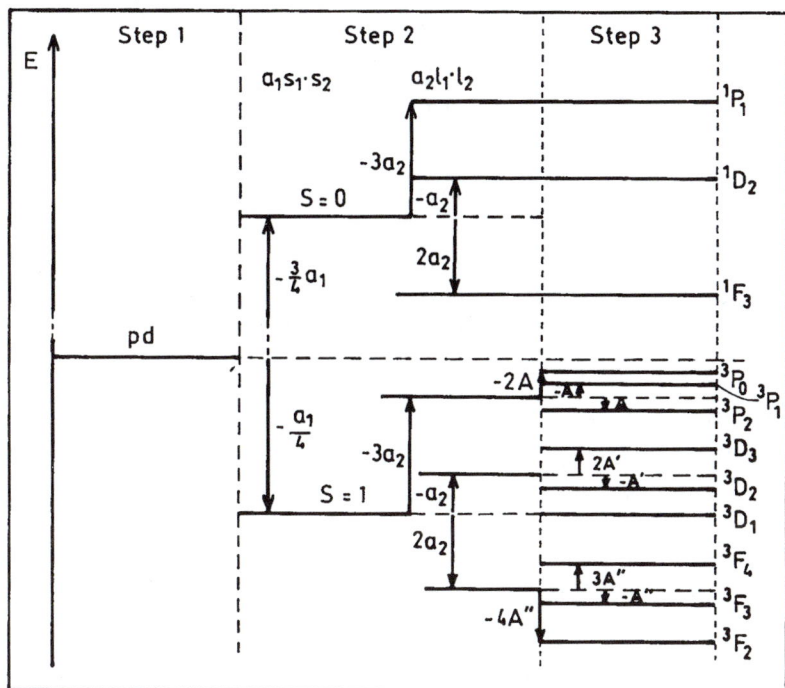

Fig. 1.29

$$(4p4d)^3P_1 \to (4p^2)^3P_0, \qquad (4p4d)^3P_1 \to (4p^2)^3P_1,$$

$$(4p4d)^3P_1 \to (4p^2)^3P_2, \qquad (4p4d)^3P_2 \to (4p^2)^3P_1,$$

$$(4p4d)^3P_2 \to (4p^2)^3P_2, \qquad (4p4d)^3P_0 \to (4p^2)^3P_1,$$

$$(4p4d)^3D_1 \to (4p^2)^3P_1, \qquad (4p4d)^3D_1 \to (4p^2)^3P_2,$$

$$(4p4d)^3D_2 \to (4p^2)^3P_1, \qquad (4p4d)^3D_2 \to (4p^2)^3P_2,$$

$$(4p4d)^3D_3 \to (4p^2)^3P_2, \qquad (4p4d)^1P_1 \to (4p^2)^1S_0,$$

$$(4p4d)^1P_1 \to (4p^2)^1D_2, \qquad (4p4d)^1D_2 \to (4p^2)^1D_2,$$

$$(4p4d)^1F_3 \to (4p^2)^1D_2.$$

1082

In the ground state of beryllium there are two $1s$ and two $2s$ electrons. The lowest excited states are those in which one of the $2s$ electrons is excited to a $2p$ state.

(a) List these states, giving all the angular momentum quantum numbers of each.

(b) Order the states according to increasing energy, indicating any degeneracies. Give a physical explanation for this ordering and estimate the magnitudes of the splitting between the various states.

(Columbia)

Solution:

(a) The electron configuration of the ground state is $1s^2 2s^2$. Pauli's principle requires $S = 0$. Thus the ground state has $S = 0$, $L = 0$, $J = 0$ and is a singlet $^1S_0$.

The lowest excited state has configuration $1s^2 2s 2p$. Pauli's principle allows for both $S = 0$ and $S = 1$. For $S = 0$, as $L = 1$, we have $J = 1$ also, and the state is $^1P_1$. For $S = 1$, as $L = 1$, $J = 2, 1, 0$ and the states are $^3P_{2,1,0}$.

(b) In order of increasing energy, we have

$$^1S_0 < {}^3P_0 < {}^3P_1 < {}^3P_2 < {}^1P_1 .$$

The degeneracies of $^3P_2$, $^3P_1$ and $^1P_1$ are 5, 3, 3 respectively. According to Hund's rule (**Problem 1008(e)**), for the same configuration, the largest S corresponds to the lowest energy; and for a less than half-filled shell, the smallest J corresponds to the smallest energy. This roughly explains the above ordering.

The energy difference between $^1S_0$ and $^1P_1$ is of the order of 1 eV. The energy splitting between the triplet and singlet states is also ~ 1 eV. However the energy splitting among the triplet levels of a state is much smaller, $\sim 10^5$–10^{-4} eV.

1083

A characteristic of the atomic structure of the noble gasses is that the highest p-shells are filled. Thus, the electronic configuration in neon, for example, is $1s^2 2s^2 2p^6$. The total angular momentum $\mathbf{J}$, total orbital angular momentum $\mathbf{L}$ and total spin angular momentum $\mathbf{S}$ of such a closed shell configuration are all zero.

(a) Explain the meaning of the symbols $1s^2 2s^2 2p^6$.

(b) The lowest group of excited states in neon corresponds to the excitation of one of the $2p$ electrons to a $3s$ orbital. The $(2p^5)$ core has orbital and spin angular momenta equal in magnitude but oppositely directed to these quantities for the electron which was removed. Thus, for its interaction with the excited electron, the core may be treated as a p-wave electron.

Assuming LS (Russell-Saunders) coupling, calculate the quantum numbers (L, S, J) of this group of states.

(c) When an atom is placed in a magnetic field H, its energy changes (from the $H = 0$ case) by ΔE:

$$\Delta E = \frac{e\hbar}{2mc} gHM \,,$$

where M can be $J, J-1, J-2, \ldots, -J$. The quantity g is known as the Laudé g-factor. Calculate g for the $L = 1$, $S = 1$, $J = 2$ state of the $1s^2 2s^2 2p^5 3s$ configuration of neon.

(d) The structure of the $1s^2 2s^2 2p^5 3p$ configuration of neon is poorly described by Russell-Saunders coupling. A better description is provided by the "pair coupling" scheme in which the orbital angular momentum $\mathbf{L}_2$ of the outer electron couples with the total angular momentum $\mathbf{J}_c$ of the core. The resultant vector $\mathbf{K}$ ($\mathbf{K} = \mathbf{J}_c + \mathbf{L}_2$) then couples with the spin $\mathbf{S}_2$ of the outer electron to give the total angular momentum $\mathbf{J}$ of the atom.

Calculate the J_c, K, J quantum numbers of the states of the $1s^2 2s^2 2p^5 3p$ configuration.

(CUSPEA)

Solution:

(a) In each group of symbols such as $1s^2$, the number in front of the letter refers to the principal quantum number n, the letter (s, p, etc.) determines the quantum number l of the orbital angular momentum (s for $l = 0$, p for $l = 1$, etc.), the superscript after the letter denotes the number of electrons in the subshell (n, l).

(b) The coupling is the same as that between a p- and an s-electron. Thus we have $l_1 = 1$, $l_2 = 0$ and so $L = 1 + 0 = 1$; $s_1 = \frac{1}{2}$, $s_2 = \frac{1}{2}$ and so

$S = \frac{1}{2} + \frac{1}{2} = 1$ or $S = \frac{1}{2} - \frac{1}{2} = 0$. Then $L = 1$, $S = 1$ give rise to $J = 2, 1$, or 0; $L = 1$, $S = 0$ give rise to $J = 1$. To summarize, the states of (L, S, J) are $(1,1,2)$, $(1,1,1)$, $(1,1,0)$, $(1,0,1)$.

(c) The g-factor is given by

$$g = 1 + \frac{J(J+1) + S(S+1) - L(L+1)}{2J(J+1)}.$$

For $(1,1,2)$ we have

$$g = 1 + \frac{6 + 2 - 2}{2 \times 6} = \frac{3}{2}.$$

(d) The coupling is between a core, which is equivalent to a p-electron, and an outer-shell p-electron, i.e. between $l_c = 1$, $s_c = \frac{1}{2}$; $l_2 = 1$, $s_2 = \frac{1}{2}$. Hence

$$J_c = \frac{3}{2}, \frac{1}{2}, L_2 = 1, S_2 = \frac{1}{2}.$$

For $J_c = \frac{3}{2}$, $L_2 = 1$, we have $K = \frac{5}{2}, \frac{3}{2}, \frac{1}{2}$.
Then for $K = \frac{5}{2}$, $J = 3, 2$; for $K = \frac{3}{2}$, $J = 2, 1$; for

$$K = \frac{1}{2}, J = 1, 0.$$

For $J_c = \frac{1}{2}$, $L_2 = 1$, we have $K = \frac{3}{2}, \frac{1}{2}$. Then for

$$K = \frac{3}{2}, J = 2, 1; \quad \text{for } K = \frac{1}{2}, J = 1, 0.$$

1084

A furnace contains atomic sodium at low pressure and a temperature of 2000 K. Consider only the following three levels of sodium:

$1s^2 2s^2 2p^6 3s$: ^2S, zero energy (ground state),
$1s^2 2s^2 2p^6 3p$: ^2P, 2.10 eV,
$1s^2 2s^2 2p^6 4s$: ^2S, 3.18 eV.

(a) What are the photon energies of the emission lines present in the spectrum? What are their relative intensities? (Give appropriate expressions and evaluate them approximately as time permits).

(b) Continuous radiation with a flat spectrum is now passed through the furnace and the absorption spectrum observed. What spectral lines are observed? Find their relative intensities.

(UC, Berkeley)

Solution:

(a) As $E_0 = 0$ eV, $E_1 = 2.10$ eV, $E_2 = 3.18$ eV, there are two electric-dipole transitions corresponding to energies

$$E_{10} = 2.10 \text{ eV}, \qquad E_{21} = 1.08 \text{ eV}.$$

The probability of transition from energy level k to level i is given by

$$A_{ik} = \frac{e^2 \omega_{ki}^3}{3\hbar^2 c^3} \frac{1}{g_k} \sum_{m_k, m_i} |\langle im_i|\mathbf{r}|km_k\rangle|^2,$$

where $\omega_{ki} = (E_k - E_i)/\hbar$, i, k being the total angular momentum quantum numbers, m_k, m_i being the corresponding magnetic quantum numbers. The intensities of the spectral lines are

$$I_{ik} \propto N_k \hbar \omega_{ki} A_{ik},$$

where the number of particles in the ith energy level $N_i \propto g_i \exp(-\frac{E_i}{kT})$. For ^2P, there are two values of $J : J = 3/2, 1/2$. Suppose the transition matrix elements and the spin weight factors of the two transitions are approximately equal. Then the ratio of the intensities of the two spectral lines is

$$\frac{I_{01}}{I_{12}} = \left(\frac{\omega_{10}}{\omega_{21}}\right)^4 \exp\left(\frac{E_{21}}{kT}\right)$$

$$= \left(\frac{2.10}{1.08}\right)^4 \exp\left(\frac{1.08}{8.6 \times 10^{-5} \times 2000}\right) = 8 \times 10^3.$$

(b) The intensity of an absorption line is

$$I_{ik} \propto B_{ik} N_k \rho(\omega_{ik}) \hbar \omega_{ik},$$

where

$$B_{ik} = \frac{4\pi^2 e^2}{3\hbar^2} \frac{1}{g_k} \sum_{m_k, m_i} |\langle im_i|\mathbf{r}|km_k\rangle|^2$$

is Einstein's coefficient. As the incident beam has a flat spectrum, $\rho(\omega)$ is constant. There are two absorption spectral lines: $E_0 \to E_1$ and $E_1 \to E_2$. The ratio of their intensities is

$$
\frac{I_{10}}{I_{21}} = \frac{B_{10}N_0\omega_{10}}{B_{21}N_1\omega_{21}} \approx \left(\frac{\omega_{10}}{\omega_{21}}\right) \exp\left(\frac{E_{10}}{kT}\right)
$$

$$
= \left(\frac{2.10}{1.08}\right) \exp\left(\frac{2.10}{8.62 \times 10^{-5} \times 2000}\right) = 4 \times 10^5 .
$$

1085

For C $(Z = 6)$ write down the appropriate electron configuration. Using the Pauli principle derive the allowed electronic states for the 4 outermost electrons. Express these states in conventional atomic notation and order in energy according to Hund's rules. Compare this with a $(2p)^4$ configuration.

(*Wisconsin*)

Solution:

The electron configuration of C is $1s^2 2s^2 2p^2$. The two $1s$ electrons form a complete shell and need not be considered. By Pauli's principle, the two electrons $2s^2$ have total spin $S = 0$, and hence total angular momentum 0. Thus we need consider only the coupling of the two p-electrons. Then the possible electronic states are $^1S_0$, $^3P_{2,1,0}$, $^1D_2$ (**Problem 1088**). According to Hund's rule, in the order of increasing energy they are $^3P_0, ^3P_1, ^3P_2, ^1D_2, ^1S_0$.

The electronic configuration of $(2p)^4$ is the same as the above but the energy order is somewhat different. Of the ^3P states, $J = 0$ has the highest energy while $J = 2$ has the lowest. The other states have the same order as in the $2s^2 2p^2$ case.

1086

The atomic number of Mg is $Z = 12$.

(a) Draw a Mg atomic energy level diagram (not necessarily to scale) illustrating its main features, including the ground state and excited states arising from the configurations in which one valence electron is in the $3s$ state and the other valence electron is in the state nl for $n = 3$, 4 and $l = 0$, 1. Label the levels with conventional spectroscopic notation. Assuming LS coupling.

(b) On your diagram, indicate the following (give your reasoning):

(1) an allowed transition,

(2) a forbidden transition,

(3) an intercombination line (if any),

(4) a level which shows (1) anomalous and (2) normal Zeeman effect, if any.

(Wisconsin)

Solution:

(a) Figure 1.30 shows the energy level diagram of Mg atom.

(b) (1) An allowed transition:

$$(3s3p)^1 P_1 \to (3s3s)^1 S_0 .$$

(2) A forbidden transition:

$$(3s4p)^1 P_1 \nrightarrow (3s3p)^1 P_1 .$$

($\Delta \pi = 0$, violating selection rule for parity)

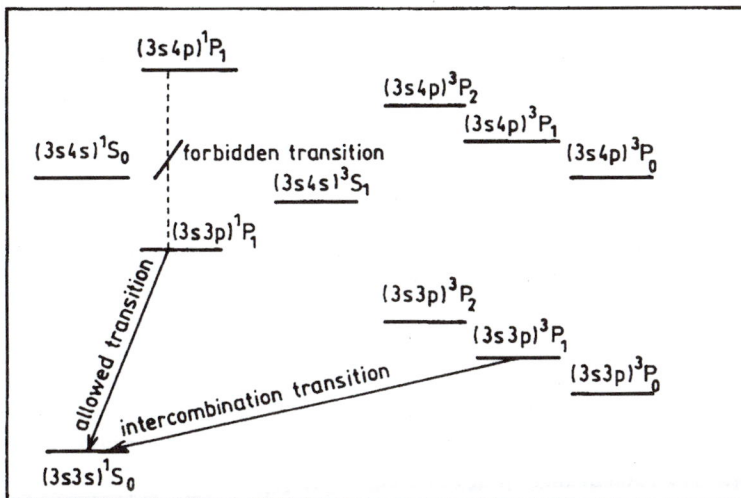

Fig. 1.30

(3) An intercombination line:

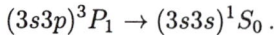

$$(3s3p)^3P_1 \to (3s3s)^1S_0 \,.$$

(4) In a magnetic field, the transition $(3s3p)^1P_1 \to (3s3s)^1S_0$ only produces three lines, which is known as normal Zeeman effect, as shown in Fig. 1.31(a). The transition $(3s4p)^3P_1 \to (3s4s)^3S_1$ produces six lines and is known as anomalous Zeeman effect. This is shown in Fig. 1.31(b). The spacings of the sublevels of $(3s3p)^1P_1$, $(3s4p)^3P_1$, and $(3s4s)^3S_1$ are $\mu_B B$, $3\mu_B B/2$ and $2\mu_B B$ respectively.

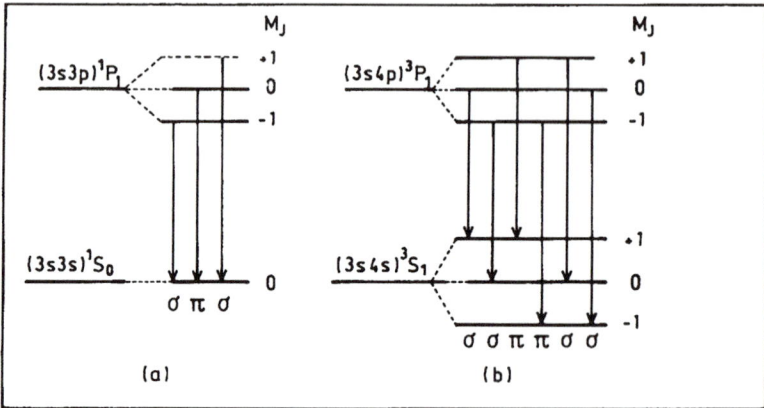

Fig. 1.31

1087

Give, in spectroscopic notation, the ground state of the carbon atom, and explain why this is the ground.

(Wisconsin)

Solution:

The electron configuration of the lowest energy state of carbon atom is $1s^2 2s^2 2p^2$, which can form states whose spectroscopic notations are $^1S_0$, $^3P_{0,1,2}$, $^1D_2$. According to Hund's rule, the ground state has the largest total spin S. But if there are more than one such states, the ground state corresponds to the largest total orbital angular momentum L among such states. If the number of electrons is less than that required to half-fill the shell, the lowest-energy state corresponds to the smallest total angular momentum J. Of the above states, $^3P_{0,1,2}$ have the largest S. As the p-shell is less than half-full, the state $^3P_0$ is the ground state.

1088

What is meant by the statement that the ground state of the carbon atom has the configuration $(1s)^2(2s)^2(2p)^2$?

Assuming that Russell-Saunders coupling applies, show that there are 5 spectroscopic states corresponding to this configuration: $^1S_0$, $^1D_2$, $^3P_1$, $^3P_2$, $^3P_0$.

(*Wisconsin*)

Solution:

The electronic configuration of the ground state of carbon being $(1s)^2$ $(2s)^2(2p)^2$ means that, when the energy of carbon atom is lowest, there are two electrons on the s-orbit of the first principal shell and two electrons each on the s- and p-orbits of the second principal shell.

The spectroscopic notations corresponding to the above electronic configuration are determined by the two equivalent electrons on the p-orbit.

For these two p-electrons, the possible combinations and sums of the values of the z-component of the orbital quantum number are as follows:

m_{l2}	m_{l1}	1	0	−1
1		2	1	0
0		1	0	−1
−1		0	−1	−2

For $m_{l1} = m_{l2}$, or $L = 2, 0$, Pauli's principle requires $m_{s1} \neq m_{s2}$, or $S = 0$, giving rise to terms $^1D_2$, $^1S_0$.

For $m_{s1} = m_{s2}$, or $S = 1$, Pauli's principle requires $m_{l1} \neq m_{l2}$, or $L = 1$, and so $J = 2,1,0$, giving rise to terms $^3P_{2,1,0}$. Hence corresponding to the electron configuration $1s^2 2s^2 2p^2$ the possible spectroscopic terms are

$$^1S_0, {}^1D_2, {}^3P_2, {}^3P_1, {}^3P_0.$$

1089

Apply the Russell-Saunders coupling scheme to obtain all the states associated with the electron configuration $(1s)^2(2s)^2(2p)^5(3p)$. Label each state by the spectroscopic notation of the angular-momentum quantum numbers appropriate to the Russell-Saunders coupling.

(Wisconsin)

Solution:

The 2p-orbit can accommodate $2(2l+1) = 6$ electrons. Hence the configuration $(1s)^2(2s)^2(2p)^5$ can be represented by its complement $(1s)^2(2s)^2$ $(2p)^1$ in its coupling with the 3p electron. In LS coupling the combination of the 2p- and 3p-electrons can be considered as follows. As $l_1 = 1$, $l_2 = 1$, $s_1 = \frac{1}{2}$, $s_2 = \frac{1}{2}$, we have $L = 2, 1, 0$; $S = 1, 0$. For $L = 2$, we have for $S = 1$: $J = 3, 2, 1$; and for $S = 0$: $J = 2$, giving rise to $^3D_{3,2,1}$, $^1D_2$. For $L = 1$, we have for $S = 1$: $J = 2, 1, 0$; and for $S = 0$: $J = 1$, giving rise to $^3P_{2,1,0}$, $^1P_1$. For $L = 0$, we have for $S = 1$: $J = 1$; for $S = 0$: $J = 0$, giving rise to $^3S_1$, $^1S_0$. Hence the given configuration has atomic states

$$^3S_1, {}^3P_{2,1,0}, {}^3D_{3,2,1}, {}^1S_0, {}^1P_1, {}^1D_2.$$

1090

The ground configuration of Sd (scandium) is $1s^2 2s^2 2p^6 3s^2 3p^6 3d4s^2$.

(a) To what term does this configuration give rise?

(b) What is the appropriate spectroscopic notation for the multiplet levels belonging to this term? What is the ordering of the levels as a function of the energy?

(c) The two lowest (if there are more than two) levels of this ground multiplet are separated by 168 cm^{-1}. What are their relative population at $T = 2000$ K?

$$h = 6.6 \times 10^{-34} \text{ J sec}, \qquad c = 3 \times 10^8 \text{ m/s}, \qquad k = 1.4 \times 10^{-23} \text{ J/K}.$$

(Wisconsin)

Solution:

(a) Outside completed shells there are one 3d-electron and two 4s-electrons to be considered. In LS coupling we have to combine $l = 2$,

$s = \frac{1}{2}$ with $L = 0$, $S = 0$. Hence $L = 2$, $S = \frac{1}{2}$, and the spectroscopic notations of the electron configuration are

$$^2D_{5/2}, ^2D_{3/2}.$$

(b) The multiplet levels are $^2D_{5/2}$ and $^2D_{3/2}$, of which the second has the lower energy according to Hund's rules as the D shell is less than half-filled.

(c) The ratio of particle numbers in these two energy levels is

$$\frac{g_1}{g_2} \exp\left(-\frac{\Delta E}{kT}\right),$$

where $g_1 = 2 \times \frac{3}{2} + 1 = 4$ is the degeneracy of $^2D_{3/2}$, $g_2 = 2 \times \frac{5}{2} + 1 = 6$ is the degeneracy of $^2D_{5/2}$, ΔE is the separation of these two energy levels. As

$$\Delta E = hc\Delta\tilde{\nu} = 6.6 \times 10^{-34} \times 3 \times 10^8 \times 168 \times 10^2$$

$$= 3.3 \times 10^{-21} \text{ J},$$

$$\frac{g_1}{g_2} \exp\left(-\frac{\Delta E}{kT}\right) = 0.6.$$

1091

Consider the case of four equivalent p-electrons in an atom or ion. (Think of these electrons as having the same radial wave function, and the same orbital angular momentum $l = 1$).

(a) Within the framework of the Russell-Saunders (LS) coupling scheme, determine all possible configurations of the four electrons; label these according to the standard spectroscopic notation, and in each case indicate the values of L, S, J and the multiplicity.

(b) Compute the Landé g-factor for all of the above states for which $J = 2$.

(*UC, Berkeley*)

Solution:

(a) The p-orbit of a principal-shell can accommodate $2(2 \times 1 + 1) = 6$ electrons and so the terms for p^n and p^{6-n} are the same. Thus the situation of four equivalent p-electrons is the same as that of 2 equivalent p-electrons. In accordance with Pauli's principle, the spectroscopic terms are (**Problem 1088**)

$$^1S_0 \quad (S = 0, L = 0, J = 0)$$

$$^1D_2 \quad (S = 0, L = 2, J = 2)$$

$$^3P_{2,1,0} \quad (S = 1, L = 1, J = 2, 1, 0).$$

(b) The Landé g-factors are given by

$$g = 1 + \frac{J(J+1) + S(S+1) - L(L+1)}{2J(J+1)}.$$

For $^1D_2$:

$$g = 1 + \frac{2 \times 3 + 0 \times 1 - 2 \times 3}{2 \times 2 \times 3} = 1,$$

For $^3P_2$:

$$g = 1 + \frac{2 \times 3 + 1 \times 2 - 1 \times 2}{2 \times 2 \times 3} = 1.5.$$

1092

For the sodium doublet give:

(a) Spectroscopic notation for the energy levels (Fig. 1.32).
(b) Physical reason for the energy difference E.
(c) Physical reason for the splitting ΔE.
(d) The expected intensity ratio

$$D_2/D_1 \quad \text{if } kT \gg \Delta E.$$

(Wisconsin)

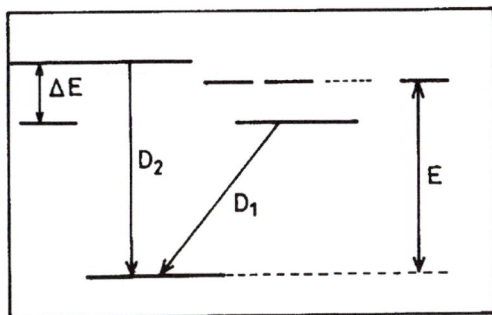

Fig. 1.32

Solution:

(a) The spectroscopic notations for the energy levels are shown in Fig. 1.33.

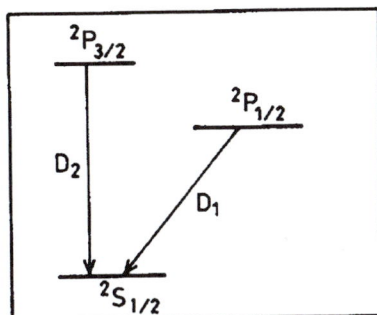

Fig. 1.33

(b) The energy difference E arises from the polarization of the atomic nucleus and the penetration of the electron orbits into the nucleus, which are different for different orbital angular momenta l.

(c) ΔE is caused by the coupling between the spin and orbit angular momentum of the electrons.

(d) When $kT \gg \Delta E$, the intensity ratio D_2/D_1 is determined by the degeneracies of $^2P_{3/2}$ and $^2P_{1/2}$:

$$\frac{D_2}{D_1} = \frac{2J_2 + 1}{2J_1 + 1} = \frac{3+1}{1+1} = 2\,.$$

1093

(a) What is the electron configuration of sodium ($Z = 11$) in its ground state? In its first excited state?

(b) Give the spectroscopic term designation (e.g. $^4S_{3/2}$) for each of these states in the LS coupling approximation.

(c) The transition between the two states is in the visible region. What does this say about kR, where k is the wave number of the radiation and R is the radius of the atom? What can you conclude about the multipolarity of the emitted radiation?

(d) What are the sodium "D-lines" and why do they form a doublet?

(Wisconsin)

Solution:

(a) The electron configuration of the ground state of Na is $1s^2 2s^2 2p^6 3s^1$, and that of the first excited state is $1s^2 2s^2 2p^6 3p^1$.

(b) The ground state: $^2S_{1/2}$.

The first excited state: $^2P_{3/2}, ^2P_{1/2}$.

(c) As the atomic radius $R \approx 1$ Å and for visible light $k \approx 10^{-4}$ Å^{-1}, we have $kR \ll 1$, which satisfies the condition for electric-dipole transition. Hence the transitions $^2P_{3/2} \rightarrow ^2 S_{1/2}$, $^2P_{1/2} \rightarrow ^2 S_{1/2}$ are electric dipole transitions.

(d) The D-lines are caused by transition from the first excited state to the ground state of Na. The first excited state is split into two energy levels $^2P_{3/2}$ and $^2P_{1/2}$ due to LS coupling. Hence the D-line has a doublet structure.

1094

Couple a p-state and an s-state electron via

(a) Russell-Saunders coupling,

(b) j, j coupling,

and identify the resultant states with the appropriate quantum numbers. Sketch side by side the energy level diagrams for the two cases and show which level goes over to which as the spin-orbit coupling is increased.

(Wisconsin)

Solution:

We have $s_1 = s_2 = 1/2$, $l_1 = 1$, $l_2 = 0$.

(a) In LS coupling, $\mathbf{L} = \mathbf{l_1} + \mathbf{l_2}$, $\mathbf{S} = \mathbf{s_1} + \mathbf{s_2}$, $\mathbf{J} = \mathbf{L} + \mathbf{S}$. Thus $L = 1, S = 1, 0$.

For $S = 1$, $J = 2$, 1, 0, giving rise to $^3P_{2,1,0}$.

For $S = 0$, $J = 1$, giving rise to $^1P_1$.

(b) In jj coupling, $\mathbf{j_1} = \mathbf{l_1} + \mathbf{s_1}$, $\mathbf{j_2} = \mathbf{l_2} + \mathbf{s_2}$, $\mathbf{J} = \mathbf{j_1} + \mathbf{j_2}$. Thus $j_1 = \frac{3}{2}, \frac{1}{2}$, $j_2 = \frac{1}{2}$.

Hence the coupled states are

$$\left(\frac{3}{2}, \frac{1}{2}\right)_2, \left(\frac{3}{2}, \frac{1}{2}\right)_1, \left(\frac{1}{2}, \frac{1}{2}\right)_1, \left(\frac{1}{2}, \frac{1}{2}\right)_0,$$

where the subscripts indicate the values of J.

The coupled states are shown in Fig. 1.34.

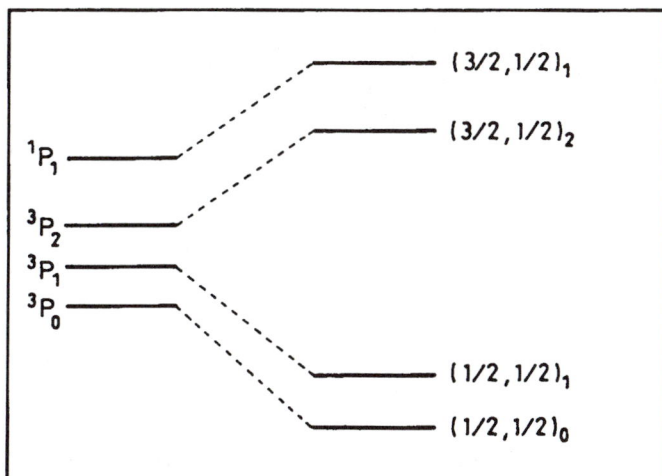

Fig. 1.34

1095

(a) State the ground state configuration of a carbon atom, and list the levels (labeled in terms of Russel-Saunders coupling) of this configuration.

(b) Which is the ground state level? Justify your answer.

(Wisconsin)

Solution:

(a) The electronic configuration of the ground state of carbon is $1s^2$ $2s^2 2p^2$. The corresponding energy levels are $^1S_0$, $^2P_{2,1,0}$, $^1D_2$.

(b) According to Hund's rules, the ground state is $^3P_0$.

1096

For each of the following atomic radiative transitions, indicate whether the transition is allowed or forbidden under the electric-dipole radiation selection rules. For the forbidden transitions, cite the selection rules which are violated.

(a) He: $(1s)(1p)$ $^1P_1 \to (1s)^2$ $^1S_0$
(b) C: $(1s)^2(2s)^2(2p)(3s)$ $^3P_1 \to (1s)^2(2s)^2(2p)^2$ $^3P_0$
(c) C: $(1s)^2(2s)^2(2p)(3s)$ $^3P_0 \to (1s)^2(2s)^2(2p)^2$ $^3P_0$
(d) Na: $(1s)^2(2s)^2(2p)^6(4d)$ $^2D_{5/2} \to (1s)^2(2s)^2(2p)^6(3p)$ $^2P_{1/2}$
(e) He: $(1s)(2p)$ $^3P_1 \to (1s)^2$ $^1S_0$

(Wisconsin)

Solution:

The selection rules for single electric-dipole transition are

$$\Delta l = \pm 1, \qquad \Delta j = 0, \pm 1.$$

The selection rules for multiple electric-dipole transition are

$$\Delta S = 0, \qquad \Delta L = 0, \pm 1, \qquad \Delta J = 0, \pm 1 (0 \not\leftrightarrow 0).$$

(a) Allowed electric-dipole transition.
(b) Allowed electric-dipole transition.
(c) Forbidden as the total angular momentum J changes from 0 to 0 which is forbidden for electric-dipole transition.
(d) Forbidden as it violates the condition $\Delta J = 0, \pm 1$.
(e) Forbidden as it violates the condition $\Delta S = 0$.

1097

Consider a hypothetical atom with an electron configuration of two identical p-shell electrons outside a closed shell.

(a) Assuming LS (Russell-Saunders) coupling, identify the possible levels of the system using the customary spectroscopic notation, $^{(2S+1)}L_J$.

(b) What are the parities of the levels in part (a)?

(c) In the independent-particle approximation these levels would all be degenerate, but in fact their energies are somewhat different. Describe the physical origins of the splittings.

(Wisconsin)

Solution:

(a) The electronic configuration is p^2. The two p-electrons being equivalent, the possible energy levels are **(Problem 1088)**

$$^{1}S_0, \,^{3}P_{2,1,0}, \,^{1}D_2 \,.$$

(b) The parity of an energy level is determined by the sum of the orbital angular momentum quantum numbers: parity $\pi = (-1)^{\Sigma l}$. Parity is even or odd depending on π being $+1$ or -1. The levels $^{1}S_0$, $^{3}P_{2,1,0}$, $^{1}D_2$ have $\Sigma l = 2$ and hence even parity.

(c) See **Problem 1079(a)**.

1098

What is the ground state configuration of potassium (atomic number 19).

(UC, Berkeley)

Solution:

The ground state configuration of potassium is $1s^2 2s^2 2p^6 3s^2 3p^6 4s^1$.

1099

Consider the ^{17}O isotope ($I = 5/2$) of the oxygen atom. Draw a diagram to show the fine-structure and hyperfine-structure splittings of the levels described by $(1s^2 2s^2 2p^4)^3P$. Label the states by the appropriate angular-momentum quantum numbers.

(*Wisconsin*)

Solution:

The fine and hyperfine structures of the ^3P state of ^{17}O is shown in Fig. 1.35.

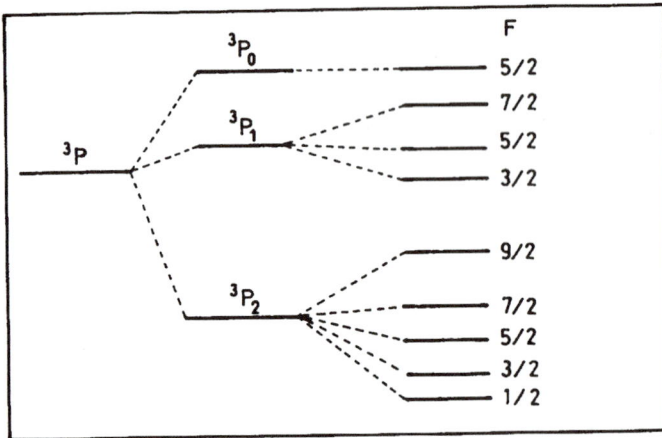

Fig. 1.35

1100

Consider a helium atom with a $1s3d$ electronic configuration. Sketch a series of energy-level diagrams to be expected when one takes successively into account:

(a) only the Coulomb attraction between each electron and the nucleus,
(b) the electrostatic repulsion between the electrons,
(c) spin-orbit coupling,
(d) the effect of a weak external magnetic field.

(*Wisconsin*)

Solution:

The successive energy-level splittings are shown in Fig. 1.36.

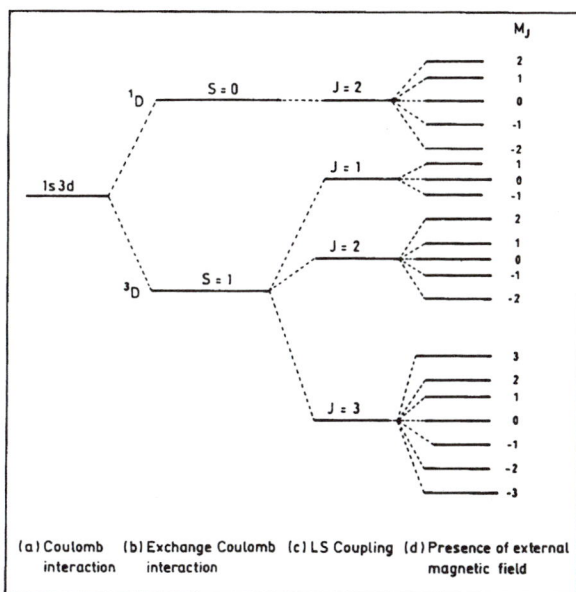

Fig. 1.36

1101

Sodium chloride forms cubic crystals with four Na and four Cl atoms per cube. The atomic weights of Na and Cl are 23.0 and 35.5 respectively. The density of NaCl is 2.16 gm/cc.

(a) Calculate the longest wavelength for which X-rays can be Bragg reflected.

(b) For X-rays of wavelength 4 Å, determine the number of Bragg reflections and the angle of each.

(UC, Berkeley)

Solution:

(a) Let V be the volume of the unit cell, N_A be Avogadro's number, ρ be the density of NaCl. Then

$$V \rho N_A = 4(23.0 + 35.5),$$

giving

$$V = \frac{4 \times 58.5}{2.16 \times 6.02 \times 10^{23}} = 1.80 \times 10^{-22} \text{ cm}^3,$$

and the side length of the cubic unit cell

$$d = ^3\sqrt{V} = 5.6 \times 10^{-8} \text{ cm} = 5.6 \text{ Å}.$$

Bragg's equation $2d \sin\theta = n\lambda$, then gives

$$\lambda_{max} = 2d = 11.2 \text{ Å}.$$

(b) For $\lambda = 4$ Å,

$$\sin\theta = \frac{\lambda n}{2d} = 0.357n.$$

Hence

$$\text{for } n = 1 : \sin\theta = 0.357, \theta = 20.9°,$$

$$\text{for } n = 2 : \sin\theta = 0.714, \theta = 45.6°$$

For $n \geq 3$: $\sin\theta > 1$, and Bragg reflection is not allowed.

1102

(a) 100 keV electrons bombard a tungsten target ($Z = 74$). Sketch the spectrum of resulting X-rays as a function of $1/\lambda$ (λ = wavelength). Mark the K X-ray lines.

(b) Derive an approximate formula for λ as a function of Z for the K X-ray lines and show that the Moseley plot ($\lambda^{-1/2}$ vs. Z) is (nearly) a straight line.

(c) Show that the ratio of the slopes of the Moseley plot for K_α and K_β (the two longest-wavelength K-lines) is $(27/32)^{1/2}$.

(Wisconsin)

Solution:

(a) The X-ray spectrum consists of two parts, continuous and characteristic. The continuous spectrum has the shortest wavelength determined by the energy of the incident electrons:

$$\lambda_{min} = \frac{hc}{E} = \frac{12.4}{100} \text{Å} = 0.124 \text{ Å}.$$

The highest energy for the K X-ray lines of W is 13.6×74^2 eV $= 74.5$ keV, so the K X-ray lines are superimposed on the continuous spectrum as shown as Fig. 1.37.

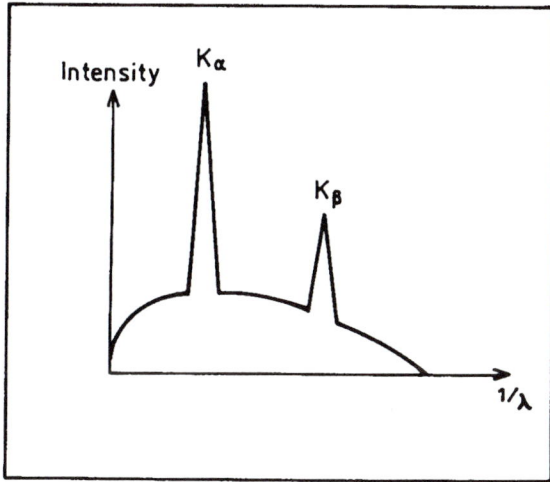

Fig. 1.37

(b) The energy levels of tungsten atom are given by

$$E_n = -\frac{RhcZ^{*2}}{n^2} \, ,$$

where Z^* is the effective nuclear charge.

The K lines arise from transitions to ground state ($n \to 1$):

$$\frac{hc}{\lambda} = -\frac{RhcZ^{*2}}{n^2} + RhcZ^{*2} \, ,$$

giving

$$\lambda = \frac{n^2}{(n^2 - 1)RZ^{*2}} \, ,$$

or

$$\lambda^{-\frac{1}{2}} = Z^* \sqrt{\left(\frac{n^2 - 1}{n^2}\right) R} \approx Z \sqrt{\left(\frac{n^2 - 1}{n^2}\right) R} \, . \qquad (n = 1, 2, 3, \dots)$$

Hence the relation between $\lambda^{-1/2}$ and Z is approximately linear.

(c) K_α lines are emitted in transitions $n = 2$ to $n = 1$, and K_β lines, from $n = 3$ to $n = 1$. In the Moseley plot, the slope of the K_α curve is $\sqrt{\frac{3}{4}R}$ and that of K_β is $\sqrt{\frac{8}{9}R}$, so the ratio of the two slopes is

$$\frac{\sqrt{(3/4)R}}{\sqrt{(8/9)R}} = \sqrt{\frac{27}{32}} \, .$$

1103

(a) If a source of continuum radiation passes through a gas, the emergent radiation is referred to as an absorption spectrum. In the optical and ultra-violet region there are absorption lines, while in the X-ray region there are absorption edges. Why does this difference exist and what is the physical origin of the two phenomena?

(b) Given that the ionization energy of atomic hydrogen is 13.6 eV, what would be the energy E of the radiation from the $n = 2$ to $n = 1$ transition of boron ($Z = 5$) that is 4 times ionized? (The charge of the ion is $+4e$.)

(c) Would the K_α fluorescent radiation from neutral boron have an energy E_k greater than, equal to, or less than E of part (b)? Explain why.

(d) Would the K absorption edge of neutral boron have an energy E_k greater than, equal to, or less than E_k of part (c)? Explain why.

(Wisconsin)

Solution:

(a) Visible and ultra-violet light can only cause transitions of the outer electrons because of their relatively low energies. The absorption spectrum consists of dark lines due to the absorption of photons of energy equal to the difference in energy of two electron states. On the other hand, photons with energies in the X-ray region can cause the ejection of inner electrons from the atoms, ionizing them. This is because in the normal state the outer orbits are usually filled. Starting from lower frequencies in the ultraviolet the photons are able to eject only the loosely bound outer electrons. As the frequency increases, the photons suddenly become sufficiently energetic to eject electrons from an inner shell, causing the absorption coefficient to increase suddenly, giving rise to an absorption edge. As the frequency is increased further, the absorption coefficient decreases approximately as ν^{-3} until the frequency becomes great enough to allow electron ejection from the next inner shell, giving rise to another absorption edge.

(b) The energy levels of a hydrogen-like atom are given by

$$E_n = -\frac{Z^2 e^2}{2n^2 a_0} = -\frac{Z^2}{n^2} E_0 \,,$$

where E_0 is the ionization energy of hydrogen. Hence

$$E_2 - E_1 = -Z^2 E_0 \left(\frac{1}{2^2} - \frac{1}{1^2} \right) = 5^2 \times 13.6 \times \frac{3}{4}$$

$$= 255 \text{ eV} \,.$$

(c) Due to the shielding by the orbital electrons of the nuclear charge, the energy E_k of K_α emitted from neutral Boron is less than that given in (b).

(d) As the K absorption edge energy E_k correspond to the ionization energy of a K shell electron, it is greater than the energy given in (c).

1104

For Zn, the X-ray absorption edges have the following values in keV:

$$K \ 9.67, L_I \ 1.21, L_{II} \ 1.05, L_{III} \ 1.03 \,.$$

Determine the wavelength of the K_α line.

If Zn is bombarded by 5-keV electrons, determine

(a) the wavelength of the shortest X-ray line, and

(b) the wavelength of the shortest characteristic X-ray line which can be emitted.

Note: The K level corresponds to $n = 1$, the three L-levels to the different states with $n = 2$. The absorption edges are the lowest energies for which X-rays can be absorbed by ejection of an electron from the corresponding level. The K_α line corresponds to a transition from the lowest L level.

(UC, Berkeley)

Solution:

The K_α series consists of two lines, $K_{\alpha 1}(L_{III} \to K)$, $K_{\alpha 2}(L_{II} \to K)$:

$$E_{K_{\alpha_1}} = K_{L_{III}} - E_K = 9.67 - 1.03 = 8.64 \text{ keV},$$

$$E_{K_{\alpha_2}} = K_{L_{II}} - E_K = 9.67 - 1.05 = 8.62 \text{ keV}.$$

Hence

$$\lambda_{K_{\alpha_1}} = \frac{hc}{E_{K_{\alpha_1}}} = \frac{12.41}{8.64} = 1.436 \text{ Å},$$

$$\lambda_{K_{\alpha_2}} = \frac{hc}{E_{K_{\alpha_2}}} = 1.440 \text{ Å}.$$

(a) The minimum X-ray wavelength that can be emitted by bombarding the atoms with 5-keV electrons is

$$\lambda_{\min} = \frac{hc}{E_{\max}} = \frac{12.41}{5} = 2.482 \text{ Å}.$$

(b) It is possible to excite electrons on energy levels other than the K level by bombardment with 5-keV electrons, and cause the emission of characteristic X-rays when the atoms de-excite. The highest-energy X-rays have energy $0 - E_I = 1.21$ keV, corresponding to a wavelength of 10.26 Å.

1105

The characteristic K_α X-rays emitted by an atom of atomic number Z were found by Morseley to have the energy $13.6 \times (1 - \frac{1}{4})(Z - 1)^2$ eV.

(a) Interpret the various factors in this expression.
(b) What fine structure is found for the K_α transitions? What are the pertinent quantum numbers?
(c) Some atoms go to a lower energy state by an Auger transition. Describe the process.

(*Wisconsin*)

Solution:

(a) In this expression, 13.6 eV is the ground state energy of hydrogen atom, i.e., the binding energy of an $1s$ electron to unit nuclear charge, the factor $(1 - \frac{1}{4})$ arises from difference in principal quantum number between the states $n = 2$ and $n = 1$, and $(Z - 1)$ is the effective nuclear charge. The K_α line thus originates from a transition from $n = 2$ to $n = 1$.

(b) The K_α line actually has a doublet structure. In LS coupling, the $n = 2$ state splits into three energy levels: $^2S_{1/2}$, $^2P_{1/2}$, $^2P_{3/2}$, while the $n = 1$ state is still a single state $^2S_{1/2}$. According to the selection rules $\Delta L = \pm 1$, $\Delta J = 0, \pm 1 (0 \leftrightarrow\!\!\!/ 0)$, the allowed transitions are

$$K_{\alpha 1}: \quad 2^2P_{3/2} \to 1^2S_{1/2},$$

$$K_{\alpha 2}: \quad 2^2P_{1/2} \to 1^2S_{1/2}.$$

(c) The physical basis of the Auger process is that, after an electron has been removed from an inner shell an electron from an outer shell falls to the vacancy so created and the excess energy is released through ejection of another electron, rather than by emission of a photon. The ejected electron is called Auger electron. For example, after an electron has been removed from the K shell, an L shell electron may fall to the vacancy so created and the difference in energy is used to eject an electron from the L shell or another outer shell. The latter, the Auger electron, has kinetic energy

$$E = -E_L - (-E_k) - E_L = E_k - 2E_L ,$$

where E_k and E_L are the ionization energies of K and L shells respectively.

1106

The binding energies of the two $2p$ states of niobium $(Z = 41)$ are 2370 eV and 2465 eV. For lead $(Z = 82)$ the binding energies of the $2p$ states are 13035 eV and 15200 eV. The $2p$ binding energies are roughly proportional to $(Z - a)^2$ while the splitting between the $2P_{1/2}$ and the $2P_{3/2}$ goes as $(Z - a)^4$. Explain this behavior, and state what might be a reasonable value for the constant a.

(Columbia)

Solution:

The $2p$ electron moves in a central potential field of the nucleus shielded by inner electrons. Taking account of the fine structure due to ls coupling, the energy of a $2p$ electron is given by

$$E = -\frac{1}{4}Rhc(Z - a_1)^2 + \frac{1}{8}Rhc\alpha^2(Z - a_2)^4 \left(\frac{3}{8} - \frac{1}{j + \frac{1}{2}} \right)$$

$$= -3.4(Z - a_1)^2 + 9.06 \times 10^{-5}(Z - a_2)^4 \left(\frac{3}{8} - \frac{1}{j + \frac{1}{2}} \right),$$

as $Rhc = 13.6$ eV, $\alpha = 1/137$. Note that $-E$ gives the binding energy and that $^2P_{3/2}$ corresponds to lower energy according to Hund's rule. For Nb, we have $95 = 9.06 \times 10^{-5}(41 - a_2)^4 \times 0.5$, or $a_2 = 2.9$, which then gives $a_1 = 14.7$. Similarly we have for Pb: $a_1 = 21.4$, $a_2 = -1.2$.

1107

(a) Describe carefully an experimental arrangement for determining the wavelength of the characteristic lines in an X-ray emission spectrum.

(b) From measurement of X-ray spectra of a variety of elements, Moseley was able to assign an atomic number Z to each of the elements. Explain explicitly how this assignment can be made.

(c) Discrete X-ray lines emitted from a certain target cannot in general be observed as absorption lines in the same material. Explain why, for example, the K_α lines cannot be observed in the absorption spectra of heavy elements.

(d) Explain the origin of the continuous spectrum of X-rays emitted when a target is bombarded by electrons of a given energy. What feature of this spectrum is inconsistent with classical electromagnetic theory?

(Columbia)

Solution:

(a) The wavelength can be determined by the method of crystal diffraction. As shown in the Fig. 1.38, the X-rays collimated by narrow slits S_1, S_2, fall on the surface of crystal C which can be rotated about a vertical axis. Photographic film P forms an arc around C. If the condition $2d\sin\theta = n\lambda$, where d is the distance between neighboring Bragg planes and n is an integer, is satisfied, a diffraction line appears on the film at A. After rotating the crystal, another diffraction line will appear at A' which is symmetric to A. As $4\theta = \text{arc}AA'/CA$, the wavelength λ can be obtained.

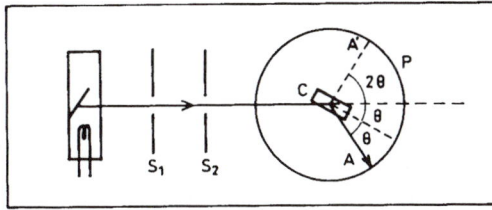

Fig. 1.38

(b) Each element has its own characteristic X-ray spectrum, of which the K series has the shortest wavelengths, and next to them the L series, etc. Moseley discovered that the K series of different elements have the same structure, only the wavelengths are different. Plotting $\sqrt{\tilde{\nu}}$ versus the atomic number Z, he found an approximate linear relation:

$$\tilde{\nu} = R(Z-1)^2 \left(\frac{1}{1^2} - \frac{1}{2^2} \right),$$

where $R = R_H c$, R_H being the Rydberg constant and c the velocity of light in free space.

Then if the wavelength or frequency of K_α of a certain element is found, its atomic number Z can be determined.

(c) The K_α lines represent the difference in energy between electrons in different inner shells. Usually these energy levels are all occupied and transitions cannot take place between them by absorbing X-rays with energy equal to the energy difference between such levels. The X-rays can only ionize the inner-shell electrons. Hence only absorption edges, but not absorption lines, are observed.

(d) When electrons hit a target they are decelerated and consequently emit bremsstrahlung radiation, which are continuous in frequency with the shortest wavelength determined by the maximum kinetic energy of the electrons, $\lambda = \frac{hc}{E_e}$. On the other hand, in the classical electromagnetic theory, the kinetic energy of the electrons can only affect the intensity of the spectrum, not the wavelength.

1108

In the X-ray region, as the photon energy decreases the X-ray absorption cross section rises monotonically, except for sharp drops in the cross section at certain photon energies characteristic of the absorbing material. For Zn ($Z = 30$) the four most energetic of these drops are at photon energies 9678 eV, 1236 eV, 1047 eV and 1024 eV.

(a) Identify the transitions corresponding to these drops in the X-ray absorption cross section.

(b) Identify the transitions and give the energies of Zn X-ray emission lines whose energies are greater than 5000 eV.

(c) Calculate the ionization energy of Zn^{29+} (i.e., a Zn atom with 29 electrons removed). (Hint: the ionization energy of hydrogen is 13.6 eV).

(d) Why does the result of part (c) agree so poorly with 9678 eV?

(Wisconsin)

Solution:

(a) The energies 9.768, 1.236, 1.047 and 1.024 keV correspond to the ionization energies of an $1s$ electron, a $2s$ electron, and each of two $2p$ electrons respectively. That is, they are energies required to eject the respective electrons to an infinite distance from the atom.

(b) X-rays of Zn with energies greater than 5 keV are emitted in transitions of electrons from other shells to the K shell. In particular X-rays emitted in transitions from L to K shells are

$$K_{\alpha 1}: \quad E = -1.024 - (-9.678) = 8.654 \text{ keV}, \qquad (L_{\text{III}} \to K)$$

$$K_{\alpha 2}: \quad E = -1.047 - (-9.678) = 8.631 \text{ keV}. \qquad (L_{\text{II}} \to K)$$

(c) The ionization energy of the Zn^{29+} (a hydrogen-like atom) is

$$E_{Zn} = 13.6 \ Z^2 = 11.44 \text{ keV}.$$

(d) The energy 9.678 keV corresponds to the ionization energy of the $1s$ electron in the neutral Zn atom. Because of the Coulomb screening effect of the other electrons, the effective charge of the nucleus is $Z^* < 30$. Also the farther is the electron from the nucleus, the less is the nuclear charge Z^* it interacts with. Hence the ionization energy of a $1s$ electron of the neutral Zn atom is much less than that of the Zn^{29+} ion.

1109

Sketch a derivation of the "Landé g-factor", i.e. the factor determining the effective magnetic moment of an atom in weak fields.

(Wisconsin)

Solution:

Let the total orbital angular momentum of the electrons in the atom be $\mathbf{P}_L$, the total spin angular momentum be $\mathbf{P}_S$ ($\mathbf{P}_L$ and $\mathbf{P}_S$ being all in units of $\hbar$). Then the corresponding magnetic moments are $\boldsymbol{\mu}_L = -\mu_B \mathbf{P}_L$ and $\boldsymbol{\mu}_S = -2\mu_B \mathbf{P}_S$, where μ_B is the Bohr magneton. Assume the total magnetic moment is $\boldsymbol{\mu}_J = -g\mu_B \mathbf{P}_J$, where g is the Landé g-factor. As

$$\mathbf{P}_J = \mathbf{P}_L + \mathbf{P}_S\,,$$

$$\boldsymbol{\mu}_J = \boldsymbol{\mu}_L + \boldsymbol{\mu}_S = -\mu_B(\mathbf{P}_L + 2\mathbf{P}_S) = -\mu_B(\mathbf{P}_J + \mathbf{P}_S)\,,$$

we have

$$\boldsymbol{\mu}_J = \frac{\boldsymbol{\mu}_J \cdot \mathbf{P}_J}{P_J^2}\mathbf{P}_J$$

$$= -\mu_B\frac{(\mathbf{P}_J + \mathbf{P}_S) \cdot \mathbf{P}_J}{P_J^2}\mathbf{P}_J$$

$$= -\mu_B\frac{P_J^2 + \mathbf{P}_S \cdot \mathbf{P}_J}{P_J^2}\mathbf{P}_J$$

$$= -g\mu_B\mathbf{P}_J\,,$$

giving

$$g = \frac{P_J^2 + \mathbf{P}_S \cdot \mathbf{P}_J}{P_J^2} = 1 + \frac{\mathbf{P}_S \cdot \mathbf{P}_J}{P_J^2}\,.$$

As

$$\mathbf{P}_L \cdot \mathbf{P}_L = (\mathbf{P}_J - \mathbf{P}_S) \cdot (\mathbf{P}_J - \mathbf{P}_S) = P_J^2 + P_S^2 - 2\mathbf{P}_J \cdot \mathbf{P}_S\,,$$

we have

$$\mathbf{P}_J \cdot \mathbf{P}_S = \frac{1}{2}(P_J^2 + P_S^2 - P_L^2)\,.$$

Hence

$$g = 1 + \frac{P_J^2 + P_S^2 - P_L^2}{2P_J^2}$$

$$= 1 + \frac{J(J+1) + S(S+1) - L(L+1)}{2J(J+1)}\,.$$

1110

In the spin echo experiment, a sample of a proton-containing liquid (e.g. glycerin) is placed in a steady but spatially inhomogeneous magnetic field of a few kilogauss. A pulse (a few microseconds) of a strong (a few gauss) radiofrequency field is applied perpendicular to the steady field. Immediately afterwards, a radiofrequency signal can be picked up from the coil around the sample. But this dies out in a fraction of a millisecond unless special precaution has been taken to make the field very spatially homogeneous, in which case the signal persists for a long time. If a second long radiofrequency pulse is applied, say 15 milliseconds after the first pulse, then a radiofrequency signal is observed 15 milliseconds after the second pulse (the echo).

(a) How would you calculate the proper frequency for the radiofrequency pulse?

(b) What are the requirements on the spatial homogeneity of the steady field?

(c) Explain the formation of the echo.

(d) How would you calculate an appropriate length of the first radiofrequency pulse?

(Princeton)

Solution:

(a) The radiofrequency field must have sufficiently high frequency to cause nuclear magnetic resonance:

$$\hbar\omega = \gamma_p \hbar H_0(\mathbf{r}),$$

or

$$\omega = \gamma_p \langle H_0(\mathbf{r}) \rangle,$$

where γ_p is the gyromagnetic ratio, and $\langle H_0(\mathbf{r}) \rangle$ is the average value of the magnetic field in the sample.

(b) Suppose the maximum variation of H_0 in the sample is $(\Delta H)_m$. Then the decay time is $\frac{1}{\gamma_p(\Delta H)_m}$. We require $\frac{1}{\gamma_p}(\Delta H)_m > \tau$, where τ is the time interval between the two pulses. Thus we require

$$(\Delta H)_m < \frac{1}{\gamma_p \tau}.$$

(c) Take the z-axis along the direction of the steady magnetic field H_0. At $t = 0$, the magnetic moments are parallel to H_0 (Fig. 1.39(a)). After introducing the first magnetic pulse H_1 in the x direction, the magnetic moments will deviate from the z direction (Fig. 1.39(b)). The angle θ of the rotation of the magnetic moments can be adjusted by changing the width of the magnetic pulse, as shown in Fig. 1.39(c) where $\theta = 90°$.

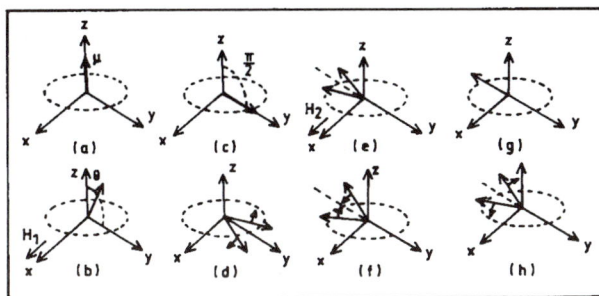

Fig. 1.39

The magnetic moments also processes around the direction of H_0. The spatial inhomogeneity of the steady magnetic field H_0 causes the processional angular velocity $\omega = \gamma_P H_0$ to be different at different points, with the result that the magnetic moments will fan out as shown in Fig. 1.39(d). If a second, wider pulse is introduced along the x direction at $t = \tau$ (say, at $t = 15$ ms), it makes all the magnetic moments turn $180°$ about the x-axis (Fig. 1.39(e)). Now the order of procession of the magnetic moments is reversed (Fig. 1.39(f)). At $t = 2\tau$, the directions of the magnetic moments will again become the same (Fig. 1.39(g)). At this instant, the total magnetic moment and its rate of change will be a maximum, producing a resonance signal and forming an echo wave (Fig. 1.40). Afterwards the magnetic moments scatter again and the signal disappears, as shown in Fig. 1.39(h).

(d) The first radio pulse causes the magnetic moments to rotate through an angle θ about the x-axis. To enhance the echo wave, the rotated magnetic moments should be perpendicular to H_0, i.e., $\theta \approx \pi/2$. This means that

$$\gamma_P H_1 t \approx \pi/2,$$

i.e., the width of the first pulse should be $t \approx \frac{\pi}{2\gamma_P H_1}$.

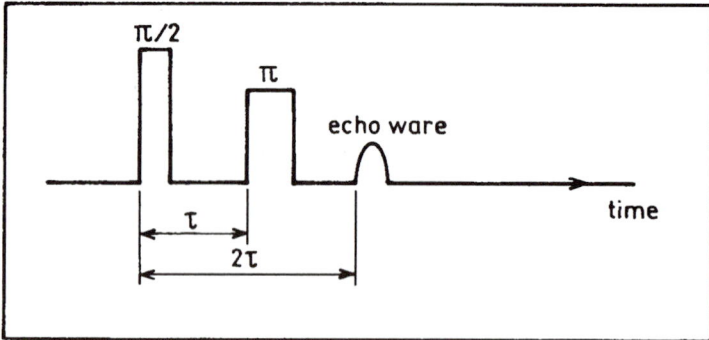

Fig. 1.40

1111

Choose only ONE of the following spectroscopes:

Continuous electron spin resonance

Pulsed nuclear magnetic resonance

Mössbauer spectroscopy

(a) Give a block diagram of the instrumentation required to perform the spectroscopy your have chosen.

(b) Give a concise description of the operation of this instrument.

(c) Describe the results of a measurement making clear what quantitative information can be derived from the data and the physical significance of this quantitative information.

(SUNY, Buffalo)

Solution:

(1) **Continuous electron spin resonance**

(a) The experimental setup is shown in Fig. 1.41.

(b) *Operation.* The sample is placed in the resonant cavity, which is under a static magnetic field B_0. Fixed-frequency microwaves B_1 created in the klystron is guided to the T-bridge. When the microwave power is distributed equally to the arms 1 and 2, there is no signal in the wave detector. As B_0 is varied, when the resonance condition is satisfied, the sample absorbs power and the balance between 1 and 2 is disturbed. The absorption

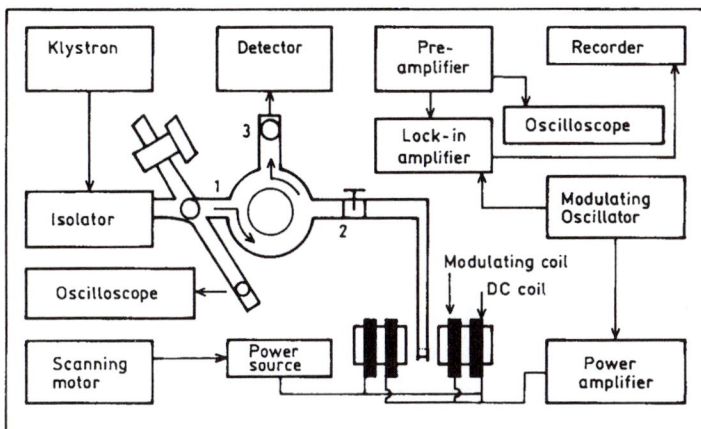

Fig. 1.41

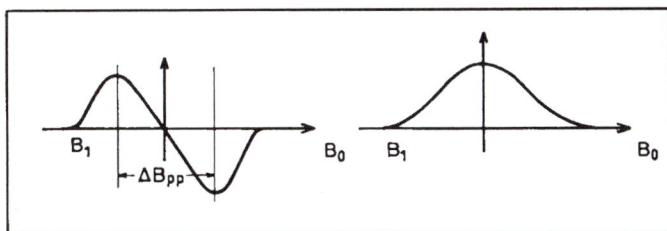

Fig. 1.42

signal is transmitted to the wave detector through arm 3, to be displayed or recorded.

(c) *Data analysis.* The monitor may show two types of differential graph (Fig. 1.42), Gaussian or Lagrangian, from which the following information may be obtained.

(i) The g-factor can be calculated from B_0 at the center and the microwave frequency.

(ii) The line width can be found from the peak-to-peak distance ΔB_{pp} of the differential signal.

(iii) The relaxation time T_1 and T_2 can be obtained by the saturation method, where T_1 and T_2 (Lorenzian profile) are given by

$$T_2(\text{spin–spin}) = \frac{1.3131 \times 10^{-7}}{g\Delta B_{pp}^0},$$

$$T_1 = \frac{0.9848 \times 10^{-7}\Delta B_{PP}^0}{gB_1^2}\left(\frac{1}{s} - 1\right).$$

In the above g is the Landé factor, ΔB_{PP}^0 is the saturation peak-to-peak distance (in gauss), B_1 is the magnetic field corresponding to the edge of the spectral line, and s is the saturation factor.

(iv) The relative intensities.

By comparing with the standard spectrum, we can determine from the g-factor and the line profile to what kind of paramagnetic atoms the spectrum belongs. If there are several kinds of paramagnetic atoms present in the sample, their relative intensities give the relative amounts. Also, from the structure of the spectrum, the nuclear spin I may be found.

(2) **Pulsed nuclear magnetic resonance**

(a) Figure 1.43 shows a block diagram of the experimental setup.

(b) *Operation.* Basically an external magnetic field is employed to split up the spin states of the nuclei. Then a pulsed radiofrequency field is introduced perpendicular to the static magnetic field to cause resonant transitions between the spin states. The absorption signals obtained from the same coil are amplified, Fourier-transformed, and displayed on a monitor screen.

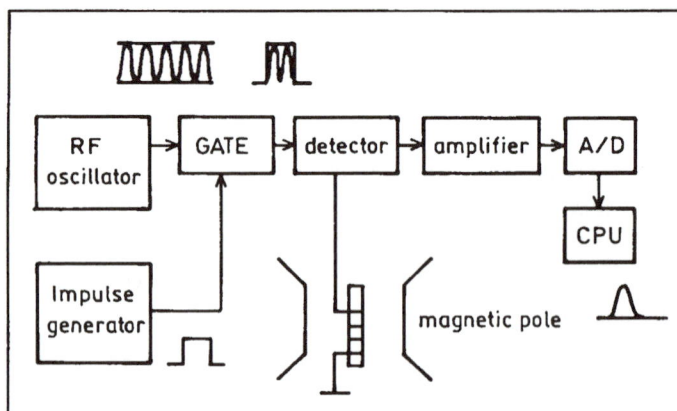

Fig. 1.43

(c) Information that can be deduced are positions and number of absorption peaks, integrated intensities of absorption peaks, the relaxation times T_1 and T_2.

The positions of absorption peaks relate to chemical displacement. From the number and integrated intensities of the peaks, the structure of the compound may be deduced as different kinds of atom have different ways of compounding with other atoms. For a given way of compounding, the integrated spectral intensity is proportional to the number of atoms. Consequently, the ratio of atoms in different combined forms can be determined from the ratio of the spectral intensities. The number of the peaks relates to the coupling between nuclei.

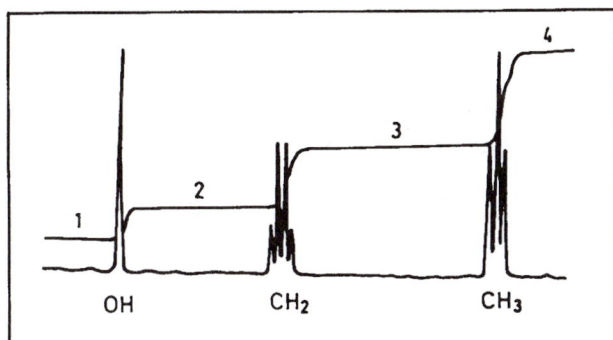

Fig. 1.44

For example Fig. 1.44 shows the nuclear magnetic resonance spectrum of H in alcohol. Three groups of nuclear magnetic resonance spectra are seen. The single peak on the left arises from the combination of H and O. The 4 peaks in the middle are the nuclear magnetic resonance spectrum of H in CH_2, and the 3 peaks on the right are the nuclear magnetic resonance spectrum of H in CH_3. The line shape and number of peaks are related to the coupling between CH_2 and CH_3. Using the horizontal line 1 as base line, the relative heights of the horizontal lines 2, 3, 4 give the relative integrated intensities of the three spectra, which are exactly in the ratio of 1:2:3.

(3) **Mössbauer spectroscopy**

(a) Figure 1.45 shows a block diagram of the apparatus.

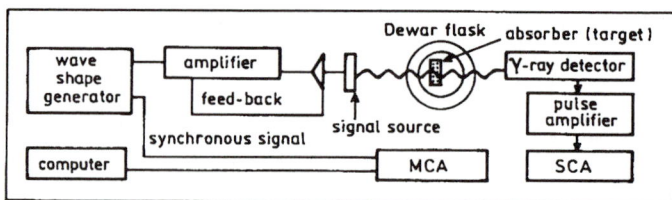

Fig. 1.45

(b) *Operation.* The signal source moves towards the fixed absorber with a velocity v modulated by the signals of the wave generator. During resonant absorption, the γ-ray detector behind the absorber produces a pulse signal, which is stored in the multichannel analyser MCA. Synchronous signals establish the correspondence between the position of a pulse and the velocity v, from which the resonant absorption curve is obtained.

(c) Information that can be obtained are the position δ of the absorption peak (Fig. 1.46), integrated intensity of the absorbing peak A, peak width Γ.

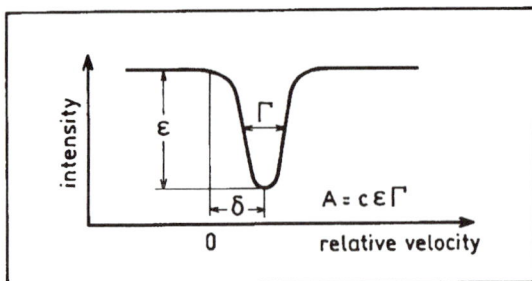

Fig. 1.46

Besides the effect of interactions among the nucleons inside the nucleus, nuclear energy levels are affected by the crystal structure, the orbital electrons and atoms nearby. In the Mössbauer spectrum the isomeric shift δ varies with the chemical environment. For instance, among the isomeric shifts of Sn^{2+}, Sn^{4+} and the metallic β-Sn, that of Sn^{2+} is the largest, that of β-Sn comes next, and that of the Sn^{4+} is the smallest.

The lifetime of an excited nuclear state can be determined from the width Γ of the peak by the uncertainty principle $\Gamma\tau \sim \hbar$.

The Mössbaur spectra of some elements show quadrupole splitting, as shown in Fig. 1.47. The quadrupole moment $Q = 2\Delta/e^2q$ of the nucleus can be determined from this splitting, where q is the gradient of the electric field at the site of the nucleus, e is the electronic charge.

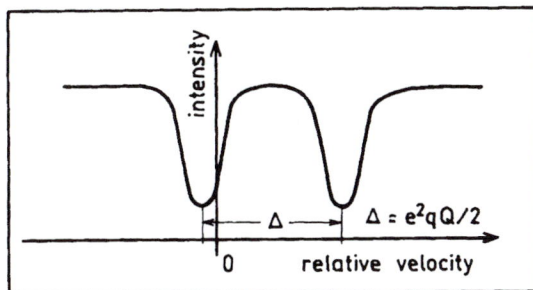

Fig. 1.47

1112

Pick ONE phenomenon from the list below, and answer the following questions about it:

(1) What is the effect? (e.g., "The Mössbauer effect is ...")

(2) How can it be measured?

(3) Give several sources of noise that will influence the measurement.

(4) What properties of the specimen or what physical constants can be measured by examining the effect?

Pick one:

(a) Electron spin resonance. (b) Mössbauer effect. (c) The Josephson effect. (d) Nuclear magnetic resonance. (e) The Hall effect.

(SUNY, Buffalo)

Solution:

(a) (b) (d) Refer to **Problem 1111**.

(c) *The Josephson effect*: Under proper conditions, superconducting electrons can cross a very thin insulation barrier from one superconductor into another. This is called the Josephson effect (Fig. 1.48). The

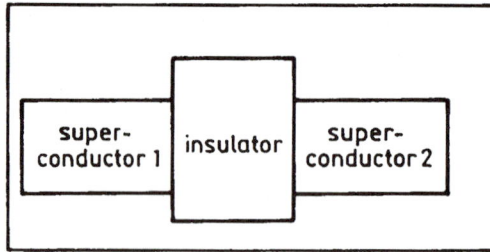

Fig. 1.48

Josephson effect is of two kinds, direct current Josephson effect and alternating current Josephson effect.

The direct current Josephson effect refers to the phenomenon of a direct electric current crossing the Josephson junction without the presence of any external electric or magnetic field. The superconducting current density can be expressed as $J_s = J_c \sin \varphi$, where J_c is the maximum current density that can cross the junction, φ is the phase difference of the wave functions in the superconductors on the two sides of the insulation barrier.

The alternating current Josephson effect occurs in the following situations:

1. When a direct current voltage is introduced to the two sides of the Josephson junction, a radiofrequency current $J_s = J_c \sin(\frac{2e}{\hbar} V t + \varphi_0)$ is produced in the Josephson junction, where V is the direct current voltage imposed on the two sides of the junction.

2. If a Josephson junction under an imposed bias voltage V is exposed to microwaves of frequency ω and the condition $V = n\hbar\omega/2e$ $(n = 1, 2, 3, \ldots)$ is satisfied, a direct current component will appear in the superconducting current crossing the junction.

Josephson effect can be employed for accurate measurement of $e/\hbar$. In the experiment the Josephson junction is exposed to microwaves of a fixed frequency. By adjusting the bias voltage V, current steps can be seen on the I–V graph, and $e/\hbar$ determined from the relation $\Delta V = \hbar\omega/2e$, where ΔV is the difference of the bias voltages of the neighboring steps.

The Josephson junction can also be used as a sensitive microwave detector. Furthermore, $\Delta V = \hbar\omega/2e$ can serve as a voltage standard.

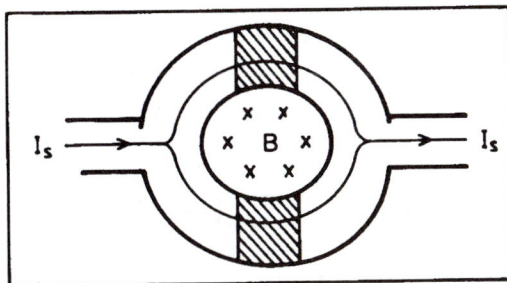

Fig. 1.49

Making use of the modulation effect on the junction current of the magnetic field, we can measure weak magnetic fields. For a ring structure consisting of two parallel Josephson junctions as shown in Fig. 1.49 ("double-junction quantum interferometer"), the current is given by

$$I_s = 2I_{s0} \sin \varphi_0 \cos \left(\frac{\pi \Phi}{\phi_0} \right) ,$$

where I_{s0} is the maximum superconducting current which can be produced in a single Josephson junction, $\phi_0 = \frac{h}{2e}$ is the magnetic flux quantum, Φ is the magnetic flux in the superconducting ring. Magnetic fields as small as 10^{-11} gauss can be detected.

(e) **Hall effect.** When a metallic or semiconductor sample with electric current is placed in a uniform magnetic field which is perpendicular to the current, a steady transverse electric field perpendicular to both the current and the magnetic field will be induced across the sample. This is called the Hall effect. The uniform magnetic field $\mathbf{B}$, electric current density $\mathbf{j}$, and the Hall electric field $\mathbf{E}$ have a simple relation: $\mathbf{E} = R_H \mathbf{B} \times \mathbf{j}$, where the parameter R_H is known as the Hall coefficient.

As shown in Fig. 1.50, a rectangular parallelepiped thin sample is placed in a uniform magnetic field $\mathbf{B}$. The Hall coefficient R_H and the electric conductivity σ of the sample can be found by measuring the Hall voltage V_H, magnetic field B, current I, and the dimensions of the sample:

$$R_H = \frac{V_H d}{IB}, \qquad \sigma = \frac{Il}{Ubd},$$

where U is the voltage of the current source. From the measured R_H and σ, we can deduce the type and density N of the current carriers in a semiconductor, as well as their mobility μ.

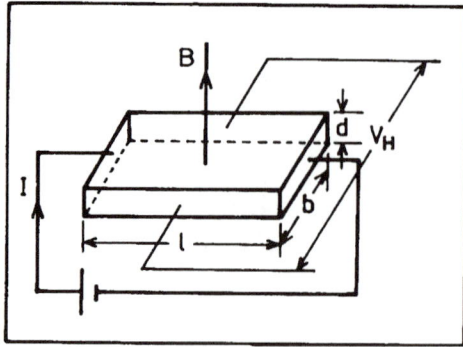

Fig. 1.50

The Hall effect arises from the action of the Lorentz force on the current carriers. In equilibrium, the magnetic force on the current carriers is balanced by the force due to the Hall electric field:

$$q\mathbf{E} = q\mathbf{v} \times \mathbf{B},$$

giving

$$\mathbf{E} = \mathbf{v} \times \mathbf{B} = \frac{1}{Nq}\mathbf{j} \times \mathbf{B}.$$

Hence $R_H = \frac{1}{Nq}$, where q is the charge of current carriers ($|q| = e$), from which we can determine the type of the semiconductor (p or n type in accordance with R_H being positive or negative). The carrier density and mobility are given by

$$N = \frac{1}{qR_H},$$

$$\mu = \frac{\sigma}{Ne} = \sigma|R_H|.$$

1113

State briefly the importance of each of the following experiments in the development of atomic physics.

(a) Faraday's experiment on electrolysis.
(b) Bunsen and Kirchhoff's experiments with the spectroscope.

(c) J. J. Thomson's experiments on e/m of particles in a discharge.

(d) Geiger and Marsdens experiment on scattering of α-particles.

(e) Barkla's experiment on scattering of X-rays.

(f) The Frank-Hertz experiment.

(g) J. J. Thomson's experiment on e/m of neon ions.

(h) Stern-Gerlach experiment.

(i) Lamb-Rutherford experiment.

(*Wisconsin*)

Solution:

(a) Faraday's experiment on electrolysis was the first experiment to show that there is a natural unit of electric charge $e = F/N_a$, where F is the Faraday constant and N_a is Avogadro's number. The charge of any charged body is an integer multiple of e.

(b) Bunsen and Kirchhoff analyzed the Fraunhofer lines of the solar spectrum and gave the first satisfactory explanation of their origin that the lines arose from the absorption of light of certain wavelengths by the atmospheres of the sun and the earth. Their work laid the foundation of spectroscopy and resulted in the discovery of the elements rubidium and cesium.

(c) J. J. Thomson discovered the electron by measuring directly the e/m ratio of cathode rays. It marked the beginning of our understanding of the atomic structure.

(d) Geiger and Marsden's experiment on the scattering of α-particles formed the experimental basis of Rutherfold's atomic model.

(e) Barkla's experiment on scattering of X-rays led to the discovery of characteristic X-ray spectra of elements which provide an important means for studying atomic structure.

(f) The Frank-Hertz experiment on inelastic scattering of electrons by atoms established the existence of discrete energy levels in atoms.

(g) J. J. Thomson's measurement of the e/m ratio of neon ions led to the discovery of the isotopes ^{20}Ne and ^{22}Ne.

(h) The Stern-Gerlach experiment provided proof that there exist only certain permitted orientations of the angular momentum of an atom.

(i) The Lamb-Rutherford experiment provided evidence of interaction of an electron with an electromagnetic radiation field, giving support to the theory of quantum electrodynamics.

1114

In a Stern-Gerlach experiment hydrogen atoms are used.

(a) What determines the *number* of lines one sees? What features of the apparatus determine the magnitude of the *separation* between the lines?

(b) Make an *estimate* of the separation between the two lines if the Stern-Gerlach experiment is carried out with H atoms. Make any reasonable assumptions about the experimental setup. For constants which you do not know by heart, state where you would look them up and what units they should be substituted in your formula.

(Wisconsin)

Solution:

(a) A narrow beam of atoms is sent through an inhomogeneous magnetic field having a gradient $\frac{dB}{dz}$ perpendicular to the direction of motion of the beam. Let the length of the magnetic field be L_1, the flight path length of the hydrogen atoms after passing through the magnetic field be L_2 (Fig. 1.51).

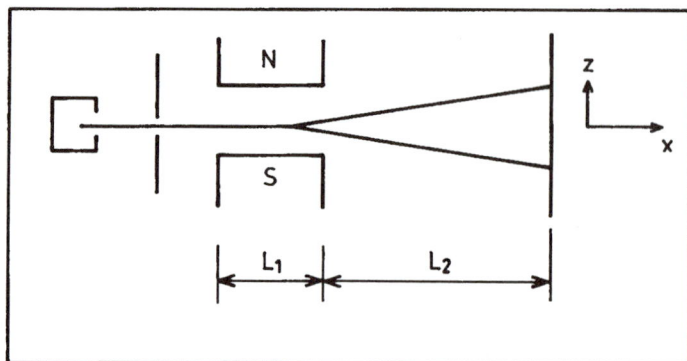

Fig. 1.51

The magnetic moment of ground state hydrogen atom is $\mu = g\mu_B J = 2\mu_B J$. In the inhomogeneous magnetic field the gradient $\frac{\partial B}{\partial z}\mathbf{i}_z$ exerts a force on the magnetic moment $F_z = 2\mu_B M_J(\frac{\partial B}{\partial z})$. As $J = \frac{1}{2}$, $M_J = \pm\frac{1}{2}$ and the beam splits into two components.

After leaving the magnetic field an atom has acquired a transverse velocity $\frac{F_z}{m} \cdot \frac{L_1}{v}$ and a transverse displacement $\frac{1}{2} \frac{F_z}{m} (\frac{L_1}{v})^2$, where m and v are respectively the mass and longitudinal velocity of the atom. When the beam strikes the screen the separation between the lines is

$$\frac{\mu_B L_1}{mv^2} (L_1 + 2L_2) \left(\frac{1}{2} + \frac{1}{2}\right) \frac{\partial B}{\partial z} .$$

(b) Suppose $L_1 = 0.03$ m, $L_2 = 0.10$ m, $dB/dz = 10^3$ T/m, $v = 10^3$ m/s. We have

$$d = \frac{0.927 \times 10^{-23} \times 0.03}{1.67 \times 10^{-27} \times 10^6} \times (0.03 + 2 \times 0.10) \times 10^3$$

$$= 3.8 \times 10^{-2} \text{ m} = 3.8 \text{ cm} .$$

1115

Give a brief description of the Stern-Gerlach experiment and answer the following questions:

(a) Why must the magnetic field be inhomogeneous?

(b) How is the inhomogeneous field obtained?

(c) What kind of pattern would be obtained with a beam of hydrogen atoms in their ground state? Why?

(d) What kind of pattern would be obtained with a beam of mercury atoms (ground state $^1S_0$)? Why?

(*Wisconsin*)

Solution:

For a brief description of the Stern-Gerlach experiment see **Problem 1114**.

(a) The force acting on the atomic magnetic moment μ in an inhomogeneous magnetic field is

$$F_z = -\frac{d}{dz}(\mu B \cos\theta) = -\mu \frac{dB}{dz} \cos\theta ,$$

where θ is the angle between the directions of μ and **B**. If the magnetic field were uniform, there would be no force and hence no splitting of the atomic beam.

(b) The inhomogeneous magnetic field can be produced by non-symmetric magnetic poles such as shown in Fig. 1.52.

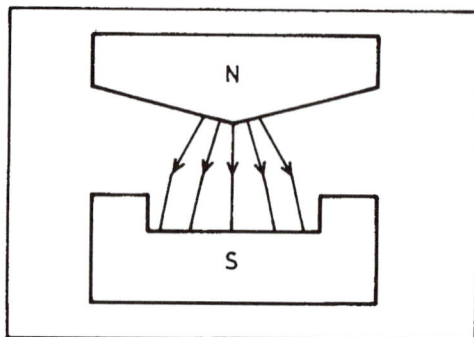

Fig. 1.52

(c) The ground state of hydrogen atom is $^2S_{1/2}$. Hence a beam of hydrogen atoms will split into two components on passing through an inhomogeneous magnetic field.

(d) As the total angular momentum J of the ground state of Hg is zero, there will be no splitting of the beam since $(2J + 1) = 1$.

1116

The atomic number of aluminum is 13.

(a) What is the electronic configuration of Al in its ground state?

(b) What is the term classification of the ground state? Use standard spectroscopic notation (e.g. $^4S_{1/2}$) and explain all superscripts and subscripts.

(c) Show by means of an energy-level diagram what happens to the ground state when a very strong magnetic field (Paschen-Back region) is applied. Label all states with the appropriate quantum numbers and indicate the relative spacing of the energy levels.

(Wisconsin)

Solution:

(a) The electronic configuration of the ground state of Al is

$$(1s)^2(2s)^2(2p)^6(3s)^2(3p)^1 .$$

(b) The spectroscopic notation of the ground state of Al is $^2P_{1/2}$, where the superscript 2 is the multiplet number, equal to $2S + 1$, S being the total spin quantum number, the subscript $1/2$ is the total angular momentum quantum number, the letter P indicates that the total orbital angular momentum quantum number $L = 1$.

(c) In a very strong magnetic field, LS coupling will be destroyed, and the spin and orbital magnetic moments interact separately with the external magnetic field, causing the energy level to split. The energy correction in the magnetic field is given by

$$\Delta E = -(\boldsymbol{\mu}_L + \boldsymbol{\mu}_s) \cdot \mathbf{B} = (M_L + 2M_s)\mu_B B ,$$

where

$$M_L = 1, 0, -1, \qquad M_S = 1/2, -1/2 .$$

The ^2P energy level is separated into 5 levels, the spacing of neighboring levels being $\mu_B B$. The split levels and the quantum numbers (L, S, M_L, M_S) are shown in Fig. 1.53.

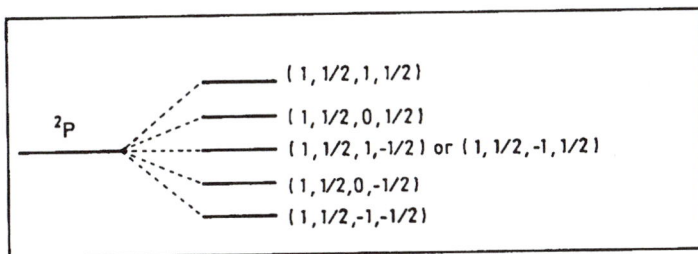

Fig. 1.53

1117

A heated gas of neutral lithium $(Z = 3)$ atoms is in a magnetic field. Which of the following states lie lowest. Give brief physical reasons for your answers.

(a) $3\ ^2P_{1/2}$ and $2\ ^2S_{1/2}$.
(b) $5\ ^2S_{1/2}$ and $5\ ^2P_{1/2}$.

(c) $5\,^2P_{3/2}$ and $5\,^2P_{1/2}$.

(d) Substates of $5^2P_{3/2}$.

(*Wisconsin*)

Solution:

The energy levels of an atom will be shifted in an external magnetic field B by

$$\Delta E = M_J g \mu_B B \,,$$

where g is the Landé factor, M_J is the component of the total angular momentum along the direction of the magnetic field B. The shifts are only $\sim 5 \times 10^{-5}$ eV even in a magnetic field as strong as 1 T.

(a) $3^2P_{1/2}$ is higher than $2^2S_{1/2}$ (energy difference ~ 1 eV), because the principal quantum number of the former is larger. Of the $^2S_{1/2}$ states the one with $M_J = -\frac{1}{2}$ lies lowest.

(b) The state with $M_J = -1/2$ of $^2S_{1/2}$ lies lowest. The difference of energy between ^2S and ^2P is mainly caused by orbital penetration and is of the order ~ 1 eV.

(c) Which of the states $^2P_{3/2}$ and $^2P_{1/2}$ has the lowest energy will depend on the intensity of the external magnetic field. If the external magnetic field would cause a split larger than that due to LS-coupling, then the state with $M_J = -3/2$ of $^2P_{3/2}$ is lowest. Conversely, $M_J = -1/2$ of $^2P_{1/2}$ is lowest.

(d) The substate with $M_J = -3/2$ of $^2P_{3/2}$ is lowest.

1118

A particular spectral line corresponding to a $J = 1 \to J = 0$ transition is split in a magnetic field of 1000 gauss into three components separated by 0.0016 Å. The zero field line occurs at 1849 Å.

(a) Determine whether the total spin is in the $J = 1$ state by studying the g-factor in the state.

(b) What is the magnetic moment in the excited state?

(*Princeton*)

Solution:

(a) The energy shift in an external magnetic field B is

$$\Delta E = g\mu_B B\,.$$

The energy level of $J = 0$ is not split. Hence the splitting of the line due to the transition $J = 1 \to J = 0$ is equal to the splitting of $J = 1$ level:

$$\Delta E(J = 1) = hc\Delta\tilde{\nu} = hc\frac{\Delta\lambda}{\lambda^2}\,,$$

or

$$g = \frac{hc}{\mu_B B}\frac{\Delta\lambda}{\lambda^2}\,.$$

With

$$\Delta\lambda = 0.0016 \text{ Å}\,,$$

$$\lambda = 1849 \text{ Å} = 1849 \times 10^{-8} \text{ cm}\,,$$

$$hc = 4\pi \times 10^{-5} \text{ eV} \cdot \text{cm}\,,$$

$$\mu_B = 5.8 \times 10^{-9} \text{ eV} \cdot \text{Gs}^{-1}\,,$$

$$B = 10^3 \text{ Gs}\,,$$

we find

$$g = 1\,.$$

As $J = 1$ this indicates (**Problem 1091(b)**) that $S = 0$, $L = 1$, i.e., only the orbital magnetic moment contributes to the Zeeman splitting.

(b) The magnetic moment of the excited atom is

$$\mu_J = g\mu_B P_J/\hbar = 1 \cdot \mu_B \cdot \sqrt{J(J + 1)} = \sqrt{2}\mu_B\,.$$

1119

Compare the weak-field Zeeman effect for the $(1s3s)$ $^1S_0 \to (1s2p)$ $^1P_1$ and $(1s3s)$ $^3S_1 \to (1s2p)$ $^3P_1$ transitions in helium. You may be qualitative so long as the important features are evident.

(Wisconsin)

Solution:

In a weak magnetic field, each energy level of $^3P_1$, $^3S_1$ and $^1P_1$ is split into three levels. From the selection rules ($\Delta J = 0, \pm 1$; $M_J = 0, \pm 1$), we see that the transition $(1s3s)^1S_0 \to (1s2p)^1P_1$ gives rise to three spectral lines, the transition $(1s3s)^3S_1 \to (1s2p)^3P_1$ gives rise to six spectral lines, as shown in Fig. 1.54.

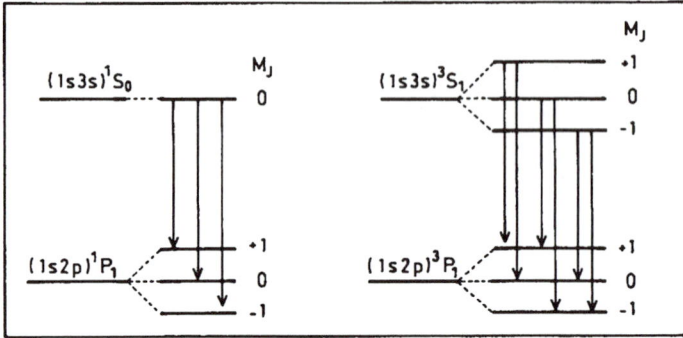

Fig. 1.54

The shift of energy in the weak magnetic field B is $\Delta E = g M_J \mu_B B$, where μ_B is the Bohr magneton, g is the Landé splitting factor given by

$$g = 1 + \frac{J(J+1) - L(L+1) + S(S+1)}{2J(J+1)} \, .$$

For the above four levels we have

Level	$(1s3s)^1S_0$	$(1s2p)^1P_1$	$(1s3s)^3S_1$	$(1s2p)^3P_1$
(JLS)	(000)	(110)	(101)	(111)
ΔE	0	$\mu_B B$	$2\mu_B B$	$3\mu_B B/2$

from which the energies of transition can be obtained.

1120

The influence of a magnetic field on the spectral structure of the prominent yellow light (in the vicinity of 6000 Å) from excited sodium vapor is being examined (Zeeman effect). The spectrum is observed for light emitted in a direction either along or perpendicular to the magnetic field.

(a) *Describe:* (i) The spectrum before the field is applied.

(ii) The change in the spectrum, for both directions of observation, after the field is applied.

(iii) What states of polarization would you expect for the components of the spectrum in each case?

(b) *Explain* how the above observations can be interpreted in terms of the characteristics of the atomic quantum states involved.

(c) If you have available a spectroscope with a resolution $(\lambda/\delta\lambda)$ of 100000 what magnetic field would be required to resolve clearly the 'splitting' of lines by the magnetic field? (Numerical estimates to a factor of two or so are sufficient. You may neglect the line broadening in the source.)

(Columbia)

Solution:

(a) The spectra with and without magnetic field are shown in Fig. 1.55.

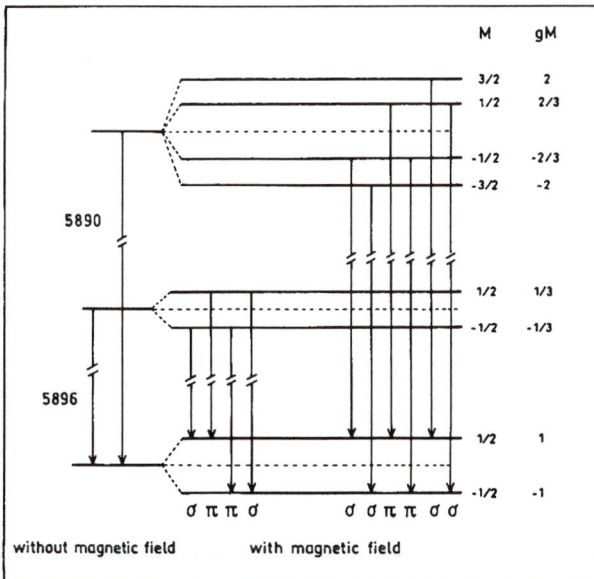

Fig. 1.55

(i) Before the magnetic field is introduced, two lines can be observed with wavelengths 5896 Å and 5890 Å in all directions.

(ii) After introducing the magnetic field, we can observe 6σ lines in the direction of the field and 10 lines, 4π lines and 6σ lines in a direction perpendicular to the field.

(iii) The σ lines are pairs of left and right circularly polarized light. The π lines are plane polarized light.

(b) The splitting of the spectrum arises from quantization of the direction of the total angular momentum. The number of split components is determined by the selection rule $(\Delta M_J = 1, 0, -1)$ of the transition, while the state of polarization is determined by the conservation of the angular momentum.

(c) The difference in wave number of two nearest lines is

$$\Delta\tilde{\nu} = \frac{|g_1 - g_2|\mu_B B}{hc} = \frac{1}{\lambda_1} - \frac{1}{\lambda_2} \approx \frac{\delta\lambda}{\lambda^2},$$

where g_1, g_2 are Landé splitting factors of the higher and lower energy levels. Hence the magnetic field strength required is of the order

$$B \sim \frac{hc\delta\lambda}{|g_1 - g_2|\mu_B\lambda^2} = \frac{12 \times 10^{-5} \times 10^8}{1 \times 6 \times 10^{-5}} \times \frac{10^{-5}}{6000} = 0.3\ T.$$

1121

Discuss qualitatively the shift due to a constant external electric field E_0 of the $n = 2$ energy levels of hydrogen. Neglect spin, but include the observed zero-field splitting W of the 2s and 2p states:

$$W = E_{2s} - E_{2p} \sim 10^{-5}\ \text{eV}.$$

Consider separately the cases $|e|E_0 a_0 \gg W$ and $|e|E_0 a_0 \ll W$, where a_0 is the Bohr radius.

(*Columbia*)

Solution:

Consider the external electric field E_0 as perturbation. Then $H' = e\mathbf{E_0} \cdot \mathbf{r}$. Nonzero matrix elements exist only between states $|200\rangle$ and $|210\rangle$

among the four $|n = 2\rangle$ states $|200\rangle, |211\rangle, |210\rangle, |21{-}1\rangle$. **Problem 1122(a)** gives

$$\langle 210|H'|200\rangle \equiv u = -3eE_0a_0 .$$

The states $|211\rangle$ and $|21-1\rangle$ remain degenerate.

(i) For $W \gg |e|E_0a_0$, or $W \gg |u|$, the perturbation is on nondegenerate states. There is nonzero energy correction only in second order calculation. The energy corrections are

$$E_+ = W + u^2/W, \ E_- = W - u^2/W .$$

(ii) For $W \ll |e|E_0a_0$, or $W \ll |u|$, the perturbation is among degenerate states and the energy corrections are

$$E_+ = -u = 3eE_0a_0, \ E_- = u = -3eE_0a_0 .$$

1122

A beam of excited hydrogen atoms in the 2s state passes between the plates of a capacitor in which a uniform electric field **E** exists over a distance L, as shown in the Fig. 1.56. The hydrogen atoms have velocity v along the x axis and the **E** field is directed along the z axis as shown.

All the $n = 2$ states of hydrogen are degenerate in the absence of the **E** field, but certain of them mix when the field is present.

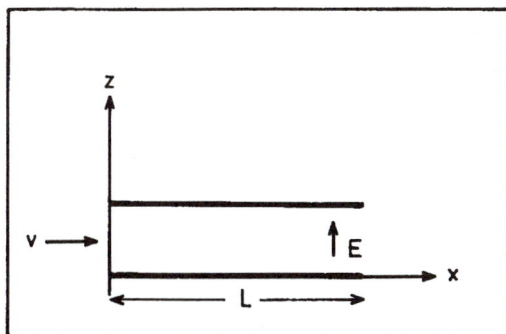

Fig. 1.56

(a) Which of the $n = 2$ states are connected in first order via the perturbation?

(b) Find the linear combination of $n = 2$ states which removes the degeneracy as much as possible.

(c) For a system which starts out in the 2s states at $t = 0$, express the wave function at time $t \le L/v$.

(d) Find the probability that the emergent beam contains hydrogen in the various $n = 2$ states.

(MIT)

Solution:

(a) The perturbation Hamiltonian $H' = eEr \cos \theta$ commutes with $\hat{l}_z = -i\hbar \frac{\partial}{\partial \varphi}$, so the matrix elements of H' between states of different m vanish. There are 4 degenerate states in the $n = 2$ energy level:

$$2s: \quad l = 0, m = 0\,,$$

$$2p: \quad l = 1, m = 0, \pm 1\,.$$

The only nonzero matrix element is that between the 2s and $2p(m = 0)$ states:

$$\langle 210|eEr\cos\theta|200\rangle = eE \int \psi_{210}(\mathbf{r})r\cos\theta\psi_{200}(\mathbf{r})d^3r$$

$$= \frac{eE}{16a^4} \int_0^\infty \int_{-1}^1 r^4 \left(2 - \frac{r}{a}\right) e^{-r/a} \cos^2\theta\, d\cos\theta\, dr$$

$$= -3eEa\,,$$

where a is the Bohr radius.

(b) The secular equation determining the energy shift

$$\begin{vmatrix} -\lambda & -3eEa & 0 & 0 \\ -3eEa & -\lambda & 0 & 0 \\ 0 & 0 & -\lambda & 0 \\ 0 & 0 & 0 & -\lambda \end{vmatrix} = 0$$

gives

$$\lambda = 3eEa\,, \qquad \Psi^{(-)} = \frac{1}{\sqrt{2}}(\Phi_{200} - \Phi_{210})\,,$$

$$\lambda = -3eEa\,, \qquad \Psi^{(+)} = \frac{1}{\sqrt{2}}(\Phi_{200} + \Phi_{210})\,,$$

$$\lambda = \quad 0\,, \qquad \Psi = \Phi_{211}, \Phi_{21-1}\,.$$

(c) Let the energy of $n = 2$ state before perturbation be E_1. As at $t = 0$,

$$\Psi(t = 0) = \Phi_{200} = \frac{1}{\sqrt{2}}\left[\frac{1}{\sqrt{2}}(\Phi_{200} - \Phi_{210}) + \frac{1}{\sqrt{2}}(\Phi_{200} + \Phi_{210})\right]$$

$$= \frac{1}{\sqrt{2}}(\Psi^{(-)} + \Psi^{(+)}),$$

we have

$$\Psi(t) = \frac{1}{\sqrt{2}}\left\{\Psi^{(-)}\exp\left[-\frac{i}{\hbar}(E_1 + 3eEa)t\right] + \Psi^{(+)}\exp\left[-\frac{i}{\hbar}(E_1 - 3eEa)t\right]\right\}$$

$$= \left[\Phi_{200}\cos\left(\frac{3eEat}{\hbar}\right) + \Phi_{210}\sin\left(\frac{3eEat}{\hbar}\right)\right]\exp\left(-\frac{i}{\hbar}E_1 t\right).$$

(d) When the beam emerges from the capacitor at $t = L/v$, the probability of its staying in $2s$ state is

$$\left|\cos\left(\frac{3eEat}{\hbar}\right)\exp\left(-\frac{i}{\hbar}E_1 t\right)\right|^2 = \cos^2\left(\frac{3eEat}{\hbar}\right) = \cos^2\left(\frac{3eEaL}{\hbar v}\right).$$

The probability of its being in $2p(m = 0)$ state is

$$\left|\sin\left(\frac{3eEat}{\hbar}\right)\exp\left(-\frac{i}{\hbar}E_1 t\right)\right|^2 = \sin^2\left(\frac{3eEat}{\hbar}\right) = \sin^2\left(\frac{3eEaL}{\hbar v}\right).$$

The probability of its being in $2p(m = \pm 1)$ state is zero.

2. MOLECULAR PHYSICS (1123–1142)

1123

(a) Assuming that the two protons of the H_2^+ molecule are fixed at their normal separation of 1.06 Å, sketch the potential energy of the electron along the axis passing through the protons.

(b) Sketch the electron wave functions for the two lowest states in H_2^+, indicating roughly how they are related to hydrogenic wave functions. Which wave function corresponds to the ground state of H_2^+, and why?

(c) What happens to the two lowest energy levels of H_2^+ in the limit that the protons are moved far apart?

(*Wisconsin*)

Solution:

(a) Take the position of one proton as the origin and that of the other proton at 1.06 Å along the x-axis. Then the potential energy of the electron is

$$V(r_1, r_2) = -\frac{e^2}{r_1} - \frac{e^2}{r_2},$$

where r_1 and r_2 are the distances of the electron from the two protons. The potential energy of the electron along the x-axis is shown in Fig. 1.57.

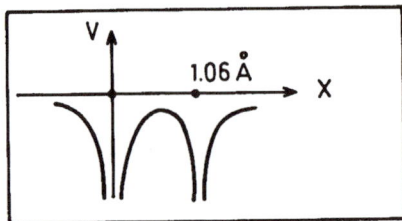

Fig. 1.57

(b) The molecular wave function of the H_2^+ has the forms

$$\Psi_S = \frac{1}{\sqrt{2}}(\Phi_{1s}(1) + \Phi_{1s}(2)),$$

$$\Psi_A = \frac{1}{\sqrt{2}}(\Phi_{1s}(1) - \Phi_{1s}(2)),$$

where $\Phi(i)$ is the wave function of an atom formed by the electron and the ith proton. Note that the energy of Ψ_S is lower than that of Ψ_A and so Ψ_S is the ground state of H_2^+; Ψ_A is the first excited state. Ψ_S and Ψ_A are linear combinations of $1s$ states of H atom, and are sketched in Fig. 1.58. The overlapping of the two hydrogenic wave functions is much larger in the case of the symmetric wave function Ψ_S and so the state is called a bonding state. The antisymmetric wave function Ψ_A is called an antibonding state. As Ψ_S has stronger binding its energy is lower.

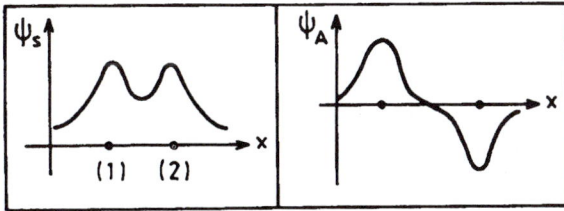

Fig. 1.58

(c) Suppose, with proton 1 fixed, proton 2 is moved to infinity, i.e. $r_2 \to \infty$. Then $\Phi(2) \sim e^{-r_2/a} \to 0$ and $\Psi_S \approx \Psi_A \approx \Phi(1)$. The system breaks up into a hydrogen atom and a non-interacting proton.

1124

Given the radial part of the Schrödinger equation for a central force field $V(r)$:

$$-\frac{\hbar^2}{2\mu} \frac{1}{r^2} \frac{d}{dr} \left(r^2 \frac{d\Psi(r)}{dr} \right) + \left[V(r) + \frac{l(l+1)\hbar^2}{2\mu r^2} \right] \Psi(r) = E\Psi(r) \,,$$

consider a diatomic molecule with nuclei of masses m_1 and m_2. A good approximation to the molecular potential is given by

$$V(r) = -2V_0 \left(\frac{1}{\rho} - \frac{1}{2\rho^2} \right) \,,$$

where $\rho = r/a$, a with a being some characteristic length parameter.

(a) By expanding around the minimum of the effective potential in the Schrödinger equation, show that for small B the wave equation reduces to that of a simple harmonic oscillator with frequency

$$\omega = \left[\frac{2V_0}{\mu a^2 (1+B)^3} \right]^{1/2} \,, \qquad \text{where } B = \frac{l(l+1)\hbar^2}{2\mu a^2 V_0} \,.$$

(b) Assuming $\hbar^2/2\mu \gg a^2 V_0$, find the rotational, vibrational and rotation-vibrational energy levels for small oscillations.

(SUNY, Buffalo)

Solution:

(a) The effective potential is

$$V_{\text{eff}} = \left[V(r) + \frac{l(l+1)\hbar^2}{2\mu r^2}\right] = -2V_0\left[\frac{a}{r} - \frac{a^2}{2r^2}(1+B)\right].$$

To find the position of minimum V_{eff}, let $\frac{dV_{\text{eff}}}{dr} = 0$, which gives $r = a(1 + B) \equiv r_0$ as the equilibrium position. Expanding V_{eff} near $r = r_0$ and neglecting terms of orders higher than $(\frac{r-r_0}{a})^2$, we have

$$V_{\text{eff}} \approx -\frac{V_0}{1+B} + \frac{V_0}{(1+B)^3 a^2}[r - (1+B)a]^2.$$

The radial part of the Schrödinger equation now becomes

$$-\frac{\hbar^2}{2\mu}\frac{1}{r^2}\frac{d}{dr}\left(r^2\frac{d\Psi(r)}{dr}\right) + \left\{-\frac{V_0}{B+1} + \frac{V_0}{(1+B)^3 a^2}[r - (1+B)a]^2\right\}\Psi(r)$$

$$= E\Psi(r),$$

or, on letting $\Psi(r) = \frac{1}{r}\chi(r)$, $R = r - r_0$,

$$-\frac{\hbar^2}{2\mu}\frac{d^2}{dR^2}\chi(R) + \frac{V_0}{(1+B)^3 a^2}R^2\chi(R) = \left(E + \frac{V_0}{1+B}\right)\chi(R),$$

which is the equation of motion of a harmonic oscillator of angular frequency

$$\omega = \left[\frac{2V_0}{\mu a^2(1+B)^3}\right]^{1/2}.$$

(b) If $\hbar^2/2\mu \gg a^2V_0$, we have

$$B = \frac{l(l+1)\hbar^2}{2\mu a^2 V_0} \gg 1, \qquad r_0 \approx Ba,$$

$$\omega \approx \sqrt{\frac{2V_0}{\mu a^2 B^3}}.$$

The vibrational energy levels are given by

$$E_v = (n + 1/2)\hbar\omega, n = 1, 2, 3 \ldots.$$

The rotational energy levels are given by

$$E_r = \frac{l(l+1)\hbar^2}{2\mu r_0} \approx \frac{l(l+1)\hbar^2}{2\mu Ba} .$$

Hence, the vibration-rotational energy levels are given by

$$E = E_v + E_r \approx \left(n + \frac{1}{2}\right)\hbar\omega + \frac{l(l+1)\hbar^2}{2\mu Ba} .$$

1125

A beam of hydrogen molecules travel in the z direction with a kinetic energy of 1 eV. The molecules are in an excited state, from which they decay and dissociate into two hydrogen atoms. When one of the dissociation atoms has its final velocity perpendicular to the z direction its kinetic energy is always 0.8 eV. Calculate the energy released in the dissociative reaction.

(*Wisconsin*)

Solution:

A hydrogen molecule of kinetic energy 1 eV moving with momentum $\mathbf{p_0}$ in the z direction disintegrates into two hydrogen atoms, one of which has kinetic energy 0.8 eV and a momentum $\mathbf{p_1}$ perpendicular to the z direction. Let the momentum of the second hydrogen atom be $\mathbf{p_2}$, its kinetic energy be E_2. As $\mathbf{p_0} = \mathbf{p_1} + \mathbf{p_2}$, the momentum vectors are as shown in Fig. 1.59.

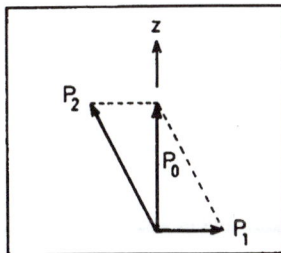

Fig. 1.59

We have

$$p_0 = \sqrt{2m(H_2)E(H_2)}$$

$$= \sqrt{2 \times 2 \times 938 \times 10^6 \times 1} = 6.13 \times 10^4 \text{ eV/c},$$

$$p_1 = \sqrt{2m(H)E(H)}$$

$$= \sqrt{2 \times 938 \times 10^6 \times 0.8} = 3.87 \times 10^4 \text{ eV/c}.$$

The momentum of the second hydrogen atom is then

$$p_2 = \sqrt{p_0^2 + p_1^2} = 7.25 \times 10^4 \text{ eV/c},$$

corresponding to a kinetic energy of

$$E_2 = \frac{p_2^2}{2m(H)} = 2.80 \text{ eV}.$$

Hence the energy released in the dissociative reaction is $0.8 + 2.8 - 1 = 2.6$ eV.

1126

Interatomic forces are due to:

(a) the mutual electrostatic polarization between atoms.
(b) forces between atomic nuclei.
(c) exchange of photons between atoms.

(CCT)

Solution:

The answer is (a).

1127

Which of the following has the smallest energy-level spacing?

(a) Molecular rotational levels,
(b) Molecular vibrational levels,
(c) Molecular electronic levels.

(CCT)

Solution:

The answer is (a). $\Delta E_e > \Delta E_v > \Delta E_r$.

1128

Approximating the molecule ${}^1_1\text{H}\ {}^{17}_{35}\text{Cl}$ as a rigid dumbbell with an internuclear separation of 1.29×10^{-10} m, calculate the frequency separation of its far infrared spectral lines. ($h = 6.6 \times 10^{-34}$ J sec, 1 amu = 1.67×10^{-27} kg).

(Wisconsin)

Solution:

The moment of inertia of the molecule is

$$I = \mu r^2 = \frac{m_{Cl} m_H}{m_{Cl} + m_H} r^2 = \frac{35}{36} \times 1.67 \times 10^{-27} \times (1.29 \times 10^{-10})^2$$

$$= 2.7 \times 10^{-47} \text{ kg} \cdot m^2$$

The frequency of its far infrared spectral line is given by

$$\nu = \frac{hcBJ(J+1) - hcBJ(J-1)}{h} = 2cBJ,$$

where $B = \hbar^2/(2Ihc)$. Hence

$$\nu = \frac{\hbar^2}{Ih} J, \text{ and so } \Delta\nu = \frac{\hbar^2}{hI} = \frac{h}{4\pi^2 I} = \frac{6.6 \times 10^{-34}}{4\pi^2 \times 2.7 \times 10^{-47}} = 6.2 \times 10^{11} \text{ Hz}.$$

1129

(a) Recognizing that a hydrogen nucleus has spin 1/2 while a deuterium nucleus has spin 1, enumerate the possible nuclear spin states for H_2, D_2 and HD molecules.

(b) For each of the molecules H_2, D_2 and HD, discuss the rotational states of the molecule that are allowed for each nuclear spin state.

(c) Estimate the energy difference between the first two rotational levels for H_2. What is the approximate magnitude of the contribution of the nuclear kinetic energy? The interaction of the two nuclear spins? The interaction of the nuclear spin with the orbital motion?

(d) Use your answer to (c) above to obtain the distribution of nuclear spin states for H_2, D_2 and HD at a temperature of 1 K.

Solution:

(a) As $s(p) = \frac{1}{2}$, $s(d) = 1$, and $\mathbf{S} = \mathbf{s}_1 + \mathbf{s}_2$, the spin of H_2 is 1 or 0, the spin of D_2 is 2, 1 or 0, and the spin of HD is 1/2 or 3/2.

(b) The two nuclei of H_2 are identical, so are the nuclei of D_2. Hence the total wave functions of H_2 and D_2 must be antisymmetric with respect to exchange of particles, while there is no such rule for DH. The total wave function may be written as $\Psi_T = \Psi_e \Psi_v \Psi_r \Psi_s$, where Ψ_e, Ψ_v, Ψ_r, and Ψ_s are the electron wave function, nuclear vibrational wave function, nuclear rotational wave function, and nuclear spin wave function respectively. For the ground state, the Ψ_e, Ψ_v are exchange-symmetric. For the rotational states of H_2 or D_2, a factor $(-1)^J$ will occur in the wave function on exchanging the two nuclei, where J is the rotational quantum number. The requirement on the symmetry of the wave function then gives the following:

H_2 :   For $S = 1$ (Ψ_s symmetric), $J = 1, 3, 5, \ldots$;

        for $S = 0$ (Ψ_s anitsymmetric), $J = 0, 2, 4, \ldots$.

D_2 :   For $S = 0, 2, J = 0, 2, 4, \ldots$;   for $S = 1, J = 1, 3, 5, \ldots$.

HD : $S = \dfrac{1}{2}, \dfrac{3}{2}$; $J = 1, 2, 3, \ldots$ (no restriction) .

(c) For H_2, take the distance between the two nuclei as $a \approx 2a_0 \approx 1$ Å $a_0 = \frac{\hbar^2}{m_e e^2}$ being the Bohr radius. Then $I = 2m_p a_0^2 = \frac{1}{2} m_p a^2$ and the energy difference between the first two rotational states is

$$\Delta E = \frac{\hbar^2}{2I} \times [1 \times (1+1) - 0 \times (0+1)] \approx \frac{2\hbar^2}{m_p a^2} \approx \frac{m_e}{m_p} E_0 \,,$$

where

$$E_0 = \frac{2\hbar^2}{m_e a^2} = \frac{e^2}{2a_0} \,.$$

is the ionization potential of hydrogen. In addition there is a contribution from the nuclear vibrational energy: $\Delta E_v \approx \hbar\omega$. The force between the nuclei is $f \approx e^2/a^2$, so that $K = |\nabla f| \approx \frac{2e^2}{a^3}$, giving

$$\Delta E_0 = \hbar\omega \approx \sqrt{\frac{K}{m_p}} = \sqrt{\frac{2e^2\hbar^2}{m_p a^3}} = \sqrt{\frac{m_e}{m_p}}\frac{e^2}{2e_0} = \sqrt{\frac{m_e}{m_p}}E_0 \,.$$

Hence the contribution of the nuclear kinetic energy is of the order of $\sqrt{\frac{m_e}{m_p}}E_0$.

The interaction between the nuclear spins is given by

$$\Delta E \approx \mu_N^2/a^3 \approx \left(\frac{e\hbar}{2m_p c}\right)^2 \frac{1}{8a_0^3} = \frac{1}{16}\left(\frac{\hbar}{m_p c}\right)^2 \left(\frac{m_e e^2}{\hbar^2}\right)^2 \frac{e^2}{2a_0}$$

$$= \frac{1}{16}\left(\frac{m_e}{m_p}\right)^2 \left(\frac{e^2}{\hbar c}\right)^2 E_0 = \frac{1}{16}\left(\frac{m_e}{m_p}\right)^2 \alpha^2 E_0 \,,$$

where $\alpha = \frac{1}{137}$ is the fine structure constant, and the interaction between nuclear spin and electronic orbital angular momentum is

$$\Delta E \approx \mu_N \mu_B/a_0^3 \approx \frac{1}{2}\left(\frac{m_e}{m_p}\right)\alpha^2 E_0 \,.$$

(d) For H_2, the moment of inertia is $I = \mu a^2 = \frac{1}{2}m_p a^2 \approx 2m_p a_0^2$, so the energy difference between states $l = 0$ and $l = 1$ is

$$\Delta E_{H_2} = \frac{\hbar^2}{2I} \times (2 - 0) = \frac{2m_e}{m_p}E_0 \,.$$

For D_2, as the nuclear mass is twice that of H_2,

$$\Delta E_{D_2} = \frac{1}{2}\Delta E_{H_2} = \frac{m_e}{m_p}E_0 \,.$$

As $kT = 8.7 \times 10^{-5}$ eV for $T = 1$ K, $\Delta E \approx \frac{E_0}{2000} = 6.8 \times 10^{-3}$ eV, we have $\Delta E \gg kT$ and so for both H_2 and D_2, the condition $\exp(-\Delta E/kT) \approx 0$ is satisfied. Then from Boltzmann's distribution law, we know that the H_2 and D_2 molecules are all on the ground state.

The spin degeneracies $2S + 1$ are for H_2, $g_{s=1} : g_{s=0} = 3 : 1$; for D_2, $g_{s=2} : g_{s=1} : g_{s=0} = 5 : 3 : 1$; and for HD, $g_{s=2/3} : g_{s=1/2} = 2 : 1$. From the population ratio g_2/g_1, we can conclude that most of H_2 is in the state of $S = 1$; most of D_2 is in the states of $S = 2$ and $S = 1$, the relative ratio being 5:3. Two-third of HD is in the state $S = 3/2$ and one-third in $S = 1/2$.

1130

Consider the (homonuclear) molecule $^{14}N_2$. Use the fact that a nitrogen nucleus has spin $I = 1$ in order to derive the result that the ratio of intensities of adjacent rotational lines in the molecular spectrum is 2:1.

(Chicago)

Solution:

As nitrogen nucleus has spin $I = 1$, the total wave function of the molecule must be symmetric. On interchanging the nuclei a factor $(-1)^J$ will occur in the wave function. Thus when the rotational quantum number J is even, the energy level must be a state of even spin, whereas a rotational state with odd J must be associated with an antisymmetric spin state. Furthermore, we have

$$\frac{g_S}{g_A} = \frac{(I+1)(2I+1)}{I(2I+1)} = (I+1)/I = 2:1$$

where g_S is the degeneracy of spin symmetric state, g_A is the degeneracy of spin antisymmetric state. As a homonuclear molecule has only Raman spectrum for which $\Delta J = 0, \pm 2$, the symmetry of the wave function does not change in the transition. The same is true then for the spin function. Hence the ratio of intensities of adjacent rotational lines in the molecular spectrum is 2 : 1.

1131

Estimate the lowest neutron kinetic energy at which a neutron, in a collision with a molecule of gaseous oxygen, can lose energy by exciting molecular rotation. (The bond length of the oxygen molecule is 1.2 Å).

(Wisconsin)

Solution:

The moment of inertia of the oxygen molecule is

$$I = \mu r^2 = \frac{1}{2}mr^2,$$

where r is the bond length of the oxygen molecule, m is the mass of oxygen atom.

The rotational energy levels are given by

$$E_J = \frac{h^2}{8\pi^2 I}J(J+1), \qquad J = 0,1,2,\dots.$$

To excite molecular rotation, the minimum of the energy that must be absorbed by the oxygen molecule is

$$E_{\min} = E_1 - E_0 = \frac{h^2}{4\pi^2 I} = \frac{h^2}{2\pi^2 mr^2} = \frac{2(\hbar c)^2}{mc^2 r^2}$$

$$= \frac{2 \times (1.97 \times 10^{-5})^2}{16 \times 938 \times 10^6 \times (1.2 \times 10^{-8})^2} = 3.6 \times 10^{-4} \text{ eV}.$$

As the mass of the neutron is much less than that of the oxygen molecule, the minimum kinetic energy the neutron must possess is 3.6×10^{-4} eV.

1132

(a) Using hydrogen atom ground state wave functions (including the electron spin) write wave functions for the hydrogen molecule which satisfy the Pauli exclusion principle. Omit terms which place both electrons on the same nucleus. Classify the wave functions in terms of their total spin.

(b) Assuming that the only potential energy terms in the Hamiltonian arise from Coulomb forces discuss qualitatively the energies of the above states at the normal internuclear separation in the molecule and in the limit of very large internuclear separation.

(c) What is meant by an "exchange force"?

(Wisconsin)

Solution:

Figure 1.60 shows the configuration of a hydrogen molecule. For convenience we shall use atomic units in which a_0 (Bohr radius) $= e = \hbar = 1$.

(a) The Hamiltonian of the hydrogen molecule can be written in the form

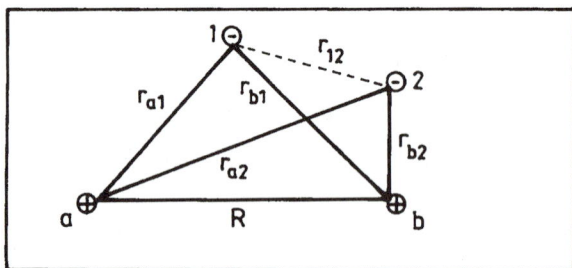

Fig. 1.60

$$\hat{H} = -\frac{1}{2}(\nabla_1^2 + \nabla_2^2) + \frac{1}{r_{12}} - \left(\frac{1}{r_{a1}} + \frac{1}{r_{a2}} + \frac{1}{r_{b1}} + \frac{1}{r_{b2}}\right) + \frac{1}{R}.$$

As the electrons are indistinguishable and in accordance with Pauli's principle the wave function of the hydrogen molecule can be written as

$$\Psi_S = [\Psi(r_{a1})\Psi(r_{b2}) + \Psi(r_{a2})\Psi(r_{b1})]\chi_0$$

or

$$\Psi_A = [\Psi(r_{a1})\Psi(r_{b2}) - \Psi(r_{a2})\Psi(r_{b1})]\chi_1,$$

where χ_0, χ_1 are spin wave functions for singlet and triplet states respectively, $\psi(r) = \frac{\lambda^{3/2}}{\sqrt{\pi}}e^{-\lambda r}$, the parameter λ being 1 for ground state hydrogen atom.

(b) When the internuclear separation is very large the molecular energy is simply the sum of the energies of the atoms.

If two electrons are to occupy the same spatial position, their spins must be antiparallel as required by Pauli's principle. In the hydrogen molecule the attractive electrostatic forces between the two nuclei and the electrons tend to concentrate the electrons between the nuclei, forcing them together and thus favoring the singlet state. When two hydrogen atoms are brought closer from infinite separation, the repulsion for parallel spins causes the triplet-state energy to rise and the attraction for antiparallel spins causes the singlet-state energy to fall until a separation of $\sim 1.5a_0$ is reached, thereafter the energies of both states will rise. Thus the singlet state has lower energy at normal internuclear separation.

(c) The contribution of the Coulomb force between the electrons to the molecular energy consists of two parts, one is the Coulomb integral arising from the interaction of an electron at location 1 and an electron at location 2. The other is the exchange integral arising from the fact that part of the time electron 1 spends at location 1 and electron 2 at location 2 and part of the time electron 1 spends at location 2 and electron 2 at location 1. The exchange integral has its origin in the identity of electrons and Pauli's principle and has no correspondence in classical physics. The force related to it is called exchange force.

The exchange integral has the form

$$\varepsilon = \iint d\tau_1 d\tau_2 \frac{1}{r_{12}} \psi^*(r_{a1})\psi(r_{b1})\psi(r_{a2})\psi^*(r_{b2}).$$

If the two nuclei are far apart, the electrons are distinguishable and the distinction between the symmetry and antisymmetry of the wave functions vanishes; so does the exchange force.

1133

(a) Consider the ground state of a dumbbell molecule: mass of each nucleus $= 1.7 \times 10^{-24}$ gm, equilibrium nuclear separation $= 0.75$ Å. Treat the nuclei as distinguishable. Calculate the energy difference between the first two rotational levels for this molecule. Take $\hbar = 1.05 \times 10^{-27}$ erg.sec.

(b) When forming H_2 from atomic hydrogen, 75% of the molecules are formed in the ortho state and the others in the para state. What is the difference between these two states and where does the 75% come from?

(*Wisconsin*)

Solution:

(a) The moment of inertia of the molecule is

$$I_0 = \mu r^2 = \frac{1}{2}mr^2,$$

where r is the distance between the nuclei. The rotational energy is

$$E_J = \frac{\hbar^2}{2I_0}J(J+1),$$

with

$$J = \begin{cases} 0, 2, 4, \dots & \text{for para-hydrogen}, \\ 1, 3, 5, \dots & \text{for ortho-hydrogen}. \end{cases}$$

As
$$\frac{\hbar^2}{2I} = \frac{\hbar^2}{mr^2} = \frac{(\hbar c)^2}{mc^2 r^2} = \frac{1973^2}{9.4 \times 10^8 \times 0.75^2} = 7.6 \times 10^{-3} \text{ eV},$$
the difference of energy between the rotational levels $J = 0$ and $J = 1$ is
$$\Delta E_{0,1} = \frac{\hbar^2}{I_0} = 1.5 \times 10^{-2} \text{ eV}.$$

(b) The two nuclei of hydrogen molecule are protons of spin $\frac{1}{2}$. Hence the H_2 molecule has two nuclear spin states $I = 1, 0$. The states with total nuclear spin $I = 1$ have symmetric spin function and are known as ortho-hydrogen, and those with $I = 0$ have antisymmetric spin function and are known as para-hydrogen.

The ratio of the numbers of ortho H_2 and para H_2 is given by the degeneracies $2I + 1$ of the two spin states:
$$\frac{\text{degeneracy of ortho } H_2}{\text{degeneracy of para } H_2} = \frac{3}{1}.$$
Thus 75% of the H_2 molecules are in the ortho state.

1134

A $^7N_{14}$ nucleus has nuclear spin $I = 1$. Assume that the diatomic molecule N_2 can rotate but does not vibrate at ordinary temperatures and ignore electronic motion. Find the relative abundance of ortho and para molecules in a sample of nitrogen gas. (Ortho = symmetric spin state; para = antisymmetric spin state), What happens to the relative abundance as the temperature is lowered towards absolute zero?

(SUNY, Buffalo)

Solution:

The $^7N_{14}$ nucleus is a boson of spin $I = 1$, so the total wave function of a system of such nuclei must be symmetric. For the ortho-nitrogen, which has symmetric spin, the rotational quantum number J must be an even number for the total wave function to be symmetric. For the para-nitrogen, which has antisymmetric spin, J must be an odd number.

The rotational energy levels of N_2 are
$$E_J = \frac{\hbar^2}{2H} J(J + 1), \qquad J = 0, 1, 2, \ldots .$$
where H is its moment of inertia. Statistical physics gives

$$\frac{\text{population of para-nitrogen}}{\text{population of ortho-nitrogen}} = \frac{\displaystyle\sum_{\text{even } J} (2J+1)\exp\left[-\frac{\hbar^2}{2HkT}J(J+1)\right]}{\displaystyle\sum_{\text{odd } J} (2J+1)\exp\left[-\frac{\hbar^2}{2HkT}J(J+1)\right]}$$

$$\times \frac{I+1}{I},$$

where I is the spin of a nitrogen nucleus.

If $\hbar^2/HRT \ll 1$, the sums can be approximated by integrals:

$$\sum_{\text{even } J} (2J+1)\exp\left[-\frac{\hbar^2}{2HkT}J(J+1)\right]$$

$$= \sum_{m=0}^{\infty} (4m+1)\exp\left[-\frac{\hbar^2}{2HkT}2m(2m+1)\right]$$

$$= \frac{1}{2}\int_0^\infty \exp\left(-\frac{\hbar^2 x}{2HkT}\right) dx = \frac{HkT}{\hbar^2},$$

where $x = 2m(2m+1)$;

$$\sum_{\text{odd } J} (2J+1)\exp\left[-\frac{\hbar^2}{2HkT}J(J+1)\right]$$

$$= \sum_{m=0}^{\infty} (4m+3)\exp\left[-\frac{\hbar^2}{2HkT}(2m+1)(2m+2)\right]$$

$$= \frac{1}{2}\int_0^\infty \exp\left(-\frac{\hbar^2 y}{2HkT}\right) dy = \frac{HkT}{\hbar^2}\exp\left(-\frac{\hbar^2}{HkT}\right),$$

where $y = (2m+1)(2m+2)$.

Hence

$$\frac{\text{population of para-nitrogen}}{\text{population of ortho-nitrogen}} = \frac{I+1}{I}\exp\left(\frac{\hbar^2}{HkT}\right) \approx \frac{I+1}{I}$$

$$= \frac{1+1}{1} = 2:1.$$

For $T \to 0$, $\hbar^2/HkT \gg 1$, then

$$\sum_{\text{even } J} (2J+1) \exp\left[-\frac{\hbar^2}{2HkT} J(J+1)\right]$$

$$= \sum_{m=0}^{\infty} (4m+1) \exp\left[-\frac{\hbar^2}{2HkT} 2m(2m+1)\right] \approx 1,$$

$$\sum_{\text{odd } J} (2J+1) \exp\left[-\frac{\hbar^2}{2HkT} J(J+1)\right]$$

$$= \sum_{m=0}^{\infty} (4m+3) \exp\left[-\frac{\hbar^2}{2HkT} (2m+1)(2m+2)\right]$$

$$\approx 3 \exp\left[-\frac{\hbar^2}{HkT}\right],$$

retaining the lowest order terms only. Hence

$$\frac{\text{population of para-nitrogen}}{\text{population of ortho-nitrogen}} \approx \frac{I+1}{3I} \exp\left(\frac{\hbar^2}{HkT}\right) \to \infty,$$

which means that the N_2 molecules are all in the para state at 0 K.

1135

In HCl a number of absorption lines with wave numbers (in cm^{-1}) 83.03, 103.73, 124.30, 145.03, 165.51, and 185.86 have been observed. Are these vibrational or rotational transitions? If the former, what is the characteristic frequency? If the latter, what J values do they correspond to, and what is the moment of inertia of HCl? In that case, estimate the separation between the nuclei.

(Chicago)

Solution:

The average separation between neighboring lines of the given spectrum is 20.57 cm^{-1}. The separation between neighboring vibrational lines is of the order of 10^{-1} eV $= 10^3$ cm^{-1}. So the spectrum cannot originate from transitions between vibrational energy levels, but must be due to transitions between rotational levels.

The rotational levels are given by

$$E = \frac{\hbar^2}{2I}J(J+1),$$

where J is the rotational quantum number, I is the moment of inertia of the molecule:

$$I = \mu R^2 = \frac{m_{Cl}m_H}{m_{Cl} + m_H}R^2 = \frac{35}{36}m_H R^2,$$

μ being the reduced mass of the two nuclei forming the molecule and R their separation. In a transition $J' \to J' - 1$, we have

$$\frac{hc}{\lambda} = \frac{\hbar^2}{2I}[J'(J'+1) - (J'-1)J'] = \frac{\hbar^2 J'}{I},$$

or

$$\tilde{\nu} = \frac{1}{\lambda} = \frac{\hbar J'}{2\pi I c}.$$

Then the separation between neighboring rotational lines is

$$\Delta\tilde{\nu} = \frac{\hbar}{2\pi I c},$$

giving

$$R = \left[\frac{\hbar c}{2\pi \left(\frac{35}{36}\right) m_H c^2 \Delta\tilde{\nu}}\right]^{\frac{1}{2}}$$

$$= \left[\frac{19.7 \times 10^{-12}}{2\pi \left(\frac{35}{36}\right) \times 938 \times 20.57}\right]^{\frac{1}{2}} = 1.29 \times 10^{-8} \text{ cm} = 1.29 \text{ Å}.$$

As $J' = \frac{\tilde{\nu}}{\Delta\tilde{\nu}}$, the given lines correspond to $J' = 4, 5, 6, 7, 8, 9$ respectively.

1136

When the Raman spectrum of nitrogen ($^{14}N^{14}N$) was measured for the first time (this was before Chadwick's discovery of the neutron in 1932), scientists were very puzzled to find that the nitrogen nucleus has a spin of $I = 1$. Explain

(a) how they could find the nuclear spin $I = 1$ from the Raman spectrum;

(b) why they were surprised to find $I = 1$ for the nitrogen nucleus. Before 1932 one thought the nucleus contained protons and electrons.

(Chicago)

Solution:

(a) For a diatomic molecule with identical atoms such as $(^{14}N)_2$, if each atom has nuclear spin I, the molecule can have symmetric and antisymmetric total nuclear spin states in the population ratio $(I + 1)/I$. As the nitrogen atomic nucleus is a boson, the total wave function of the molecule must be symmetric. When the rotational state has even J, the spin state must be symmetric. Conversely when the rotational quantum number J is odd, the spin state must be antisymmetric. The selection rule for Raman transitions is $\Delta J = 0, \pm 2$, so Raman transitions always occur according to $J_{even} \to J_{even}$ or J_{odd} to J_{odd}. This means that as J changes by one successively, the intensity of Raman lines vary alternately in the ratio $(I + 1)/I$. Therefore by observing the intensity ratio of Raman lines, I may be determined.

(b) If a nitrogen nucleus were made up of 14 protons and 7 electrons (nuclear charge = 7), it would have a half-integer spin, which disagrees with experiments. On the other hand, if a nitrogen nucleus is made up of 7 protons and 7 neutrons, an integral nuclear spin is expected, as found experimentally.

1137

A molecule which exhibits one normal mode with normal coordinate Q and frequency Ω has a polarizability $\alpha(Q)$. It is exposed to an applied incident field $E = E_0 \cos \omega_0 t$. Consider the molecule as a classical oscillator.

(a) Show that the molecule can scatter radiation at the frequencies ω_0 (Rayleigh scattering) and $\omega_0 \pm \Omega$ (first order Raman effect).

(b) For which $\alpha(Q)$ shown will there be no first order Raman scattering?

(c) Will O_2 gas exhibit a first order vibrational Raman effect? Will O_2 gas exhibit a first order infrared absorption band? Explain your answer briefly.

(Chicago)

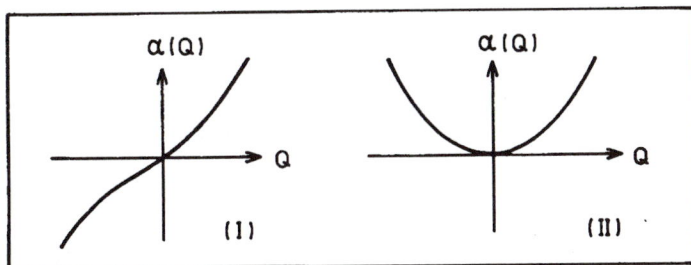

Fig. 1.61

Solution:

(a) On expanding $\alpha(Q)$ about $Q = 0$,

$$\alpha(Q) = \alpha_0 + \left(\frac{d\alpha}{dQ}\right)_{Q=0} Q + \frac{1}{2}\left(\frac{d^2\alpha}{dQ^2}\right)_{Q=0} Q^2 + \cdots.$$

and retaining only the first two terms, the dipole moment of the molecule can be given approximately as

$$P = \alpha E \approx \left[\alpha_0 + \left(\frac{d\alpha}{dQ}\right)_{Q=0} Q \cos \Omega t\right] E_0 \cos \omega_0 t$$

$$= \alpha_0 E_0 \cos \omega_0 t + QE_0 \left(\frac{d\alpha}{dQ}\right)_{Q=0} \left\{\frac{1}{2}[\cos(\omega_0 + \Omega)t + \cos(\omega_0 - \Omega)t]\right\}.$$

As an oscillating dipole radiates energy at the frequency of oscillation, the molecule not only scatters radiation at frequency ω_0 but also at frequencies $\omega_0 \pm \Omega$.

(b) The first order Raman effect arises from the term involving $(\frac{d\alpha}{dQ})_{Q=0}$. Hence in case (II) where $(\frac{d\alpha}{dQ})_{Q=0} = 0$ there will be no first order Raman effect.

(c) There will be first order Raman effect for O_2, for which there is a change of polarizability with its normal coordinate such that $(\frac{d\alpha}{dQ})_{Q=0} \neq 0$.

However, there is no first order infrared absorption band, because as the charge distribution of O_2 is perfectly symmetric, it has no intrinsic electric dipole moment, and its vibration and rotation cause no electric dipole moment change.

1138

Figure 1.62 shows the transmission of light through HCl vapor at room temperature as a function of wave number (inverse wavelength in units of cm^{-1}) decreasing from the left to the right.

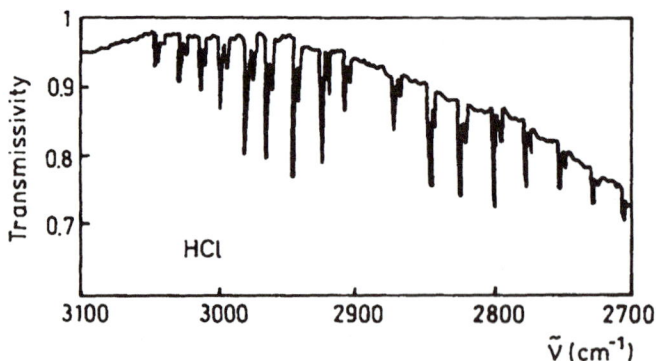

Fig. 1.62

Explain all the features of this transmission spectrum and obtain quantitative information about HCl. Sketch an appropriate energy level diagram labeled with quantum numbers to aid your explanation. Disregard the slow decrease of the top baseline for $\lambda^{-1} < 2900$ cm^{-1} and assume that the top baseline as shown represents 100% transmission. The relative magnitudes of the absorption lines are correct.

(Chicago)

Solution:

Figure 1.62 shows the vibration-rotational spectrum of the molecules of hydrogen with two isotopes of chlorine, H^{35}Cl and H^{37}Cl, the transition energy being

$$E_{v,k} = (v + 1/2)h\nu_0 + \frac{\hbar^2 k(k+1)}{2I} \, ,$$

where v, k are the vibrational and rotational quantum numbers respectively. The selection rules are $\Delta v = \pm 1, \Delta k = \pm 1$.

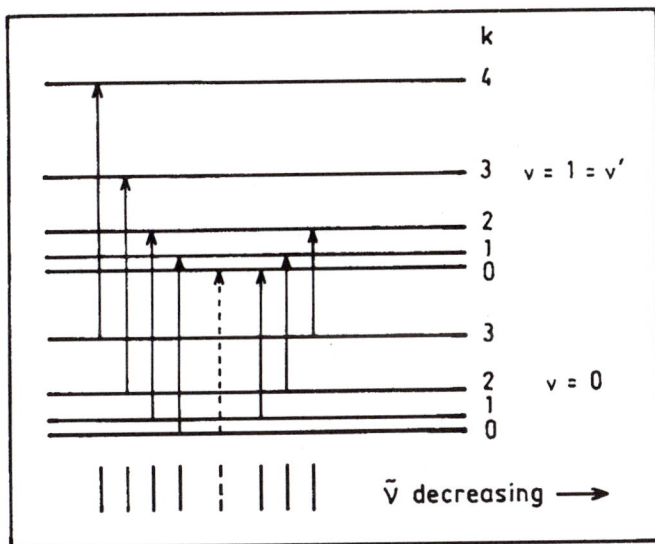

Fig. 1.63

The "missing" absorption line at the center of the spectrum shown in Fig. 1.63 corresponds to $k = 0 \to k' = 0$. This forbidden line is at $\lambda^{-1} = 2890$ cm^{-1}, or $\nu_0 = c\lambda^{-1} = 8.67 \times 10^{13}$ s^{-1}.

From the relation

$$\nu_0 = \frac{1}{2\pi}\sqrt{\frac{K}{\mu}} \, ,$$

where K is the force constant, $\mu = \frac{35}{36} m_H = 1.62 \times 10^{-24}$ g is the reduced mass of HCl, we obtain $K = 4.8 \times 10^5$ erg cm$^{-2} = 30$ eV Å^{-2}.

Figure 1.64 shows roughly the potential between the two atoms of HCl. Small oscillations in r may occur about r_0 with a force constant $K = \frac{d^2V}{dr^2}|_{r=r_0}$. From the separation of neighboring rotational lines $\Delta\tilde{\nu} = 20.5$ cm^{-1}, we can find the equilibrium atomic separation (**Problem 1135**)

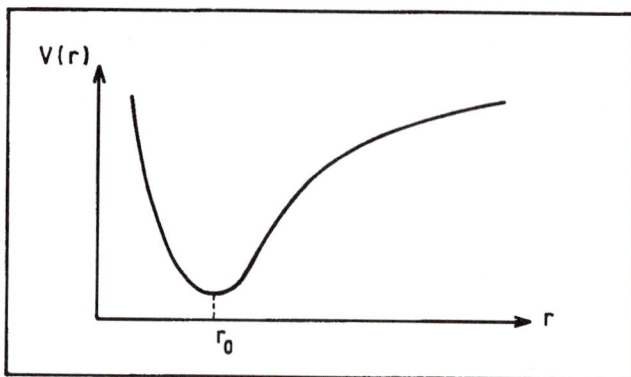

Fig. 1.64

$$r_0 = \left[\frac{\hbar c}{2\pi \left(\dfrac{36}{37} \right) m_H c^2 \Delta \tilde{\nu}} \right]^{\frac{1}{2}} = 1.30 \times 10^{-8} \text{ cm}$$

$$= 1.30 \text{ Å}.$$

The Isotope ratio can be obtained from the intensity ratio of the two series of spectra in Fig. 1.62. For $H^{35}Cl$, $\mu = \frac{35}{36} m_H$, and for $H^{37}Cl$, $\mu = \frac{37}{38} m_H$. As the wave number of a spectral line $\tilde{\nu} \propto \frac{1}{\mu}$, the wave number of a line of $H^{37}Cl$ is smaller than that of the corresponding line of $H^{35}Cl$. We see from Fig. 1.62 that the ratio of the corresponding spectral intensities is 3:1, so the isotope ratio of ^{35}Cl to ^{37}Cl is 3:1.

1139

(a) Using the fact that electrons in a molecule are confined to a volume typical of the molecule, estimate the spacing in energy of the excited states of the electrons (E_{elect}).

(b) As nuclei in a molecule move they distort electronic wave functions. This distortion changes the electronic energy. The nuclei oscillate about positions of minimum total energy, comprising the electron energy and the repulsive Coulomb energy between nuclei. Estimate the frequency and therefore the energy of these vibrations (E_{vib}) by saying that a nucleus is in a harmonic oscillator potential.

(c) Estimate the deviations from the equilibrium sites of the nuclei.

(d) Estimate the energy of the rotational excitations (E_{rot}).

(e) Estimate the ratio of $E_{\text{elect}} : E_{\text{vib}} : E_{\text{rot}}$ in terms of the ratio of electron mass to nuclear mass, m_e/m_n.

(Columbia)

Solution:

(a) The uncertainty principle $pd \approx \hbar$ gives the energy spacing between the excited states as $E_{\text{elect}} = \frac{p^2}{2m_e} \approx \frac{\hbar^2}{2m_e d^2}$, where d, the linear size of the molecule, is of the same order of magnitude as the Bohr radius $a_0 = \frac{\hbar^2}{m_e e^2}$.

(b) At equilibrium, the Coulomb repulsion force between the nuclei is $f \approx \frac{e^2}{d^2}$, whose gradient is $K \approx \frac{f}{d} \approx \frac{e^2}{d^3}$. The nuclei will oscillate about the equilibrium separation with angular frequency

$$\omega = \sqrt{\frac{K}{m}} \approx \sqrt{\frac{m_e}{m}} \sqrt{\frac{e^2 a_0}{m_e d^4}} = \sqrt{\frac{m_e}{m}} \frac{\hbar}{m_e d^2} \,,$$

where m is the reduced mass of the atomic nuclei.

Hence

$$E_{\text{vib}} = \hbar\omega \approx \sqrt{\frac{m_e}{m}} E_{\text{elect}} \,.$$

(c) As

$$E_{\text{vib}} = \frac{1}{2} m\omega^2 (\Delta x)^2 = \hbar\omega \,,$$

we have

$$\Delta x \approx \left(\frac{m_e}{m}\right)^{\frac{1}{4}} d \,.$$

(d) The rotational energy is of the order $E_{\text{rot}} \approx \frac{\hbar^2}{2I}$. With $I \approx md^2$, we have

$$E_{\text{rot}} \approx \frac{m_e}{m} E_{\text{elect}} \,.$$

(e) As $m \approx m_n$, the nuclear mass, we have

$$E_{\text{elect}} : E_{\text{vib}} : E_{\text{rot}} \approx 1 : \sqrt{\frac{m_e}{m_n}} : \frac{m_e}{m_n} \,.$$

1140

Sketch the potential energy curve $V(r)$ for the HF molecule as a function of the distance r between the centers of the nuclei, indicating the dissociation energy on your diagram.

(a) What simple approximation to $V(r)$ can be used near its minimum to estimate vibrational energy levels? If the zero-point energy of HF is 0.265 eV, use your approximation (without elaborate calculations) to estimate the zero-point energy of the DF molecule (D = deuteron, F = ^{19}F).

(b) State the selection rule for electromagnetic transitions between vibrational levels in HF within this approximation, and briefly justify your answer. What is the photon energy for these transitions?

(*Wisconsin*)

Solution:

(a) Figure 1.65 shows $V(r)$ and the dissociation energy E_d for the HF molecule. Near the minimum potential point r_0, we may use the approximation

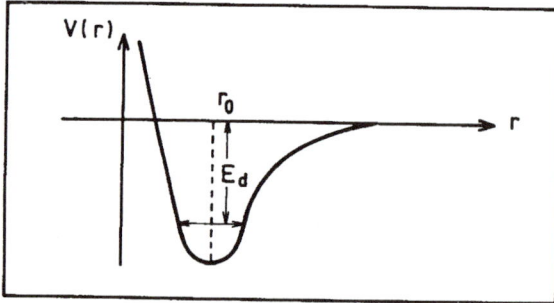

Fig. 1.65

$$V(r) \approx \frac{1}{2}k(r - r_0)^2 .$$

Thus the motion about r_0 is simple harmonic with angular frequency $\omega_0 = \sqrt{\frac{k}{\mu}}$, μ being the reduced mass of the nuclei. The zero-point energy is $E_0 = \frac{1}{2}\hbar\omega_0$.

As their electrical properties are the same, DF and HF have the same potential curve. However their reduced masses are different:

$$\mu(DF) = \frac{m(D)m(F)}{m(D) + m(F)} = \frac{2 \times 19}{2 + 19}u = 1.81u \,,$$

$$\mu(HF) = \frac{m(H)m(F)}{m(H) + m(F)} = \frac{1 \times 19}{1 + 19}u = 0.95u \,.$$

where u is the nucleon mass.

Hence

$$\frac{E_0(HF)}{E_0(DF)} = \sqrt{\frac{\mu(DF)}{\mu(HF)}}$$

and the zero-point energy of DF is

$$E_0(DF) = \sqrt{\frac{\mu(HF)}{\mu(DF)}} E_0(HF) = 0.192 \text{ eV} \,.$$

(b) In the harmonic oscillator approximation, the vibrational energy levels are given by

$$E_\nu = (\nu + 1/2)\hbar\omega, \quad \nu = 0, 1, 2, 3 \dots .$$

The selection rule for electromagnetic transitions between these energy levels is

$$\Delta\nu = \pm 1, \pm 2, \pm 3, \dots ,$$

while the selection rule for electric dipole transitions is

$$\Delta\nu = \pm 1 \,.$$

In general, the electromagnetic radiation emitted by a moving charge consists of the various electric and magnetic multipole components, each with its own selection rule $\Delta\nu$ and parity relationship between the initial and final states. The lowest order perturbation corresponds to electric dipole transition which requires $\Delta\nu = \pm 1$ and a change of parity.

For purely vibrational transitions, the energy of the emitted photon is approximately $\hbar\omega_0 \sim 0.1$ to 1 eV.

1141

Diatomic molecules such as HBr have excitation energies composed of electronic, rotational, and vibrational terms.

(a) Making rough approximations, estimate the magnitudes of these three contributions to the energy, in terms of fundamental physical constants such as $M, m_e, e, \ldots$, where M is the nuclear mass.

(b) For this and subsequent parts, assume the molecule is in its electronic ground state. What are the selection rules that govern radiative transitions? Justify your answer.

(c) An infrared absorption spectrum for gaseous HBr is shown in Fig. 1.66. (Infrared absorption involves no electronic transitions.) Use it to determine the moment of interia I and the vibrational frequency ω_0 for HBr.

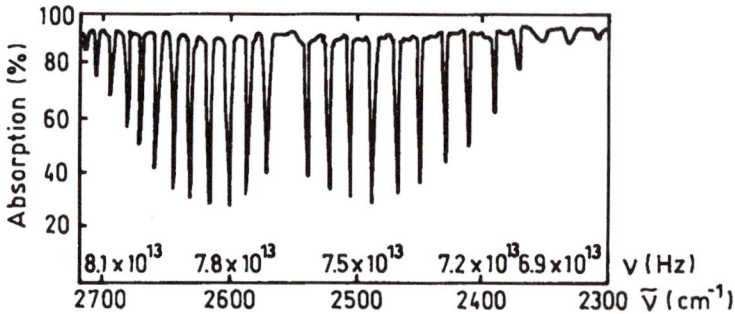

Fig. 1.66

(d) Note that the spacing between absorption lines increases with increasing energy. Why?

(e) How does this spectrum differ from that of a homonuclear molecule such as H_2 or D_2?

(*Princeton*)

Solution:

(a) Let a denote the linear dimension of the diatomic molecule. As the valence electron moves in an orbit of linear dimension a, the uncertainty of momentum is $\Delta p \approx \hbar/a$ and so the order of magnitude of the zero-point energy is

$$E_e \approx \frac{(\Delta p)^2}{m_e} \approx \frac{\hbar^2}{m_e a^2} \, .$$

A harmonic oscillator with mass m and coefficient of stiffness k is used as model for nuclear vibration. A change of the distance between the two nuclei will considerably distort the electronic wave function and thus relate to a change of the electronic energy, i.e. $ka^2 \approx E_e$.

Hence

$$E_{\text{vib}} \approx \hbar\omega \approx \hbar\sqrt{\frac{k}{M}} = \sqrt{\frac{m_e}{M}}\sqrt{\frac{\hbar^2}{m_e a^2}}\sqrt{ka^2} \approx \left(\frac{m_e}{M}\right)^{\frac{1}{2}} E_e$$

The molecular rotational energy levels are obtained by treating the molecule as a rotator of moment of inertia $I \approx Ma^2$. Thus

$$E_{\text{rot}} \approx \frac{\hbar^2}{I} \approx \frac{m_e}{M}\frac{\hbar^2}{m_e a^2} \approx \frac{m_e}{M}E_e .$$

(b) The selection rules for radiative transitions are $\Delta J = \pm 1, \Delta v = \pm 1$, where J is the rotational quantum number, v is the vibrational quantum number. As the electrons remain in the ground state, there is no transition between the electronic energy levels. The transitions that take place are between the rotational or the vibrational energy levels.

(c) From Fig. 1.66 we can determine the separation of neighboring absorption lines, which is about $\Delta\tilde{\nu} = 18$ cm^{-1}. As (**Problem 1135**) $\Delta\tilde{\nu} = 2B$, where $B = \frac{\hbar}{4\pi Ic}$, the moment of inertia is

$$I = \frac{\hbar}{2\pi c\Delta\tilde{\nu}} = 3.1 \times 10^{-40} \text{ g cm}^2 .$$

Corresponding to the missing spectral line in the middle we find the vibrational frequency $\nu_0 = 3 \times 10^{10} \times 2560 = 7.7 \times 10^{13}$ Hz.

(d) Actually the diatomic molecule is not exactly equivalent to a harmonic oscillator. With increasing vibrational energy, the average separation between the nuclei will become a little larger, or B_v a little smaller:

$$B_v = B_e - \left(v + \frac{1}{2}\right)\alpha_e ,$$

where B_e is the value B when the nuclei are in the equilibrium positions, $\alpha_e > 0$ is a constant. A transition from E to $E'(E < E')$ produces an absorption line of wave number

$$\tilde{\nu} = \frac{E' - E}{hc} = \frac{1}{hc}[(E'_{\text{vib}} + E'_{\text{rot}}) - (E_{\text{vib}} + E_{\text{rot}})]$$

$$= \tilde{\nu}_0 + B'J'(J'+1) - BJ(J+1) .$$

where $B' < B$. For the R branch, $J' = J + 1$, we have

$$\tilde{\nu}_R = \tilde{\nu}_0 + (B' + B)J' + (B' - B)J'^2,$$

and hence the spectral line separation

$$\Delta\tilde{\nu} = (B' + B) + (B' - B)(2J' + 1),$$

where $J' = 1, 2, 3 \ldots$. Hence, when the energy of spectral lines increases, i.e., J' increases, $\Delta\tilde{\nu}$ will decrease.

For the P branch, $J' = J - 1$,

$$\tilde{\nu}_P = \tilde{\nu}_0 - (B' + B)J + (B' - B)J^2,$$

$$\Delta\tilde{\nu} = (B' + B) - (B' - B)(2J + 1),$$

where $J = 1, 2, 3 \ldots$. Thus $\Delta\tilde{\nu}$ will decrease with increasing spectral line energy.

(e) Molecules formed by two identical atoms such as H_2 and D_2 have no electric dipole moment, so the vibration and rotation of these molecules do not relate to absorption or emission of electric-dipole radiation. Hence they are transparent in the infrared region.

1142

In a recent issue of Science Magazine, G. Zweig discussed the idea of using free quarks (if they should exist) to catalyze fusion of deuterium. In an ordinary negative deuterium molecule (ded) the two deuterons are held together by an electron, which spends most of its time between the two nuclei. In principle a neutron can tunnel from one proton to the other, making a tritium plus p + energy, but the separation is so large that the rate is negligible. If the electron is replaced with a massive quark, charge $-4e/3$, the separation is reduced and the tunneling rate considerably increased. After the reaction, the quark generally escapes and captures another deuteron to make a dQ atom, charge $-e/3$. The atom decays radiatively to the ground state, then captures another deuteron in a large-n orbit. This again decays down to the ground state. Fusion follows rapidly and the quark is released again.

(a) Suppose the quark is much more massive than the deuteron. What is the order of magnitude of the separation of the deuterons in the ground state of the dQd molecule?

(b) Write down an expression for the order of magnitude of the time for a deuteron captured at large radius (large n) in dQ to radiatively settle to the ground state. Introduce symbols like mass, charge, etc. as needed; do not evaluate the expression.

(c) Write down the expression for the probability of finding the neutron-proton separation in a deuteron being $r \geq r_0$, with $r_0 \gg 10^{-13}$ cm. Again, introduce symbols like deuteron binding energy as needed, and do not evaluate the expression.

(d) As a simple model for the tunneling rate suppose that if the neutron reaches a distance $r \geq r_0$ from the proton it certainly is captured by the other deuteron. Write down an order of magnitude expression for the halflife of dQd (but do not evaluate it).

(Princeton)

Solution:

(a) The dQd molecule can be considered as H_2^+ ion with the replacements $m_e \to m$, the deuteron mass, nuclear charge $e \to$ quark charge $-\frac{4}{3}e$.

Then by analogy with H_2^+ ion, the Hamiltonian for the dQd molecule can be written as

$$H = \frac{p_1^2}{2m} + \frac{p_2^2}{2m} - \frac{4e^2}{3r_1} - \frac{4e^2}{3r_2} + \frac{e^2}{r_{12}},$$

where $r_{12} = |\mathbf{r}_1 - \mathbf{r}_2|$, $\mathbf{r}_1, \mathbf{r}_2$ being the radius vectors of the deuterons from the massive quark.

Assume the wave function of the ground state can be written as

$$\Psi(\mathbf{r}_1, \mathbf{r}_2) = \Psi_{100}(\mathbf{r}_1)\Psi_{100}(\mathbf{r}_2),$$

where

$$\Psi_{100}(\mathbf{r}) = \frac{1}{\sqrt{\pi}} a^{-3/2} \exp\left(-\frac{r}{a}\right),$$

with

$$a = \frac{3\hbar^2}{4me^2}.$$

The average separation of the deuterons in the ground state is

$$\bar{r}_{12} = \frac{1}{\pi^2 a^6} \iint r_{12} \exp\left[-\frac{2(r_1 + r_1)}{a}\right] d\mathbf{r}_1 d\mathbf{r}_1 = \frac{8}{5}a = \frac{6\hbar^2}{5me^2}.$$

(b) A hydrogen-like atom of nuclear charge Ze has energy $-\frac{Z^2 e^4 m_e}{2\hbar^2 n^2}$. By analogy the dQd molecule has ground state energy

$$E = -2 \times \left(\frac{4}{3}\right)^2 \frac{me^4}{2\hbar^2} + \frac{e^2}{r_{12}}$$

$$= -\frac{4}{3}\frac{e^2}{a} + \frac{5}{8}\frac{e^2}{a} = -\frac{17}{24}\frac{e^2}{a} \,.$$

When n is very large, the molecule can be considered as a hydrogen-like atom with dQ as nucleus (charge $= -\frac{4e}{3} + e = -\frac{e}{3}$) and the second d taking the place of orbital electron (charge $= +e$). Accordingly the energy is

$$E_n = -\frac{4}{6}\frac{e^2}{a} - \left(\frac{1}{3}\right)^2 \frac{me^4}{2\hbar^2 n^2} = -\frac{4}{6}\frac{e^2}{a} - \frac{1}{6}\frac{e^2}{a'}\frac{1}{n^2} \,,$$

where $a' = \frac{3\hbar^2}{me^2}$. Hence when the system settles to the ground state, the energy emitted is

$$\Delta E = E_n - E_0 = -\frac{4e^2}{6a} + \frac{17}{24}\frac{e^2}{a} - \frac{e^2}{6a'}\frac{1}{n^2} \approx \frac{e^2}{24a} \,.$$

The emitted photons have frequency

$$\omega = \frac{\Delta E}{\hbar} \approx \frac{e^2}{24\hbar a} \,.$$

The transition probability per unit time is given by

$$A_{n1} = \frac{4e^2 \omega^3}{3\hbar c^3} |\mathbf{r}_{1n}|^2 \,,$$

and so the time for deuteron capture is of the order

$$\tau = 1/A_{n1} = \frac{3\hbar c^3}{4e^2 \omega^3 |\mathbf{r}_{1n}|^2} \,.$$

The wave function of the excited state is

$$\Psi_\pm = \frac{1}{\sqrt{2}}[\Psi_{100}(\mathbf{r}_1)\Psi_{nlm}(\mathbf{r}_2) \pm \Psi_{100}(\mathbf{r}_2)\Psi_{nlm}(\mathbf{r}_1)] \,,$$

which only acts on one d. As

$$\langle \Psi_{100} | \mathbf{r} | \Psi_{100} \rangle = 0 \,,$$

we have

$$\mathbf{r}_{1n} = \frac{1}{\sqrt{2}} \langle \Psi_{nlm} | \mathbf{r} | \Psi_{100} \rangle$$

and hence

$$\tau = \frac{3\hbar c^3}{2e^2 \omega^3 |\langle \Psi_{nlm} | \mathbf{r} | \Psi_{100} \rangle|^2} \,.$$

(c) In a deuteron the interaction potential between the proton and neutron can be taken to be that shown in Fig. 1.67, where W is the binding energy and $a \approx 10^{-13}$ cm.

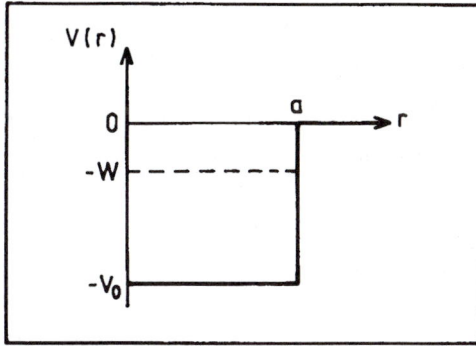

Fig. 1.67

The radial part of the wave function can be shown to satisfy the equation

$$R'' + \frac{1}{r} R' + \frac{M}{\hbar^2} [-W - V(r)] R = 0 \,,$$

where M is the mass of the neutron. Let $rR = u$. The above becomes

$$u'' - \frac{M}{\hbar^2} [W + V(r)] u = 0 \,.$$

As $V = -V_0$ for $0 \le r \le a$ and $V = 0$ otherwise, we have

$$\begin{cases} u'' - \dfrac{M}{\hbar^2} [W - V_0] u = 0, & (r \le a) \,, \\[2mm] u'' - \dfrac{MW}{\hbar^2} u = 0, & (r \ge a) \,. \end{cases}$$

The boundary conditions are $u|_{r=0} = 0$ and $u|_{r\to\infty}$ finite. Satisfying these the solutions are

$$u = \begin{cases} A\sin(k_1 r), & (r \le a) \\ B\exp(-k_2 r), & (r \ge a) \end{cases}$$

where $k_1 = \sqrt{\frac{M}{\hbar^2}(V_0 - W)}$, $k_2 = \sqrt{\frac{MW}{\hbar^2}}$. Continuity of the wave function at $r = a$ further requires

$$u = \begin{cases} A\sin(k_1 r), & (r \le a) \\ A\sin(k_1 a)\exp[-k_2(a - r)]. & (r \ge a) \end{cases}$$

Continuity of the first derivative of the wave function at $r = a$ gives

$$\cot(k_1 a) = -k_2/k_1 \,.$$

Hence the probability of finding $r \ge r_0$ is

$$P = \frac{\int_{r_0}^{\infty} r^2 R^2(r)dr}{\int_0^{\infty} r^2 R^2(r)dr} = \frac{\sin^2(k_1 a)\exp[2k_2(a - r_0)]}{ak_2 - \dfrac{k_2}{2k_1}\sin(2k_1 a) + \sin^2(k_1 a)}$$

$$\approx \frac{\sin^2(k_1 a)}{ak_2}\exp(-2k_2 r_0)\,,$$

as $r_0 \gg a$.

A rough estimate of the probability can be obtained by putting $u \approx C\exp(-k_2 r)$, for which

$$P = \frac{\int_{r_0}^{\infty} \exp(-2k_2 r)dr}{\int_0^{\infty} \exp(-2k_2 r)dr} = \exp(-2k_2 r_0)\,.$$

(d) The neutron has radial velocity

$$v = \frac{p}{M} = \sqrt{\frac{2(V_0 - W)}{M}}$$

in the potential well. The transition probability per unit time is

$$\lambda = \frac{vP}{a}\,,$$

and so the halflife of dQd is given by

$$\tau = \frac{\ln 2}{\lambda} = \frac{a\ln 2}{vP} \approx a\ln 2\sqrt{\frac{M}{2(V_0 - W)}}\exp(2k_2 r_0)\,.$$

PART II

NUCLEAR PHYSICS

1. BASIC NUCLEAR PROPERTIES (2001–2023)

2001

Discuss 4 independent arguments against electrons existing inside the nucleus.

(Columbia)

Solution:

First argument - Statistics. The statistical nature of nuclei can be deduced from the rotational spectra of diatomic molecules. If a nucleus (A,Z) were to consist of A protons and (A-Z) electrons, the spin of an odd-odd nucleus or an odd-even nucleus would not agree with experimental results, Take the odd-odd nucleus ^{14}N as example. An even number of protons produce an integer spin while an odd number of electrons produce a half-integer spin, so the total spin of the ^{14}N nucleus would be a half-integer, and so it is a fermion. But this result does not agree with experiments. Hence, nuclei cannot be composed of protons and electrons.

Second argument - Binding energy. The electron is a lepton and cannot take part in strong interactions which bind the nucleons together. If electrons existed in a nucleus, it would be in a bound state caused by Coulomb interaction with the binding energy having an order of magnitude

$$E \approx -\frac{Ze^2}{r},$$

where r is the electromagnetic radius of the nucleus, $r = 1.2A^{1/3}$fm. Thus

$$E \approx -Z\left(\frac{e^2}{\hbar c}\right)\frac{\hbar c}{r} = -\frac{197Z}{137 \times 1.2A^{1/3}} \approx -1.20\frac{Z}{A^{1/3}} \text{ MeV}.$$

Note that the fine structure constant

$$\alpha = \frac{e^2}{\hbar c} = \frac{1}{137}.$$

Suppose $A \approx 124$, $Z \approx A/2$. Then $E \approx -15$ MeV, and the de Brogile wavelength of the electron would be

$$\lambda = \hbar/p = c\hbar/cp = 197/15 = 13 \text{ fm}.$$

As $\lambda > r$ the electron cannot be bound in the nucleus.

Third argument - Nuclear magnetic moment. If the nucleus consists of neutrons and protons, the nuclear magnetic moment is the sum of the contributions of the two kinds of nucleons. While different coupling methods give somewhat different results, the nuclear magnetic moment should be of the same order of magnitude as that of a nucleon, μ_N. On the other hand, if the nucleus consisted of protons and electrons, the nuclear magnetic moment should be of the order of magnitude of the magnetic moment of an electron, $\mu_e \approx 1800\mu_N$. Experimental results agree with the former assumption, and contradict the latter.

Fourth argument - β-decay. Nucleus emits electrons in β-decay, leaving behind a daughter nucleus. So this is a two-body decay, and the electrons emitted should have a monoenergetic spectrum. This conflicts with the continuous β energy spectrum found in such decays. It means that, in a β-decay, the electron is accompanied by some third, neutral particle. This contracts the assumption that there were only protons and electrons in a nucleus.

The four arguments above illustrate that electrons do not exist in the nucleus.

2002

The size of the nucleus can be determined by (a) electron scattering, (b) energy levels of muonic atoms, or (c) ground state energies of the isotopic spin multiplet . Discuss what physical quantities are measured in two and only two of these three experiments and how these quantities are related to the radius of the nucleus.

(SUNY, Buffalo)

Solution:

(a) It is the nuclear form factor that is measured in electron scattering experiments:

$$F(q^2) = \frac{(d\sigma)_{\text{exp}}}{(d\sigma)_{\text{point}}} ,$$

where $(d\sigma)_{\text{exp}}$ is the experimental value, $(d\sigma)_{\text{point}}$ is the theoretical value obtained by considering the nucleus as a point. With first order Born approximation, we have

$$F(q^2) = \int \rho(\mathbf{r}) e^{i\mathbf{q}\cdot\mathbf{r}} d^3\mathbf{r} \,.$$

Assuming $\rho(\mathbf{r}) = \rho(r)$ and $\mathbf{q}\cdot\mathbf{r} \ll 1$, we have

$$F(q^2) \approx \int \rho(r) \left[1 + \frac{1}{2}(i\mathbf{q}\cdot\mathbf{r})^2 \right] d^3\mathbf{r} = 1 - \frac{1}{2}\int \rho(r)(\mathbf{q}\cdot\mathbf{r})^2 d^3\mathbf{r}$$

$$= 1 - \frac{1}{c2}\int \rho(r) q^2 r^2 \cdot 4\pi r^2 dr \int_0^\pi \frac{1}{2}\cos^2\theta \sin\theta d\theta$$

$$= 1 - \frac{1}{6} q^2 \langle r^2 \rangle$$

with $\langle r^2 \rangle = \int \rho(r) r^2 d^3\mathbf{r}$.

By measuring the angular distribution of elastically scattered electrons, we can deduce $F(q^2)$, and so obtain the charge distribution $\rho(r)$ as a function of r, which gives a measure of the nuclear size.

(b) We can measure the energy differences between the excited states and the ground state of the muonic atom. As the mass of a muon is $m_\mu \approx 210 m_e$, the first radius of the muonic atom is $a_\mu \approx (1/210) a_0$, where a_0 is the Bohr radius, so that the energy levels of muonic atom are more sensitive to its nuclear radius. Consider for example the s state, for which the Hamiltonian is

$$H = -\frac{1}{2m_\mu}\nabla^2 + V(r)\,.$$

If the nucleus can be considered as a point charge, then $V(r) = V_0(r) = -e^2/r$, r being the distance of the muon from the nucleus.

If on the other hand we consider the nuclear charge as being uniformly distributed in a sphere of radius R, then

$$V(r) = \begin{cases} -\dfrac{e^2}{2R^3}(3R^2 - r^2), & 0 < r \le R, \\[2ex] -\dfrac{e^2}{r}, & r > R. \end{cases}$$

To obtain the energy shift of the ground state, ΔE, caused by the finite size of the nucleus, we take

$$H' = H - H_0 = V(r) - V_0(r) = \begin{cases} -\dfrac{e^2}{2R^3}(3R^2 - r^2) + \dfrac{e^2}{r}, & 0 < r \le R, \\[2ex] 0, & r > R, \end{cases}$$

as perturbation. Then

$$\Delta E = \langle \Phi_0 | H' | \Phi_0 \rangle = 4\pi \int_0^R |\Phi_0|^2 H' r^2 dr \, ,$$

where $\Phi_0 = \left(\frac{1}{\pi a_\mu^3} \right)^{1/2} e^{-\frac{r}{a_\mu}}$. As $R \sim 10^{-12}$ cm, $a_\mu \sim 10^{-10}$ cm, we can take $\frac{R}{a_\mu} \ll 1$ and hence $e^{-2r/a_\mu} \approx \left(1 - \frac{2r}{a_\mu} \right)$. Then $\Delta E = \frac{2}{5} \left(\frac{e^2}{2a_\mu} \right) \left(\frac{R}{a_\mu} \right)^2$, neglecting terms of order $\left(\frac{R}{a_\mu} \right)^3$ and higher.

We can measure the energy of the X-rays emitted in the transition from the first excited state to the ground state,

$$E_X = (E_1 - E_0) - \frac{2}{5} \left(\frac{e^2}{a_\mu} \right) \left(\frac{R}{a_\mu} \right)^2 \, ,$$

where E_1 and E_0 are eigenvalues of H_0, i.e. E_1 is the energy level of the first excited state and E_0 is the energy level of the ground state (for a point-charge nucleus). If the difference between E_X and $(E_1 - E_0)$, is known, R can be deduced.

(c) The nuclear structures of the same isotopic spin multiplet are the same so that the mass difference in the multiplet arises from electromagnetic interactions and the proton-neutron mass difference. Thus **(Problem 2009)**

$$\Delta E \equiv [M(Z, A) - M(Z - 1, A)]c^2$$

$$= \Delta E_e - (m_n - m_p)c^2$$

$$= \frac{3e^2}{5R}[Z^2 - (Z - 1)^2] - (m_n - m_p)c^2 \, ,$$

from which R is deduced

It has been found that $R \approx R_0 A^{\frac{1}{3}}$ with $R_0 = 1.2 - 1.4$ fm.

2003

To study the nuclear size, shape and density distribution one employs electrons, protons and neutrons as probes.

(a) What are the criteria in selecting the probe? Explain.

(b) Compare the advantages and disadvantages of the probes mentioned above.

(c) What is your opinion about using photons for this purpose?

(SUNY, Buffalo)

Solution:

(a) The basic criterion for selecting probes is that the de Broglie wavelength of the probe is less than or equal to the size of the object being studied. Thus $\lambda = h/p \leq d_n$, or $p \geq h/d_n$, where d_n is the linear size of the nucleus. For an effective study of the nuclear density distribution we require $\lambda \ll d_n$.

(b) Electrons are a suitable probe to study the nuclear electromagnetic radius and charge distribution because electrons do not take part in strong interactions, only in electromagnetic interactions. The results are therefore easy to analyze. In fact, many important results have been obtained from electron-nucleus scatterings, but usually a high energy electron beam is needed. For example, take a medium nucleus. As $d_n \approx 10^{-13}$ cm, we require

$$p_e \approx \hbar/d_n \approx 0.2 \text{ GeV}/c, \qquad \text{or} \qquad E_e \approx pc = 0.2 \text{ GeV}.$$

Interactions between protons and nuclei can be used to study the nuclear structure, shape and distribution. The advantage is that proton beams of high flux and suitable parameters are readily available. The disadvantage is that both electromagnetic and strong interactions are present in proton-nucleus scatterings and the results are rather complex to analyse.

Neutrons as a probe are in principle much 'neater' than protons, However, it is much more difficult to generate neutron beams of high energy and suitable parameters. Also detection and measurements are more difficult for neutrons.

(c) If photons are used as probe to study nuclear structure, the high energy photons that must be used to interest with nuclei would show a hadron-like character and complicate the problem.

2004

Consider a deformed nucleus (shape of an ellipsoid, long axis 10% longer than short axis). If you compute the electric potential at the first Bohr radius, what accuracy can you expect if you treat the nucleus as a point charge? Make reasonable estimate; do not get involved in integration.

(Wisconsin)

Solution:

Assume the charge distribution in the nucleus is uniform, ellipsoidal and axially symmetric. Then the electric dipole moment of the nucleus is zero, and the potential can be written as

$$V = V_p + V_q ,$$

where $V_p = Q/r$ is the potential produced by the nucleus as a point charge, $V_q = MQ/r^3$, M being the electric quadrupole moment.

For the ellipsoid nucleus, let the long axis be $a = (1 + \varepsilon)R$, the short axis be $b = (1 - \varepsilon/2)R$, where ε is the deformed parameter, and R is the nuclear radius. As $a : b = 1.1$, we have $\frac{3\varepsilon}{2} = 0.1$, or $\varepsilon = 0.2/3$, and so

$$M = \frac{2}{5}(a^2 - b^2) = \frac{2}{5}(a - b)(a + b) = \frac{1.22}{15}R^2 .$$

For a medium nucleus, take $A \sim 125$, for which $R = 1.2A^{1/3} = 6$ fm. Then

$$\Delta V = \frac{V_q}{V_p} = \frac{M}{r^2} = \frac{1.22}{15}\frac{R^2}{r^2} = \frac{1.22}{15} \times \left(\frac{6 \times 10^{-13}}{0.53 \times 10^{-8}}\right) \approx 1 \times 10^{-9} ,$$

at the first Bohr radius $r = 0.53 \times 10^{-8}$ cm. Thus the relative error in the potential if we treat the nucleus as a point charge is about 10^{-9} at the first Bohr orbit.

2005

The precession frequency of a nucleus in the magnetic field of the earth is 10^{-1}, 10^1, 10^3, 10^5 sec^{-1}.

(Columbia)

Solution:

The precession frequency is given by

$$\omega = \frac{geB}{2m_Nc} .$$

With $g = 1$, $e = 4.8 \times 10^{-10}$ esu, $c = 3 \times 10^{10}$ cm/s, $B \approx 0.5$ Gs, $m_N \approx 10^{-23}$g for a light nucleus, $\omega = \frac{4.8 \times 0.5 \times 10^{-10}}{2 \times 10^{-23} \times 3 \times 10^{10}} = 0.4 \times 10^3 s^{-1}$.

Hence the answer is 10^3 s^{-1}.

2006

Given the following information for several light nuclei (1 amu = 931.5 MeV) in Table 2.1.

(a) What are the approximate magnetic moments of the neutron, $^3H_1$, $^3He_2$, and $^6Li_3$?

(b) What is the maximum-energy β-particle emitted when $^3H_1$ decays to $^3He_2$?

(c) Which reaction produces more energy, the fusion of $^3H_1$ and $^3He_2$ or $^2H_1$ and $^4He_2$?

(Wisconsin)

Table 2.1

Nuclide	J^π	Nuclide mass (amu)	magnetic moment (μ_N)
$^1H_1$	$1/2^+$	1.00783	$+2.79$
$^2H_1$	1^+	2.01410	$+0.86$
$^3H_1$	$1/2^+$	3.01605	—
$^3He_2$	$1/2^+$	3.01603	—
$^4He_2$	0^+	4.02603	0
$^6Li_3$	1^+	6.01512	—

Solution:

The nuclear magnetic moment is given by $\mu = g\mu_N \mathbf{J}$, where $\mathbf{J}$ is the nuclear spin, g is the Landé factor, μ_N is the nuclear magneton. Then from the table it is seen that

$$g(^1H_1) = 2 \times 2.79 = 5.58, \qquad g(^2H_1) = 0.86, \qquad g(^4He_2) = 0.$$

When two particles of Landé factors g_1 and g_2 combine into a new particle of Landé factor g, (assuming the orbital angular momentum of relative motion is zero), then

$$g = g_1 \frac{J(J+1) + j_1(j_1+1) - j_2(j_2+1)}{2J(J+1)}$$

$$+ g_2 \frac{J(J+1) + j_2(j_2+1) - j_1(j_1+1)}{2J(J+1)},$$

where J is the spin of the new particle, j_1 and j_2 are the spins of the constituent particles.

^{2}H$_1$ is the combination of a neutron and ^{1}H$_1$, with $J = 1$, $j_1 = j_2 = 1/2$. Let $g_1 = g(n)$, $g_2 = g(^1\text{H}_1)$. Then $\frac{1}{2}g_1 + \frac{1}{2}g_2 = g(^2\text{H}_1)$, or

$$g(n) = g_1 = 2(0.86 - 2.79) = -3.86 \,.$$

According to the single-particle shell model, the magnetic moment is due to the last unpaired nucleon. For ^{3}H, $j = 1/2$, $l = 0$, $s = 1/2$, same as for ^{1}H. Thus, $g(^3\text{H}) = g(^1\text{H})$. Similarly ^{3}He has an unpaired n so that $g(^3\text{He}) = g(n)$. Hence

$$\mu(^3\text{H}) = 2.79\mu_N, \qquad \mu(^3\text{He}) = -1.93\mu_N \,.$$

^{6}Li$_3$ can be considered as the combination of ^{4}He$_2$ and ^{2}H$_1$, with $J = 1$, $j_1 = 0$, $j_2 = 1$. Hence

$$g = \left(\frac{2-2}{2\times 2}\right) g_1 + \left(\frac{2+2}{2\times 2}\right) g_2 = g_2 \,,$$

or

$$g(^6\text{Li}_3) = g(^2\text{H}_1) = 0.86 \,.$$

(a) The approximate values of the magnetic moments of neutron, ^{3}H$_1$, ^{3}He$_2$, ^{6}Li$_3$ are therefore

$$\mu(n) = g(n)\mu_N/2 = -1.93\mu_N \,,$$

$$\mu(^3\text{H}_1) = 2.79\mu_N \,,$$

$$\mu(^3\text{He}_2) = -1.93\mu_N \,,$$

$$\mu(^6\text{Li}) = g(^6\text{Li}_3)\mu_N \times 1 = 0.86\mu_N \,.$$

(b) The β-decay from ^{3}H$_1$ to ^{3}He is by the interaction

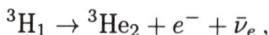

$$^3\text{H}_1 \rightarrow {}^3\text{He}_2 + e^- + \bar{\nu}_e \,,$$

where the decay energy is

$$Q = m(^3\text{H}_1) - m(^3\text{He}_2) = 3.01605 - 3.01603 = 0.00002 \text{ amu}$$

$$= 2 \times 10^{-5} \times 938 \times 10^3 \text{ keV} = 18.7 \text{ keV} \,.$$

Hence the maximum energy of the β-particle emitted is 18.7 keV.

(c) The fusion reaction of $^3H_1$ and $^3He_2$,

$$^3H_1 + {}^3He_2 \rightarrow {}^6Li_3 \,,$$

releases an energy

$$Q = m(^3H_1) + m(^3He_2) - m(^6Li_3) = 0.01696 \text{ amu} = 15.9 \text{ MeV} \,.$$

The fusion reaction of $^2H_1$ and $^4He_2$,

$$^2H_1 + {}^4He_2 \rightarrow {}^6Li_3 \,,$$

releases an energy

$$Q' = m(^2H_1) + m(^4He_2) - m(^6Li_3) = 0.02501 \text{ amu} = 23.5 \text{ MeV} \,.$$

Thus the second fusion reaction produces more energy.

2007

To penetrate the Coulomb barrier of a light nucleus, a proton must have a minimum energy of the order of
  (a) 1 GeV.
  (b) 1 MeV.
  (c) 1 KeV.

(CCT)

Solution:

The Coulomb barrier of a light nucleus is $V = Q_1 Q_2 / r$. Let $Q_1 \approx Q_2 \approx e$, $r \approx 1$ fm. Then

$$V = e^2/r = \frac{\hbar c}{r} \left(\frac{e^2}{\hbar c} \right) = \frac{197}{1} \cdot \frac{1}{137} = 1.44 \text{ MeV} \,.$$

Hence the answer is (b).

2008

What is the density of nuclear matter in ton/cm^3?
  (a) 0.004.
  (b) 400.
  (c) 10^9.

(CCT)

Solution:

The linear size of a nucleon is about 10^{-13} cm, so the volume per nucleon is about 10^{-39} cm^3. The mass of a nucleon is about 10^{-27} kg $= 10^{-30}$ ton, so the density of nuclear matter is $\rho = m/V \approx 10^{-30}/10^{-39} = 10^9$ton/cm^3. Hence the answer is (c).

2009

(a) Calculate the electrostatic energy of a charge Q distributed uniformly throughout a sphere of radius R.

(b) Since $^{27}_{14}$Si and $^{27}_{13}$Al are "mirror nuclei", their ground states are identical except for charge. If their mass difference is 6 MeV, estimate their radius (neglecting the proton-neutron mass difference).

(Wisconsin)

Solution:

(a) The electric field intensity at a point distance r from the center of the uniformly charged sphere is

$$
E(r) = \begin{cases} \dfrac{Qr}{R^3} & \text{for } r < R, \\[2mm] \dfrac{Q}{r^2} & \text{for } r > R. \end{cases}
$$

The electrostatic energy is

$$
\begin{aligned}
W &= \int_0^\infty \frac{1}{8\pi} E^2 d\tau \\
&= \frac{Q^2}{8\pi} \left[\int_0^R \left(\frac{r}{R^3} \right)^2 4\pi r^2 dr + \int_R^\infty \frac{1}{r^4} 4\pi r^2 dr \right] \\
&= \frac{Q^2}{2} \left(\int_0^R \frac{r^4}{R^6} dr + \int_R^\infty \frac{1}{r^2} dr \right) \\
&= \frac{Q^2}{2} \left(\frac{1}{5R} + \frac{1}{R} \right) \\
&= \frac{3Q^2}{5R}.
\end{aligned}
$$

(b) The mass difference between the mirror nuclei $^{27}_{14}Si$ and $^{27}_{13}Al$ can be considered as due to the difference in electrostatic energy:

$$\Delta W = \frac{3e^2}{5R}(Z_1^2 - Z_2^2).$$

Thus

$$R = \frac{3e^2}{5\Delta W}(14^2 - 13^2) = \frac{3\hbar c}{5\Delta W}\left(\frac{e^2}{\hbar c}\right)(14^2 - 13^2)$$

$$= \frac{3 \times 1.97 \times 10^{-11}}{5 \times 6} \times \frac{1}{137} \times (14^2 - 13^2)$$

$$= 3.88 \times 10^{-11} \text{ cm}$$

$$= 3.88 \text{ fm}.$$

2010

The nucleus $^{27}_{14}Si$ decays to its "mirror" nucleus $^{27}_{13}Al$ by positron emission. The maximum (kinetic energy$+m_e c^2$) energy of the positron is 3.48 MeV. Assume that the mass difference between the nuclei is due to the Coulomb energy. Assume the nuclei to be uniformly charged spheres of charge Ze and radius R. Assuming the radius is given by $r_0 A^{1/3}$, use the above data to estimate r_0.

(Princeton)

Solution:

$$^{27}_{14}Si \rightarrow {}^{27}_{13}Al + \beta^+ + \nu.$$

If the recoil energy of the nucleus is neglected, the maximum energy of the positron equals roughly the mass difference between the nuclei minus $2m_e c^2$. The Coulomb energy of a uniformly charged sphere is (**Problem 2009**)

$$E_e = \frac{3e^2 Z^2}{5R} = \frac{3e^2}{5r_0}Z^2 A^{-1/3}.$$

For $^{27}_{14}Si$ and $^{27}_{13}Al$,

$$E_e = \frac{3e^2}{5r_0}27^{-\frac{1}{3}}(14^2 - 13^2) = \frac{27e^2}{5r_0} = 3.48 + 1.02 = 4.5 \text{ MeV},$$

or

$$r_0 = \frac{27e^2}{5 \times 4.5} = \frac{27\hbar c}{5 \times 4.5}\left(\frac{e^2}{\hbar c}\right) = \frac{27 \times 1.97 \times 10^{-11}}{5 \times 4.5 \times 137}$$

$$= 1.73 \times 10^{-13} \text{ cm} = 1.73 \text{ fm}.$$

2011

The binding energy of $^{90}_{40}\text{Zr}_{50}$ is 783.916 MeV. The binding energy of $^{90}_{39}\text{Y}_{51}$ is 782.410 MeV. Estimate the excitation energy of the lowest $T = 6$ isospin state in ^{90}Zr.

(*Princeton*)

Solution:

The energy difference between two members of the same isospin multiplet is determined by the Coulomb energies and the neutron-proton mass difference. Thus (**Problem 2009**)

$$\Delta E = E(A, Z+1) - E(A, Z) = \Delta E_e - (m_n - m_p)c^2$$

$$= \frac{3e^2}{5R}(2Z+1) - 0.78 = \frac{3(2Z+1)c\hbar\alpha}{5R} - 0.78$$

$$= \frac{3(2 \times 39 + 1) \times 197}{5 \times 1.2 \times 90^{1/3} \times 137} - 0.78$$

$$= 11.89 \text{ MeV}$$

using $R = 1.2A^{1/3}$ fm.

Hence the excitation energy of the $T = 6$ state of ^{90}Zr is

$$E = -782.410 + 11.89 + 783.916 = 13.40 \text{ MeV}.$$

2012

The masses of a set of isobars that are members of the same isospin multiplet can be written as the expectation value of a mass operator with the form

$$M = a + bT_z + cT_z^2,$$

where a, b, c are constants and T_z is the operator for the z component of the isotopic spin.

(a) Derive this formula.

(b) How large must the isospin be in order to test it experimentally?

(*Princeton*)

Solution:

(a) Members of the same isospin multiplet have the same spin-parity J^p because of the similarity of their structures. Their mass differences are determined by the Coulomb energies and the neutron-proton mass difference. Let the nuclear mass number be A, neutron number be N, then $A = Z + N = 2Z - (Z - N) = 2Z - 2T_z$. As (**Problem 2009**)

$$M = \frac{3e^2 Z^2}{5R} + (m_p - m_n)T_z + M_0$$

$$= B\left(\frac{A}{2} + T_z\right)^2 + CT_z + M_0$$

$$= \frac{BA^2}{4} + BAT_z + BT_z^2 + CT_z + M_0$$

$$= M_0 + \frac{BA^2}{4} + (C + BA)T_z + BT_z^2$$

$$= a + bT_z + cT_z^2$$

with $a = M_0 + BA^2/4$, $b = C + BA$, $c = B$.

The linear terms in the formula arise from the neutron-proton mass difference and the Coulomb energy, while the quadratic term is mainly due to the Coulomb energy.

(b) There are three constants a,b,c in the formula, so three independent linear equations are needed for their determination. As there are $(2T + 1)$ multiplets of an isospin T, in order to test the formula experimentally we require at least $T = 1$.

2013

Both nuclei $^{14}_{7}$N and $^{12}_{6}$C have isospin $T = 0$ for the ground state. The lowest $T = 1$ state has an excitation energy of 2.3 MeV in the case of

$^{14}_{7}$N and about 15.0 MeV in the case of $^{12}_{6}$C. Why is there such a marked difference? Indicate also the basis on which a value of T is ascribed to such nuclear states. (Consider other members of the $T = 1$ triplets and explain their relationship in terms of systematic nuclear properties.)

(Columbia)

Solution:

The excited states with $T = 1$ of $^{12}_{6}$C form an isospin triplet which consists of $^{12}_{5}$B, $^{12}_{6}$C and $^{12}_{7}$N. $^{12}_{5}$B and $^{12}_{7}$N have $|T_3| = 1$, so they are the ground states of the triplet. Likewise, $^{14}_{6}$C and $^{14}_{8}$O are the ground states of the isospin triplet of the $T = 1$ excited states of $^{14}_{7}$N. The binding energies $M - A$ are given in the table below.

Elements	M-A (MeV)
$^{12}_{6}C$	0
$^{12}_{5}B$	13.370
$^{14}_{7}N$	2.864
$^{14}_{6}C$	3.020

The energy difference between two nuclei of an isospin multiplet is

$$\Delta E = [M(Z, A) - M(Z - 1, A)]c^2$$

$$= \frac{3e^2}{5R}(2Z - 1) - (m_n - m_p)c^2$$

$$= \frac{3e^2}{5R_0 A^{1/3}}(2Z - 1) - 0.78$$

$$= \frac{3\hbar c}{5R_0 A^{1/3}}\left(\frac{e^2}{\hbar c}\right)(2Z - 1) - 0.78$$

$$= \frac{3 \times 197}{5 \times 137 R_0 A^{1/3}}(2Z - 1) - 0.78 \text{ MeV}.$$

Taking $R_0 \approx 1.4$ fm and so

$$M(^{14}N, T = 1) - M(^{14}C, T = 1) = 2.5 \text{ MeV}/c^2,$$

$$M(^{12}C, T = 1) - M(^{12}B, T = 1) = 2.2 \text{ MeV}/c^2,$$

we have

$$M(^{14}N, T = 1) - M(^{14}N, T = 0)$$
$$= M(^{14}N, T = 1) - M(^{14}C, T = 1)$$
$$+ M(^{14}C, T = 1) - M(^{14}N, T = 0)$$
$$= 2.5 + 3.02 - 2.86$$
$$= 2.66 \text{ MeV}/c^2,$$

$$M(^{12}C, T = 1) - M(^{12}C, T = 0)$$
$$= M(^{12}C, T = 1) - M(^{12}B, T = 1)$$
$$+ M(^{12}B, T = 1) - M(^{12}C, T = 0)$$
$$= 2.2 + 13.37$$
$$= 15.5 \text{ MeV}/c^2,$$

which are in agreement with the experiment values 2.3 MeV and 15.0 MeV. The large difference between the excitation energies of ^{12}C and ^{14}N is due to the fact that the ground state of ^{12}C is of an α-group structure and so has a very low energy.

The nuclei of an isospin multiplet have similar structures and the same J^p. The mass difference between two isospin multiplet members is determined by the difference in the Coulomb energy of the nuclei and the neutron-proton mass difference. Such data form the basis of isospin assignment. For example ^{14}O, $^{14}N^*$ and ^{14}C belong to the same isospin multiplet with $J^p = 0^+$ and ground states ^{14}C and ^{14}O, $^{14}N^*$ being an exciting state. Similarly $^{12}C^*$, ^{12}C and ^{12}B form an isospin multiplet with $J^p = 1^+$, of which ^{14}N and ^{12}B are ground states while $^{12}C^*$ is an excited state.

2014

(a) Fill in the missing entries in the following table giving the properties of the ground states of the indicated nuclei. The mass excess $\Delta M_{Z,A}$ is defined so that

$$M_{Z,A} = A(931.5 \text{ MeV}) + \Delta M_{Z,A},$$

where $M_{Z,A}$ is the atomic mass, A is the mass member, T and T_z are the quantum members for the total isotopic spin and the third component of isotopic spin. Define your convention for T_z.

(b) The wave function of the isobaric analog state (IAS) in ^{81}Kr is obtained by operating on the ^{81}Br ground state wave function with the isospin upping operator T.

(i) What are J^π, T, and T_z for the IAS in ^{81}Kr?

(ii) Estimate the excitation energy of the IAS in ^{81}Kr.

(iii) Now estimate the decay energy available for decay of the IAS in ^{81}Kr by emission of a

$$\text{neutron}, \qquad \gamma\text{-ray}, \qquad \alpha\text{-particle}, \qquad \beta^+\text{-ray}.$$

(iv) Assuming sufficient decay energy is available for each decay mode in (iii), indicate selection rules or other factors which might inhibit decay by that mode.

(Princeton)

Isotopes Z		T_z	T	J^p	Mass excess (MeV)
n	0				8.07
^{1}H	1				7.29
^{4}He	2				2.43
^{77}Se	34			$1/2^-$	-74.61
^{77}Br	35			$3/2^-$	-73.24
^{77}Kr	36			$7/2^+$	-70.24
^{80}Br	35			1^+	-76.89
^{80}Kr	36				-77.90
^{81}Br	35			$3/2^-$	-77.98
^{81}Kr	36			$7/2^+$	-77.65
^{81}Rb	37			$3/2^-$	-77.39

Solution:

(a) The table is completed as shown in the next page.

(b) (i) The isobasic analog state (IAS) is a highly excited state of a nucleus with the same mass number but with one higher atomic number, i.e. a state with the same A, the same T, but with T_z increased by 1. Thus for ^{81}Br, $|T, T_z\rangle = |11/2, -11/2\rangle$, the quantum numbers of the IAS in ^{81}Kr are $T = 11/2$, $T_z = -9/2$, $J^p[^{81}\text{Kr(IAS)}] = J^p(^{81}\text{Br}) = 3/2^-$.

Isotopes Z		T_z	T	J^p	Mass excess (MeV)
n	0	$-1/2$	1/2	$1/2^+$	8.07
^{1}H	1	1/2	1/2	$1/2^+$	7.29
^{4}He	2	0	0	0^+	2.43
^{77}Se	34	$-9/2$	9/2	$1/2^-$	-74.61
^{77}Br	35	$-7/2$	7/2	$3/2^-$	-73.24
^{77}Kr	36	$-5/2$	5/2	$7/2^+$	-70.24
^{80}Br	35	-5	5	1^+	-76.89
^{80}Kr	36	-4	4	0^+	-77.90
^{81}Br	35	$-11/2$	11/2	$3/2^-$	-77.98
^{81}Kr	36	$-9/2$	9/2	$7/2^+$	-77.65
^{81}Rb	37	$-7/2$	7/2	$3/2^-$	-77.39

(ii) The mass difference between ^{81}Br and ^{81}Kr(IAS) is due to the difference between the Coulomb energies of the nuclei and the neutron-proton mass difference:

$$\Delta M_{^{81}\text{Kr}(IAS)} = \Delta M_{^{81}\text{Br}} + \frac{3}{5} \times \frac{(2Z-1)e^2}{R_0 A^{1/3}} - [m(n) - M(^1\text{H})]$$

$$= \Delta M_{^{81}\text{Br}} + 0.719\left(\frac{2Z-1}{A^{\frac{1}{3}}}\right) - 0.78 \text{ MeV},$$

as $R_0 = 1.2$ fm, $m_n - m_p = 0.78$ MeV. With $Z = 36$, $A = 81$, $\Delta M_{^{81}\text{Br}} = -77.98$ MeV, we have $\Delta M_{^{81}\text{Kr}(IAS)} = -67.29$ MeV.

Hence the excitation energy of ^{81}Kr(IAS) from the ground state of ^{81}Kr is

$$\Delta E = -67.29 - (-77.65) = 10.36 \text{ MeV}.$$

(iii) For the neutron-decay ^{81}Kr(IAS)$\rightarrow n + {}^{80}$Kr,

$$Q_1 = \Delta M_{^{81}\text{Kr}(IAS)} - \Delta(n) - \Delta M_{^{80}\text{Kr}}$$

$$= -67.29 - 8.07 + 77.90 = 2.54 \text{ MeV}.$$

For the γ-decay ^{81}Kr$(IAS) \rightarrow {}^{81}$Kr $+ \gamma$,

$$Q_2 = \Delta M_{^{81}\text{Kr}(IAS)} - \Delta M_{^{81}\text{Kr}} = -67.29 - (-77.65) = 10.36 \text{ MeV}.$$

For the α-decay ^{81}Kr$(IAS) \rightarrow \alpha + {}^{77}$Se,

$$Q_3 = \Delta M_{81\text{Kr}(IAS)} - \Delta M_\alpha - \Delta M_{77\text{Se}}$$

$$= -67.29 - 2.43 - (-74.61) = 4.89 \text{ MeV}.$$

For the β^+-decay $^{81}\text{Kr}(IAS) \to {}^{81}\text{Br} + \beta^+ + \nu_e$,

$$Q_4 = \Delta M_{81\text{Kr}(IAS)} - \Delta M_{81\text{Br}} - 2m_e$$

$$= -67.29 - (-77.98) - 1.02 = 9.67 \text{ MeV}.$$

(iv)

In the interaction $\quad {}^{81}\text{Kr}(IAS) \to \quad {}^{81}\text{Kr} \quad + \quad n$

$$T: \qquad\qquad 11/2 \qquad\quad 4 \qquad \frac{1}{2}$$

$\Delta T \neq 0$. As strong interaction requires conservation of T and T_z, the interaction is inhibited.

In the interaction $\quad {}^{81}\text{Kr}(IAS) \to \quad {}^{81}\text{Kr} + \gamma$

$$J^p: \qquad\qquad\qquad \frac{3}{2}^- \qquad\qquad \frac{7}{2}^+$$

we have $\Delta J = \left| \frac{3}{2} - \frac{7}{2} \right| = 2$, $\Delta P = -1$; so it can take place through $E3$ or $M2$ type transition.

The interaction $\quad {}^{81}\text{Kr}(IAS) \to \quad {}^{77}\text{Se} \quad + \quad \alpha$

$$T: \qquad\qquad 11/2 \qquad\quad 9/2 \qquad 0$$

$$T_z: \qquad\qquad -9/2 \qquad -9/2 \qquad 0$$

is inhibited as isospin is not conserved.

The interaction $\quad {}^{81}\text{Kr}(IAS) \to \quad {}^{81}\text{Br} \quad + \quad \beta^+ + \nu_e$

$$J^p: \qquad\qquad\quad 3/2^- \qquad\quad \frac{3^-}{2}$$

is allowed, being a mixture of the Fermi type and Gamow–Teller type interactions.

2015

Isospin structure of magnetic dipole moment.

The magnetic dipole moments of the free neutron and free proton are $-1.913\mu_N$ and $+2.793\mu_N$ respectively. Consider the nucleus to be a collection of neutrons and protons, having their free moments.

(a) Write down the magnetic moment operator for a nucleus of A nucleons.

(b) Introduce the concept of isospin and determine the isoscalar and isovector operators. What are their relative sizes?

(c) Show that the sum of magnetic moments in nuclear magnetons of two $T = 1/2$ mirror nuclei is

$$J + (\mu_p + \mu_n - 1/2)\left\langle \sum_{i=l}^{A} \sigma_z^{(i)} \right\rangle ,$$

where J is the total nuclear spin and $\sigma_z^{(i)}$ is the Pauli spin operator for a nucleon.

(*Princeton*)

Solution:

(a) The magnetic moment operator for a nucleus of A nucleons is

$$\mu = \sum_{i=l}^{A} (g_l^i l_i + g_s^i S_i) ,$$

where for neutrons: $g_l = 0$, $g_s = 2\mu_n$; for protons: $g_l = 1$, $g_s = 2\mu_p$ and S is the spin operator $\frac{1}{2}\sigma$.

(b) Charge independence has been found to hold for protons and neutrons such that, if Coulomb forces are ignored, the $p-p$, $p-n$, $n-n$ forces are identical provided the pair of nucleons are in the same spin and orbital motions. To account for this, isospin T is introduced such that p and n have the same T while the z component T_z in isospin space is $T_z = \frac{1}{2}$ for p and $T_z = -\frac{1}{2}$ for n. There are four independent operators in isospin space:

scalar operator: unit matrix $I = \begin{pmatrix} 1 & 0 \\ 0 & 1 \end{pmatrix}$;

vector operators: Pauli matrices, $\tau_1 = \begin{pmatrix} 0 & 1 \\ 1 & 0 \end{pmatrix}$, $\tau_2 = \begin{pmatrix} 0 & -i \\ i & 0 \end{pmatrix}$,

$\tau_3 = \begin{pmatrix} 1 & 0 \\ 0 & -1 \end{pmatrix}$.

Let the wave functions of proton and neutron be $\psi_p = \begin{pmatrix} 1 \\ 0 \end{pmatrix}$, $\psi_n = \begin{pmatrix} 0 \\ 1 \end{pmatrix}$ respectively, and define $\tau_{\pm} = \tau_1 \pm i\tau_2$, $T = \tau/2$. Then

$$T_3\Psi_p = \frac{1}{2}\Psi_p, \qquad \tau_3\Psi_p = \Psi_p,$$

$$T_3\Psi_n = -\frac{1}{2}\Psi_n, \qquad \tau_3\Psi_n = -\Psi_n,$$

$$T_+\Psi_n = \Psi_p, \qquad T_-\Psi_p = \Psi_n.$$

(c) The mirror nucleus is obtained by changing all the protons of a nucleus into neutrons and all the neutrons into protons. Thus mirror nuclei have the same T but opposite T_z. In other words, for mirror nuclei, if the isospin quantum numbers of the first nucleus are $\left(\frac{1}{2}, \frac{1}{2}\right)$, then those of the second are $\left(\frac{1}{2}, -\frac{1}{2}\right)$.

For the first nucleus, the magnetic moment operator is

$$\mu_1 = \sum_{i=1}^{A}(g_l^i l_1^i + g_s^i S_1^i).$$

We can write

$$g_l = \frac{1}{2}(1 + \tau_3), \qquad g_s = (1 + \tau_3)\mu_p + (1 - \tau_3)\mu_n,$$

since $g_l\psi_p = \psi_p$, $g_l\psi_n = 0$, etc. Then

$$\mu_1 = \sum_{i=1}^{A}\frac{(1 + \tau_3^i)}{2}l_1^i + \left[\sum_{i=1}^{A}(1 + \tau_3^i)\mu_p + \sum_{i=1}^{A}(1 - \tau_3^i)\mu_n\right]S_l^i$$

$$= \frac{1}{2}\sum_{i=1}^{A}(l_1^i + S_1^i) + \left(\mu_p + \mu_n - \frac{1}{2}\right)\sum_{i=1}^{A}S_1^i$$

$$+ \frac{1}{2}\sum_{i=1}^{A}\tau_3^i[l_1^i + 2(\mu_p - \mu_n)S_1^i].$$

Similarly for the other nucleus we have

$$\mu_2 = \frac{1}{2}\sum_{i=1}^{A}(l_2^i + S_2^i) + \left(\mu_p + \mu_n - \frac{1}{2}\right)\sum_{i=1}^{A}S_2^i + \frac{1}{2}\sum_{i=1}^{A}\tau_3^i[l_2^i + 2(\mu_p - \mu_n)S_2^i].$$

As $\boldsymbol{J}^i = \sum_{i=1}^{A}(\boldsymbol{l}^i + \boldsymbol{S}^i)$, the mirror nuclei have $\boldsymbol{J}^1 = \boldsymbol{J}^2$ but opposite T_3 values, where $T_3 = \frac{1}{2}\sum_{i=1}^{A}\tau_3^i$, $\boldsymbol{S} = \frac{1}{2}\boldsymbol{\sigma}$.

The observed magnetic moment is $\mu = \langle \mu_z \rangle = \langle JJ_zTT_3|\mu_z|JJ_zTT_3\rangle$. Then for the first nucleus:

$$\mu_1 = \left\langle JJ_z\frac{1}{2}\frac{1}{2}\left|\frac{J_z}{2} + \left(\mu_p + \mu_n - \frac{1}{2}\right) \times \frac{1}{2}\sum_{i=1}^{A}(\sigma_1^i)_z\right.\right.$$

$$\left.\left. + \frac{1}{2}\sum_{i=1}^{A}\tau_3^i[l_{1z}^i + 2(\mu_p - \mu_n)S_{1z}^i]\right|JJ_z\frac{1}{2}\frac{1}{2}\right\rangle$$

$$= \frac{J_z}{2} + \frac{1}{2}\left(\mu_p + \mu_n - \frac{1}{2}\right)\left\langle \sum_{i=1}^{A}(\sigma_1^i)_z\right\rangle$$

$$+ \left\langle JJ_z\frac{1}{2}\frac{1}{2}\left|\frac{1}{2}\sum_{i=1}^{A}\tau_3^i[l_{1z}^i + 2(\mu_p - \mu_n)S_{1z}^i]\right|JJ_z\frac{1}{2}\frac{1}{2}\right\rangle,$$

and for the second nucleus:

$$\mu_2 = \frac{J_z}{2} + \frac{1}{2}\left(\mu_p + \mu_n - \frac{1}{2}\right)\left\langle \sum_{i=1}^{A}(\sigma_1^i)_z\right\rangle$$

$$+ \left\langle JJ_z\frac{1}{2}-\frac{1}{2}\left|\frac{1}{2}\sum_{i=1}^{A}\tau_3^i[l_{2z}^i + 2(\mu_p - \mu_n)S_{2z}^i]\right|JJ_z\frac{1}{2}-\frac{1}{2}\right\rangle.$$

The sum of the magnetic moments of the mirror nuclei is

$$\mu_1 + \mu_2 = J_z + \left(\mu_p + \mu_n - \frac{1}{2}\right)\left\langle \sum_{i=1}^{A}\sigma_z^i\right\rangle,$$

as the last terms in the expression for μ_1 and μ_2 cancel each other.

2016

Hard sphere scattering:

Show that the classical cross section for elastic scattering of point particles from an infinitely massive sphere of radius R is isotropic.

(MIT)

Solution:

In classical mechanics, in elastic scattering of a point particle from a fixed surface, the emergent angle equals the incident angle. Thus if a particle moving along the $-z$ direction impinges on a hard sphere of radius R at a surface point of polar angle θ, it is deflected by an angle $\alpha = 2\theta$. As the impact parameter is $b = R\sin\theta$, the differential scattering cross section is

$$\frac{d\sigma}{d\Omega} = \frac{2\pi bdb}{2\pi \sin \alpha d\alpha} = \frac{R^2 \sin\theta \cos\theta d\theta}{4\sin\theta \cos\theta d\theta} = \frac{R^2}{4},$$

which is independent of θ, showing that the scattering is isotropic.

2017

A convenient model for the potential energy V of a particle of charge q scattering from an atom of nuclear charge Q is $V = qQe^{-\alpha r}/r$. Where α^{-1} represents the screening length of the atomic electrons.

(a) Use the Born approximation

$$f = -\frac{1}{4\pi}\int e^{-i\Delta \mathbf{k}\cdot\mathbf{r}}\frac{2m}{\hbar^2}V(r)d^3\mathbf{r}$$

to calculate the scattering cross section σ.

(b) How should α depend on the nuclear charge Z?

(Columbia)

Solution:

(a) In Born approximation

$$f = -\frac{m}{2\pi\hbar^2}\int V(\mathbf{r})e^{-i\mathbf{q}\cdot\mathbf{r}}d^3\mathbf{r},$$

where $\mathbf{q} = \mathbf{k} - \mathbf{k}_0$ is the momentum transferred from the incident particle to the outgoing particle. We have $|\mathbf{q}| = 2k_0\sin\frac{\theta}{2}$, where θ is the angle between the incident and outgoing particles. As $V(\mathbf{r})$ is spherically symmetric,

$$f(\theta) = -\frac{m}{2\pi\hbar^2}\int_0^\infty \int_0^{2\pi}\int_0^\pi V(r)e^{-i\Delta kr\cos\theta}\sin\theta r^2 drd\varphi d\theta$$

$$= -\frac{2m}{\hbar^2 \Delta k}\int_0^\infty V(r)\sin(\Delta kr)rdr$$

$$= -\frac{2mQq}{\hbar^2}\cdot\frac{1}{\alpha^2 + (\Delta k)^2}.$$

The differential cross section is

$$d\sigma = |f(\theta)|^2 d\Omega = \frac{4m^2 Q^2 q^2}{\hbar^4} \cdot \frac{d\Omega}{[\alpha^2 + (\Delta k^2)]^2}$$

$$= \frac{m^2 Q^2 q^2}{4\hbar^4 k_0^4} \cdot \frac{d\Omega}{\left(\frac{\alpha^2}{4k_0^2} + \sin^2 \frac{\theta}{2}\right)^2},$$

and the total cross-section is

$$\sigma = \int d\sigma = \frac{m^2 Q^2 q^2}{4\hbar^4 k_0^4} \int_0^{2\pi} \int_0^\pi \frac{\sin\theta d\theta d\varphi}{\left(\frac{\alpha^2}{4k_0^2} + \sin^2 \frac{\theta}{2}\right)^2}$$

$$= \frac{16\pi m^2 Q^2 q^2}{\hbar^4 \alpha^2 (4k_0^2 + \alpha^2)}.$$

(b) α^{-1} gives a measure of the size of atoms. As Z increases, the number of electrons outside of the nucleus as well as the probability of their being near the nucleus will increase, enhancing the screening effect. Hence α is an increasing function of Z.

2018

Consider the scattering of a 1-keV proton by a hydrogen atom.

(a) What do you expect the angular distribution to look like? (Sketch a graph and comment on its shape).

(b) Estimate the total cross section. Give a numerical answer in cm^2, m^2 or barns, and a reason for your answer.

(*Wisconsin*)

Solution:

The differential cross section for elastic scattering is (**Problem 2017**)

$$\frac{d\sigma}{d\Omega} = \frac{m^2 q^2 Q^2}{4\hbar^4 k_0^4} \cdot \frac{1}{\left(\frac{\alpha^2}{4k_0^4} + \sin \frac{\theta}{2}\right)^2}.$$

For proton and hydrogen nuclues $Q = q = e$. The screening length can be taken to be $\alpha^{-1} \approx R_0$, R_0 being the Bohr radius of hydrogen atom. For an incident proton of 1 keV; The wave length is

$$\lambda_0 = \frac{\hbar}{\sqrt{2\mu E}} = \frac{c\hbar}{\sqrt{2\mu c^2 E}} = \frac{197}{\sqrt{1 \times 938 \times 10^{-3}}} = 203 \text{ fm}.$$

With $\alpha^{-1} \approx R_0 = 5.3 \times 10^4$ fm, $\frac{\alpha^2}{4k_0^2} = \left(\frac{\lambda_0}{2\alpha^{-1}}\right)^2 \ll 1$ and so

$$\frac{d\sigma}{d\Omega} \approx \frac{m^2 e^4}{4\hbar^2 k_0^2} \frac{1}{\sin^4 \frac{\theta}{2}},$$

which is the Rurthford scattering formula.

The scattering of 1-keV protons from hydrogen atom occurs mainly at small angles (see Fig. 2.1). The probability of large angle scattering (near head-on collision) is very small, showing that hydrogen atom has a very small nucleus.

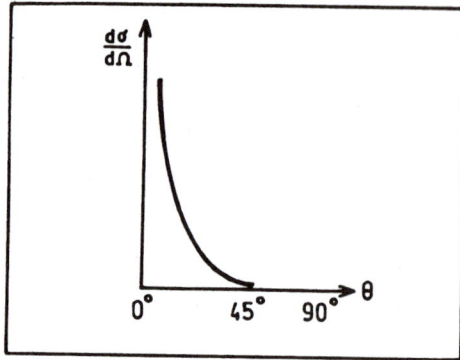

Fig. 2.1

(b) As given in **Problem 2017**,

$$\sigma = \frac{16\pi m^2 e^4}{\hbar^4 \alpha^2 (4k_0^2 + \alpha^3)} \approx \frac{16\pi m^2 e^4}{\hbar^4 \alpha^2 4k_0^2}$$

$$= 4\pi \left[\frac{mc^2 R_0 \lambda_0}{\hbar c} \left(\frac{e^2}{\hbar c}\right)\right]^2 = 4\pi \left(\frac{938 \times 5.3 \times 10^4 \times 203}{197 \times 137}\right)^2$$

$$= 1.76 \times 10^{12} \text{ fm}^2 = 1.76 \times 10^{-14} \text{ cm}^2.$$

2019

(a) At a center-of-mass energy of 5 MeV, the phase describing the elastic scattering of a neutron by a certain nucleus has the following values: $\delta_0 = 30^0$, $\delta_1 = 10^0$. Assuming all other phase shifts to be negligible, plot $d\sigma/d\Omega$ as a function of scattering angle. Explicitly calculate $d\sigma/d\Omega$ at 30^0, 45^0 and 90^0. What is the total cross section σ?

(b) What does the fact that all of the phase shifts δ_2, $\delta_3 \ldots$ are negligible imply about the range of the potential? Be as quantitative as you can.

(Columbia)

Solution:

(a) The differential cross section is given by

$$\frac{d\sigma}{d\Omega} = \frac{1}{k^2} \left| \sum_{l=0}^{\infty} (2l+1) e^{i\delta_l} \sin \delta_l P_l(\cos\theta) \right|^2 .$$

Supposing only the first and second terms are important, we have

$$\frac{d\sigma}{d\Omega} \approx \frac{1}{k^2} |e^{i\delta_0} \sin \delta_0 + 3 e^{i\delta_1} \sin \delta_1 \cos\theta|^2$$

$$= \frac{1}{k^2} |(\cos \delta_0 \sin \delta_0 + 3 \cos \delta_1 \sin \delta_1 \cos\theta) + i(\sin^2 \delta_0 + 3 \sin^2 \delta_1 \cos\theta)|^2$$

$$= \frac{1}{k^2} [\sin^2 \delta_0 + 9 \sin^2 \delta_1 \cos^2\theta + 6 \sin \delta_0 \sin \delta_1 \cos(\delta_1 - \delta_0) \cos\theta]$$

$$= \frac{1}{k^2} [0.25 + 0.27 \cos^2\theta + 0.49 \cos\theta] ,$$

where k is the wave number of the incident neutron in the center-of-mass frame. Assume that the mass of the nucleus is far larger than that of the neutron m_n. Then

$$k^2 \approx \frac{2 m_n E}{\hbar^2} = \frac{2 m_n c^2 E}{(\hbar c)^2} = \frac{2 \times 938 \times 5}{197^2 \times 10^{-30}}$$

$$= 2.4 \times 10^{29} \ m^{-2} = 2.4 \times 10^{25} \ cm^{-2} .$$

The differential cross section for other angles are given in the following table. The data are plotted in Fig. 2.2 also.

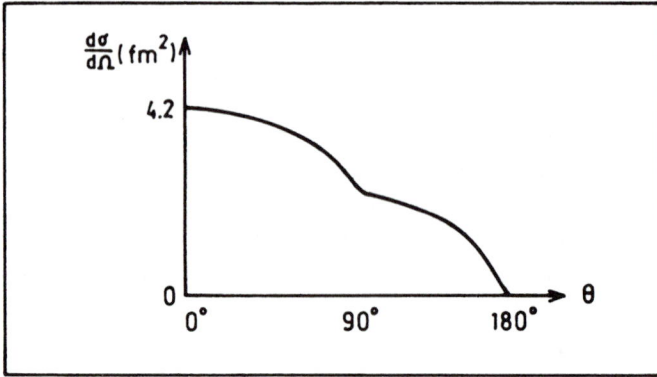

Fig. 2.2

θ	0^0	30^0	45^0	90^0	180^0
$k^2 \dfrac{d\sigma}{d\Omega}$	1	0.88	0.73	0.25	0
$\dfrac{d\sigma}{d\Omega} \times 10^{26}$ (cm^2)	4.2	3.7	3.0	1.0	0

The total cross section is

$$\sigma = \int \frac{d\sigma}{d\Omega} d\Omega = \frac{2\pi}{k^2} \int_0^\pi (0.25 + 0.49 \cos\theta + 0.27 \cos^2\theta) \sin\theta d\theta$$

$$= \frac{4\pi}{k^2} \left(0.25 + \frac{1}{3} \times 0.27 \right) = 1.78 \times 10^{-25} \text{ cm}^2 \approx 0.18 \, b \,.$$

(b) The phase shift δ_l is given by

$$\delta_l \approx -\frac{2m_n k}{\hbar^2} \int_0^\infty V(r) J_l^2(kr) r^2 dr \,,$$

where J_l is a spherical Bessel function. As the maximum of $J_l(x)$ occurs nears $x = l$, for higher l values J_l in the region of potential $V(r)$ is rather small and can be neglected. In other words, $\delta_2, \delta_3 \ldots$ being negligible means that the potential range is within $R \approx 1/k$. Thus the range of the potential is $R \approx (2.4 \times 10^{25})^{-1/2} = 2 \times 10^{-13}$ cm = 2 fm.

2020

Neutrons of 1000 eV kinetic energy are incident on a target composed of carbon. If the inelastic cross section is 400×10^{-24} cm^2, what upper and lower limits can you place on the elastic scattering cross section?

(Chicago)

Solution:

At 1 keV kinetic energy, only s-wave scattering is involved. The phase shift δ must have a positive imaginary part for inelastic process to take place. The elastic and inelastic cross sections are respectively given by

$$\sigma_e = \pi \lambdabar^2 |e^{2i\delta} - 1|^2 \,,$$

$$\sigma_{in} = \pi \lambdabar^2 (1 - |e^{2i\delta}|^2) \,.$$

The reduced mass of the system is

$$\mu = \frac{m_n m_c}{m_c + m_n} \approx \frac{12}{13} m_n \,.$$

For $E = 1000$ eV,

$$\lambdabar = \frac{\hbar}{\sqrt{2\mu E}} = \frac{\hbar c}{\sqrt{2\mu c^2 E}}$$

$$= \frac{197}{\sqrt{2 \times \frac{12}{13} \times 940 \times 10^{-3}}} = 150 \text{ fm} \,,$$

$$\pi \lambdabar^2 = 707 \times 10^{-24} \text{ cm}^2 \,.$$

As

$$1 - |e^{2i\delta}|^2 = \frac{\sigma_{in}}{\pi \lambdabar^2} = \frac{400}{707} = 0.566 \,,$$

we have

$$|e^{2i\delta}| = \sqrt{1 - 0.566} = 0.659 \,,$$

or

$$e^{2i\delta} = \pm 0.659 \,.$$

Hence the elastic cross section

$$\sigma_e = \pi \lambdabar^2 |e^{2i\delta} - 1|^2$$

has maximum and minimum values

$$(\sigma_e)_{\max} = 707 \times 10^{-24}(-0.659 - 1)^2 = 1946 \times 10^{-24} \text{ cm}^2 \,,$$

$$(\sigma_e)_{\min} = 707 \times 10^{-24}(0.659 - 1)^2 = 82 \times 10^{-24} \text{ cm}^2 \,.$$

2021

The study of the scattering of high energy electrons from nuclei has yielded much interesting information about the charge distributions in nuclei and nucleons. We shall here consider a simple version in which the electrons are supposed to have zero spin. We also assume that the nucleus, of charge Ze, remains fixed in space (i.e., its mass is assumed infinite). Let $\rho(\mathbf{x})$ denote the charge density in the nucleus. The charge distribution is assumed to be spherically symmetric but otherwise arbitrary.

Let $f_c\,(\mathbf{p}_i, \mathbf{p}_f)$, where $\mathbf{p}_i$ is the initial and $\mathbf{p}_f$ the final momentum, be the scattering amplitude in the first Born approximation for the scattering of an electron from a point-nucleus of charge Ze. Let $f(\mathbf{p}_i, \mathbf{p}_f)$ be the scattering amplitude of an electron from a real nucleus of the same charge. Let $\mathbf{q} = \mathbf{p}_i - \mathbf{p}_f$ denote the momentum transfer. The quantity F defined by

$$f(\mathbf{p}_i, \mathbf{p}_f) = F(q^2) f_c(\mathbf{p}_i, \mathbf{p}_f)$$

is called the form factor. It is easily seen that F, in fact, depends on $\mathbf{p}_i$ and $\mathbf{p}_f$ only through the quantity q^2.

(a) The form factor $F(q^2)$ and the Fourier transform of the charge density $\rho(\mathbf{x})$ are related in a very simple manner. State and derive this relationship within the framework of the nonrelativistic Schrödinger theory. The assumption that the electrons are "nonrelativistic" is here made so that the problem will be simplified. However, on careful consideration it will probably be clear that the assumption is irrelevant: the same result applies in the "relativistic" case of the actual experiment. It is also the case that the neglect of the electron spin does not affect the essence of what we are here concerned with.

(b) Figure 2.3 shows some experimental results pertaining to the form factor for the proton, and we shall regard our theory as applicable to these data. On the basis of the data shown, compute the root-mean-square (charge) radius of the proton. Hint: Note that there is a simple relationship between the root-mean-square radius and the derivative of $F(q^2)$ with respect to q^2, at $q^2 = 0$. Find this relationship, and then compute.

(UC, Berkeley)

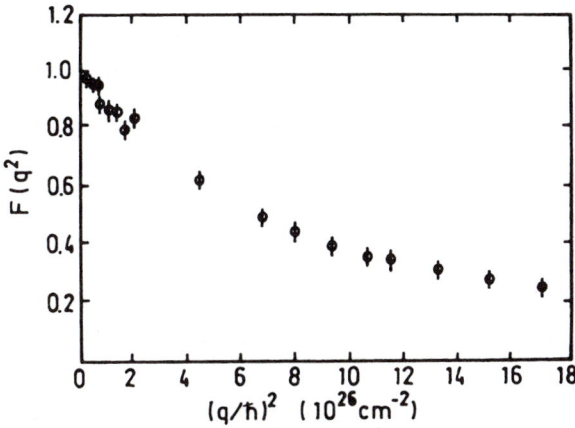

Fig. 2.3

Solution:

(a) In the first Born approximation, the scattering amplitude of a high energy electron from a nucleus is

$$f(\mathbf{p}_i, \mathbf{p}_f) = -\frac{m}{2\pi\hbar^2} \int V(\mathbf{x}) e^{i\mathbf{q}\cdot\mathbf{x}/\hbar} d^3x.$$

For a nucleus of spherically symmetric charge distribution, the potential at position $\mathbf{x}$ is

$$V(x) = \int \frac{\rho(r)Ze}{|\mathbf{x}-\mathbf{r}|} d^3r.$$

Thus

$$f(\mathbf{p}_i, \mathbf{p}_f) = -\frac{m}{2\pi\hbar^2} \int d^3x\, e^{i\mathbf{q}\cdot\mathbf{x}/\hbar} \int d^3r \frac{\rho(r)Ze}{|\mathbf{x}-\mathbf{r}|}$$

$$= -\frac{m}{2\pi\hbar^2} \int d^3r \rho(r) e^{i\mathbf{q}\cdot\mathbf{r}/\hbar} \int d^3x \frac{Ze}{|\mathbf{x}-\mathbf{r}|} e^{i\mathbf{q}\cdot(\mathbf{x}-\mathbf{r})/\hbar}$$

$$= -\frac{m}{2\pi\hbar^2} \int d^3r \rho(r) e^{i\mathbf{q}\cdot\mathbf{r}/\hbar} \int d^3x' \frac{Ze}{x'} e^{i\mathbf{q}\cdot\mathbf{x}'/\hbar}.$$

On the other hand, for a point nucleus we have $V(\mathbf{x}) = \frac{Ze}{x}$ and so

$$f_c(\mathbf{p}_i, \mathbf{p}_f) = -\frac{m}{2\pi\hbar^2} \int \frac{Ze}{x} e^{i\mathbf{q}\cdot\mathbf{x}/\hbar} d^3x.$$

Comparing the two equations above we obtain

$$f(\mathbf{p}_i, \mathbf{p}_f) = f_c(\mathbf{p}_i, \mathbf{p}_f) \int d^3 r \rho(r) e^{i\mathbf{q} \cdot \mathbf{r}/\hbar}$$

and hence

$$F(q^2) = \int d^3 r \rho(r) e^{i\mathbf{q} \cdot \mathbf{r}/\hbar} .$$

(b) When $q \approx 0$,

$$F(q^2) = \int \rho(r) e^{i\mathbf{q} \cdot \mathbf{r}/\hbar} d^3 r$$

$$\approx \int \rho(r) \left[1 + i\mathbf{q} \cdot \mathbf{r}/\hbar - \frac{1}{2}(\mathbf{q} \cdot \mathbf{r})^2/\hbar^2 \right] d^3 r$$

$$= \int \rho(r) d^3 r - \frac{1}{2} \int (\rho(r) q^2 r^2 \cos^2 \theta/\hbar^2) \cdot r^2 \sin \theta dr d\theta d\varphi$$

$$= F(0) - \frac{2\pi}{3} \frac{q^2}{\hbar^2} \int r^4 \rho(r) dr ,$$

i.e.,

$$F(q^2) - F(0) = -\frac{2\pi}{3} \frac{q^2}{\hbar^2} \int r^4 \rho(r) dr .$$

Note that $\frac{i}{\hbar} \int \rho(r) \mathbf{q} \cdot \mathbf{r} d^3 r = 0$ as $\int_0^\pi \cos \theta \sin \theta d\theta = 0$. The mean-square radius $\langle r^2 \rangle$ is by definition

$$\langle r^2 \rangle = \int d^3 r \rho(r) r^2 = 4\pi \int \rho(r) r^4 dr$$

$$= -6\hbar^2 \frac{F(q^2) - F(0)}{q^2} = -6\hbar^2 \left(\frac{\partial F}{\partial q^2} \right)_{q^2 = 0} .$$

From Fig. 2.3,

$$-\hbar^2 \left(\frac{\partial F}{\partial q^2} \right)_{q^2 = 0} \approx -\frac{0.8 - 1.0}{2 - 0} \times 10^{-26} = 0.1 \times 10^{-26} \text{ cm}^2$$

Hence $\langle r^2 \rangle = 0.6 \times 10^{-26}$ cm^2, or $\sqrt{\langle r^2 \rangle} = 0.77 \times 10^{-13}$ cm, i.e., the root-mean-square proton radius is 0.77 fm.

2022

The total (elastic+inelastic) proton-neutron cross section at center-of-mass momentum $p = 10$ GeV/c is $\sigma = 40$ mb.

(a) Disregarding nucleon spin, set a lower bound on the elastic center-of-mass proton-neutron forward differential cross-section.

(b) Assume experiments were to find a violation of this bound. What would this mean?

Solution:

(a) The forward $p - n$ differential cross section is given by

$$\frac{d\sigma}{d\Omega}\Big|_{0^\circ} = |f(0)|^2 \geq |\text{Im} f(0)|^2 = \left(\frac{k}{4\pi}\sigma_t\right)^2 ,$$

where the relation between $\text{Im} f(0)$ and σ_t is given by the optical theorem. As $k = p/\hbar$ we have

$$\frac{d\sigma}{d\Omega}\Big|_{0^\circ} \geq \left(\frac{pc}{4\pi\hbar c}\sigma_t\right)^2 = \left(\frac{10^4 \times 40 \times 10^{-27}}{4\pi \times 1.97 \times 10^{-11}}\right)^2$$

$$= 2.6 \times 10^{-24} \text{ cm}^2 = 2.6 \text{ } b .$$

(b) Such a result would mean a violation of the optical theorem, hence of the unitarity of the S-matrix, and hence of the probabilistic interpretation of quantum theory.

2023

When a 300-GeV proton beam strikes a hydrogen target (see Fig. 2.4), the elastic cross section is maximum in the forward direction. Away from the exact forward direction, the cross section is found to have a (first) minimum.

(a) What is the origin of this minimum? Estimate at what laboratory angle it should be located.

(b) If the beam energy is increased to 600 GeV, what would be the position of the minimum?

(c) If the target were lead instead of hydrogen, what would happen to the position of the minimum (beam energy= 300 GeV)?

(d) For lead, at what angle would you expect the second minimum to occur?

(*Chicago*)

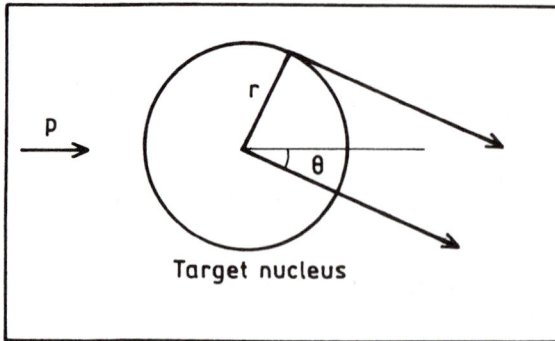

Fig. 2.4

Solution:

(a) The minimum in the elastic cross section arises from the destructive interference of waves resulting from scattering at different impact parameters. The wavelength of the incident proton, $\lambda = \frac{h}{p} = \frac{2\pi\hbar c}{pc} = \frac{2\pi \times 1.97 \times 10^{-11}}{300 \times 10^3} = 4.1 \times 10^{-16}$ cm, is much smaller than the size $\sim 10^{-13}$ cm of the target proton. The first minimum of the diffraction pattern will occur at an angle θ such that scattering from the center and scattering from the edge of the target proton are one-half wavelength out of phase, i.e.,

$$r\theta_{min} = \lambda/2 = 2.1 \times 10^{-16} \text{ cm}.$$

Thus, if $r = 1.0 \times 10^{-13}$ cm, the minimum occurs at 2.1×10^{-3} rad.

(b) If $E \rightarrow 600$ GeV/c, then $\lambda \rightarrow \lambda/2$ and $\theta_{min} \rightarrow \theta_{min}/2$ i.e., the minimum will occur at $\theta_{min} = 1.05 \times 10^{-3}$ rad.

(c) For $Pb : A = 208$, $r = 1.1 \times 208^{\frac{1}{3}} = 6.5$ fm, and we may expect the first minimum to occur at $\theta_{min} = 3.2 \times 10^{-4}$ rad.

(d) At the second minium, scattering from the center and scattering from the edge are $3/2$ wavelengths out of phase. Thus the second minimum will occur at $\theta_{min} = 3 \times 3.2 \times 10^{-4} = 9.6 \times 10^{-4}$ rad.

2. NUCLEAR BINDING ENERGY, FISSION AND FUSION (2024–2047)

2024

The semiempirical mass formula relates the mass of a nucleus, $M(A, Z)$, to the atomic number Z and the atomic weight A. Explain and justify each of the terms, giving approximate values for the magnitudes of the coefficients or constants in each term.

(Columbia)

Solution:

The mass of a nucleus, $M(Z, A)$, is

$$M(Z, A) = ZM(^1\text{H}) + (A - Z)m_n - B(Z, A),$$

where $B(Z, A)$ is the binding energy of the nucleus, given by the liquid-drop model as

$$B(Z, A) = B_v + B_s + B_e + B_a + B_p = a_v A - a_s A^{2/3} - a_e Z^2 A^{-1/3}$$

$$- a_a \left(\frac{A}{2} - Z\right)^2 A^{-1} + a_p \delta A^{-1/2},$$

where B_v, B_s, B_e are respectively the volume and surface energies and the electrostatic energy between the protons.

As the nuclear radius can be given as $r_0 A^{-1/3}$, r_0 being a constant, B_v, which is proportional to the volume of the nucleus, is proportional to A. Similarly the surface energy is proportional to $A^{2/3}$. The Coulomb energy is proportional to Z^2/R, and so to $Z^2 A^{-1/3}$.

Note that B_s arises because nucleus has a surface, where the nucleons interact with only, on the average, half as many nucleons as those in the interior, and may be considered as a correction to B_v.

B_a arises from the symmetry effect that for nuclides with mass number A, nuclei with $Z = \frac{A}{2}$ is most stable. A departure from this condition leads to instability and a smaller binding energy.

Lastly, neutrons and protons in a nucleus each have a tendency to exist in pairs. Thus nuclides with proton number and neutron number being even-even are the most stable; odd-odd, the least stable; even-odd or odd-even, intermediate in stability. This effect is accounted for by the pairing energy $B_p = a_p \delta A^{-1/2}$, where

$$\delta = \begin{cases} 1 & \text{for even-even nucleus,} \\ 0 & \text{for odd-even or even-odd nucleus,} \\ -1 & \text{for odd-odd nucleus.} \end{cases}$$

The values of the coefficients can be determined by a combination of theoretical calculations and adjustments to fit the experimental binding energy values. These have been determined to be

$$a_v = 15.835 \text{ MeV}, \qquad a_s = 18.33 \text{ MeV}, \qquad a_e = 0.714 \text{ MeV},$$

$$a_a = 92.80 \text{ MeV}, \qquad a_p = 11.20 \text{ MeV}.$$

2025

The nuclear binding energy may be approximated by the empirical expression

$$\text{B.E.} = a_1 A - a_2 A^{2/3} - a_3 Z^2 A^{-1/3} - a_4 (A - 2Z)^2 A^{-1}.$$

(a) Explain the various terms in this expression.

(b) Considering a set of isobaric nuclei, derive a relationship between A and Z for naturally occurring nuclei.

(c) Use a Fermi gas model to estimate the magnitude of a_4. You may assume $A \neq 2Z$ and that the nuclear radius is $R = R_0 A^{1/3}$.

(Princeton)

Solution:

(a) The terms in the expression represent volume, surface, Coulomb and symmetry energies, as explained in **Problem 2024** (where $a_a = 4a_4$).

(b) For isobaric nuclei of the same A and different Z, the stable nuclides should satisfy

$$\frac{\partial(B.E.)}{\partial Z} = -2A^{-1/3}a_3 Z + 4a_4 A^{-1}(A - 2Z) = 0,$$

giving

$$Z = \frac{A}{2 + \frac{a_3}{2a_4}A^{2/3}}.$$

With $a_3 = 0.714$ MeV, $a_4 = 23.20$ MeV,

$$Z = \frac{A}{2 + 0.0154 A^{2/3}} .$$

(c) A fermi gas of volume V at absolute temperature $T = 0$ has energy

$$E = \frac{2V}{h^3} \cdot \frac{4\pi}{5} \cdot \frac{p_0^5}{2m}$$

and particle number

$$N = \frac{2V}{h^3} \cdot \frac{4\pi}{3} \cdot p_0^3 ,$$

where we have assumed that each phase cell can accommodate two particles (neutrons or protons) of opposite spins. The limiting momentum is then

$$p_0 = h \left(\frac{3}{8\pi} \cdot \frac{N}{V} \right)^{\frac{1}{3}}$$

and the corresponding energy is

$$E = \frac{3}{40} \left(\frac{3}{\pi} \right)^{\frac{2}{3}} \frac{h^2}{m} V^{-\frac{2}{3}} N^{\frac{5}{3}} .$$

For nucleus (A, Z) consider the neutrons and protons as independent gases in the nuclear volume V. Then the energy of the lowest state is

$$E = \frac{3}{40} \left(\frac{3}{\pi} \right)^{2/3} \frac{h^2}{m} \frac{N^{5/3} + Z^{5/3}}{V^{2/3}}$$

$$= \frac{3}{40} \left(\frac{9}{4\pi^2} \right)^{2/3} \frac{h^2}{mR_0^2} \frac{N^{5/3} + Z^{5/3}}{A^{2/3}}$$

$$= C \frac{N^{5/3} + Z^{5/3}}{A^{2/3}} ,$$

where $V = \frac{4\pi}{3} R_0^3 A$, $R_0 \approx 1.2\,\text{fm}$, $C = \frac{3}{40} \left(\frac{9}{4\pi^2} \right)^{2/3} \frac{1}{mc^2} \left(\frac{hc}{R_0} \right)^2 = \frac{3}{40} \left(\frac{9}{4\pi^2} \right)^{\frac{2}{3}}$
$\times \frac{1}{940} \left(\frac{1238}{1.2} \right)^2 = 31.7\,\text{MeV}$.

For stable nuclei, $N + Z = A$, $N \approx Z$. Let $N = \frac{1}{2} A(1 + \varepsilon/A)$, $Z = \frac{1}{2} A(1 - \varepsilon/A)$, where $\frac{\varepsilon}{A} \ll 1$. As

$$\left(1 + \frac{\varepsilon}{A} \right)^{5/3} = 1 + \frac{5\varepsilon}{3A} + \frac{5\varepsilon^2}{9A^2} + \cdots ,$$

$$\left(1 - \frac{\varepsilon}{A} \right)^{5/3} = 1 - \frac{5\varepsilon}{3A} + \frac{5\varepsilon^2}{9A^2} - \cdots ,$$

we have

$$N^{\frac{5}{3}} + Z^{\frac{5}{3}} \approx 2 \left(\frac{A}{2} \right)^{\frac{5}{3}} \left(1 + \frac{5\varepsilon^2}{9A^2} \right)$$

and

$$E \approx 2^{-2/3} CA \left[1 + \frac{5\varepsilon^2}{9A^2} \right] = 2^{-2/3} CA + \frac{5}{9} \times 2^{-2/3} C \frac{(N-Z)^2}{A} .$$

The second term has the form $a_4 \frac{(N-Z)^2}{A}$ with

$$a_4 = \frac{5}{9} \times 2^{-2/3} C \approx 11 \text{ MeV} .$$

The result is smaller by a factor of 2 from that given in **Problem 2024**, where $a_4 = a_a/4 = 23.20$ MeV. This may be due to the crudeness of the model.

2026

The greatest binding energy per nucleon occurs near ^{56}Fe and is much less for ^{238}U. Explain this in terms of the semiempirical nuclear binding theory. State the semiempirical binding energy formula (you need not specify the values of the various coefficients).

(Columbia)

Solution:

The semiempirical formula for the binding energy of nucleus (A, Z) is

$$B(Z, A) = B_v + B_s + B_e + B_a + B_p = a_v A - a_s A^{2/3} - a_e Z^2 A^{-1/3}$$

$$- a_a \left(\frac{A}{2} - Z \right)^2 A^{-1} + a_p \delta A^{-1/2} .$$

The mean binding energy per nucleon is then

$$\varepsilon = B/A = a_v - a_s A^{-1/3} - a_e Z^2 A^{-4/3} - a_a \left(\frac{1}{2} - \frac{Z}{A} \right)^2 + a_p \delta A^{-3/2} .$$

Consider the five terms that contribute to ε. The contribution of the pairing energy (the last term) for the same A may be different for different combinations of Z, N, though it generally decreases with increasing A. The contribution of the volume energy, which is proportional to A, is a constant. The surface energy makes a negative contribution whose absolute value decreases with increasing A. The Coulomb energy also makes a negative contribution whose absolute value increases with A as Z and A increase together. The symmetry energy makes a negative contribution too, its absolute value increasing with A because Z/A decreases when A increases. Adding together these terms, we see that the mean binding energy increases with A at first, reaching a flat maximum at $A \sim 50$ and then decreases gradually, as shown in Fig. 2.5.

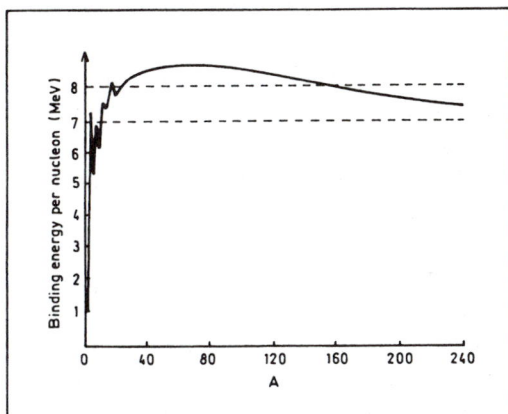

Fig. 2.5

2027

Draw a curve showing binding energy per nucleon as a function of nuclear mass. Give values in MeV, as accurately as you can. Where is the maximum of the curve? From the form of this curve explain nuclear fission and estimate the energy release per fission of ^{235}U. What force is principally responsible for the form of the curve in the upper mass region?

(Wisconsin)

Solution:

Figure 2.5 shows the mean binding energy per nucleon as a function of nuclear mass number A. The maximum occurs at $A \sim 50$. As A increases from 0, the curve rises sharply for $A < 30$, but with considerable fluctuations. Here the internucleon interactions have not reached saturation and there are not too many nucleons present so that the mean binding energy increases rapidly with the mass number. But because of the small number of nucleons, the pairing and symmetry effects significantly affect the mean binding energy causing it to fluctuate.

When $A > 30$, the mean binding energy goes beyond 8 MeV. As A increases further, the curve falls gradually. Here, with sufficient number of nucleons, internucleon forces become saturated and so the mean binding energy tends to saturate too. As the number of nucleons increases further, the mean binding energy decreases slowly because of the effect of Coulomb repulsion.

In nuclear fission a heavy nucleus dissociates into two medium nuclei. From the curve, we see that the products have higher mean binding energy. This excess energy is released. Suppose the fission of ^{235}U produces two nuclei of $A \sim 117$. The energy released is $235 \times (8.5 - 7.6) = 210$ MeV.

2028

Is the binding energy of nuclei more nearly proportional to $A(= N + Z)$ or to A^2? What is the numerical value of the coefficient involved (state units). How can this A dependence be understood? This implies an important property of nucleon-nucleon forces. What is it called? Why is a neutron bound in a nucleus stable against decay while a lambda particle in a hypernucleus is not?

(Wisconsin)

Solution:

The nuclear binding energy is more nearly proportional to A with a coefficient of 15.6 MeV. Because of the saturation property of nuclear forces, a nucleon can only interact with its immediate neighbors and hence with only a limited number of other nucleons. For this reason the binding energy is proportional to A, rather than to A^2, which would be the case if the nucleon interacts with all nucleons in the nuclues. Nuclear forces are therefore short-range forces.

The underlying cause of a decay is for a system to transit to a state of lower energy which is, generally, also more stable. A free neutron decays according to

$$n \to p + e + \bar{\nu}$$

and releases an energy

$$Q = m_n - m_p - m_e = 939.53 - 938.23 - 0.51 = 0.79 \text{ MeV}.$$

The decay of a bound neutron in a nucleus $^A X_N$ will result in a nucleus $^A X_{N-1}$. If the binding energy of $^A X_{N-1}$ is lower than that of $^A X_N$ and the difference is larger than 0.79 MeV, the decay would increase the system's energy and so cannot take place. Hence neutrons in many non-β-radioactive nuclei are stable. On the other hand, the decay energy of a Λ^0-particle, 37.75 MeV, is higher than the difference of nuclear binding energies between the initial and final systems, and so the Λ-particle in a hypernucleus will decay.

2029

Figure 2.5 shows a plot of the average binding energy per nucleon E vs. the mass number A. In the fission of a nucleus of mass number A_0 (mass M_0) into two nuclei A_1 and A_2 (masses M_1 and M_2), the energy released is

$$Q = M_0 c^2 - M_1 c^2 - M_2 c^2.$$

Express Q in terms of $\varepsilon(A)$ and A. Estimate Q for symmetric fission of a nucleus with $A_0 = 240$.

(Wisconsin)

Solution:

The mass of a nucleus of mass number A is

$$M = Z m_p + (A - Z) m_n - B/c^2,$$

where Z is its charge number, m_p and m_n are the proton and neutron masses respectively, B is the binding energy. As $Z_0 = Z_1 + Z_2$, $A_0 = A_1 + A_2$, and so $M_0 = M_1 + M_2 + (B_1 + B_2)/c^2 - B_0/c^2$, we have

$$Q = M_0 c^2 - M_1 c^2 - M_2 c^2 = B_1 + B_2 - B_0.$$

The binding energy of a nucleus is the product of the average binding energy and the mass number:

$$B = \varepsilon(A) \times A.$$

Hence

$$Q = B_1 + B_2 - B_0 = A_1\varepsilon(A_1) + A_2\varepsilon(A_2) - A_0\varepsilon(A_0).$$

With $A_0 = 240$, $A_1 = A_2 = 120$ in a symmetric fission, we have from Fig. 2.5

$$\varepsilon(120) \approx 8.5 \text{ MeV}, \qquad \varepsilon(240) \approx 7.6 \text{ MeV}.$$

So the energy released in the fission is

$$Q = 120\varepsilon(120) + 120\varepsilon(120) - 240\varepsilon(240) \approx 216 \text{ (MeV)}.$$

2030

(a) Construct an energy-versus-separation plot which can be used to explain nuclear fission. Describe qualitatively the relation of the features of this plot to the liquid-drop model.

(b) Where does the energy released in the fission of heavy elements come from?

(c) What prevents the common elements heavier than iron but lighter than lead from fissioning spontaneously?

(Wisconsin)

Solution:

(a) Nuclear fission can be explained using the curve of specific binding energy ε vs. nuclear mass number A (Fig. 2.5). As A increases from 0, the binding energy per nucleon E, after reaching a broad maximimium, decreases gradually. Within a large range of A, $\varepsilon \approx 8$ MeV/nucleon. The approximate linear dependence of the binding energy on A, which shows the saturation of nuclear forces (**Problems 2028**), agrees with the liquid-drop model.

(b) As a heavy nucleus dissociates into two medium nuclei in fission, the specific binding energy increases. The nuclear energy released is the difference between the binding energies before and after the fission:

$$Q = A_1\varepsilon(A_1) + A_2\varepsilon(A_2) - A\varepsilon(A),$$

where A, A_1 and A_2 are respectively the mass numbers of the nuclei before and after fission, $\varepsilon(A_i)$ being the specific binding energy of nucleus A_i.

(c) Although the elements heavier than iron but lighter than lead can release energy in fission if we consider specific binding energies alone, the Coulomb barriers prevent them from fissioning spontaneously. This is because the fission barriers of these nuclei are so high that the probability of penetration is very small.

2031

Stable nuclei have N and Z which lie close to the line shown roughly in Fig. 2.6.

(a) Qualitatively, what features determine the shape of this curve.

(b) In heavy nuclei the number of protons is considerably less than the number of neutrons. Explain.

(c) $^{14}O(Z = 8, N = 6)$ has a lifetime of 71 sec. Give the particles in the final state after its decay.

(*Wisconsin*)

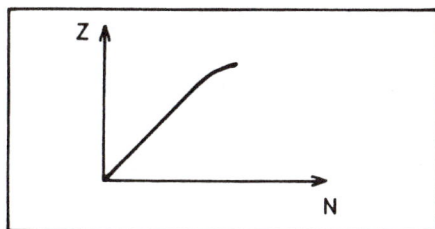

Fig. 2.6

Solution:

(a) Qualitatively, Pauli's exclusion principle allows four nucleons, 2 protons of opposite spins and 2 neutrons of opposite spins, to occupy the same energy level, forming a tightly bound system. If a nucleon is added, it would have to go to the next level and would not be so lightly bound. Thus the most stable nuclides are those with $N = Z$.

From binding energy considerations (**Problem 2025**), A and Z of a stable nuclide satisfy

$$Z = \frac{A}{2 + 0.0154A^{2/3}},$$

or, as $A = N + Z$,

$$N = Z(1 + 0.0154A^{2/3}).$$

This shows that for light nuclei, $N \approx Z$, while for heavy nuclei, $N > Z$, as shown in Fig. 2.6.

(b) For heavy nuclei, the many protons in the nucleus cause greater Coulomb repulsion. To form a stable nucleus, extra neutrons are needed to counter the Coulomb repulsion. This competes with the proton-neutron symmetry effect and causes the neutron-proton ratio in stable nuclei to increase with A. Hence the number of protons in heavy nuclei is considerably less than that of neutrons.

(c) As the number of protons in ^{14}O is greater than that of neutrons, and its half life is 71 s, the decay is a β^+ decay

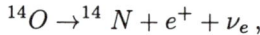

$$^{14}O \rightarrow^{14} N + e^+ + \nu_e,$$

the decay products being ^{14}N, e^+, and electron-neutrino. Another possible decay process is by electron capture. However, as the decay energy of ^{14}O is very large, $(E_{\max} > 4 \text{ MeV})$, the branching ratio of electron capture is very small.

2032

The numbers of protons and neutrons are roughly equal for stable lighter nuclei; however, the number of neutrons is substantially greater than the number of protons for stable heavy nuclei. For light nuclei, the energy required to remove a proton or a neutron from the nucleus is roughly the same; however, for heavy nuclei, more energy is required to remove a proton than a neutron. Explain these facts, assuming that the specific nuclear forces are exactly equal between all pairs of nucleons.

(Columbia)

Solution:

The energy required to remove a proton or a neutron from a stable nucleus (Z, A) is

$$S_p = B(Z, A) - B(Z - 1, A - 1),$$

or

$$S_n = B(Z, A) - B(Z, A - 1).$$

respectively, where B is the binding energy per nucleon of a nuclues. In the liquid-drop model (**Problem 2024**), we have

$$B(Z, A) = a_v A - a_s A^{2/3} - a_c Z^2 A^{-1/3} - a_a \left(\frac{A}{2} - Z\right)^2 A^{-1} + a_p \delta A^{-1/2}.$$

Hence

$$S_p - S_n = -a_c (2Z - 1)(A - 1)^{-\frac{1}{3}} + a_a (A - 2Z)(A - 1)^{-1},$$

where $a_c = 0.714$ MeV, $a_a = 92.8$ MeV. For stable nuclei (**Problem 2025**),

$$Z = \frac{A}{2 + \frac{2a_c}{a_a} A^{2/3}} \approx \frac{A}{2}\left(1 - \frac{a_c}{a_a} A^{2/3}\right),$$

and so

$$S_p - S_n \approx \frac{a_c}{A - 1}\left[A^{5/3} - (A - 1)^{5/3} + \frac{a_c}{a_a} A^{5/3}(A - 1)^{2/3}\right].$$

For heavy nuclei, $A \gg 1$ and $S_p - S_n \approx 5.5 \times 10^{-3} A^{4/3}$. Thus $S_p - S_n$ increases with A, i.e., to dissociate a proton from a heavy nucleus needs more energy than to dissociate a neutron.

2033

All of the heaviest naturally-occurring radioactive nuclei are basically unstable because of the Coulomb repulsion of their protons. The mechanism by which they decrease their size is alpha-decay. Why is alpha-decay favored over other modes of disintegration (like proton-, deuteron-, or triton-emission, or fission)? Discuss briefly in terms of
   (a) energy release, and
   (b) Coulomb barrier to be penetrated.

(Wisconsin)

Solution:

(a) A basic condition for a nucleus to decay is that the decay energy is larger than zero. For heavy nuclei however, the decay energy of proton-, deuteron- or triton-emission is normally less than zero. Take the isotopes and isotones of $^{238}_{95}$Am as an example. Consider the ten isotopes of Am. The proton-decay energies are between -3.9 MeV and -5.6 MeV, the deuteron-decay energies are between -7.7 MeV and -9.1 MeV, the triton-decay energies are between -7.6 MeV and -8.7 MeV, while the α-decay energies are between 5.2 MeV and 6.1 MeV. For the three isotones of $^{238}_{95}Am$, the proton-, deuteron- and triton-decay energies are less than zero while their α-decay energies are larger than zero. The probability for fission of a heavy nucleus is less than that for α-decay also because of its much lower probability of penetrating the Coulomb barrier. Therefore α-decay is favored over other modes of disintegration for a heavy nucleus.

(b) Figure 2.7 shows the Coulomb potential energy of a nucleus of charge Z_1e and a fragment of charge Z_2e.

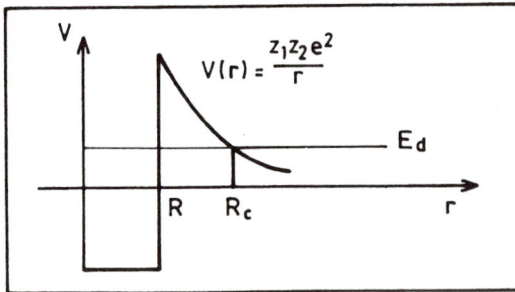

Fig. 2.7

Suppose a nucleus is to break up into two fragments of charges Z_1e and Z_2e. The probability of penetrating the Coulomb barrier by a fragment of energy E_d is

$$\exp\left(-\frac{2}{\hbar}\int_R^{R_c}\left[2\mu\left(\frac{Z_1Z_2e^2}{r}-E_d\right)\right]^{1/2}dr\right) = \exp(-G),$$

where μ is the reduced mass of the system,

$$R_c = \frac{Z_1Z_2e^2}{E_d},$$

and

$$G = \frac{2\sqrt{2\mu E_d}}{\hbar}\int_R^{R_c}\left(\frac{R_c}{r}-1\right)^{1/2}dr.$$

Integrating we have

$$\int_R^{R_c} \sqrt{\frac{R_c}{r} - 1}\, dr = R_c \int_1^{R_c/R} \frac{1}{p^2} \sqrt{p-1}\, dp$$

$$= R_c \left[-\frac{1}{p}\sqrt{p-1} + \tan^{-1}\sqrt{p-1} \right]_1^{R_c/R}$$

$$\approx R_c \left[\frac{\pi}{2} - \left(\frac{R}{R_c}\right)^{\frac{1}{2}} \right]$$

taking $\frac{R_c}{R} \gg 1$, and hence

$$G \approx \frac{2R_c\sqrt{2\mu E_d}}{\hbar} \left[\frac{\pi}{2} - \left(\frac{R}{R_c}\right)^{1/2} \right] \approx \frac{2Z_1 Z_2 e^2 \sqrt{2\mu}}{\hbar\sqrt{E_d}} \left[\frac{\pi}{2} - \left(\frac{R}{R_c}\right)^{1/2} \right].$$

For fission, though the energy release is some 50 times larger than that of α-decay, the reduced mass is 20 times larger and $Z_1 Z_2$ is 5 times larger. Then the value of G is 4 times larger and so the barrier penetrating probability is much lower than that for α-decay.

2034

Instability ('radioactivity') of atomic nuclei with respect to α-particle emission is a comparatively common phenomenon among the very heavy nuclei but proton-radioactivity is virtually nonexistent. Explain, with such relevant quantitative arguments as you can muster, this striking difference.

(Columbia)

Solution:

An explanation can be readily given in terms of the disintegration energies. In the α-decay of a heavy nucleus (A, Z) the energy release given by the liquid-drop model (**Problem 2024**) is

$$E_d = M(A, Z) - M(A - 4, Z - 2) - M(4, 2)$$

$$= -B(A, Z) + B(A - 4, Z - 2) + B(4, 2)$$

$$= -a_s[A^{2/3} - (A - 4)^{2/3}] - a_c[Z^2 A^{-\frac{1}{3}} - (Z - 2)^2 (A - 4)^{-\frac{1}{3}}]$$

$$- a_a \left[\left(\frac{A}{2} - Z \right)^2 A^{-1} - \left(\frac{A-4}{2} - Z + 2 \right)^2 (A-4)^{-1} \right]$$

$$+ B(4,2) - 4a_v .$$

For heavy nuclei, $\frac{2}{Z} \ll 1$, $\frac{4}{A} \ll 1$, and the above becomes

$$E_d \approx \frac{8}{3} a_s A^{-1/3} + 4 a_c Z A^{-\frac{1}{3}} \left(1 - \frac{Z}{3A} \right) - a_a \left(1 - \frac{2Z}{A} \right)^2 + 28.3 - 4a_v$$

$$= 48.88 A^{-1/3} + 2.856 Z A^{-1/3} \left(1 - \frac{Z}{3A} \right)$$

$$- 92.80 \left(1 - \frac{2Z}{A} \right)^2 - 35.04 \text{ MeV} .$$

For stable nuclei we have (**Problem 2025**)

$$Z = \frac{A}{2 + 0.0154 A^{2/3}} .$$

E_d is calculated for such nuclei and plotted as the dashed wave in Fig. 2.8.

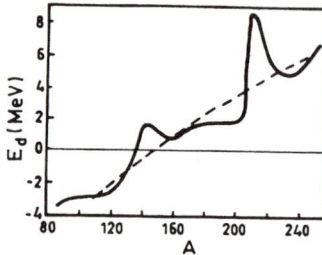

Fig. 2.8

For α-decay to take place, we require $E_d > 0$. It is seen that E_d increases generally with A and is positve when $A \geq 150$. Thus only heavy nuclei have α-decays. The actual values of E_d for naturally occurring nuclei are shown as the solid curve in the figure. It intersects the $E_d = 0$ line at $A \approx 140$, where α-radioactive isotopes $^{147}_{62}Sm$, $^{144}_{60}Nd$ are actually observed. For the proton-decay of a heavy nucleus, we have

$$M(A, Z) - M(A - 1, Z - 1) - M(0, 1)$$
$$= -B(A, Z) + B(A - 1, Z - 1) + B(0, 1)$$
$$\approx -B(A, Z) + B(A - 1, Z - 1) = -\varepsilon < 0,$$

where ε is the specific binding energy and is about 7 MeV for heavy nuclei. As the decay energy is negative, proton-decay cannot take place. However, this consideration is for stable heavy nuclei. For those nuclei far from stability curve, the neutron-proton ratio may be much smaller so that the binding energy of the last proton may be negative and proton-emission may occur. Quite different from neutron-emission, proton-emission is not a transient process but similar to α-decay; it has a finite half-life due to the Coulomb barrier. As the proton mass is less than the α-particle mass and the height of the Coulomb barrier it has to penetrate is only half that for the α-particle, the half-life against p-decay should be much less than that against α-decay. All proton-emitters should also have β^+-radioactivity and orbital-electron capture, and their half-lives are related to the probabilities of such competing proceses. Instances of proton-radioactivity in some isomeric states have been observed experimentally.

2035

(a) Derive argument for why heavy nuclei are α-radioactive but stable against neutron-emission.

(b) What methods and arguments are used to determine nuclear radii?

(c) What are the properties that identify a system of nucleons in its lowest energy state? Discuss the nonclassical properties.

(d) The fission cross sections of the following uranium ($Z = 92$) isotopes for thermal neutrons are shown in the table below.

Isotope	σ (barns)
^{230}U	20
^{231}U	300
^{232}U	76
^{233}U	530
^{234}U	0
^{235}U	580
^{236}U	0

The fast-neutron fission cross sections of the same isotopes are all of the order of a few barns, and the even-odd periodicity is much less pronounced. Explain these facts.

Solution:

(a) The reason why heavy nuclei only are α-radioactive has been discussed in **Problems 2033** and **2034**. For ordinary nuclei near the β-stability curve, the binding energy of the last neutron is positive so that no neutron-radioactivity exists naturally. However, for neutron-rich isotopes far from the β-stability curve, the binding energy may be negative for the last neutron, and so neutron-emission may occur spontaneously. As there is no Coulomb barrier for neutrons, emission is a transient process. Also, certain excited states arising from β-decays may emit neutrons. In such cases, as the neutron-emission follows a β-decay the emitted neutrons are called delayed neutrons. The half-life against delayed-neutron emission is the same as that against the related β-decay.

(b) There are two categories of methods for measuring nuclear radii. The first category makes use of the range of the strong interaction of nuclear forces by studying the scattering by nuclei of neutrons, protons or α-particles, particularly by measuring the total cross-section of intermediate-energy neutrons. Such methods give the nuclear radius as

$$R = R_0 A^{1/3}, \qquad R_0 \approx (1.4 \sim 1.5) \text{ fm}.$$

The other category of methods makes use of the Coulomb interaction between charged particles and atomic nuclei or that among particles within a nucleus to get the electromagnetic nuclear radius. By studying the scattering between high energy electrons and atomic nuclei, the form factors of the nuclei may be deduced which gives the electromagnetic nuclear radius. Assuming mirror nuclei to be of the same structure, their mass difference is caused by Coulomb energy difference and the mass difference between neutron and proton. We have **(Problem 2010)**

$$\Delta E = \frac{3}{5}\frac{e^2}{R}(2Z - 1) - (m_n - m_p)c^2$$

for the energy difference between the ground states of the mirror nuclei, which then gives the electromagnetic nuclear radius R. A more precise method is to study the deviation of μ-mesic atom from Bohr's model of hydrogen atom (**problem 1062**). Because the Bohr radius of the mesic atom is much smaller than that of the hydrogen atom, the former is more sensitive to the value of the electromagnetic nuclear radius, which, by this method, is

$$R = R_0 A^{1/3}, \qquad R_0 \approx 1.1 \text{ fm}.$$

High-energy electron scattering experiments show that charge distribution within a nucleus is nonuniform.

(c) The ground state of a system of nucleons is identified by its spin, parity and isospin quantum numbers.

Spin and parity are determined by those of the last one or two unpaired nucleons. For the ground state of an even-even nucleus, $J^p = 0^+$. For an even-odd nucleus, the nuclear spin and parity are determined by the last nucleon, and for an odd-odd nucleus, by the spin-orbit coupling of the last two nucleons.

The isospin of the nuclear ground state is $I = \frac{1}{2}|N - Z|$.

(d) There is a fission barrier of about 6 MeV for uranium so that spontaneous fission is unlikely and external inducement is required. At the same time, there is a tendency for neutrons in a nucleus to pair up so that isotopes with even numbers of neutrons, N, have higher binding energies. When an uranium isotope with an odd number of neutrons captures a neutron and becomes an isotope of even N, the excitation energy of the compound nucleus is large, sufficient to overcome the fission barrier, and fission occurs. On the other hand, when an even-N uranium isotope captures a neutron to become an isotope of odd N, the excitation energy of the compound nucleus is small, not sufficient to overcome the fission barrier, and fission does not take place. For example, in $^{235}U + n \rightarrow ^{236}U^*$ the excitation energy of the compound nucleus $^{236}U^*$ is 6.4 MeV, higher than the fission barrier of ^{236}U of 5.9 MeV, so the probability of this reaction results in a fission is large. In $^{238}U + n \rightarrow ^{239}U^*$, the excitation energy is only 4.8 MeV, lower than the fission barrier of 6.2 MeV of ^{239}U, and so the probability for fission is low. Such nuclides require neutrons of higher energies to achieve fission. When the neutron energy is higher than a certain threshold, fission cross section becomes large and fission may occur.

Thermal neutrons, which can cause fission when captured by odd-N uranium isotopes, have long wavelengths and hence large capture cross sections. Thus the cross sections for fission induced by thermal neutrons are large, in hundreds of barns, for uranium isotopes of odd N. They are small for isotope of even N.

If a fast neutron is captured by an uranium isotope the excitation energy of the compound nucleus is larger than the fission barrier and fission occurs irrespective of whether the isotope has an even or an odd number of neutrons. While fast neutrons have smaller probability of being captured their fission cross section, which is of the order of a few barns, do not change with the even-odd periodicity of the neutron number of the uranium isotope.

2036

The semiempirical mass formula modified for nuclear-shape eccentricity suggests a binding energy for the nucleus $^A_Z X$:

$$B = \alpha A - \beta A^{2/3} \left(1 + \frac{2}{5}\varepsilon^2\right) - \gamma Z^2 A^{-\frac{1}{3}} \left(1 - \frac{1}{5}\varepsilon^2\right),$$

where α, β, $\gamma = 14, 13, 0.6$ MeV and ε is the eccentricity.

(a) Briefly interpret this equation and find a limiting condition involving Z and A such that a nucleus can undergo prompt (unhindered) spontaneous fission. Consider $^{240}_{94}Pu$ as a specific example.

(b) The discovery of fission shape isomers and the detection of spontaneous fission of heavy isotopes from their ground state suggest a more complicated nuclear potential energy function $V(\varepsilon)$. What simple nuclear excitations can account for the two sets of states of $^{240}_{94}Pu$ shown below (Fig. 2.9). Discuss similarities and differences between the two. What are the implications for $V(\varepsilon)$? Draw a rough sketch of $V(\varepsilon)$.

(Princeton)

Solution:

(a) In the mass formula, the first term represents volume energy, the second term surface energy, in which the correction $\frac{2}{5}\varepsilon^2$ is for deformation from spherical shape of the nucleus, the third term, the Coulomb energy, in which the correction $\frac{1}{5}\varepsilon^2$ is also for nucleus deformation. Consequent to nuclear shape deformation, the binding energy is a function of the eccentricity ε. The limiting condition for stability is $\frac{dB}{d\varepsilon} = 0$. We have

$$\frac{dB}{d\varepsilon} = -\frac{4\beta}{5}A^{2/3}\varepsilon + \gamma\frac{Z^2}{A^{1/3}} \cdot \frac{2}{5}\varepsilon = \frac{2}{5}\varepsilon A^{2/3}\left(\frac{\gamma Z^2}{A} - 2\beta\right).$$

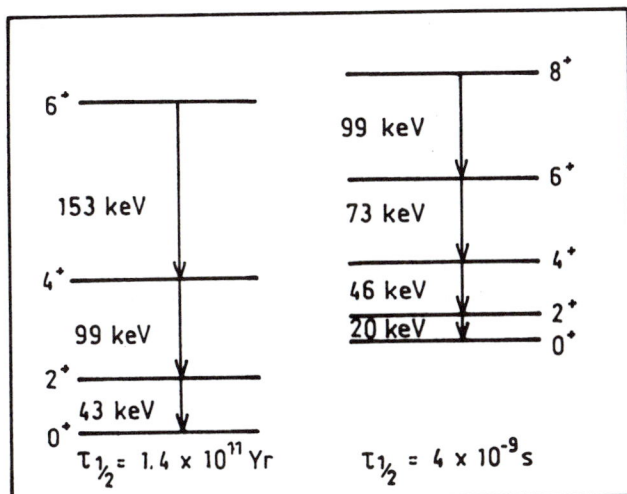

Fig. 2.9

If $\frac{dB}{d\varepsilon} > 0$, nuclear binding energy increases with ε so the deformation will keep on increasing and the nucleus becomes unstable. If $\frac{dB}{d\varepsilon} < 0$, binding energy decreases as ε increases so the nuclear shape will tend to that with a lower ε and the nucleus is stable. So the limiting condition for the nucleus to undergo prompt spontaneous fission is $\frac{d\beta}{d\varepsilon} > 0$, or

$$\frac{Z^2}{A} \geq \frac{2\beta}{\gamma} = 43.3 \,.$$

For ^{240}Pu, $\frac{Z^2}{A} = 36.8 < 43.3$ and so it cannot undergo prompt spontaneous fission; it has a finite lifetime against spontaneous fission.

(b) The two sets of energy levels of ^{240}Pu (see Fig. 2.9) can be interpreted in terms of collective rotational excitation of the deformed nucleus, as each set satisfies the rotational spectrum relation for the $K = 0$ rotational band

$$E_I = \frac{\hbar^2}{2M}[I(I+1)] \,.$$

Both sets of states show characteristics of the rotational spectrums of even-even nuclei; they differ in that the two rotational bands correspond to different rotational moments of inertia M. The given data correspond to

$\frac{\hbar^2}{2J} \approx 7$ MeV for the first set, $\frac{\hbar^2}{2J} \approx 3.3$ MeV for the second set. The different moments of inertia suggest different deformations. Use of a liquid-drop shell model gives a potential $V(\varepsilon)$ in the form of a two-peak barrier, as shown in Fig. 2.10. The set of states with the longer lifetime corresponds to the ground-state rotational band at the first minimum of the two-peak potential barrier. This state has a thicker fission barrier to penetrate and hence a longer lifetime ($T_{1/2} = 1.4 \times 10^{11}$ yr for ^{240}Pu). The set of rotational band with the shorter lifetime occurs at the second minimum of the potential barrier. In this state the fission barrier to penetrate is thinner, hence the shorter lifetime ($T_{1/2} = 4 \times 10^{-9}$s for ^{240}Pu). The difference between the two rotational bands arises from the different deformations; hence the phenomenon is referred to as nuclear shape isomerism.

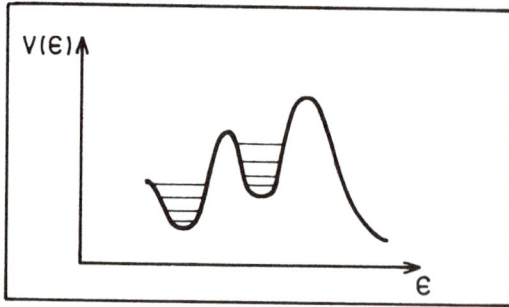

Fig. 2.10

2037

Assume a uranium nucleus breaks up spontaneously into two roughly equal parts. Estimate the reduction in electrostatic energy of the nuclei. What is the relationship of this to the total change in energy? (Assume uniform charge distribution; nuclear radius= $1.2 \times 10^{-13} A^{1/3}$ cm)

(*Columbia*)

Solution:

Uranium nucleus has $Z_0 = 92$, $A_0 = 236$, and radius $R_0 = 1.2 \times 10^{-13} A_0^{1/3}$ cm. When it breaks up into to two roughly equal parts, each part has

$$Z = \frac{1}{2}Z_0, \ A = \frac{1}{2}A_0, \ R = 1.2 \times 10^{-13}A^{1/3} \text{ cm}.$$

The electrostatic energy of a sphere of a uniformly distributed charge Q is $\frac{3}{5}Q^2/R$, where R is the radius. Then for uranium fission, the electrostatic energy reduction is

$$\Delta E = \frac{3}{5}\left[\frac{(Z_0 e)^2}{R_0} - 2 \times \frac{(Ze)^2}{R}\right]$$

$$= \frac{3 \times Z_0^2 e^2}{5}\frac{1}{R_0}\left[1 - \frac{1}{2^{2/3}}\right] = 0.222 \times \frac{Z_0^2}{R_0}\left(\frac{e^2}{\hbar c}\right)\hbar c$$

$$= \frac{0.222 \times 92^2}{1.2 \times 10^{-13} \times 236^{\frac{1}{3}}} \times \frac{1}{137} \times 1.97 \times 10^{-11}$$

$$= 364 \text{ MeV}.$$

This reduction is the source of the energy released in uranium fission. However, to calculate the actual energy release, some other factors should also be considered such as the increase of surface energy on fission.

2038

Estimate (order of magnitude) the ratio of the energy released when 1 g of uranium undergoes fission to the energy released when 1 g of TNT explodes.

(Columbia)

Solution:

Fission is related to nuclear forces whose interaction energy is about 1 MeV/nucleon. TNT explosion is related to electromagnetic forces whose interaction energy is about 1 eV/molecule. As the number of nucleons in 1 g of uranium is of the same order of magnitude as the number of molecules in 1 g of TNT, the ratio of energy releases should be about 10^6.

A more precise estimate is as follows. The energy released in the explosion of 1 g of TNT is about 2.6×10^{22} eV. The energy released in the fission of a uranium nucleus is about 210 MeV. Then the fission of 1 g of uranium releases an energy $\frac{6.023 \times 10^{23}}{238} \times 210 = 5.3 \times 10^{23}$ MeV. Hence the ratio is about 2×10^7.

2039

The neutron density $\rho(\mathbf{x}, t)$ inside a block of U^{235} obeys the differential equation

$$\frac{\partial \rho(x, t)}{\partial t} = A\nabla^2 \rho(\mathbf{x}, t) + B\rho(\mathbf{x}, t),$$

where A and B are positive constants. Consider a block of U^{235} in the shape of a cube of side L. Assume that those neutrons reaching the cube's surface leave the cube immediately so that the neutron density at the U^{235} surface is always zero.

(a) Briefly describe the physical processes which give rise to the $A\nabla^2\rho$ and the $B\rho$ terms. In particular, explain why A and B are both positive.

(b) There is a critical length L_0 for the sides of the U^{235} cube. For $L > L_0$, the neutron density in the cube is unstable and increases exponentially with time — an explosion results. For $L < L_0$, the neutron density decreases with time — there is no explosion. Find the critical length L_0 in terms of A and B.

(Columbia)

Solution:

(a) The term $B\rho(\mathbf{x}, t)$, which is proportional to the neutron density, accounts for the increase of neutron density during nuclear fission. $B\rho(\mathbf{x}, t)$ represents the rate of increase of the number of neutrons, in a unit volume at location $\mathbf{x}$ and at time t, caused by nuclear fission. It is proportional to the number density of neutrons which induce the fission. As the fission of U^{235} increases the neutron number, B is positive. The term $A\nabla^2\rho(\mathbf{x}, t)$ describes the macroscopic motion of neutrons caused by the nonuniformity of neutron distribution. As the neutrons generally move from locations of higher density to locations of lower density, A is positive too.

(b) Take a vertex of the cube as the origin, and its three sides as the x-, y- and z-axes. Let $\rho(\mathbf{x}, t) = f(x, y, z)e^{-\alpha t}$. Then the differential equation becomes

$$A\nabla^2 f(x, y, z) + (\alpha + B)f(x, y, z) = 0$$

with boundary condition

$$f(x, y, z)|_{i=0, L} = 0, \qquad i = x, y, z.$$

Try a solution of the form $f = X(x)Y(y)Z(z)$. Substitution gives

$$\frac{1}{X}\frac{d^2X}{dx^2} + \frac{1}{Y}\frac{d^2Y}{dy^2} + \frac{1}{Z}\frac{d^2Z}{dz^2} + k_x^2 + k_y^2 + k_z^2 = 0,$$

where we have rewritten $\frac{\alpha+B}{A} = k_x^2 + k_y^2 + k_z^2$. The boundary condition becomes

$$X(x) = 0 \text{ at } x = 0, L; \quad Y(y) = 0 \text{ at } y = 0, L; \quad Z(z) = 0 \text{ at } z = 0, L.$$

The last differentiation equation can be separated into 3 equations:

$$\frac{d^2X}{dx^2} + k_x^2 X = 0, \quad \text{etc.}$$

The solutions of these equations are

$$X = C_{xi} \sin\left(\frac{n_{xi}\pi}{L}x\right),$$

$$Y = C_{yj} \sin\left(\frac{n_{yj}\pi}{L}y\right),$$

$$Z = C_{zk} \sin\left(\frac{n_{zk}\pi}{L}x\right),$$

with $n_{xi}, n_{yj}, n_{zk} = \pm 1, \pm 2, \pm 3 \ldots$ and C_{xi}, C_{yj}, C_{zk} being arbitrary constants. Thus

$$f(x,y,z) = \sum_{ijk} C_{ijk} \sin\left(\frac{n_{xi}\pi}{L}x\right) \sin\left(\frac{n_{yj}\pi}{L}y\right) \sin\left(\frac{n_{zk}\pi}{L}z\right),$$

with

$$\frac{\alpha+B}{A} = \left(\frac{\pi}{L}\right)^2 (n_{xi}^2 + n_{yj}^2 + n_{zk}^2), \qquad C_{ijk} = C_{zi}C_{yi}C_{zk}.$$

If $\alpha < 0$, the neutron density will increase exponentially with time, leading to instability and possible explosion. Hence the crital length L_0 is given by

$$\alpha = \frac{A\pi^2}{L_0^2}(n_{xi}^2 + n_{yj}^2 + n_{zk}^2) - B = 0,$$

or

$$L_0 = \pi\sqrt{\frac{A}{B}(n_{xi}^2 + n_{yj}^2 + n_{zk}^2)}.$$

In particular, for $n_{xi} = n_{yj} = n_{zk} = 1$,

$$L_0 = \pi \sqrt{\frac{3A}{B}} \, .$$

2040

The half-life of U^{235} is $10^3, 10^6, 10^9, 10^{12}$ years.

(Columbia)

Solution:

10^9 years. (Half-life of U^{235} is 7×10^8 years)

2041

Number of fission per second in a 100-MW reactor is: 10^6, 10^{12}, 10^{18}, 10^{24}, 10^{30}.

(Columbia)

Solution:

Each fission of uranium nucleus releases about 200 MeV $= 320 \times 10^{-13}$ J. So the number of fissions per second in a 100-MW reactor is

$$N = \frac{100 \times 10^6}{320 \times 10^{-13}} = 3 \times 10^{18} \, .$$

Hence the answer is 10^{18}.

2042

Explain briefly the operation of a "breeder" reactor. What physical constant of the fission process is a prerequisite to the possibility of "breeding"? What important constraint is placed on the choice of materials in the reactor? In particular, could water be used as a moderator?

(Wisconsin)

Solution:

A breeder reactor contains a fissionable material and a nonfissionable one that can be made fissionable by absorbing a neutron. For example,

^{235}U and ^{238}U. Suppose 3 neutrons are emitted per fission. One is needed to induce a fission in another fuel atom and keep the chain reaction going. If the other two neutrons can be used to convert two nonfissionable atoms into fissionable ones, then two fuel atoms are produced when one is consumed, and the reactor is said to be a breeder.

In the example, neutrons from the fission of ^{235}U may be used to convert ^{238}U to fissionable ^{239}Pu:

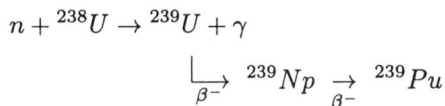

$$n + {}^{238}U \rightarrow {}^{239}U + \gamma$$
$$\phantom{n + {}^{238}U \rightarrow} \xrightarrow[\beta^-]{} {}^{239}Np \xrightarrow[\beta^-]{} {}^{239}Pu$$

A prerequisite to breeding is that η, the number of neutrons produced per neutron absorbed in the fuel, should be larger than 2. In the example, this is achieved by the use of fast neutrons and so no moderator is needed.

2043

(a) Describe briefly the type of reaction on which a nuclear fission reactor operates.

(b) Why is energy released, and roughly how much per reaction?

(c) Why are the reaction products radioactive?

(d) Why is a "moderator" necessary? Are light or heavy elements preferred for moderators, and why?

(Wisconsin)

Solution:

(a) In nuclear fission a heavy nucleus disassociates into two medium nuclei. In a reactor the fission is induced. It takes place after a heavy nucleus captures a neutron. For example

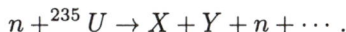

$$n + {}^{235}U \rightarrow X + Y + n + \cdots.$$

(b) The specific binding energy of a heavy nucleus is about 7.6 MeV per nucleon, while that of a medium nucleus is about 8.5 MeV per nucleon. Hence when a fission occurs, some binding energies will be released. The energy released per fission is about 210 MeV.

(c) Fission releases a large quantity of energy, some of which is in the form of excitation energies of the fragments. Hence fission fragments are in general highly excited and decay through γ emission. In addition, the neutron-to-proton ratios of the fragments, which are similar to that of the original heavy nucleus, are much larger than those of stable nuclei of the same mass. So the fragments are mostly unstable neutron-rich isotopes having strong β^- radioactivity.

(d) For reactors using ^{235}U, fission is caused mainly by thermal neutrons. However, fission reaction emits fast neutrons; so some moderator is needed to reduce the speed of the neutrons. Lighter nuclei are more suitable as moderator because the energy lost by a neutron per neutron-nucleus collision is larger if the nucleus is lighter.

2044

Give the three nuclear reactions currently considered for controlled thermonuclear fusion. Which has the largest cross section? Give the approximate energies released in the reactions. How would any resulting neutrons be used?

(Wisconsin)

Solution:

Reactions often considered for controlled thermonuclear fusion are

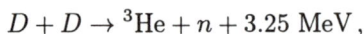

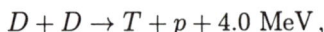

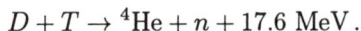

$$D + D \rightarrow {}^3\text{He} + n + 3.25 \text{ MeV},$$

$$D + D \rightarrow T + p + 4.0 \text{ MeV},$$

$$D + T \rightarrow {}^4\text{He} + n + 17.6 \text{ MeV}.$$

The cross section of the last reaction is the largest.

Neutrons resulting from the reactions can be used to induce fission in a fission-fusion reactor, or to take part in reactions like $^6\text{Li} + n \rightarrow {}^4\text{He} + T$ to release more energy.

2045

Discuss thermonuclear reactions. Give examples of reactions of importance in the sun, the H bomb and in controlled fusion attempts. Estimate roughly in electron volts the energy release per reaction and give the characteristic of nuclear forces most important in these reactions.

(Wisconsin)

Solution:

The most important thermonuclear reactions in the sun are the proton-proton chain

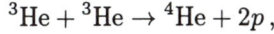

$$p + p \rightarrow d + e^+ + \nu_e \,,$$

$$d + p \rightarrow {}^3\text{He} + \gamma \,,$$

$${}^3\text{He} + {}^3\text{He} \rightarrow {}^4\text{He} + 2p \,,$$

the resulting reaction being

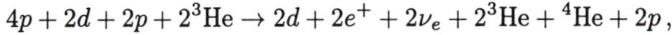

$$4p + 2d + 2p + 2\,{}^3\text{He} \rightarrow 2d + 2e^+ + 2\nu_e + 2\,{}^3\text{He} + {}^4\text{He} + 2p \,,$$

or

$$4p \rightarrow {}^4\text{He} + 2e^+ + 2\nu_e \,.$$

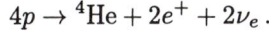

The energy released in this reaction is roughly

$$Q = [4M({}^1\text{H}) - M({}^4\text{He})]c^2 = 4 \times 1.008142 - 4.003860$$

$$= 0.02871 \text{ amu} = 26.9 \text{ MeV} \,.$$

The explosive in a H bomb is a mixture of deuterium, tritium and lithium in some condensed form. H bomb explosion is an uncontrolled thermonuclear reaction which releases a great quantity of energy at the instant of explosion. The reaction chain is

$${}^6\text{Li} + n \rightarrow {}^4\text{He} + t \,,$$

$$D + t \rightarrow {}^4\text{He} + n \,,$$

with the resulting reaction

$${}^6\text{Li} + d \rightarrow 2\,{}^4\text{He} \,.$$

The energy released per reaction is

$$Q = [M({}^6\text{Li}) + M({}^2\text{H}) - 2M({}^4\text{He})]c^2$$

$$= 6.01690 + 2.01471 - 2 \times 4.00388$$

$$= 0.02385 \text{ amu} = 22.4 \text{ MeV} \,.$$

An example of possible controlled fusion is

$$t + d \rightarrow {}^4\text{He} + n\,,$$

where the energy released is

$$Q = [M({}^3\text{H}) + M({}^2\text{H}) - M({}^4\text{He}) - M(n)]c^2$$

$$= 3.01695 + 2.01471 - 4.00388 - 1.00896$$

$$= 0.01882 \text{ amu} = 17.65 \text{ MeV}\,.$$

The most important characteristic of nuclear forces in these reactions is saturation, which means that a nucleon interacts only with nucleons in its immediate neighborhood. So while the nuclear interactions of a nucleon in the interior of a nucleus are saturated, the interactions of a nucleon on the surface of the nucleus are not. Then as the ratio of the number of nucleons on the nucleus surface to that of those in the interior is larger for lighter nuclei, the mean binding energy per nucleon for a lighter nucleus is smaller than for a heavier nucleus. In other words nucleons in lighter nuclei are combined more loosely. However, because of the effect of the Coulomb energy of the protons, the mean binding energies of very heavy nuclei are less than those of medium nuclei.

2046

For some years now, R. Davis and collaborators have been searching for solar neutrinos, in a celebrated experiment that employs as detector a large tank of C_2Cl_4 located below ground in the Homestake mine. The idea is to look for argon atoms (A^{37}) produced by the inverse β-decay reaction $Cl^{37}(\nu, e^-)Ar^{37}$. This reaction, owing to threshold effects, is relatively insensitive to low energy neutrinos, which constitute the expected principal component of neutrinos from the sun. It is supposed to respond to a smaller component of higher energy neutrinos expected from the sun. The solar constant (radiant energy flux at the earth) is $\sim 1 \ kW/m^2$.

(a) Outline the principal sequence of nuclear processes presumed to account for energy generation in the sun. What is the slow link in the chain? Estimate the mean energy of the neutrinos produced in this chain.

What is the expected number flux at the earth of the principal component of solar neutrinos?

(b) Outline the sequence of minor nuclear reactions that is supposed to generate the higher energy component of the neutrino spectrum, the component being looked for in the above experiment. Briefly discuss the experiment itself, and the findings to date.

(Princeton)

Solution:

(a) The principal sequence of nuclear processes presumed to generate solar energy is

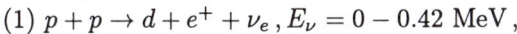

(1) $p + p \rightarrow d + e^+ + \nu_e$, $E_\nu = 0 - 0.42$ MeV,
(2) $d + p \rightarrow {}^3\text{He} + \gamma$,
(3) ${}^3\text{He} + {}^3\text{He} \rightarrow {}^4\text{He} + 2p$,

The resulting reaction being $4p \rightarrow {}^4\text{He} + 2e^+ + 2\nu_e + 26.7$ MeV.

The reaction (1) is the slow link. About 25 MeV of the energy changes into thermal energy in the sequence, the rest being taken up by the neutrinos. So the mean energy of a neutrino is

$$E_\nu \approx (26.7 - 25)/2 \approx 0.85 \text{ MeV}.$$

As each 25 MeV of solar energy arriving on earth is accompanied by 2 neutrions, the number flux of solar neutrinos at the earth is

$$I = 2\left(\frac{1 \times 10^3}{25 \times 1.6 \times 10^{-13}}\right) = 5 \times 10^{14} \ m^{-2}s^{-1}.$$

(b) The minor processes in the sequence are

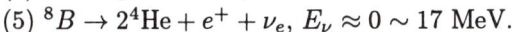

(1) ${}^3\text{He} + {}^4\text{He} \rightarrow {}^7Be + \gamma$,
(2) ${}^7Be + e^- \rightarrow {}^7\text{Li} + \nu_e$, $E_\nu = 0.478$ MeV(12%) and 0.861 MeV (88%),
(3) ${}^7\text{Li} + p \rightarrow 2{}^4\text{He}$,
(4) ${}^7Be + p \rightarrow {}^8B + \gamma$,
(5) ${}^8B \rightarrow 2{}^4\text{He} + e^+ + \nu_e$, $E_\nu \approx 0 \sim 17$ MeV.

The high energy neutrinos produced in the ${}^8B$ decay are those being measured in the experiment

In the experiment of Davis et al, a tank of 390000 liters of C_2Cl_4 was placed in a mine 1.5 kilometers below ground, to reduce the cosmic-ray background. The threshold energy for the reaction between solar neutrino and Cl, $\nu_e + {}^{37}Cl \rightarrow e^- + {}^{37}Ar$, is 0.814 MeV. The Ar gas produced then decays by electron capture, $e^- + {}^{37}Ar \rightarrow \nu_e + {}^{37}Cl$, the energy of the Auger electron emitted following this process being 2.8 keV. The half-life of Ar against the decay is 35 days. When the Ar gas produced, which had accumulated in the tank for several months, was taken out and its radioactivity measured with a proportional counter, the result was only one-third of what had been theoretically expected. This was the celebrated case of the "missing solar neutrinos". Many possible explanations have been proposed, such as experimental errors, faulty theories, or "neutrinos oscillation", etc.

2047

In a crude, but not unreasonable, approximation, a neutron star is a sphere which consists almost entirely of neutrons which form a nonrelativistic degenerate Fermi gas. The pressure of the Fermi gas is counterbalanced by gravitational attraction.

(a) Estimate the radius of such a star to within an order of magnitude if the mass is 10^{33} g. Since only a rough numerical estimate is required, you need to make only reasonable simplifying assumptions like taking a uniform density, and estimate integrals you cannot easily evaluate, etc. (Knowing the answer is not enough here; you must derive it.)

(b) In the laboratory, neutrons are unstable, decaying according to $n \rightarrow p + e + \nu + 1$ MeV with a lifetime of 1000 s. Explain briefly and qualitatively, but precisely, why we can consider the neutron star to be made up almost entirely of neutrons, rather than neutrons, protons, and electrons.

(Columbia)

Solution:

(a) Let R be the radius of the neutron star. The gravitational potential energy is

$$V_g = -\int_0^R \frac{4}{3}\pi r^3 \rho \left(\frac{G}{r}\right) 4\pi r^2 \rho \, dr = -\frac{3}{5}\frac{GM^2}{R},$$

where $\rho = \frac{3M}{4\pi R^3}$ is the density of the gas, M being its total mass, G is the gravitational constant. When R increases by ΔR, the pressure P of the gas does an external work $\Delta W = P\Delta V = 4\pi P R^2 \Delta R$. As $\Delta W = -\Delta V_g$, we have

$$P = \frac{3GM^2}{20\pi R^4} .$$

The pressure of a completely degenerate Fermi gas is

$$P = \frac{2}{5} N E_f ,$$

where $N = \frac{\rho}{M_n}$ is the neutron number density, M_n being the neutron mass,

$$E_f = \frac{\hbar^2}{2M_n} \left(\frac{9\pi}{4} \frac{M}{M_n R^3} \right)^{2/3}$$

is the limiting energy. Equating the expressions for P gives

$$R = \left(\frac{9\pi}{4} \right)^{\frac{2}{3}} \frac{\hbar^2}{GM_n^3} \left(\frac{M_n}{M} \right)^{\frac{1}{3}}$$

$$= \left(\frac{9\pi}{4} \right)^{\frac{2}{3}} \times \frac{(1.05 \times 10^{-34})^2}{6.67 \times 10^{-11} \times (1.67 \times 10^{-27})^3} \times \left(\frac{1.67 \times 10^{-27}}{10^{30}} \right)^{\frac{1}{3}}$$

$$= 1.6 \times 10^4 \ m .$$

(b) Let d be the distance between neighboring neutrons. As $\frac{M}{M_n} \approx \left(\frac{2R}{d} \right)^3$, $d \approx 2R \left(\frac{M_n}{M} \right)^{\frac{1}{3}} = 4 \times 10^{-15}$ m. If electrons existed in the star, the magnitude of their mean free path would be of the order of d, and so the order of magnitude of the kinetic energy of an electron would be $E \approx cp \sim c\hbar/d \sim 50$ MeV. Since each neutron decay only gives out 1 MeV, and the neutron's kinetic energy is less than $E_f \approx 21$ MeV, it is unlikely that there could be electrons in the neutron star originating from the decay of neutrons, if energy conservation is to hold. Furthermore, because the neutrons are so close together, e and p from a decay would immediately recombine. Thus there would be no protons in the star also.

3. THE DEUTERON AND NUCLEAR FORCES (2048–2058)

2048

If the nuclear force is charge independent and a neutron and a proton form a bound state, then why is there no bound state for two neutrons? What information does this provide on the nucleon-nucleon force?

(Wisconsin)

Solution:

A system of a neutron and a proton can form either singlet or triplet spin state. The bound state is the triplet state because the energy level of the singlet state is higher. A system of two neutrons which are in the same energy level can form only singlet spin state, and no bound state is possible. This shows the spin dependency of the nuclear force.

2049

A deuteron of mass M and binding energy $B(B \ll Mc^2)$ is disintegrated into a neutron and a proton by a gamma ray of energy E_γ. Find, to lowest order in B/Mc^2, the minimum value of $(E_\gamma - B)$ for which the reaction can occur.

(Wisconsin)

Solution:

In the disintegration of the deuteron, $E_\gamma - B$ is smallest when E_γ is at threshold, at which the final particles are stationary in the center-of-mass system. In this case the energy of the incident photon in the center-of-mass system of the deuteron is $E^* = (m_n + m_p)c^2$.

Let M be the mass of the deuteron. As $E^2 - p^2c^2$ is Lorentz-invariant and $B = (m_n + m_p - M)c^2$, we have

$$(E_\gamma + Mc^2) - E_\gamma^2 = (m_n + m_p)^2 c^4 ,$$

i.e.,

$$2E_\gamma Mc^2 = [(m_n + m_p)^2 - M^2]c^4 = (B + 2Mc^2)B ,$$

or

$$E_\gamma - B = \frac{B^2}{2Mc^2} ,$$

which is the minimum value of $E_\gamma - B$ for the reaction to occur.

2050

According to a simple-minded picture, the neutron and proton in a deuteron interact through a square well potential of width $b = 1.9 \times 10^{-15}$ m and depth $V_0 = 40$ MeV in an $l = 0$ state.

(a) Calculate the probability that the proton moves within the range of the neutron. Use the approximation that $m_n = m_p = M$, $kb = \frac{\pi}{2}$, where $k = \sqrt{\frac{M(V_0 - \varepsilon)}{\hbar^2}}$ and ε is the binding energy of the deuteron.

(b) Find the mean-square radius of the deuteron.

Solution:

The interaction may be considered as between two particles of mass M, so the reduced mass is $\mu = \frac{1}{2}M$. The potential energy is

$$V(r) = \begin{cases} -V_0, & r < b, \\ 0, & r > b, \end{cases}$$

where r is the distance between the proton and the neutron. The system's energy is $E = -\varepsilon$.

For $l = 0$ states, let the wave function be $\Psi = u(r)/r$. The radial Schrödinger equation

$$u'' + \frac{2\mu}{\hbar^2}(E - V)u = 0$$

can be written as

$$u'' + k^2 u = 0, \qquad r \leq b,$$

$$u'' - k_1^2 u = 0, \qquad r > b,$$

where

$$k = \sqrt{\frac{M(V_0 - \varepsilon)}{\hbar^2}},$$

$$k_1 = \sqrt{\frac{M\varepsilon}{\hbar^2}}.$$

With the boundary condition $\psi = 0$ at $r = 0$ and $\psi = $ finite at $r = \infty$, we get $u(r) = A\sin(kr)$, $r \leq b$; $Be^{-k_1(r-b)}$, $r > b$.

The continuity of $\psi(r)$ and that of $\psi'(r)$ at $r = b$ require

$$A\sin(kb) = B,$$

$$kA\cos(kb) = -k_1 B,$$

which give

$$\cot(kb) = -\frac{k_1}{k} = -\sqrt{\frac{\varepsilon}{V_0 - \varepsilon}}.$$

If we take the approximation $kb = \frac{\pi}{2}$, then $A \approx B$ and $\cot(kb) \approx 0$. The latter is equivalent to assuming $V_0 \gg \varepsilon$, which means there is only one found state.

To normalize, consider

$$1 = \int_0^\infty |\psi(r)|^2 4\pi r^2 dr$$

$$= 4\pi A^2 \int_0^b \sin^2(kr)dr + 4\pi B^2 \int_b^\infty e^{-2k_1(r-b)}d\gamma$$

$$\approx 2\pi A^2 b \left(1 + \frac{1}{bk_1}\right).$$

Thus

$$A \approx B \approx \left[2\pi b \left(1 + \frac{1}{bk_1}\right)\right]^{-\frac{1}{2}}.$$

(a) The probability of the proton moving within the range of the force of the neutron is

$$P = 4\pi A^2 \int_0^b \sin^2(kr)dr = \left(1 + \frac{1}{k_1 b}\right)^{-1}.$$

As

$$k = \frac{\sqrt{M(V_0 - \varepsilon)}}{\hbar} \approx \frac{\pi}{2b},$$

i.e.

$$\varepsilon \approx V_0 - \frac{1}{Mc^2}\left(\frac{\pi \hbar c}{2b}\right)^2$$

$$= 40 - \frac{1}{940}\left(\frac{\pi \times 1.97 \times 10^{-13}}{2 \times 1.9 \times 10^{-15}}\right)^2 = 11.8 \text{ MeV},$$

and

$$k_1 = \frac{\sqrt{Mc^2\varepsilon}}{\hbar c} = \frac{\sqrt{940 \times 11.8}}{1.97 \times 10^{-13}} = 5.3 \times 10^{14} \ m^{-1},$$

we have

$$P = \left(1 + \frac{1}{5.3 \times 10^{14} \times 1.9 \times 10^{-15}}\right)^{-1} = 0.50 .$$

(b) The mean-square radius of the deuteron is

$$\overline{r^2} = \langle \Psi|r^2|\Psi \rangle_{r<b} + \langle \Psi|r^2|\Psi \rangle_{r>b}$$

$$= 4\pi A^2 \left[\int_0^b \sin^2(kr) r^2 dr + \int_b^\infty e^{-2k_1(r-b)} r^2 dr \right]$$

$$= \frac{b^2}{1 + \frac{1}{k_1 b}} \left[\left(\frac{1}{3} + \frac{4}{\pi^2}\right) + \frac{1}{k_1 b} + \frac{1}{(k_1 b)^2} + \frac{1}{2(k_1 b)^3} \right]$$

$$\approx \frac{b^2}{2} \left(\frac{1}{3} + \frac{4}{\pi^2} + 2.5\right) = 5.8 \times 10^{-30} m^2 .$$

Hence

$$(\overline{r^2})^{\frac{1}{2}} = 2.4 \times 10^{-15} \ m .$$

2051

(a) A neutron and a proton can undergo radioactive capture at rest: $p + n \to d + \gamma$. Find the energy of the photon emitted in this capture. Is the recoil of the deuteron important?

(b) Estimate the energy a neutron incident on a proton at rest must have if the radioactive capture is to take place with reasonable probability from a p-state ($l = 1$). The radius of the deuteron is $\sim 4 \times 10^{-13}$ cm.

$m_p = 1.00783$ amu, $m_n = 1.00867$ amu, $m_d = 2.01410$ amu, 1 amu = 1.66×10^{-24} g $= 931$ MeV, 1 MeV $= 1.6 \times 10^{-13}$ joule $= 1.6 \times 10^{-6}$ erg, $\hbar = 1.05 \times 10^{-25}$ erg.s.

(Wisconsin)

Solution:

(a) The energy released in the radioactive capture is

$$Q = [m_p + m_n - m_d]c^2 = 1.00783 + 1.00867 - 2.01410 \text{ amu} = 2.234 \text{ MeV} .$$

This energy appears as the kinetic energies of the photon and recoil deuteron. Let their respective momenta be $\mathbf{p}$ and $-\mathbf{p}$. Then

$$Q = pc + \frac{p^2}{2m_d},$$

or

$$(pc)^2 + 2m_d c^2 (pc) - 2m_d c^2 Q = 0.$$

Solving for pc we have

$$pc = m_d c^2 \left(-1 + \sqrt{1 + \frac{2Q}{m_d c^2}} \right).$$

As $Q/m_d c^2 \ll 1$, we can take the approximation

$$p \approx m_d c \left(-1 + 1 + \frac{Q}{m_d c^2} \right) \approx \frac{Q}{c}.$$

Thus the kinetic energy of the recoiling deuteron is

$$E_{\text{recoil}} = \frac{p^2}{2m_d} = \frac{Q^2}{2m_d c^2} = \frac{2.234^2}{2 \times 2.0141 \times 931} = 1.33 \times 10^{-3} \text{ MeV}.$$

Since

$$\frac{\Delta E_{\text{recoil}}}{E_\gamma} = \frac{1.34 \times 10^{-3}}{2.234} = 6.0 \times 10^{-4},$$

the recoiling of the deuteron does not significantly affect the energy of the emitted photon, its effect being of the order 10^{-4}.

(b) Let the position vectors of the neutron and proton be $\mathbf{r}_1$, $\mathbf{r}_2$ respectively. The motion of the system can be treated as that of a particle of mass $\mu = \frac{m_p m_n}{m_p + m_n}$, position vector $\mathbf{r} = \mathbf{r}_1 - \mathbf{r}_2$, having momentum $\mathbf{p}' = \mu \dot{\mathbf{r}}$ and kinetic energy $T' = \frac{p'^2}{2\mu}$ in the center-of-mass frame. The laboratory energy is

$$T = T' + \frac{1}{2}(m_p + m_n)\dot{\mathbf{R}}^2,$$

where $\dot{\mathbf{R}} = (m_n \dot{\mathbf{r}}_1 + m_p \dot{\mathbf{r}}_2)/(m_n + m_p)$.

To a good approximation we can take $m_p \simeq m_n$. Initially $\dot{\mathbf{r}}_2 = 0$, so that $\dot{\mathbf{R}} = \frac{1}{2}\dot{\mathbf{r}}_1$, $T = \frac{m_n}{2}\dot{\mathbf{r}}_1^2 = \frac{p^2}{2m_n}$, where $\mathbf{p} = m_n \dot{\mathbf{r}}_1$ is the momentum of the neutron in the laboratory. Substitution in the energy equation gives

$$\frac{p^2}{2m_n} = \frac{p'^2}{m_n} + \frac{p^2}{4m_n},$$

or

$$p^2 = 4p'^2 .$$

The neutron is captured into the p-state, which has angular momentum eigenvalue $\sqrt{1(1+1)}\hbar$. Using the deuteron radius a as the radius of the orbit, we have $p'a \approx \sqrt{2}\hbar$ and hence the kinetic energy of the neutron in the laboratory

$$T = \frac{p^2}{2m_n} = \frac{2p'^2}{m_n} = \frac{4}{m_n c^2} \left(\frac{\hbar c}{a}\right)^2 = \frac{4}{940} \left(\frac{1.97 \times 10^{-11}}{4 \times 10^{-13}}\right)^2 = 10.32 \text{ MeV} .$$

2052

Consider the neutron-proton capture reaction leading to a deuteron and photon, $n + p \rightarrow d + \gamma$. Suppose the initial nucleons are unpolarized and that the center of mass kinetic energy T in the initial state is very small (thermal). Experimental study of this process provides information on s-wave proton-neutron scattering, in particular on the singlet scattering length a_s. Recall the definition of scattering length in the terms of phase shift: $k \cot \delta \rightarrow -1/a_s$, as $k \rightarrow 0$. Treat the deuteron as being a pure s-state .

(a) Characterize the leading multipolarity of the reaction (electric dipole? magnetic dipole? etc.?). Give your reason.

(b) Show that the capture at low energies occurs from a spin singlet rather than spin triplet initial state.

(c) Let B be the deuteron binding energy and let $m = m_p = m_n$ be the nucleon mass. How does the deuteron spatial wave function vary with neutron-proton separation r for large r?

(d) In the approximation where the neutron-proton force is treated as being of very short range, the cross section σ depends on T, B, a_s, m and universal parameters in the form $\sigma = \sigma_0(T, B, m)f(a_s, B, m)$, where f would equal unity if $a_s = 0$. Compute the factor f for $a_s \neq 0$.

(Princeton)

Solution:

(a) As the center-of-mass kinetic energy of the $n - p$ system is very small, the only reaction possible is s-wave capture with $l = 0$. The possible initial states are $^1S_0$ state: $s_p + s_n = 0$. As $P(^1S_0) = 1$, we have $J^P = 0^+$; $^3S_1$ state: $s_p + s_n = 1$. As $P(^3S_1) = 1$, we have $J^P = 1^+$. The final state is a deuteron, with $J^P = 1^+$, and thus $S = 1, l = 0, 2$ (**Problem 2058(b)**). The initial states have $l = 0$. Hence there are two possible transitions with $\Delta l = 0, 2$ and no change of parity. Therefore the reactions are of the $M1, E2$ types.

(b) Consider the two transitions above: $^1S_0 \rightarrow {}^3S_1$, and $^3S_1 \rightarrow {}^3S_1$. As both the initial and final states of each case have $l = 0$, only those interaction terms involving spin in the Hamiltonian can cause the transition. For such operators, in order that the transition matrix elements do not vanish the spin of one of the nucleons must change during the process. Since

$$\text{for} \quad {}^3S_1 \rightarrow {}^3S_1, \quad \Delta l = 0, \quad \Delta S = 0,$$

$$\text{for} \quad {}^1S_0 \rightarrow {}^3S_1, \quad \Delta l = 0, \quad \Delta S \neq 0,$$

the initial state which satisfies the transition requirement is the spin-singlet $^1S_0$ state of the $n - p$ system.

(c) Let the range of neutron-proton force be a. The radial part of the Schrödinger equation for the system for s waves is

$$\frac{d^2u}{dr^2} + \frac{2\mu}{\hbar^2}(T - V)u = 0,$$

where $u = rR(r), R(r)$ being the radial spatial wave function, $\mu = \frac{m}{2}$, and V can be approximated by a rectangular potential well of depth B and width a:

$$V = \begin{cases} -B & \text{for} \quad 0 \leq r \leq a, \\ 0 & \text{for} \quad a < r. \end{cases}$$

The solution for large r gives the deuteron spatial wave function as

$$R(r) = \frac{A}{r}\sin(kr + \delta)$$

where $k = \frac{\sqrt{mT}}{\hbar}$, A and δ are constants.

(d) The solutions of the radial Schrödinger equation for s waves are

$$u = \begin{cases} A\sin(kr + \delta), & \text{with} \quad k = \dfrac{\sqrt{mT}}{\hbar}, & \text{for } r \geq a, \\ \\ A'\sin Kr, & \text{with} \quad K = \dfrac{\sqrt{m(T + B)}}{\hbar}, & \text{for } r \leq a. \end{cases}$$

The continuity of the wave function and its first derivative at $r = a$ gives

$$\tan(ka + \delta) = \frac{k}{K} \tan Ka,\tag{1}$$

and hence

$$\delta = \arctan\left(\frac{k}{K} \tan Ka\right) - ka.\tag{2}$$

The scattering cross section is then

$$\sigma = \frac{4\pi}{k^2} \sin^2 \delta.$$

Consider the case of $k \to 0$. We have $\delta \to \delta_0$, $K \to K_0 = \frac{\sqrt{mB}}{\hbar}$, and, by definition, $a_s = -\frac{\tan \delta_0}{k}$.

With $k \to 0$, Eq. (1) gives

$$ka + \tan \delta_0 \approx \frac{k}{K_0} \tan K_0 a(1 - ka \tan \delta_0) \approx \frac{k}{K_0} \tan K_0 a,$$

or

$$ka - ka_s \approx \frac{k}{K_0} \tan K_0 a,$$

i.e.,

$$a_s \approx -a\left(\frac{\tan K_0 a}{K_0 a} - 1\right).$$

If $a_s = -\frac{\tan \delta_0}{k} \to 0$, then $\delta_0 \to 0$ also (k is small but finite). The corresponding scattering cross section is

$$\sigma_0 = \frac{4\pi}{k^2} \sin^2 \delta_0 \approx \frac{4\pi}{k^2} \delta_0^2 = \frac{4\pi}{k^2} k^2 a_s^2 = 4\pi a^2 \left(\frac{\tan K_0 a}{K_0 a} - 1\right)^2.$$

Hence

$$f(a_s, B, m) = \frac{\sigma}{\sigma_0} \approx \frac{\sin^2[\arctan(\frac{k}{K} \tan Ka) - ka]}{k^2 a^2 (\frac{\tan K_0 a}{K_0 a} - 1)^2}$$

$$\approx \frac{\sin^2[\arctan(\frac{k}{K} \tan Ka) - ka]}{k^2 a_s^2}.$$

2053

The only bound two-nucleon configuration that occurs in nature is the deuteron with total angular momentum $J = 1$ and binding energy -2.22 MeV.

(a) From the above information alone, show that the $n - p$ force must be spin dependent.

(b) Write down the possible angular momentum states for the deuteron in an LS coupling scheme. What general liner combinations of these states are possible? Explain.

(c) Which of the states in (b) are ruled out by the existence of the quadrupole moment of the deuteron? Explain. Which states, in addition, are ruled out if the deuteron has pure isospin $T = 0$?

(d) Calculate the magnetic moment of the deuteron in each of the allowed states in part (c), and compare with the observed magnetic moment $\mu_d = 0.875\mu_N$, μ_N being the nuclear magneton.

(NOTE: $\mu_p = 2.793\mu_N$ and $\mu_n = -1.913\mu_N$)

The following Clebsch–Gordan coefficients may be of use:

[Notation; $\langle J_1 J_2 M_1 M_2 | J_{\text{TOT}} M_{\text{TOT}} \rangle$]

$$\langle 2, 1; 2, -1 | 1, 1 \rangle = (3/5)^{1/2},$$

$$\langle 2, 1; 1, 0 | 1, 1 \rangle = -(3/10)^{1/2},$$

$$\langle 2, 1; 0, 1 | 1, 1 \rangle = (1/10)^{1/2}.$$

(Princeton)

Solution:

(a) The spin of naturally occurring deuteron is $J = 1$. As $\mathbf{J} = \mathbf{s}_n + \mathbf{s}_p + \mathbf{l}_p$, we can have

for $|\mathbf{s}_n + \mathbf{s}_p| = 1$, $l = 0, 1, 2$, possible states ${}^3S_1 \cdot {}^3P_1, {}^3D_1$,

for $|\mathbf{s}_n + \mathbf{s}_p| = 0$, $l = 1$, possible state ${}^1P_1$.

However, as no stable singlet state ${}^1S_0$, where n, p have antiparallel spins and $l = 0$, is found, this means that when n, p interact to form $S = 1$ and $S = 0$ states, one is stable and one is not, indicating the spin dependence of nuclear force.

(b) As shown above, in LS coupling the possible configurations are $^3S_1$, $^3D_1$ of even party and $^3P_1$, $^1P_1$ of odd parity.

As the deuteron has a definite parity, only states of the same parity can be combined. Thus

$$\Psi(n,p) = a\,^3S_1 + b\,^3D_1 \text{ or } c\,^3P_1 + d\,^1P_1\,,$$

where a, b, c, d are constants, are the general linear combinations possible.

(c) $l = 1$ in the P state corresponds to a translation of the center of mass of the system, and does not give rise to an electric quadrupole moment. So the existence of an electric quadrupole moment of the deuteron rules out the combination of P states. Also, in accordance with the generalized Pauli's principle, the total wave function of the $n-p$ system must be antisymmetric. Thus, in

$$\Psi(n,p) = \Psi_l(n,p)\Psi_s(n,p)\Psi_T(n,p)\,,$$

where l, s, T label the space, spin and isospin wave functions, as $T = 0$ and so the isospin wave function is exchange antisymmetric, the combined space and spin wave function must be exchange symmetric. It follows that if $l = 1$, then $S = 0$, if $l = 0, 2$ then $S = 1$. This rules out the $^3P_1$ state. Hence, considering the electric quadrupole moment and the isospin, the deuteron can only be a mixed state of $^3S_1$ and $^3D_1$.

(d) For the $^3S_1$ state, $l = 0$, and the orbital part of the wave function has no effect on the magnetic moment; only the spin part does. As $S = 1$, the n and p have parallel spins, and so

$$\mu(^3S_1) = \mu_p + \mu_n = (2.793 - 1.913)\mu_N = 0.88\mu_N\,.$$

For the $^3D_1$ state, when $m = 1$, the projection of the magnetic moment on the z direction gives the value of the magnetic moment. Expanding the total angular momentum $|1,1\rangle$ in terms of the D states we have

$$|1,1\rangle = \sqrt{\frac{3}{5}}|2,2,1,-1\rangle - \sqrt{\frac{3}{10}}|2,1,1,0\rangle + \sqrt{\frac{1}{10}}|2,0,1,1\rangle\,.$$

The contribution of the D state to the magnetic moment is therefore

$$\mu(^3D_1) = \left[\frac{3}{5}(g_l m_{l1} + g_s m_{s1}) + \frac{3}{10}(g_l m_{l2} + g_s m_{s2})\right.$$

$$\left. + \frac{1}{10}(g_l m_{l3} + g_s m_{s3})\right]\mu_N$$

$$= \left[\left(\frac{3}{5}m_{l1} + \frac{3}{10}m_{l2} + \frac{1}{10}m_{l3}\right) \times \frac{1}{2}\right.$$

$$\left. + \left(\frac{3}{5}m_{s1} + \frac{3}{10}m_{s2} + \frac{1}{10}m_{s3}\right) \times 0.88\right]\mu_N$$

$$= 0.31\mu_N .$$

Note that g_l is 1 for p and 0 for n, g_s is 5.5855 for p and -3.8256 for n, and so g_l is $\frac{1}{2}$ and g_s is 0.88 for the system (**Problem 2056**).

As experimentally $\mu_d = 0.857\mu_N$, the deuteron must be a mixed state of S and D. Let the proportion of D state be x, and that of S state be $1 - x$. Then

$$0.88(1 - x) + 0.31x = 0.857 ,$$

giving $x \approx 0.04$, showing that the deuteron consists of 4% $^3D_1$ state and 96% $^3S_1$ state.

2054

Consider a nonrelativistic two-nucleon system. Assume the interaction is charge independent and conserves parity.

(a) By using the above assumptions and the Pauli principle, show that $\mathbf{S}^2$, the square of the two-nucleon spin, is a good quantum number.

(b) What is the isotopic spin of the deuteron? Justify your answer!

(c) Specify all states of a two-neutron system with total angular momentum $J \leq 2$. Use the notation $^{2S+1}X_J$ where X gives the orbital angular momentum.

(SUNY Buffalo)

Solution:

(a) Let the total exchange operator of the system be $P = P'P_{12}$, where P' is the space reflection, or parity, operator, P_{12} is the spin exchange operator

$$P_{12} = \frac{1}{2}(1 + \boldsymbol{\sigma}_1 \cdot \boldsymbol{\sigma}_2) = \mathbf{S}^2 - 1,$$

where $\boldsymbol{\sigma}_i = 2\mathbf{s}_i (i = 1, 2)$, $\mathbf{S} = \mathbf{s}_1 + \mathbf{s}_2$, using units where $\hbar = 1$. Pauli's principle gives $[P, H] = 0$, and conservation of parity gives $[P', H] = 0$. As

$$0 = [P, H] = [P'P_{12}, H] = P'[P_{12}, H] + [P', H]P_{12}$$

$$= P'[P_{12}, H] = P'[\mathbf{S}^2 - 1, H] = P'[\mathbf{S}^2, H],$$

we have $[\mathbf{S}^2, H] = 0$, and so $\mathbf{S}^2$ is a good quantum number.

(b) The isospin of the nuclear ground state always takes the smallest possible value. For deuteron,

$$\mathbf{T} = \mathbf{T}_p + \mathbf{T}_n, \qquad T_z = T_{pz} + T_{nz} = \frac{1}{2} - \frac{1}{2} = 0.$$

For ground state $T = 0$.

(c) As $\mathbf{S} = \mathbf{s}_1 + \mathbf{s}_2$ and $s_1 = s_2 = \frac{1}{2}$ the quantum number S can be 1 or 0. The possible states with $J \leq 2$ are

$$S = 0, \qquad l = 0: \quad {}^1S_0,$$

$$S = 0, \qquad l = 1: \quad {}^1P_1,$$

$$S = 0, \qquad l = 2: \quad {}^1D_2,$$

$$S = 1, \qquad l = 0: \quad {}^3S_1,$$

$$S = 1, \qquad l = 1: \quad {}^3P_2, {}^3P_1, {}^3P_0,$$

$$S = 1, \qquad l = 2: \quad {}^3D_2, {}^3D_1,$$

$$S = 1, \qquad l = 3: \quad {}^3F_2,$$

However, a two-neutron system is required to be antisymmetric with respect to particle exchange. Thus $(-1)^{l+S+1} = -1$, or $l + S =$ even. Hence the possible states are ${}^1S_0, {}^1D_2, {}^3P_2, {}^3P_1, {}^3P_0, {}^3F_2$.

2055

Consider the potential between two nucleons. Ignoring velocity-dependent terms, derive the most general form of the potential which is consistent with applicable conservation laws including that of isotopic spin. Please list each conservation law and indicate its consequences for the potential.

(Chicago)

Solution:

(a) Momentum conservation – invariance in space translation.

This law means that the potential function depends only on the relative position between the two nucleons $x = x_1 - x_2$.

(b) Angular momentum conservation – invariance in continuous space rotation: $x' = \hat{R}x$, $J^{(i)\prime} = \hat{R}J^{(i)}$, $i = 1, 2$, where $\hat{R}$ is the rotational operator.

The invariants in the rotational transformation are 1, x^2 $J^{(i)} \cdot x$, $J^{(1)} \cdot J^{(2)}$ and $[J^{(1)} \times J^{(2)}] \cdot x$. Terms higher than first order in $J^{(1)}$ or in $J^{(2)}$ can be reduced as $J_i J_j = \delta_{ij} + i\varepsilon_{ijk}J_k$. Also $(J^{(1)} \times x) \cdot (J^{(2)} \times x) = (J^{(1)} \times x) \times J^{(2)} \cdot x = (J^{(1)} \cdot J^{(2)})x^2 - (J^{(1)} \cdot x)(J^{(2)} \cdot x)$.

(c) Parity conservation – invariance in space reflection: $x' = -x$, $J^{(i)\prime} = J^{(i)}$, $i = 1, 2$.

Since x is the only polar vector, in the potential function only terms of even power in x are possible. Other invariants are 1, x^2, $J^{(1)} \cdot J^{(2)}$, $(J^{(1)} \cdot x)(J^{(2)} \cdot x)$.

(d) Isotopic spin conservation – rotational invariance in isotopic spin space:

$$I^{(i)\prime} = R_J I^{(i)}, \qquad i = 1, 2.$$

The invariants are 1 and $I^{(1)} \cdot I^{(2)}$.

(e) Conservation of probability – Hamiltonian is hermitian: $V^+ = V$.

This implies the realness of the coefficient of the potential function, i.e., $V_{sk}(r)$, where $r = |x|$, is real. Thus in

$$V(x_1, x_2, J^{(1)}, J^{(2)}, I^{(1)}, I^{(2)}) = V_a + J^{(1)} \cdot J^{(2)} V_b,$$

where V_a and V_b are of the form

$$V_0(r) + V_1(r) J^{(1)} \cdot J^{(2)} + V_2(r) \frac{(J^{(1)} \cdot x)(J^{(2)} \cdot x)}{x^2},$$

as the coefficients $V_{sk}(r)$ ($s = a, b$; $k = 0, 1, 2$) are real functions.

(f) Time reversal (inversion of motion) invariance:

$$V = U^{-1}V^*U, \qquad U^{-1}J^*U = -\boldsymbol{J}.$$

This imposes no new restriction on V.

Note that V is symmetric under the interchange $1 \leftrightarrow 2$ between two nucleons.

2056

The deuteron is a bound state of a proton and a neutron of total angular momentum $J = 1$. It is known to be principally an $S(l = 0)$ state with a small admixture of a $D(l = 2)$ state.

(a) Explain why a P state cannot contribute.

(b) Explain why a G state cannot contribute.

(c) Calculate the magnetic moment of the pure D state $n - p$ system with $J = 1$. Assume that the n and p spins are to be coupled to make the total spin **S** which is then coupled to the orbital angular momentum **L** to give the total angular momentum **J**. Express your result in nuclear magnetons. The proton and neutron magnetic moments are 2.79 and -1.91 nuclear magnetons respectively.

(CUSPEA)

Solution:

(a) The P state has a parity opposite to that of S and D states. As parity is conserved in strong interactions states of opposite parities cannot be mixed. Hence the P state cannot contribute to a state involving S and D states

(b) The orbital angular momentum quantum number of G state is $l = 4$. It cannot be coupled with two $1/2$ spins to give $J = 1$. Hence the G state cannot contribute to a state of $J = 1$.

(c) We have $\mathbf{J} = \mathbf{L} + \mathbf{S}$,

$$\mu = \frac{[(g_L\mathbf{L} + g_s\mathbf{S}) \cdot \mathbf{J}]}{J(J+1)}\mathbf{J}\mu_0,$$

where μ_0 is the nuclear magneton. By definition,

$$\mathbf{S} = \mathbf{s}_p + \mathbf{s}_n\,,$$

$$\mu_s = \frac{[(g_p\mathbf{s}_p + g_n\mathbf{s}_n)\cdot\mathbf{S}]}{S(S+1)}\mathbf{S}\mu_0 \equiv g_s\mathbf{S}\mu_0\,,$$

or

$$g_s = \frac{g_p\mathbf{s}_p\cdot\mathbf{S} + g_n\mathbf{s}_n\cdot\mathbf{S}}{S(S+1)}\,.$$

Consider $\mathbf{s}_n = \mathbf{S} - \mathbf{s}_p$. As $\mathbf{s}_n^2 = \mathbf{S}^2 + \mathbf{s}_p^2 - 2\mathbf{S}\cdot\mathbf{s}_p$, we have

$$\mathbf{S}\cdot\mathbf{s}_p = \frac{S(S+1) + s_p(s_p+1) - s_n(s_n+1)}{2} = 1\,,$$

since $s_p = s_n = \frac{1}{2}$, $S = 1$ (for $J = 1$, $l = 2$). Similarly $\mathbf{S}\cdot\mathbf{s}_n = 1$. Hence

$$g_s = \frac{1}{2}(g_p + g_n)\,.$$

As the neutron, which is uncharged, makes no contribution to the orbital magnetic moment, the proton produces the entire orbital magnetic moment, but half the orbital angular momentum. Hence $g_L = \frac{1}{2}$.

Substitution of g_s and g_L in the expression for μ gives

$$\frac{\mu}{\mu_0} = \frac{\frac{1}{2}(\mathbf{L}\cdot\mathbf{J}) + \frac{1}{2}(g_p + g_n)(\mathbf{S}\cdot\mathbf{J})}{J(J+1)}\mathbf{J}\,.$$

As

$$\mathbf{L}\cdot\mathbf{J} = \frac{1}{2}[J(J+1) + L(L+1) - S(S+1)]$$

$$= \frac{1}{2}(1\times 2 + 2\times 3 - 1\times 2) = 3\,,$$

$$\mathbf{S}\cdot\mathbf{J} = \frac{1}{2}[J(J+1) + S(S+1) - L(L+1)]$$

$$= \frac{1}{2}(1\times 2 + 1\times 2 - 2\times 3) = -1\,,$$

$$\frac{\mu}{\mu_0} = \frac{1}{2}\left(\frac{3}{2} - \frac{g_p + g_n}{2}\right)\mathbf{J}\,.$$

with $\mu_p = g_p s_p \mu_0 = \frac{1}{2} g_p \mu_0$, $\mu_n = g_n s_n \mu_0 = \frac{1}{2} g_n \mu_0$, we have

$$\mu = \left(\frac{3}{4} - \frac{\mu_p + \mu_n}{2} \right) \mu_0 = \left(\frac{3}{4} - \frac{2.79 - 1.91}{2} \right) \mu_0 = 0.31 \mu_0 \,.$$

2057

(a) The deuteron ($^2_1$H) has $J = 1\hbar$ and a magnetic moment ($\mu = 0.857 \mu_N$) which is approximately the sum of proton and neutron magnetic moments ($\mu_p = 2.793 \mu_N$, and $\mu_n = -1.913 \mu_N$). From these facts what can one infer concerning the orbital motion and spin alignment of the neutron and proton in the deuteron?

(b) How might one interpret the lack of exact equality of μ and $\mu_n + \mu_p$?

(c) How can the neutron have a nonzero magnetic moment?

(*Wisconsin*)

Solution:

(a) As $\mu \approx \mu_n + \mu_p$, the orbital motions of proton and neutron make no contribution to the magnetic moment of the deuteron. This means that the orbital motion quantum number is $l = 0$. As $J = 1$ the spin of the deuteron is 1 and it is in the $^3S_1$ state formed by proton and neutron of parallel-spin alignment.

(b) The difference between μ and $\mu_n + \mu_p$ cannot be explained away by experimental errors. It is interpreted as due to the fact that the neutron and proton are not in a pure $^3S_1$ state, but in a mixture of $^3S_1$ and $^3D_1$ states. If a proportion of the latter of about 4% is assumed, agreement with the experimental value can be achieved.

(c) While the neutron has net zero charge, it has an inner structure. The current view is that the neutron consists of three quarks of fractional charges. The charge distribution inside the neutron is thus not symmetrical, resulting in a nonzero magnetic moment.

2058

The deuteron is a bound state of a proton and a neutron. The Hamiltonian in the center-of-mass system has the form

$$H = \frac{\mathbf{p}^2}{2\mu} + V_1(r) + \boldsymbol{\sigma}_p \cdot \boldsymbol{\sigma}_n V_2(r) + \left[\left(\boldsymbol{\sigma}_p \cdot \frac{\mathbf{x}}{r} \right) \left(\boldsymbol{\sigma}_n \cdot \frac{\mathbf{x}}{r} \right) - \frac{1}{3} (\boldsymbol{\sigma}_p \cdot \boldsymbol{\sigma}_n) \right] V_3(r) \,,$$

where $x = x_n - x_p$, $r = |x|$, σ_p and σ_n are the Pauli matrices for the spins of the proton and neutron, μ is the reduced mass, and $\mathbf{p}$ is conjugate to $\mathbf{x}$.

(a) Total angular momentum ($\mathbf{J}^2 = J(J + 1)$) and parity are good quantum numbers. Show that if $V_3 = 0$, total orbital angular momentum ($\mathbf{L}^2 = L(L+1)$) and total spin ($\mathbf{S}^2 = S(S+1)$) are good quantum numbers, where $\mathbf{S} = \frac{1}{2}(\sigma_p + \sigma_n)$. Show that if $V_3 \neq 0$, S is still a good quantum number. [It may help to consider interchange of proton and neutron spins.]

(b) The deuteron has $J = 1$ and positive parity. What are the possible values of L? What is the value of S?

(c) Assume that V_3 can be treated as a small perturbation. Show that in zeroth order ($V_3 = 0$) the wave function of the state with $J_z = +1$ is of the form $\Psi_0(r)|\alpha, \alpha\rangle$, where $|\alpha, \alpha\rangle$ is the spin state with $s_{pz} = s_{nz} = 1/2$. What is the differential equation satisfied by $\Psi_0(r)$?

(d) What is the first order shift in energy due to the term in V_3? Suppose that to first order the wave function is

$$\Psi_0(r)|\alpha, \alpha\rangle + \Psi_1(x)|\alpha, \alpha\rangle + \Psi_2(x)(|\alpha, \beta\rangle + |\beta, \alpha\rangle) + \Psi_3(x)|\beta, \beta\rangle,$$

where $|\beta\rangle$ is a state with $s_z = -\frac{1}{2}$ and Ψ_0 is as defined in part (c). By selecting out the part of the Schördinger equation that is first order in V_3 and proportional to $|\alpha, \alpha\rangle$, find the differential equation satisfied by $\Psi_1(x)$. Separate out the angular dependence of $\Psi_1(x)$ and write down a differential equation for its radial dependence.

$$(MIT)$$

Solution:

(a) We have $[\mathbf{L}^2, \sigma_p \cdot \sigma_n] = 0$, $[\mathbf{L}^2, V_i(r)] = 0$, $[\mathbf{S}^2, V_i(r)] = 0$, $[\mathbf{S}^2, \mathbf{p}^2] = 0$; $[\mathbf{S}^2, \sigma_p \cdot \sigma_n] = [\mathbf{S}^2, 2\mathbf{S}^2 - 3] = 0$ as $\mathbf{S}^2 = s_p^2 + s_n^2 + 2s_p \cdot s_n = \frac{3}{4} + \frac{3}{4} + \frac{1}{2}\sigma_p \cdot \sigma_n$;

$$\left[\mathbf{S}^2, 3\left(\sigma_p \cdot \frac{x}{r}\right)\left(\sigma_n \cdot \frac{x}{r}\right) - \sigma_p \cdot \sigma_n\right]$$

$$= \left[\mathbf{S}^2, \frac{12(\mathbf{s} \cdot \mathbf{x})^2}{r^2} - 2\mathbf{S}^2 + 3\right] = \left[\mathbf{S}^2, \frac{12(\mathbf{s} \cdot \mathbf{x})^2}{r^2}\right]$$

$$= \frac{12(\mathbf{s} \cdot \mathbf{x})}{r^2}[\mathbf{S}^2, \mathbf{s} \cdot \mathbf{x}] + [\mathbf{S}^2, \mathbf{s} \cdot \mathbf{x}]\frac{12(\mathbf{s} \cdot \mathbf{x})}{r^2} = 0$$

as

$$\frac{(\sigma_p \cdot \mathbf{x})(\sigma_n \cdot \mathbf{x})}{r} = \frac{4}{r^2}(s_p \cdot \mathbf{x})(s_n \cdot \mathbf{x}) = \frac{4}{r^2}(\mathbf{s} \cdot \mathbf{x})^2 ;$$

$$[\mathbf{L}^2, \mathbf{p}^2] = \mathbf{L}[\mathbf{L}, \mathbf{p}^2] + [\mathbf{L}, \mathbf{p}^2]\mathbf{L} = 0 \text{ as } [l_\alpha, \mathbf{p}^2] = 0.$$

Hence if $V_3 = 0$, $[\mathbf{L}^2, H] = 0$, $[\mathbf{S}^2, H] = 0$, and the total orbital angular momentum and total spin are good quantum numbers. If $V_3 \neq 0$, as $[\mathbf{S}^2, H] = 0$, S is still a good quantum number.

(b) The possible values of L are 0,2 for positive parity, and so the value of S is 1.

(c) If $V_3 = 0$, the Hamiltonian is centrally symmetric. Such a symmetric interaction potential between the proton and neutron gives rise to an S state ($L = 0$). The S state of deuteron would have an admixture of D-state if the perturbation V_3 is included.

In the case of $V_3 = 0$, $L = 0$, $S = 1$ and $S_z = 1$, so $J_z = +1$ and the wave function has a form $\Psi_0(r)|\alpha, \alpha\rangle$. Consider

$$H\Psi_0(r)|\alpha, \alpha\rangle = \left[-\frac{\nabla^2}{2\mu} + V_1(r) + (2\mathbf{S}^2 - 3)V_2(r)\right] \Psi_0(r)|\alpha, \alpha\rangle$$

$$= \left[-\frac{\nabla^2}{2\mu} + V_1(r) + V_2(r)\right] \Psi_0(r)|\alpha, \alpha\rangle$$

$$= E_c\Psi_0(r)|\alpha, \alpha\rangle$$

noting that $2S^2 - 3 = 2.1.2 - 3 = 1$. Thus $\Psi_0(r)$ satisfies

$$\left[-\frac{\nabla^2}{2\mu} + V_1(r) + V_2(r) - E_c\right] \Psi_0(r) = 0,$$

or

$$-\frac{1}{2\mu}\frac{1}{r^2}\frac{d}{dr}[r^2\Psi_0'(r)] + [V_1(r) + V_2(r) - E_c]\Psi_0(r) = 0,$$

i.e.,

$$-\frac{1}{2\mu}\Psi_0''(r) - \frac{1}{\mu r}\Psi_0'(r) + [V_1(r) + V_2(r) - E_c]\Psi_0(r) = 0.$$

(d) Now, writing S_{12} for the coefficient of $V_3(r)$,

$$H = -\frac{\nabla^2}{2\mu} + V_1(r) + (2\mathbf{S}^2 - 3)V_2(r) + S_{12}V_3(r)$$

$$= -\frac{\nabla^2}{2\mu} + V_1(r) + V_2(r) + S_{12}V_3(r),$$

so

$$HΨ = \left(-\frac{\nabla^2}{2\mu} + V_1 + V_2\right) Ψ_0(r)|\alpha, \alpha\rangle + \left(-\frac{\nabla^2}{2\mu} + V_1 + V_2\right) [Ψ_1|\alpha, \alpha\rangle$$

$$+ Ψ_2(|\alpha, \beta\rangle + |\beta, \alpha\rangle) + Ψ_3|\beta, \beta\rangle] + S_{12}V_3Ψ_0|\alpha, \alpha\rangle$$

$$= E_cΨ_0(r)|\alpha, \alpha\rangle + E_c[Ψ_1|\alpha, \alpha\rangle + Ψ_2(|\alpha, \beta\rangle$$

$$+ |\beta, \alpha\rangle) + Ψ_3|\beta, \beta\rangle] + ΔEΨ_0(r)|\alpha, \alpha\rangle,$$

where

$$S_{12}V_3Ψ_0(r)|\alpha, \alpha\rangle = [(\sigma_{pz}\cos\theta \cdot \sigma_{nz}\cos\theta)|\alpha, \alpha\rangle - \frac{1}{3}|\alpha, \alpha\rangle]V_3Ψ_0(r) + \cdots$$

$$= \left(\cos^2\theta - \frac{1}{3}\right) V_3Ψ_0(r)|\alpha, \alpha\rangle + \cdots,$$

terms not proportional to $|\alpha, \alpha\rangle$ having been neglected.

Selecting out the part of the Schrödinger equation that is first order in V_3 and proportional to $|\alpha, \alpha\rangle$, we get

$$\left(-\frac{\nabla^2}{2\mu} + V_1 + V_2\right) Ψ_1(\mathbf{x}) + \left(\cos^2\theta - \frac{1}{3}\right) V_3Ψ_0(r) = E_cΨ_1(\mathbf{x}) + ΔEΨ_0(r).$$

Thus the angular-dependent part of $Ψ_1(\mathbf{x})$ is

$$Y_{20} = 3\left(\frac{5}{16\pi}\right)^{\frac{1}{2}}\left(\cos^2\theta - \frac{1}{3}\right),$$

since for the state $|\alpha, \alpha\rangle$, $S_z = 1$ and so $L_z = 0$, i.e. the angular part of the wave function is Y_{20}. Therefore we have

$$-\frac{1}{2\mu}\frac{1}{r^2}\frac{d}{dr}\left(r^2\frac{dΨ_1(r)}{dr}\right) + V_1(r)Ψ_1(r) + V_2(r)Ψ_2(r)$$

$$+ \frac{l(l+1)}{r^2}Ψ_1(r) + \frac{1}{3}\sqrt{\frac{16\pi}{5}}V_3Ψ_0(r)$$

$$= E_cΨ_1(r) + ΔEΨ_0(r)$$

with $\Psi_1(\mathbf{x}) = \Psi_1(r)Y_{20}$, $l = 2$, or

$$-\frac{1}{2\mu}\Psi_1''(r) - \frac{1}{\mu r}\Psi_1'(r) + \left[V_1(r) + V_2(r) + \frac{6}{r^2} - E_c\right]\Psi_1(r)$$

$$+ \left(\frac{1}{3}\sqrt{\frac{16\pi}{5}}V_3 - \Delta E\right)\Psi_0(r) = 0$$

with

$$\Delta E = \left(\cos^2\theta - \frac{1}{3}\right)V_3.$$

4. NUCLEAR MODELS (2059–2075)

2059

What are the essential features of the liquid-drop, shell, and collective models of the nucleus? Indicate what properties of the nucleus are well predicted by each model, and how the model is applied.

(Columbia)

Solution:

It is an empirical fact that the binding energy per nucleon, B, of a nucleus and the density of nuclear matter are almost independent of the mass number A. This is similar to a liquid-drop whose heat of evaporation and density are independent of the drop size. Add in the correction terms of surface energy, Coulomb repulsion energy, pairing energy, symmetry energy and we get the liquid-drop model. This model gives a relationship between A and Z of stable nuclei, i.e., the β-stability curve, in agreement with experiment. Moreover, the model explains why the elements ^{43}Te, ^{61}Pm have no β-stable isobars. If we treat the nucleus's radius as a variable parameter in the mass-formula coefficients a_{surface} and a_{volume} and fit the mass to the experimental value, we find that the nuclear radius so deduced is in good agreement with those obtained by all other methods. So the specific binding energy curve is well explained by the liquid-drop model.

The existence of magic numbers indicates that nuclei have internal structure. This led to the nuclear shell model similar to the atomic model, which could explain the special stability of the magic-number nuclei. The shell model requires:

(1) the existence of an average field, which for a spherical nucleus is a central field,

(2) that each nucleon in the nucleus moves independently,

(3) that the number of nucleons on an energy level is limited by Pauli's principle,

(4) that spin-orbit coupling determines the order of energy levels.

The spin and parity of the ground state can be predicted using the shell model. For even-even nuclei the predicted spin and parity of the ground state, 0^+, have been confirmed by experiment in all cases. The prediction is based on the fact that normally the spin and parity are 0^+ when neutrons and protons separately pair up. The predictions of the spin and parity of the ground state of odd-A nuclei are mostly in agreement with experiment. Certain aspects of odd-odd nuclei can also be predicted. In particular it attributes the existence of magic numbers to full shells.

The shell model however cannot solve all the nuclear problems. It is quite successfull in explaining the formation of a nucleus by adding one or several nucleons to a full shell (spherical nucleus), because the nucleus at this stage is still approximately spherical. But for a nucleus between two closed shells, it is not spherical and the collective motion of a number of nucleons become much more important. For example, the experimental values of nuclear quadrupole moment are many times larger than the values calculated from a single particle moving in a central field for a nucleus between full shells. This led to the collective model, which, by considering the collective motion of nucleons, gives rise to vibrational and rotational energy levels for nuclides in the ranges of $60 < A < 150$ and $190 < A < 220$, $150 < A < 190$ and $A > 220$ respectively:

2060

Discuss briefly the chief experimental systematics which led to the shell model description for nuclear states. Give several examples of nuclei which correspond to closed shells and indicate which shells are closed.

(Wisconsin)

Solution:

The main experimental evidence in support of the nuclear shell model is the existence of magic numbers. When the number of the neutrons or of the protons in a nucleus is 2, 8, 20, 28, 50, 82 and 126 (for neutrons only), the nucleus is very stable. In nature the abundance of nuclides with such magic numbers are larger than those of the nearby numbers. Among all the stable nuclides, those of neutron numbers 20, 28, 50 and 82 have more isotones, those of proton numbers 8, 20, 28, 50 and 82 have more stable isotopes, than the nearby nuclides. When the number of neutrons or protons in a nuclide is equal to a magic number, the binding energy measured experimentally is quite different from that given by the liquid-drop model. The existence of such magic numbers implies the existence of shell structure inside a nucleus similar to the electron energy levels in an atom.

^4He is a double-magic nucleus; its protons and neutrons each fill up the first main shell. ^{16}O is also a double-magic nucleus, whose protons and neutrons each fill up the first and second main shells. ^{208}Pb is a double-magic nucleus, whose protons fill up to the sixth main shell, while whose neutrons fill up to the seventh main shell. Thus these nuclides all have closed shells.

2061

(a) Discuss the standard nuclear shell model. In particular, characterize the successive shells according to the single-particle terms that describe the shell, i.e., the principal quantum number n, the orbital angular momentum quantum number l, and the total angular momentum quantum number j (spectroscopic notation is useful here, e.g., $2s_{1/2}, 1p_{3/2}$, etc..). Discuss briefly some of the basic evidence in support of the shell model.

(b) Consider a nuclear level corresponding to a closed shell plus a single proton in a state with the angular momentum quantum numbers l and j. Of course $j = l \pm 1/2$. Let g_p be the empirical gyromagnetic ratio of the free proton. Compute the gyromagnetic ratio for the level in question, for each of the two cases $j = l + 1/2$ and $j = l - 1/2$.

(Princeton)

Solution:

(a) The basic ideas of the nuclear shell model are the following. Firstly we assume each nucleon moves in an average field which is the sum of the actions of the other nucleons on it. For a nucleus nearly spherically in shape, the average field is closely represented by a central field. Second, we assume that the low-lying levels of a nucleus are filled up with nucleons in accordance with Pauli's principle. As collisions between nucleons cannot cause a transition and change their states, all the nucleons can maintain

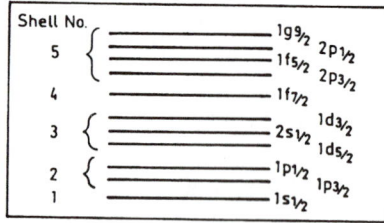

Fig. 2.11

their states of motion, i.e., they move independently in the nucleus. We can take for the average central field a Woods–Saxon potential well compatible with the characteristics of the interaction between nucleons, and obtain the energy levels by quantum mechanical methods. Considering the spin-orbital interaction, we get the single-particle energy levels (Fig. 2.11), which can be filled up with nucleons one by one. Note that each level has a degeneracy $2j + 1$. So up to the first 5 shells as shown, the total number of protons or neutrons accommodated are 2, 8, 20, 28 and 50.

The main experimental evidence for the shell model is the existence of magic numbers. Just like the electrons outside a nucleus in an atom, if the numbers of neutrons on protons in a nucleus is equal to some 'magic number' (8,20,28,50 or 82), the nucleus has greater stability, larger binding energy and abundance, and many more stable isotopes.

(b) According to the shell model, the total angular momentum of the nucleons in a closed shell is zero, so is the magnetic moment. This means that the magnetic moment and angular momentum of the nucleus are determined by the single proton outside the closed shell.

As

$$\boldsymbol{\mu}_j = \boldsymbol{\mu}_l + \boldsymbol{\mu}_s ,$$

i.e.,

$$g_j \mathbf{j} = g_l \mathbf{l} + g_s \mathbf{s} ,$$

we have

$$g_j \mathbf{j} \cdot \mathbf{j} = g_l \mathbf{l} \cdot \mathbf{j} + g_s \mathbf{s} \cdot \mathbf{j} .$$

With

$$\mathbf{l} \cdot \mathbf{j} = \frac{1}{2}(\mathbf{j}^2 + \mathbf{l}^2 - \mathbf{s}^2) = \frac{1}{2}[j(j+1) + l(l+1) - s(s+1)] ,$$

$$\mathbf{s} \cdot \mathbf{j} = \frac{1}{2}(\mathbf{j}^2 + \mathbf{s}^2 - \mathbf{l}^2) = \frac{1}{2}[j(j+1) + s(s+1) - l(l+1)] ,$$

we have

$$g_j = g_l \frac{j(j+1) + l(l+1) - s(s+1)}{2j(j+1)} + g_s \frac{j(j+1) + s(s+1) - l(l+1)}{2j(j+1)}.$$

For proton, $g_l = 1$, $g_s = g_p$, the gyromagnetic ratio for free proton ($l = 0, j = s$), $s = \frac{1}{2}$. Hence we have

$$g_j = \begin{cases} \dfrac{2j-1}{2j} + \dfrac{g_p}{2j} & \text{for } j = l + 1/2, \\[3mm] \dfrac{1}{j+1}\left(j + \dfrac{3}{2} - \dfrac{g_p}{2}\right) & \text{for } j = l - 1/2. \end{cases}$$

2062

The energy levels of the three-dimensional isotropic harmonic oscillator are given by

$$E = (2n + l + 3/2)\hbar\omega = \left(N + \frac{3}{2}\right)\hbar\omega.$$

In application to the single-particle nuclear model $\hbar\omega$ is fitted as $44A^{-\frac{1}{3}}$ MeV.

(a) By considering corrections to the oscillator energy levels relate the levels for $N \leq 3$ to the shell model single-particle level scheme. Draw an energy level diagram relating the shell model energy levels to the unperturbed oscillator levels.

(b) Predict the ground state spins and parities of the following nuclei using the shell model:

$$^3_2He, \ ^{17}_8O, \ ^{34}_{19}K, \ ^{41}_{20}Ca.$$

(c) Strong electric dipole transitions are not generally observed to connect the ground state of a nucleus to excited levels lying in the first 5 MeV of excitation. Using the single-particle model, explain this observation and predict the excitation energy of the giant dipole nuclear resonance.

(Princeton)

Solution:

(a) Using LS coupling, we have the splitting of the energy levels of a harmonic oscillator as shown in Fig. 2.12.

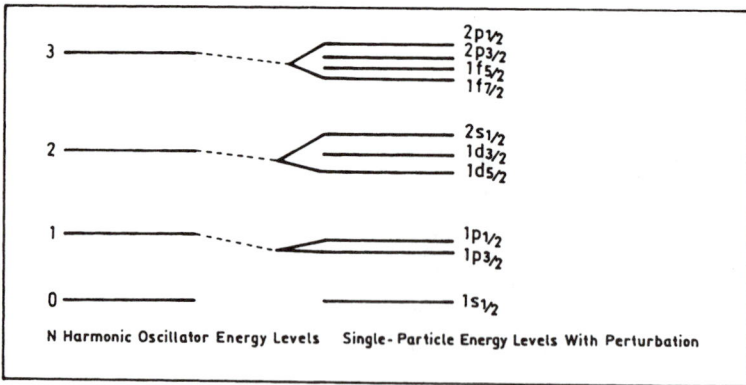

Fig. 2.12

(b) According to Fig. 2.12 we have the following:

$^{3}_{2}$He: The last unpaired nucleon is a neutron of state $1s_{\frac{1}{2}}$, so $J^{\pi} = (1/2)^{+}$.

$^{17}_{8}$O: The last unpaired nucleon is a neutron of state $1d_{5/2}$, so $J^{\pi} = (5/2)^{+}$.

$^{34}_{19}$K: The last two unpaired nucleons are a proton of state $2s_{\frac{1}{2}}$ and a neutron of state $1d_{3/2}$, so $J^{\pi} = 1^{+}$.

$^{41}_{20}$Ca: The last unpaired nucleon is a neutron of state $1f_{7/2}$, so $J^{\pi} = (7/2)^{-}$.

(c) The selection rules for electric dipole transition are

$$\Delta J = J_f - J_i = 0, 1, \qquad \Delta \pi = -1,$$

where J is the nuclear spin, π is the nuclear parity. As $\hbar \omega = 44A^{-\frac{1}{3}}$ MeV, $\hbar \omega > 5$ MeV for a nucleus. When N increases by 1, the energy level increases by $\Delta E = \hbar \omega > 5$ MeV. This means that excited states higher than the ground state by less than 5 MeV have the same N and parity as the latter. As electric dipole transition requires $\Delta \pi = -1$, such excited states cannot connect to the ground state through an electric dipole transition. However, in LS coupling the energy difference between levels of different N can be smaller than 5 MeV, especially for heavy nuclei, so that electric dipole transition may still be possible.

The giant dipole nuclear resonance can be thought of as a phenomenon in which the incoming photon separates the protons and neutrons in the nucleus, increasing the potential energy, and causing the nucleus to vibrate.

Resonant absorption occurs when the photon frequency equals resonance frequency of the nucleus.

2063

To some approximation, a medium weight nucleus can be regarded as a flat-bottomed potential with rigid walls. To simplify this picture still further, model a nucleus as a cubical box of length equal to the nuclear diameter. Consider a nucleus of iron-56 which has 28 protons and 28 neutrons. Estimate the kinetic energy of the highest energy nucleon. Assume a nuclear diameter of 10^{-12} cm.

(Columbia)

Solution:

The potential of a nucleon can be written as

$$V(x,y,z) = \begin{cases} \infty, & |x|, |y|, |z| > \dfrac{a}{2}, \\ 0, & |x|, |y|, |z| < \dfrac{a}{2}, \end{cases}$$

where a is the nuclear diameter. Assume the Schrödinger equation

$$-\frac{\hbar^2}{2m}\nabla^2\Psi(x,y,z) + V(x,y,z)\Psi(x,y,z) = E\Psi(x,y,z)$$

can be separated in the variables by letting $\Psi(x,y,z) = \Psi(x)\Psi(y)\Psi(z)$. Substitution gives

$$-\frac{\hbar^2}{2m}\frac{d^2}{dx_i^2}\Psi(x_i) + V(x_i)\Psi(x_i) = E_i\Psi(x_i),$$

with

$$V(x_i) = \begin{cases} \infty, & |x_i| > \dfrac{a}{2}, \\ 0, & |x_i| < \dfrac{a}{2}, \end{cases}$$

$i = 1, 2, 3$; $x_1 = x$, $x_2 = y$, $x_3 = z$, $E = E_1 + E_2 + E_3$.

Solving the equations we have

$$\Psi(x_i) = A_i \sin(k_i x_i) + B_i \cos(k_i x_i)$$

with $k_i = \frac{\sqrt{2mE_i}}{\hbar}$. The boundary condition $\Psi(x_i)|_{x_i=\pm\frac{a}{2}} = 0$ gives

$$\Psi(x_i) = \begin{cases} A_i \sin\left(\dfrac{n\pi}{a}x_i\right), & \text{with } n \text{ even}, \\ \\ B_i \cos\left(\dfrac{n\pi}{a}x_i\right), & \text{with } n \text{ odd}, \end{cases}$$

and hence

$$E_{xi} = \frac{k_{xi}^2\hbar^2}{2m} = \frac{\pi^2 n_{xi}^2 \hbar^2}{2ma^2}, \quad n_x = 1, 2, 3, \ldots,$$

$$E = E_0(n_x^2 + n_y^2 + n_z^2),$$

where

$$E_0 = \frac{\pi^2\hbar^2}{2ma^2} = \frac{\pi^2(c\hbar)^2}{2mc^2 \cdot a^2} = \frac{\pi^2(1.97 \times 10^{-11})^2}{2 \times 939 \times 10^{-24}} = 2.04 \text{ MeV}.$$

(n_x, n_y, n_z)	Number of states	Number of nucleons	E
(111)	1	4	$3E_0$
(211) (121) (112)	3	12	$6E_0$
(221) (122) (212)	3	12	$9E_0$
(311) (131) (113)	3	12	$11E_0$
(222)	1	4	$12E_0$
(123) (132) (231) (213) (312) (321)	6	24	$14E_0$

According to Pauli's principle, each state can accommodate one pair of neutrons and one pair of protons, as shown in the table.

For ^{56}Fe, $E_{max} = 14E_0 = 2.04 \times 14 = 28.6$ MeV.

2064

Light nuclei in the shell model.

(a) Using the harmonic-oscillator shell model, describe the expected configurations for the ground states of the light stable nuclei with $A \leq 4$, specifying also their total L, S, J and T quantum numbers and parity.

(b) For ^{4}He, what states do you expect to find at about one oscillator quantum of excitation energy?

(c) What radioactive decay modes are possible for each of these states?

(d) Which of these states do you expect to find in ^{4}H? Which do you expect to find in ^{4}Be?

(e) Which of the excited states of ^{4}He do you expect to excite in α-particle inelastic scattering? Which would you expect to be excited by proton inelastic scattering?

(Princeton)

Solution:

(a) According to Fig. 2.11 we have

$A = 1$: The stable nucleus ^{1}H has configuration: $p(1s_{1/2})^1$,

$$L = 0, \ S = 1/2, \ J^P = 1/2^+, \ T = 1/2.$$

$A = 2$: The stable nucleus ^{2}H has configuration: $p(1s_{1/2})^1, n(1s_{1/2})^1$,

$$L = 0, \ S = 1, \ J^P = 1^+, \ T = 0.$$

$A = 3$: The stable nucleus ^{3}He has configuration: $p(1s_{1/2})^2, n(1s_{1/2})^1$,

$$L = 0, \ S = 1/2, \ J^P = 1/2^+, \ T = 1/2.$$

$A = 4$: The stable nucleus ^{4}He has configuration: $p(1s_{1/2})^2, n(1s_{1/2})^2$,

$$L = 0, \ S = 0, \ J^P = 0^+, \ T = 0.$$

(b) Near the first excited state of the harmonic oscillator, the energy level is split into two levels $1p_{3/2}$ and $1p_{1/2}$ because of the LS coupling of the p state. The isospin of ^{4}He is $T_z = 0$, $T = 0$ for the ground state. So the possible excitated states are the following:

(i) When a proton (or neutron) is of $1p_{3/2}$ state, the other of $1s_{1/2}$ state, the possible coupled states are $1^-, 2^-$ $(T = 0$ or $T = 1)$.

(ii) When a proton (or neutron) is of $1p_{1/2}$ state, the other of $1s_{1/2}$ state, the possible coupled states are $0^-, 1^-$ $(T = 0$ or $1)$.

(iii) When two protons (or two neutrons) are of $1p_{1/2}$ (or $1p_{3/2}$) state, the possible coupled state is 0^+ $(T = 0)$.

(c) The decay modes of the possible states of ^{4}He are:

	J^P	T	Decay modes
Ground state:	0^+	0	*Stable*
Excited states:	0^+	0	p
	0^-	0	p, n
	2^-	0	p, n
	2^-	1	p, n
	1^-	1	$p, n\gamma$
	0^-	1	p, n
	1^-	1	$p, n\gamma$
	1^-	0	p, n, d

(d) ^{4}H has isospin $T = 1$, so it can have all the states above with $T = 1$, namely $2^-, 1^-, 0^-$.

The isospin of ^{4}Be is $T \geq 2$, and hence cannot have any of the states above.

(e) $\alpha - \alpha$ scattering is between two identical nuclei, so the total wave function of the final state is exchange symmetric and the total angular momentum is conserved

In the initial state, the two α-particles have $L = 0, 2, \ldots$

In the final state, the two α-particles are each of 0^- state, $L = 0, 2 \ldots$

Thus an α-particle can excite ^{4}He to 0^- state while a proton can excite it to 2^-, or 0^- states.

2065

Explain the following statements on the basis of physical principles:

(a) The motion of individual nucleons inside a nucleus may be regarded as independent from each other even though they interact very strongly.

(b) All the even-even nuclei have 0^+ ground state.

(c) Nuclei with outer shells partially filled by odd number of nucleons tend to have permanent deformation.

(SUNY, Buffalo)

Solution:

(a) The usual treatment is based on the assumption that the interaction among nucleons can be replaced by the action on a nucleon of the mean field produced by the other nucleons. The nucleons are considered to move independently of one another. Despite the high nucleon density inside a nucleus it is assumed that the individual interactions between nucleons do not manifest macroscopically. Since nucleons are fermions, all the low energy levels of the ground state are filled up and the interactions among nucleons cannot excite a nucleon to a higher level. We can then employ a model of moderately weak interaction to describe the strong interactions among nucleons.

(b) According to the nuclear shell model, the protons and neutrons in an even-even nucleus tend to pair off separately, i.e., each pair of neutrons or protons are in the same orbit and have opposite spins, so that the total angular momentum and total spin of each pair of nucleons are zero. It follows that the total angular momentum of the nucleus is zero. The parity of each pair of nucleons is $(-1)^{2l} = +1$, and so the total parity of the nucleus is positive. Hence for an even-even nucleus, $J^p = 0^+$.

(c) Nucleons in the outermost partially-filled shell can be considered as moving around a nuclear system of zero spin. For nucleons with $l \neq 0$, the orbits are ellipses. Because such odd nucleons have finite spins and magnetic moments, which can polarize the nuclear system, the nucleus tends to have permanent deformation.

2066

Explain the following:

(a) The binding energy of adding an extra neutron to a ^{3}He nucleus (or of adding an extra proton to a ^{3}H nucleus) to form ^{4}He is greater than 20 MeV. However neither a neutron nor a proton will bind stably to ^{4}He.

(b) Natural radioactive nuclei such as ^{232}Th and ^{238}U decay in stages, by α- and β-emissions, to isotopes of Pb. The half-lives of ^{232}Th and ^{238}U are greater than 10^9 years and the final Pb-isotopes are stable; yet the intermediate α-decay stages have much shorter half-lives – some less than 1 hour or even 1 second – and successive stages show generally a decrease in half-life and an increase in α-decay energy as the final Pb-isotope is approached.

(Columbia)

Solution:

(a) ^{4}He is a double-magic nucleus in which the shells of neutrons and protons are all full. So it is very stable and cannot absorb more neutrons or protons. Also, when a ^{3}He captures a neutron, or a ^{3}H captures a proton to form ^{4}He, the energy emitted is very high because of the high binding energy.

(b) The reason that successive stages of the decay of ^{232}Th and ^{238}U show a decrease in half-life and an increase in α-decay energy as the final Pb-isotopes are approached is that the Coulomb barrier formed between the α-particle and the daughter nucleus during α-emission obstructs the decay. When the energy of the α-particle increases, the probability of its penetrating the barrier increases, and so the half-life of the nucleus decreases. From the Geiger–Nuttall formula for α-decays

$$\log \lambda = A - BE_d^{-1/2},$$

where A and B are constants with A different for different radioactivity series, λ is the α-decay constant and E_d is the decay energy, we see that a small change in decay energy corresponds to a large change in half-life.

We can deduce from the liquid-drop model that the α-decay energy E_d increases with A. However, experiments show that for the radioactive family ^{232}Th and ^{238}U, E_d decreases as A increases. This shows that the liquid-drop model can only describe the general trend of binding energy change with A and Z, but not the fluctuation of the change, which can be explained only by the nuclear shell model.

2067

(a) What spin-parity and isospin would the shell model predict for the ground states of $^{13}_5$B, $^{13}_6$C, and $^{13}_7$N? (Recall that the $p_{3/2}$ shell lies below the $p_{1/2}$.)

(b) Order the above isobaric triad according to mass with the lowest-mass first. Briefly justify your order.

(c) Indicate how you could estimate rather closely the energy difference between the two lowest-mass members of the above triad.

(*Wisconsin*)

Solution:

(a) The isospin of the ground state of a nucleus is $I = |Z - N|/2$, where N, Z are the numbers of protons and neutrons inside the nucleus respectively. The spin-parity of the ground state of a nucleus is decided by that of the last unpaired nucleon. Thus (Fig. 2.11)

$$^{13}_5 B : J^p = \left(\frac{3}{2}\right)^- , \quad \text{as the unpaired proton is in } 1p_{\frac{3}{2}} \text{ state},$$

$$I = \frac{3}{2} ;$$

$$^{13}_6 C : J^p = \left(\frac{1}{2}\right)^- , \quad \text{as the unpaired neutron is in } 1p_{1/2} \text{ state},$$

$$I = \frac{1}{2} ;$$

$$^{13}_7 N : J^p = \left(\frac{1}{2}\right)^- , \quad \text{as the unpaired proton is in } 1p_{1/2} \text{state},$$

$$I = \frac{1}{2} .$$

(b) Ordering the nuclei with the lowest-mass first gives $^{13}_6$C, $^{13}_7$N, $^{13}_5$B. $^{13}_6$C and $^{13}_7$N belong to the same isospin doublet. Their mass difference arises from the difference in Coulomb energy and the mass difference between neutron and proton, with the former being the chiefly cause. $^{13}_7$N has one more proton than $^{13}_6$C, and so has greater Coulomb energy and hence larger mass. Whereas $^{13}_5$B has fewer protons, it has more neutrons and is far from the line of stable nuclei and so is less tightly formed. Hence it has the largest mass.

(c) Consider the two lowest-mass members of the above triad, $^{13}_6$C and $^{13}_6$N. If the nuclei are approximated by spheres of uniform charge, each will have electrostatic (Coulomb) energy $W = 3Q^2/5R$, R being the nuclear radius $R \approx 1.4A^{1/2}$ fm. Hence the mass difference is

$$[M(^{13}_7N) - M(^{13}_6C)]c^2 = \frac{3}{5R}(Q_N^2 - Q_C^2) - [M_n - M(^1H)]c^2$$

$$= \frac{3\hbar c}{5R}\left(\frac{e^2}{\hbar c}\right)(7^2 - 6^2) - 0.78$$

$$= 0.6 \times \frac{197}{137} \times \frac{49 - 36}{1.4 \times 13^{1/3}} - 0.78$$

$$= 2.62 \text{ MeV}.$$

2068

In the nuclear shell model, orbitals are filled in the order

$$1s_{1/2}, \ 1p_{3/2}, \ 1p_{1/2}, \ 1d_{5/2}, \ 2s_{1/2}, \ 1d_{3/2}, \text{ etc.}$$

(a) What is responsible for the splitting between the $p_{3/2}$ and $p_{1/2}$ orbitals?

(b) In the model, ^{16}O ($Z = 8$) is a good closed-shell nucleus and has spin and parity $J^\pi = 0^+$. What are the predicted J^π values for ^{15}O and ^{17}O?

(c) For odd-odd nuclei a range of J^π values is allowed. What are the allowed values for ^{18}F ($Z = 9$)?

(d) For even-even nuclei (e.g. for ^{18}O) J^π is always 0^+. How is this observation explained?

(Wisconsin)

Solution:

(a) The splitting between $p_{3/2}$ and $p_{1/2}$ is caused by the spin-orbit coupling of the nucleons.

(b) Each orbital can accommodate $2j + 1$ protons and $2j + 1$ neutrons. Thus the proton configuration of ^{15}O is $(1s_{1/2})^2(1p_{3/2})^4(1p_{1/2})^2$, and its neutron configuration is $(1s_{1/2})^2(1p_{3/2})^4(1p_{1/2})^1$. As the protons all pair up but the neutrons do not, the spin-parity of ^{15}O is determined by the angular momentum and parity of the unpaired neutron in the $1p_{\frac{1}{2}}$ state. Hence the spin-parity of ^{15}O of $J^p = 1/2^-$.

The proton configuration of ^{17}O is the same as that of ^{15}O, but its neutron configuration is $(1s_{1/2})^2(1p_{3/2})^4(1p_{1/2})^2(1d_{5/2})^1$. So the spin-parity of ^{17}O is that of the neutron in the $1d_{5/2}$ state, $J^p = 5/2^+$.

(c) The neutron configuration of ^{18}F is $(1s_{1/2})^2(1p_{3/2})^4(1p_{1/2})^2(1d_{5/2})^1$, its proton configuration is $(1s_{1/2})^2(1p_{3/2})^4(1p_{1/2})^2(1d_{5/2})^1$. As there are two unpaired nucleons, a range of J^p values are allowed, being decided by the neutron and proton in the $1d_{5/2}$ states. As $l_n = 2, l_p = 2$, the parity is $\pi = (-1)^{l_n+l_p} = +1$. As $j_n = 5/2, j_p = 5/2$, the possible spins are $J = 0, 1, 2, 3, 4, 5$. Thus the possible values of the spin-parity of ^{18}F are $0^+, 1^+, 2^+, 3^+, 4^+, 5^+$. (It is in fact 1^+.)

(d) For an even-even nucleus, as an even number of nucleons are in the lowest energy levels, the number of nucleons in every energy level is even. As an even number of nucleons in the same energy level have angular momenta of the same absolute value, and the angular momenta of paired nucleons are aligned oppositely because of the pairing force, the total angular momentum of the nucleons in an energy level is zero. Since all the proton shells and neutron shells have zero angular momentum, the spin of an even-even nucleus is zero. As the number of nucleons in every energy level of an even-even nucleus is even, the parity of the nucleus is positive.

2069

The single-particle energies for neutrons and protons in the vicinity of $^{208}_{82}Pb_{126}$ are given in Fig. 2.13. Using this figure as a guide, estimate or evaluate the following.

(a) The spins and parities of the ground state and the first two excited states of ^{207}Pb.

(b) The ground state quadrupole moment of ^{207}Pb.

(c) The magnetic moment of the ground state of ^{209}Pb.

(d) The spins and parities of the lowest states of $^{208}_{83}Bi$ (nearly degenerate). What is the energy of the ground state of ^{208}Bi relative to ^{208}Pb?

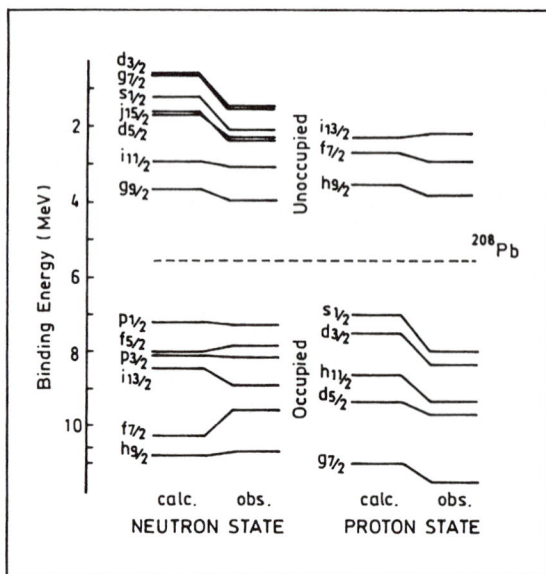

Fig. 2.13

(e) The isobaric analog state in ^{208}Bi of the ground state of ^{208}Pb is defined as

$$T_+|^{208}\text{Pb (ground state)}\rangle$$

with $T_+ = \sum_i t_+(i)$, where t_+ changes a neutron into a proton. What are the quantum numbers (spin, parity, isospin, z component of isospin) of the isobaric analog state? Estimate the energy of the isobaric analog state above the ground state of ^{208}Pb due to the Coulomb interaction.

(f) Explain why one does not observe super-allowed Fermi electron or positron emission in heavy nuclei.

(Princeton)

Solution:

(a) $^{207}_{82}$Pb consists of full shells with a vacancy for a neutron in $p_{1/2}$ level. The spin-parity of the ground state is determined by that of the unpaired neutron in $p_{1/2}$ and so is $(1/2)^-$. The first excited state is formed by a $f_{5/2}$ neutron transiting to $p_{1/2}$. Its J^p is determined by the single neutron vacancy left in $f_{5/2}$ level and is $(5/2)^-$. The second excited state is formed by a $p_{3/2}$ neutron refilling the $f_{5/2}$ vacancy (that is to say a $p_{3/2}$ neutron goes to $p_{1/2}$ directly). J^P of the nucleus in the second excited state is then determined by the single neutron vacancy in $p_{3/2}$ level and is $\left(\frac{3}{2}\right)^-$. Hence the ground and first two excited states of ^{207}Pb have $J^P = \left(\frac{1}{2}\right)^-, \left(\frac{5}{2}\right)^-, \left(\frac{3}{2}\right)^-$.

(b) The nucleon shells of $^{207}_{82}$Pb are full except there is one neutron short in $p_{1/2}$ levels. An electric quadrupole moment can arise from polarization at the nuclear center caused by motion of neutrons. But as $J = 1/2$, the electric quadrupole moment of ^{207}Pb is zero.

(c) $^{209}_{82}$Pb has a neutron in $g_{9/2}$ outside the full shells. As the orbital motion of a neutron makes no contribution to the nuclear magnetic moment, the total magnetic moment equals to that of the neutron itself:

$$\mu(^{209}Pb) = -1.91\mu_N, \mu_N \text{ being the nuclear magneton.}$$

(d) For $^{208}_{83}$Bi, the ground state has an unpaired proton and an unpaired neutron, the proton being in $h_{9/2}$, the neutron being in $p_{1/2}$. As $J = 1/2 + 9/2 = 5$ (since both nucleon spins are antiparallel to l), $l_p = 5$, $l_n = 1$ and so the parity is $(-1)^{l_p + l_n} = +$, the states has $J^P = 5^+$. The first excited state is formed by a neutron in $f_{5/2}$ transiting to $p_{1/2}$ and its spin-parity is determined by the unpaired $f_{5/2}$ neutron and $h_{9/2}$ proton. Hence $J = 5/2 + 9/2 = 7$, parity is $(-1)^{1+5} = +$, and so $J^P = 7^+$. Therefore, the two lowest states have spin-parity 5^+ and 7^+.

The energy difference between the ground states of ^{208}Bi and ^{208}Pb can be obtained roughly from Fig. 2.13. As compared with ^{208}Pb, ^{208}Bi has one more proton at $h_{9/2}$ and one less neutron at $p_{1/2}$ we have

$$\Delta E = E(Bi) - E(Pb) \approx 7.2 - 3.5 + 2\Delta \approx 3.7 + 1.5 = 5.2 \text{ MeV},$$

where $\Delta = m_n - m_p$, i.e., the ground state of ^{208}Bi is 5.2 MeV higher than that of ^{208}Pb.

(e) As T_+ only changes the third component of the isospin,

$$T_+|T, T_3\rangle = A|T, T_3 + 1\rangle.$$

Thus the isobaric analog state should have the same spin, parity and isospin, but a different third component of the isospin of the original nucleus. As ^{208}Pb has $J^P = 0^+$, $T = 22$, $T_3 = -22$, ^{208}Bi, the isobaric analog state of ^{208}Pb, has the same J^P and T but a different $T_3 = -21$. The energy difference between the two isobaric analog states is

$$\Delta E \approx \frac{6}{5}\frac{Ze^2}{R} + (m_H - m_n)c^2 = \frac{6}{5}\frac{Z\hbar c}{R}\left(\frac{e^2}{\hbar c}\right) - 0.78$$

$$= \frac{6 \times 82 \times 197}{5 \times 1.2 \times 208^{1/3} \times 137} - 0.78 = 19.1 \text{ MeV}.$$

(f) The selection rules for super-allowed Fermi transition are $\Delta J = 0$, $\Delta P = +$, $\Delta T = 0$, so the wave function of the daughter nucleus is very similar to that of the parent. As the isospin is a good quantum number super-allowed transitions occur generally between isospin multiplets. For a heavy nucleus, however, the difference in Coulomb energy between isobaric analog states can be 10 MeV or higher, and so the isobaric analogy state is highly excited. As such, they can emit nucleons rather than undergo β-decay.

2070

The simplest model for low-lying states of nuclei with N and Z between 20 and 28 involves only $f_{7/2}$ nucleons.

(a) Using this model predict the magnetic dipole moments of $^{41}_{20}Ca_{21}$ and $^{41}_{21}Sc_{20}$. Estimate crudely the electric quadrupole moments for these two cases as well.

(b) What states are expected in $^{42}_{20}Ca$ according to an application of this model? Calculate the magnetic dipole and electric quadrupole moments for these states. Sketch the complete decay sequence expected experimentally for the highest spin state.

(c) The first excited state in $^{43}_{21}Ca_{23}$ is shown below in Fig. 2.14 with a half-life of 34 picoseconds for decay to the ground state. Estimate the lifetime expected for this state on the basis of a single-particle model. The

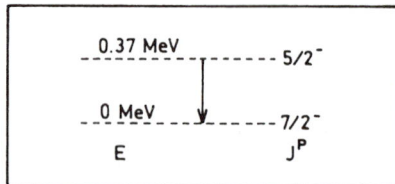

Fig. 2.14

experimental values are

$$\mu_n = -1.91\mu_N, \qquad \mu(^{41}Ca) = -1.59\mu_N$$

$$\mu_p = 2.79\mu_N, \qquad \mu(^{41}Sc) = 5.43\mu_N$$

Solution:

(a) ^{41}Ca has a neutron and ^{41}Sc has a proton outside closed shells in state $1f_{7/2}$. As closed shells do not contribute to the nuclear magnetic moment, the latter is determined by the extra-shell nucleons. The nuclear magnetic moment is given by

$$\mu = gj\mu_N,$$

where **j** is the total angular momentum, μ_N is the nuclear magneton. For a single nucleon in a central field, the g-factor is (**Problem 2061**)

$$g = \frac{(2j-1)g_l + g_s}{2j} \qquad \text{for } j = l + \frac{1}{2},$$

$$g = \frac{(2j+3)g_l - g_s}{2(j+1)} \qquad \text{for } j = l - \frac{1}{2}.$$

For neutron, $g_l = 0$, $g_s = g_n = -\frac{1.91}{\frac{1}{2}} = -3.82$. As $l = 3$ and $j = \frac{7}{2} = 3 + \frac{1}{2}$, we have for ^{41}Ca

$$\mu(^{41}Ca) = -\frac{3.82}{2j} \times j\mu_N = -1.91\mu_N.$$

For proton, $g_l = 1$, $g_s = g_p = \frac{2.79}{1/2} = 5.58$. As $j = \frac{7}{2} = 3 + \frac{1}{2}$, we have for ^{41}Sc

$$\mu(^{41}Sc) = \frac{(7-1) + 5.58}{7} \times \frac{7}{2}\mu_N = 5.79\mu_N.$$

Note that these values are only in rough agreement with the given experimental values.

The electric quadrupole moment of ^{41}Sc, which has a single proton outside closed shells, is given by

$$Q(^{41}Sc) = -e^2\langle r^2\rangle \frac{2j-1}{2(j+1)} = -\langle r^2\rangle \frac{2j-1}{2(j+1)},$$

where $\langle r^2 \rangle$ is the mean-square distance from the center and the proton charge is taken to be one. For an order-of-magnitude estimate take $\langle r^2 \rangle = (1.2 \times A^{1/3})^2 \ fm^2$. Then

$$Q(^{41}Sc) = -\frac{6}{9} \times (1.2 \times 41^{\frac{1}{3}})^2 = -1.14 \times 10^{-25} \ cm^2 \,.$$

^{41}Ca has a neutron outside the full shells. Its electric quadrupole moment is caused by the polarization of the neutron relative to the nucleus center and is

$$Q(^{41}Ca) \approx \frac{Z}{(A-1)^2}|Q(^{41}Sc)| = 1.43 \times 10^{-27} \ cm^2 \,.$$

(b) As shown in Fig. 2.15 the ground state of ^{42}Ca nucleus is 0^+. The two last neutrons, which are in $f_{7/2}$ can be coupled to form levels of $J = 7, 6, 5 \ldots, 0$ and positive parity. Taking into account the antisymmetry for identical particles, the possible levels are those with $J = 6, 4, 2, 0$. (We require $L + S = $ even, see **Problem 2054**. As $S = 0$, $J = $ even.)

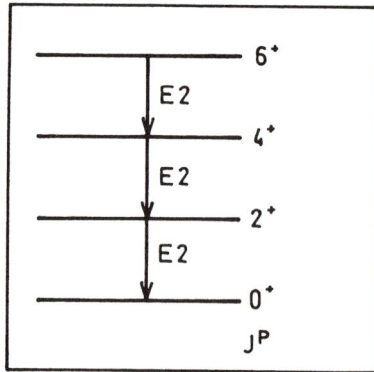

Fig. 2.15

The magnetic dipole moment μ of a two-nucleon system is given by

$$\mu = g\mathbf{J}\mu_N = (g_1\mathbf{j}_1 + g_2\mathbf{j}_2)\mu_N$$

with $\mathbf{J} = \mathbf{j}_1 + \mathbf{j}_2$. As

$$g\mathbf{J}^2 = g_1\mathbf{j}_1 \cdot \mathbf{J} + g_2\mathbf{j}_2 \cdot \mathbf{J},$$

$$\mathbf{j}_1 \cdot \mathbf{J} = \frac{1}{2}(\mathbf{J}^2 + \mathbf{j}_1^2 - \mathbf{j}_2^2),$$

$$\mathbf{j}_2 \cdot \mathbf{J} = \frac{1}{2}(\mathbf{J}^2 + \mathbf{j}_2^2 - \mathbf{j}_1^2),$$

we have

$$g\mathbf{J}^2 = \frac{1}{2}(g_1 + g_2)\mathbf{J}^2 + \frac{1}{2}(g_1 - g_2)(\mathbf{j}_1^2 - \mathbf{j}_2^2).$$

or

$$g = \frac{1}{2}(g_1 + g_2) + \frac{1}{2}(g_1 - g_2)\frac{j_1(j_1 + 1) - j_2(j_2 + 1)}{J(J + 1)}.$$

For ^{42}Ca, the two nucleons outside full shells each has $j = 7/2$. As

$$g_1 = g_2 = \frac{-3.82}{j_1}, \qquad j_1 = \frac{7}{2},$$

we have $\mu(^{42}Ca) = g_1 J \mu_N = -1.09J\mu_N$ with $J = 0, 2, 4, 6$.

The ground-state quadrupole moment of ^{42}Ca is $Q = 0$. One can get the excited state quadrupole moment using the reduced transition rate for γ-transition

$$B(E2, 2^+ \to 0^+) = \frac{e^2 Q_0^2}{16\pi}$$

where Q_0 is the intrinsic electric quadrupole moment. The first excited state 2^+ of ^{42}Ca has excitation energy 1.524 MeV and

$$B(E2 : 2^+ \to 0^+) = 81.5e^2 \; fm^4,$$

or

$$Q_0 = \sqrt{16\pi \times 81.5} = 64 \; fm^2.$$

For other states the quadrupole moments are given by

$$Q = \frac{K^2 - J(J + 1)}{(J + 1)(2J + 3)}Q_0 = -\frac{J(J + 1)Q_0}{(J + 1)(2J + 3)} = \frac{-J}{2J + 3}Q_0$$

as $K = 0$. Thus $Q = 18.3 \; fm^2$ for $J = 2$, $23.3 \; fm^2$ for $J = 4$, and $25.6 \; fm^2$ for $J = 6$.

(c) The selection rule for the γ-transition $(\frac{5}{2})^- \to (\frac{7}{2})^-$ is $(\frac{5}{2} - \frac{7}{2}) \leq L \leq \frac{5}{2} + \frac{7}{2}$, i.e. $L = 1, 2, 3, 4, 5, 6$, with the lowest order having the highest probability, for which parity is conserved. Then the most probable are magnetic dipole transition M_1 for which $\Delta P = -(-1)^{1+1} = +$, or electric quadrupole transition $E2$ for which $\Delta P = (-1)^2 = +$. According to the single-particle model (**Problem 2093**),

$$\lambda_{M1} = \frac{1.9(L+1)}{L[(2L+1)!!]^2}\left(\frac{3}{L+3}\right)^2\left(\frac{E_\gamma}{197}\right)^{2L+1} \times (1.4 \times A^{1/3})^{2L-2} \times 10^{21}$$

$$= \frac{1.9 \times 2}{3^2}\left(\frac{3}{4}\right)^2\left(\frac{0.37}{197}\right)^3 (1.4 \times 43^{1/3})^0 \times 10^{21}$$

$$= 1.57 \times 10^{12}\ s^{-1},$$

$$\lambda_{E2} = \frac{4.4(L+1)}{L[(2L+1)!!]^2}\left[\frac{3}{L+3}\right]^2\left(\frac{E_\gamma}{197}\right)^{2L+1} \times (1.4 \times A^{1/3})^{2L} \times 10^{21}$$

$$= \frac{4.4 \times 3}{2 \times (5 \times 3)^2}\left(\frac{3}{L+3}\right)^2\left(\frac{0.37}{197}\right)^5 (1.4 \times 43^{1/3})^4 \times 10^{21}$$

$$= 1.4 \times 10^8\ s^{-1}.$$

As $\lambda_{E2} \ll \lambda_{M1}$, $E2$ could be neglected, and so

$$T_{1/2} \approx \frac{\ln 2}{\lambda_{M1}} = \frac{\ln 2}{1.57 \times 10^{12}} = 4.4 \times 10^{-13}\ s.$$

This result from the single-particle model is some 20 times smaller than the experimental value. The discrepancy is probably due to γ-transition caused by change of the collective motion of the nucleons.

2071

The variation of the binding energy of a single neutron in a "realistic" potential model of the neutron-nucleus interaction is shown in Fig. 2.16.

(a) What are the neutron separation energies for $^{40}_{20}$Ca and $^{208}_{82}$Pb?

(b) What is the best neutron magic number between those for ^{40}Ca and ^{208}Pb?

(c) Draw the spectrum including spins, parities and approximate relative energy levels for the lowest five states you would expected in ^{210}Pb and explain.

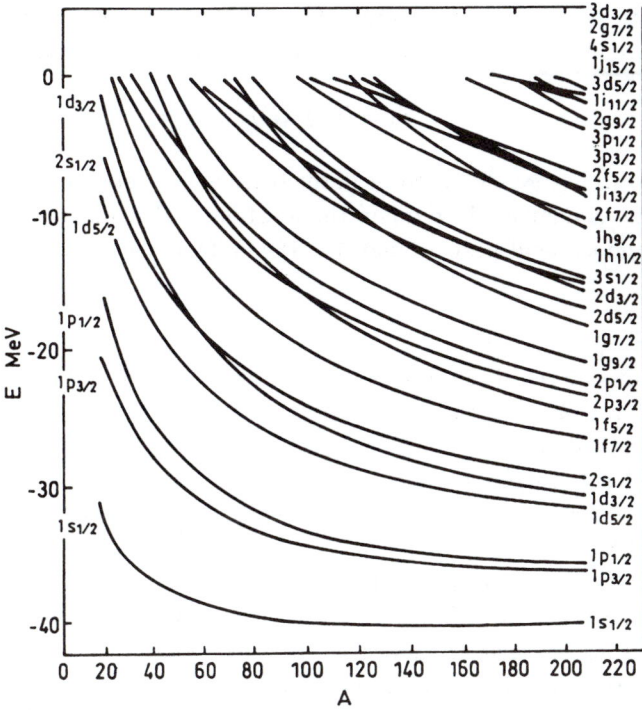

Fig. 2.16

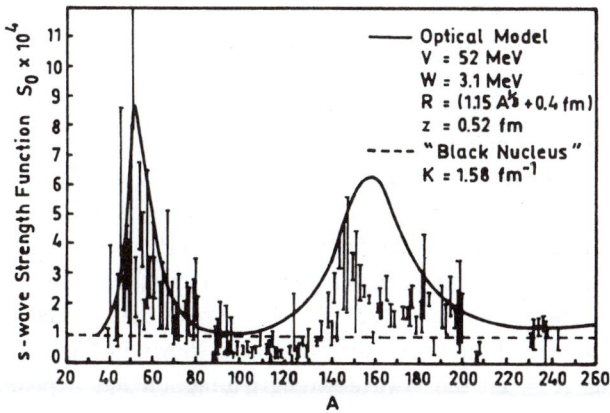

Fig. 2.17

(d) The s-wave neutron strength function S_0 is defined as the ratio of the average neutron width $\langle \Gamma_n \rangle$ to the average local energy spacing $\langle D \rangle$:

$$S_0 = \langle \Gamma_n \rangle / \langle D \rangle \,.$$

Figure 2.17 shows the variation of the thermal neutron strength function S_0 with mass number A. Explain the location of the single peak around $A \approx 50$, and the split peak around $A \approx 160$. Why is the second peak split?

(*Princeton*)

Solution:

(a) The outermost neutron of ^{40}Ca is the twentieth one. Figure 2.16 gives for $A = 40$ that the last neutron is in $1d_{3/2}$ shell with separation energy of about 13 MeV.

^{208}Pb has full shells, the last pair of neutrons being in $3p_{1/2}$ shell. From Fig. 2.16 we note that for $A = 208$, the separation energy of each neutron is about 3 MeV.

(b) The neutron magic numbers between ^{40}Ca and ^{208}Pb are 28, 50 and 82. For nuclei of $N = Z$, at the neutron magic number $N = 28$ the separation energies are about 13 MeV. At neutron number $N = 50$, the separation energies are also about 13 MeV. At N=82, the separation energies are about 12 MeV. However, for heavy nuclei, there are more neutrons than protons, so $A < 2N$. On account of this, for the nuclei of magic numbers 50 and 82, the separation energies are somewhat less than those given above. At the magic number 28 the separation energy is highest, and so this is the best neutron magic number.

(c) The last two neutrons of ^{210}Pb are in $2g_{9/2}$ shell, outside of the double-full shells. As the two nucleons are in the same orbit and will normally pair up to $J = 0$, the even-even nucleus has ground state 0^+.

The two outermost neutrons in $2g_{9/2}$ of ^{210}Pb can couple to form states of $J = 9, 8, 7 \ldots$. However a two-neutron system has isospin $T = 1$. As the antisymmetry of the total wave function requires $J + T = $ odd, the allowed J are 8, 6, 4, 2, 0 and the parity is positive. Thus the spin-parities of the lowest five states are $8^+, 6^+, 4^+, 2^+, 0^+$. Because of the residual interaction, the five states are of different energy levels as shown in Fig. 2.18.

(d) Near $A = 50$ the s-wave strength function has a peak. This is because when $A = 50$ the excitation energy of $3s$ energy level roughly equals the neutron binding energy. A calculation using the optical model gives the

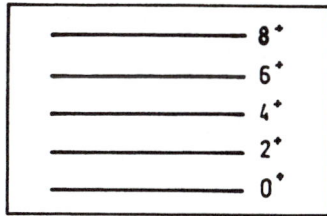

Fig. 2.18

shape of the peak as shown in Fig. 2.17 (solid curve). When $150 < A < 190$, the s-wave strength function again peaks due to the equality of excitation energy of $4s$ neutron and its binding energy. However, nuclear deformation in this region is greater, particularly near $A = 160$ to 170, where the nuclei have a tendency to deform permanently. Here the binding energies differ appreciably from those given by the single-particle model: the peak of the s-wave strength function becomes lower and splits into two smaller peaks.

2072

Figure 2.19 gives the low-lying states of ^{18}O with their spin-parity assignments and energies (in MeV) relative to the 0^+ ground state.

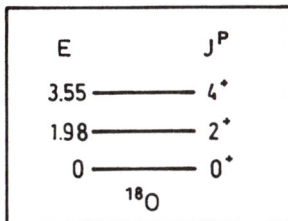

Fig. 2.19

(a) Explain why these J^p values are just what one would expect in the standard shell model.

(b) What J^p values are expected for the low-lying states of ^{19}O?

(c) Given the energies (relative to the ground state) of these ^{18}O levels, it is possible within the shell model, ignoring interconfiguration interactions, to compute the energy separations of the ^{19}O levels. However, this requires familiarity with complicated Clebsch–Gordon coefficients. To simplify matters, consider a fictitious situation where the 2^+ and 4^+ levels of ^{18}O have the energies 2 MeV and $6\frac{2}{3}$ MeV respectively. For this fictitious world, compute the energies of the low-lying ^{19}O levels.

(Princeton)

Solution:

(a) In a simple shell model, ignoring the residual interactions between nucleons and considering only the spin-orbit coupling, we have for a system of A nucleons,

$$H = \Sigma H_i \,,$$

with

$$H_i = T_i + V_i \,,$$

$$V_i = V_0^i(r) + f(r)\boldsymbol{S}_i \cdot \boldsymbol{l}_i \,,$$

$$H_i \Psi_i = E_i \Psi_i \,,$$

$$\Psi = \prod_{i=1}^{A} \psi_i \,.$$

When considering residual interactions, the difference of energy between different interconfigurations of the nucleons in the same level must be taken into account.

For ^{18}O nucleus, the two neutrons outside the full shells can fill the $1d_{5/2}$, $2s_{1/2}$ and $1d_{3/2}$ levels (see Fig. 2.16). When the two nucleons are in the same orbit, the antisymmetry of the system's total wave function requires $T + J = $ odd. As $T = 1$, J is even. Then the possible ground and excited states of ^{18}O are:

$$(1d_{5/2})^2 : \quad J = 0^+, 2^+, 4^+, \quad T = 1 \,,$$

$$(1d_{5/2}2s_{1/2}) : \quad J = 2^+, \quad T = 1 \,,$$

$$(2s_{1/2})^2 : \quad J = 0^+, \quad T = 1 \,,$$

$$(1d_{3/2})^2 : \quad J = 0^+, 2^+, \quad T = 1 \,.$$

The three low-lying states of ^{18}O as given in Fig. 2.19, $0^+, 2^+, 4^+$, should then correspond to the configuration $(1d_{5/2})^2$. However, when considering the energies of the levels, using only the $(d_{5/2})^2$ configuration does not agree well with experiment. One must also allow mixing the configurations $1d_{5/1}, 2s_{1/2}, 1d_{3/2}$, which gives fairly good agreement with the experimental values, as shown in Fig. 2.20.

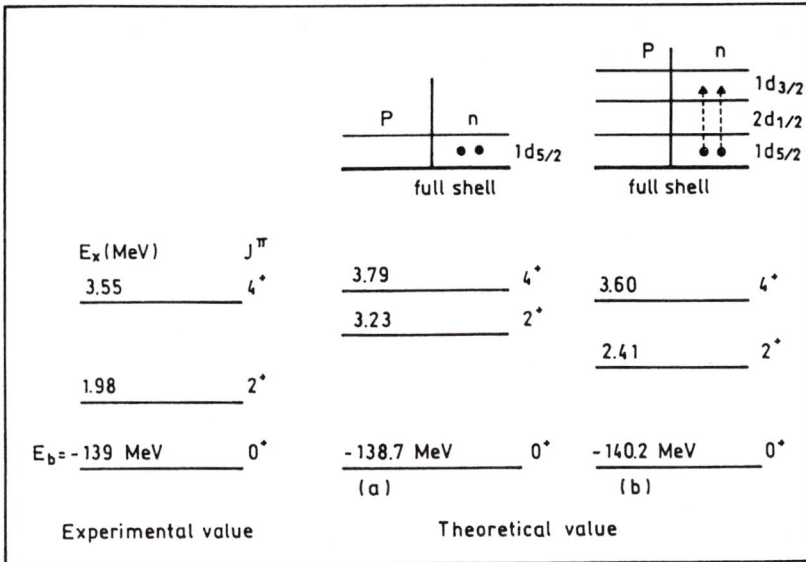

Fig. 2.20

(b) To calculate the lowest levels of ^{19}O using the simple shell model and ignoring interconfiguration interactions, we consider the last unpaired neutron. According to Fig. 2.16, it can go to $1d_{5/2}, 2s_{1/2}$, or $1d_{3/2}$. So the ground state is $\left(\frac{5}{2}\right)^+$, the first excited state $\left(\frac{1}{2}\right)^+$, and the second excited state $\left(\frac{3}{2}\right)^+$.

If interconfiguration interactions are taken into account, the three neutrons outside the full shells can go into the $1d_{5/2}$ and $2s_{1/2}$ orbits to form the following configurations:

$$[(d_{5/2})^3]_{5/2,m}, \quad [(d_{5/2})^2 s_{1/2}]_{5/2,m}, \quad [d_{5/2}(s_{1/2})_0^2]_{5/2,m}, J^P = \left(\frac{5}{2}\right)^+,$$

$$[(d_{5/2})_0^2 s_{1/2}]_{1/2,m}, J^P = \left(\frac{1}{2}\right)^+,$$

$$[(d_{5/2})^3]_{3/2,m}, \quad [(d_{5/2})_2^2 s_{1/2}]_{3/2,m}, J^P = \left(\frac{3}{2}\right)^+.$$

Moreover, states with $J^P = \frac{7^+}{2}, \frac{9^+}{2}$ are also possible.

(c) In the fictitious case the lowest excited states of ^{18}O are $0^+, 2^+, 4^+$ with energies $0, 2, 6\frac{2}{3}$ MeV as shown in Fig. 2.21.

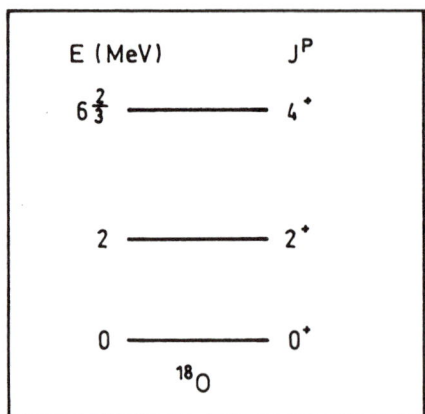

E (MeV) J^P

$6\frac{2}{3}$ ———— $4^\bullet$

2 ———— $2^\bullet$

0 ———— $0^\bullet$

^{18}O

Fig. 2.21

This fictitious energy level structure corresponds to the rotational spectrum of an even-even nucleus, for in the latter we have

$$\frac{E_2}{E_1} = \frac{J_2(J_2+1)}{J_1(J_1+1)} = \frac{4(4+1)}{2(2+1)} = \frac{6\frac{2}{3}}{2}.$$

Taking this assumption as valid, one can deduce the moment of inertia I of ^{18}O. If this assumption can be applied to ^{19}O also, and if the moments of inertia of ^{19}O, ^{18}O can be taken to be roughly equal, then one can estimate the energy levels of ^{19}O. As $E_J = \frac{\hbar^2}{2I}J(J+1)$, we have for ^{18}O

$$\frac{\hbar^2}{2I} = \frac{E_J}{J(J+1)} = \frac{2}{2(2+1)} = \frac{1}{3} \text{ MeV}.$$

Assume that I is the same for ^{19}O. From (b) we see that the three lowest rotational levels of ^{19}O correspond to $J = \frac{5}{2}, \frac{7}{2}, \frac{9}{2}$. Hence

$$E_{5/2} = 0, \text{ being the ground state of } {}^{19}O,$$

$$E_{7/2} = \frac{1}{3} \left[\frac{7}{2} \left(\frac{7}{2} + 1 \right) - \frac{5}{2} \left(\frac{5}{2} + 1 \right) \right] = 2\frac{1}{3} \text{ MeV},$$

$$E_{9/2} = \frac{1}{3} \times \frac{1}{4} (9 \times 11 - 5 \times 7) = 5\frac{1}{3} \text{ MeV}.$$

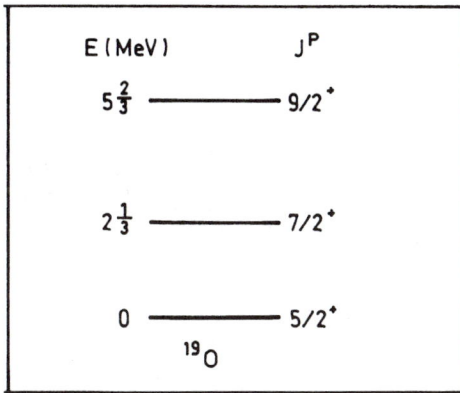

Fig. 2.22

2073

The following nonrelativistic Hamiltonians can be used to describe a system of nucleons:

$$H_0 = \sum_i \frac{p_i^2}{2m} + \frac{1}{2} m\omega_0^2 r_i^2,$$

$$H_1 = H_0 - \sum_i \beta \hat{l}_i \cdot s_i,$$

$$H_2 = H_1 - \sum_i \frac{1}{2} m\omega^2 (2z_i^2 - x_i^2 - y_i^2),$$

where $\hbar\omega_0 \gg \beta \gg \hbar\omega$.

(a) For each Hamiltonian H_0, H_1, H_2, identify the exactly and approximately conserved quantities of the system. For the ground state of each model, give the appropriate quantum numbers for the last filled single-particle orbital when the number n of identical nucleons is 11, 13 and 15.

(b) What important additional features should be included when the low-lying states of either spherical or deformed nucleons are to be described?

(c) The known levels of Aluminum 27, $^{27}_{13}Al_{14}$, below 5 MeV are shown in Fig. 2.23. Which states correspond to the predictions of the spherical and of the deformed models?

(Princeton)

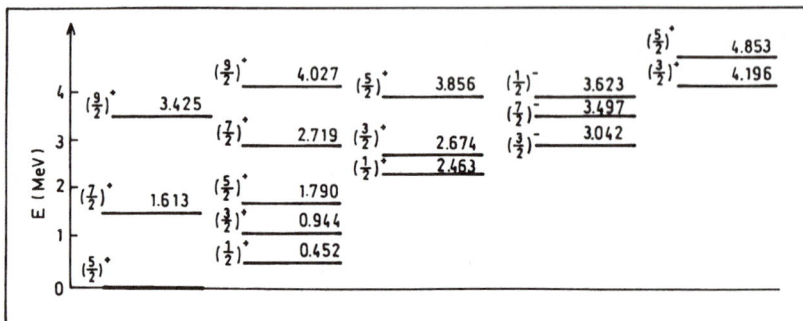

Fig. 2.23

Solution:

(a) For H_0 the exactly conserved quantities are energy E, orbital angular momentum L, total spin S, total angular momentum J, and parity.

For H_1 the exactly conserved quantities are E, J and parity, the approximately conserved ones are L and S.

For H_2 the exactly conserved quantities are E, the third component of the total angular momentum J_z, and parity, the approximately conserved ones are J, L, S.

As H_0 is an isotropic harmonic oscillator field, $E_N = \left(N + \frac{3}{2}\right)\hbar\omega$. The low-lying states are as follows (Figs. 2.12 and 2.16):

$N = 0$ gives the ground state $1s_{1/2}$.

$N = 1$ gives the p states, $1p_{3/2}$ and $1p_{1/2}$ which are degenerate.

$N = 2$ gives $2s$ and $1d$ states, $1d_{5/2}$, $2s_{1/2}$, $1d_{3/2}$, which are degenerate.

When the number of identical nucleons is $n = 11, 13, 15$, the last filled nucleons all have $N = 2$.

H_1 can be rewritten as

$$H_1 = H_0 - \sum_i \beta(\mathbf{l}_i \cdot \mathbf{s}_i) = H_0 - \sum_i \frac{1}{2}\beta[j_i(j_i + 1) - l_i(l_i + 1) - s_i(s_i + 1)].$$

The greater is j_i, the lower is the energy. For this Hamiltonian, some of the degeneracy is lost: $1p_{3/2}$ and $1p_{1/2}$ are separated, so are $1d_{3/2}$ and $1d_{5/2}$. 11 or 13 identical nucleons can fill up to the $1d_{5/2}$ state, while for $n = 15$, the last nucleon well go into the $2s_{1/2}$ state.

H_2 can be rewritten as

$$H_2 = H_1 - \sum_i \frac{1}{2}m\omega^2 r_i^2(3\cos^2\theta - 1),$$

which corresponds to a deformed nucleus. For the Hamiltonain, $1p_{3/2}$, $1d_{3/2}$, and $1d_{5/2}$ energy levels are split further:

$1d_{5/2}$ level is split into $\left(\frac{1}{2}\right)^+, \left(\frac{3}{2}\right)^+, \left(\frac{5}{2}\right)^+$,

$1d_{3/2}$ level is split into $\left(\frac{1}{2}\right)^+, \left(\frac{3}{2}\right)^+$,

$1p_{3/2}$ level is split into $\left(\frac{1}{2}\right)^-, \left(\frac{3}{2}\right)^-$,

Let the deformation parameter be ε. The order of the split energy levels well depend on ε. According to the single-particle model of deformed nuclei, when $\varepsilon \approx 0.3$ (such as for ^{27}Al), the orbit of the last nucleon is

$\left(\frac{3}{2}\right)^+$ of the $1d_{5/2}$ level if $n = 11$,

$\left(\frac{5}{2}\right)^+$ of the $1d_{5/2}$ level if $n = 13$,

$\left(\frac{1}{2}\right)^+$ of the $2s_{1/2}$ level if $n = 15$.

(b) For a spherical nucleus, when considering the ground and low excited states, pairing effect and interconfiguration interactions are to be included. For a deformed nucleus, besides the above, the effect of the deforming field on the single-particle energy levels as well as the collective vibration and rotation are to be taken into account also.

(c) ^{27}Al is a deformed nucleus with $\varepsilon \approx 0.3$. The configurations of the 14 neutrons and 13 protons in a spherical nucleus are

$$n : (1s_{1/2})^2(1p_{3/2})^4(1p_{1/2})^2(1d_{5/2})^6,$$

$$n : (1s_{1/2})^2(1p_{3/2})^4(1p_{1/2})^2(1d_{5/2})^5.$$

The ground state is given by the state of the last unpaired nucleon $(1d_{5/2})$: $J^p = \left(\frac{5}{2}\right)^+$.

If the nucleus is deformed, not only are energy levels like $1p_{3/2}$, $1d_{5/2}$, $1d_{3/2}$ split, the levels become more crowded and the order changes. Strictly speaking, the energy levels of ^{27}Al are filled up in the order of single-particle energy levels of a deformed nucleus. In addition, there is also collective motion, which makes the energy levels very complicated. Comparing the energy levels with theory, we have, corresponding to the levels of a spherical nucleus of the same J^p, the levels,

$$\text{ground state :} \quad J^p = \left(\frac{5}{2}\right)^+ , E = 0 ,$$

$$\text{excited states :} \quad J^p = \left(\frac{1}{2}\right)^+ , E = 2.463 \text{ MeV} ,$$

$$J^p = \left(\frac{3}{2}\right)^+ , E = 4.156 \text{ MeV} ;$$

corresponding to the single-particle energy levels of a deformed nucleus the levels

$$\text{ground state :} \quad K^p = \left(\frac{5}{2}\right)^+ , E = 0 ,$$

$$\text{excited states :} \quad K^p = \left(\frac{1}{2}\right)^+ , E = 0.452 \text{ MeV} ,$$

$$K^p = \left(\frac{1}{2}\right)^+ , E = 2.463 \text{ MeV} ,$$

$$K^p = \left(\frac{1}{2}\right)^- , E = 3.623 \text{ MeV} ,$$

$$K^p = \left(\frac{3}{2}\right)^+ , E = 4.196 \text{ MeV} ,$$

Also, every K^p corresponds to a collective-rotation energy band of the nucleus given by

$$E_J = \frac{\hbar^2}{2I}[J(J+1) - K(K+1)],$$

where $K \neq 1/2, J = K, K+1, \ldots$.

$$E_J = \frac{\hbar^2}{2I}\left[J(J+1) - \frac{3}{4} + a - a(-1)^{J+1/2}\left(J + \frac{1}{2}\right)\right],$$

where $K = 1/2, J = K, K+1, \ldots$.

For example, for rotational bands $\left(\frac{5}{2}\right)^+ (0), \left(\frac{7}{2}\right)^+ (1.613), \left(\frac{9}{2}\right)^+ (3.425)$, we have $K = \frac{5}{2}$,

$$\left(\frac{\hbar^2}{2I}\right)[(K+1)(K+2) - K(K+1)] = 1.613 \text{ MeV},$$

$$\left(\frac{\hbar^2}{2I}\right)[(K+2)(K+3) - K(K+1)] = 3.425 \text{ MeV}.$$

giving $\frac{\hbar^2}{2I} \approx 0.222$ MeV. For rotational bands $\left(\frac{1}{2}\right)^+ (0.452), \left(\frac{3}{2}\right)^+ (0.944),$ $\left(\frac{5}{2}\right)^+ (1.790), \left(\frac{7}{2}\right)^+ (2.719), \left(\frac{9}{2}\right)^+ (4.027)$, we have

$$\frac{\hbar^2}{2I} \approx 0.150 \text{ MeV}, \qquad a \approx -3.175 \times 10^2.$$

Similarly for $\left(\frac{1}{2}\right)^- (3.623), \left(\frac{7}{2}\right)^- (3.497)$ and $\left(\frac{3}{2}\right)^- (3.042)$ we have

$$\frac{\hbar^2}{2I} \approx 0.278 \text{ MeV}, \qquad a \approx 5.092.$$

2074

A recent model for collective nuclear states treats them in terms of interacting bosons. For a series of states that can be described as symmetric superposition of S and D bosons (i.e. of spins 0 and 2 respectively), what are the spins of the states having $N_d = 0, 1, 2$ and 3 bosons? If the energy of the S bosons is E_s and the energy of the D bosons is E_d, and there is a residual interaction between pairs of D bosons of constant strength α, what is the spectrum of the states with $N_s + N_d = 3$ bosons?

(Princeton)

Solution:

When $N_d = 0$, spin is 0,
$N_d = 1$, spin is 2,
$N_d = 2$, spin is 4,2,0,
$N_d = 3$, spin is 6, 4, 2, 0.
For states of $N_s + N_d = 3$, when

$$N_d = 0 : \quad N_s = 3, \qquad E = 3E_s \,,$$

$$N_d = 1 : \quad N_s = 2, \qquad E = E_d + 2E_s \,,$$

$$N_d = 2 : \quad N_s = 1, \qquad E = 2E_d + E_s + \alpha \,,$$

$$N_d = 3 : \quad N_s = 0, \qquad E = 3E_d + 3\alpha \,.$$

2075

A simplified model of the complex nuclear interaction is the pairing force, specified by a Hamiltonian of the form

$$H = -g \begin{pmatrix} 1 & 1 & \cdot & \cdot & 1 \\ 1 & 1 & \cdot & \cdot & 1 \\ \cdot & \cdot & \cdot & \cdot & \cdot \\ \cdot & \cdot & \cdot & \cdot & \cdot \\ 1 & 1 & \cdot & \cdot & 1 \end{pmatrix} \,,$$

in the two-identical-particle space for a single j orbit, with the basic states given by $(-1)^{j-m}|jm\rangle|j-m\rangle$. This interaction has a single outstanding eigenstate. What is its spin? What is its energy? What are the spins and energies of the rest of the two-particle states?

(*Princeton*)

Solution:

Suppose H is a $(j + \frac{1}{2}) \times (j + \frac{1}{2})$ matrix. The eigenstate can be written in the form

$$\Psi^{N=2} = \left(j + \frac{1}{2}\right)^{-1/2} \begin{pmatrix} 1 \\ 1 \\ \vdots \\ 1 \\ 1 \end{pmatrix},$$

where the column matrix has rank $\left(j + \frac{1}{2}\right) \times 1$. Then

$$\hat{H}\Psi^{N=2} = -g\left(j + \frac{1}{2}\right)\Psi^{N=2}.$$

Thus the energy eigenvalue of $\Psi^{N=2}$ is $-g\left(j + \frac{1}{2}\right)$. As the pairing force acts on states of $J = 0$ only, the spin is zero.

As the sum of the energy eigenvalues equals the trace of the $\hat{H}$ matrix, $-g\left(j + \frac{1}{2}\right)$, and H is a negative quantity, all the eigenstates orthogonal to $\Psi^{N=2}$ have energy eigenvalues zero, the corresponding angular momenta being $J = 2, 4, 6 \ldots$, etc.

5. NUCLEAR DECAYS (2076–2107)

2076

In its original (1911) form the Geiger–Nuttall law expresses the general relationship between α-particle range (R_α) and decay constant (λ) in natural α-radioactivity as a linear relation between $\log \lambda$ and $\log R$. Subsequently this was modified to an approximate linear relationship between $\log \lambda$ and some power of the α-particle energy, $E^x(\alpha)$.

Explain how this relationship between decay constant and energy is explained quantum-mechanically. Show also how the known general features of the atomic nucleus make it possible to explain the extremely rapid dependence of λ on $E(\alpha)$. (For example, from $E(\alpha) = 5.3$ MeV for Po^{210} to $E(\alpha) = 7.7$ MeV for Po^{214}, λ increases by a factor of some 10^{10}, from a half-life of about 140 days to one of 1.6×10^{-4} sec.)

(Columbia)

Solution:

α-decay can be considered as the transmission of an α-particle through the potential barrier of the daughter nucleus. Similar to that shown in

Fig. 2.7, where R is the nuclear radius, r_1 is the point where the Coulomb repulsive potential $V(r) = Zze^2/r$ equals the α-particle energy E. Using a three-dimensional potential and neglecting angular momentum, we can obtain the transmission coefficient T by the W.K.B. method:

$$T = e^{-2G},$$

where

$$G = \frac{1}{\hbar} \int_R^{r_1} (2m|E - V|)^{1/2} dr,$$

with $V = zZe^2/r$, $E = zZe^2/r_1$, $z = 2$, Ze being the charge of the daughter nucleus. Integration gives

$$G = \frac{1}{\hbar}(2mzZe^2 r_1)^{1/2} \left[\arccos\left(\frac{R}{r_1}\right) - \left(\frac{R}{r_1} - \frac{R^2}{r_1^2}\right)^{1/2} \right]$$

$$\xrightarrow{\frac{R}{r_1} \to 0} \frac{1}{\hbar}(2mzZe^2 r_1)^{1/2} \left[\frac{\pi}{2} - \left(\frac{R}{r_1}\right)^{1/2} \right].$$

Suppose the α-particle has velocity v_0 in the potential well. Then it collides with the walls $\frac{v_0}{R}$ times per unit time and the probability of decay per unit time is $\lambda = v_0 T/R$. Hence

$$\ln \lambda = -\frac{\sqrt{2m}BR\pi}{\hbar} \left(E^{-\frac{1}{2}} - \frac{2}{\pi}B^{-\frac{1}{2}} \right) + \ln\frac{v_0}{R},$$

where $B = zZe^2/R$. This is a linear relationship between $\log \lambda$ and $E^{-1/2}$ for α-emitters of the same radioactive series.

For $_{84}$Po,

$$\log_{10} \frac{T(^{210}Po)}{T(^{214}Po)} = 0.434[\ln \lambda(^{214}Po) - \ln \lambda(^{210}Po)]$$

$$= 0.434 \times \sqrt{2mc^2}zZ \left(\frac{e^2}{\hbar c}\right) \left(\frac{1}{\sqrt{E_{210}}} - \frac{1}{\sqrt{E_{214}}}\right)$$

$$= \frac{0.434 \times \sqrt{8 \times 940} \times 2 \times (84 - 2)}{137} \left(\frac{1}{\sqrt{5 \cdot 3}} - \frac{1}{\sqrt{7 \cdot 7}}\right)$$

$$\approx 10.$$

Thus the life-times differ by 10 orders of magnitude.

2077

The half-life of a radioactive isotope is strongly dependent on the energy liberated in the decay. The energy dependence of the half-life, however, follows quite different laws for α- and β-emitters.

(a) Derive the specific law for α-emitters.

(b) Indicate why the law for β-emitters is different by discussing in detail the difference between the two processes.

(Columbia)

Solution:

(a) For a quantum-mechanical derivation of the Geiger–Nuttall law for α-decays see **Problem 2076**.

(b) Whereas α-decay may be considered as the transmission of an α-particle through a Coulomb potential barrier to exit the daughter nucleus, β-decay is the result of the disintegration of a neutron in the nucleus into a proton, which remains in the nucleus, an electron and an antineutrino, which are emitted. Fermi has obtained the β-particle spectrum using a method similar to that for γ-emission. Basically the transition probability per unit time is given by Fermi's golden rule No. 2,

$$\omega = \frac{2\pi}{\hbar}|H_{fi}|^2\rho(E)\,,$$

where E is the decay energy, H_{fi} is the transition matrix element and $\rho(E) = \frac{dN}{dE}$ is the number of final states per unit energy interval.

For decay energy E, the number of states of the electron in the momentum interval p_e and $p_e + dp_e$ is

$$dN_e = \frac{V4\pi p_e^2 dp_e}{(2\pi\hbar)^3}\,,$$

where V is the volume of normalization. Similarly for the antineutrino we have

$$dN_\nu = \frac{4\pi p_\nu^2 dp_\nu}{(2\pi\hbar)^3}\,,$$

and so $dN = dN_e dN_\nu$. However p_e and p_ν are not independent. They are related through $E_e = \sqrt{p_e^2 c^2 + m_e^2 c^4}$, $E_\nu = p_\nu c$ by $E = E_e + E_\nu$. We can write $p_\nu = \frac{E - E_e}{c}$, and for a given E_e, $dp_\nu = \frac{dE_\nu}{c} = \frac{dE}{c}$. Thus

$$\frac{dN}{dE} = \int \frac{dN_e dN_\nu}{dE} = \frac{V^2}{4\pi^4\hbar^6 c^3}\int_0^{p_{\max}} (E - E_e)^2 p_e^2 dp_e\,,$$

where p_{max} corresponds to the end-point energy of the β-particle spectrum $E_0 \approx E$, and hence

$$\lambda = \frac{2\pi}{\hbar}|H_{fi}|^2 \frac{dN}{dE} = \frac{g^2|M_{fi}|^2}{2\pi^3\hbar^7 c^3} \int_0^{p_{max}} (E - \sqrt{p_e^2 c^2 + m_e^2 c^4})^2 p_e^2 dp_e \,,$$

where $M_{fi} = \frac{V H_{fi}}{g}$ and g is the coupling constant.

In terms of the kinetic energy T, as

$$E_e = T + m_e c^2 = \sqrt{p_e^2 c^2 + m_e^2 c^4}\,, \qquad E = T_0 + m_e c^2\,,$$

the above integral can be written in the form

$$\int_0^{T_0} (T + m_e c^2)(T^2 + 2m_e c^2 T)^{\frac{1}{2}} (T_0 - T)^2 dT\,.$$

This shows that for β-decays

$$\lambda \sim T_0^5\,,$$

which is the basis of the Sargent curve.

This relation is quite different from that for α-decays,

$$\lambda \sim \exp\left(-\frac{c}{\sqrt{E}}\right)\,,$$

where E is the decay energy and C is a constant.

2078

Natural gold $^{197}_{79}\text{Au}$ is radioactive since it is unstable against α-decay with an energy of 3.3 MeV. Estimate the lifetime of $^{197}_{79}\text{Au}$ to explain why gold does not burn a hole in your pocket.

(Princeton)

Solution:

The Geiger–Nuttall law

$$\log_{10} \lambda = C - DE_\alpha^{-1/2}\,,$$

where C, D are constants depending on Z, which can be calculated using quantum theory, E_α is the α-particle energy, can be used to estimate the life-time of ^{197}Au. For a rough estimate, use the values of C, D for Pb, $C \approx 52$, $D \approx 140$ (MeV)$^{\frac{1}{2}}$. Thus

$$\lambda \approx 10^{(52-140E^{-1/2})} \approx 10^{-25} \ s^{-1}$$

and so

$$T_{1/2} = \frac{1}{\lambda} \ln 2 \approx 6.9 \times 10^{24} \ s \approx 2.2 \times 10^{17} \ yr \,.$$

Thus the number of decays in a human's lifetime is too small to worry about.

2079

The half-life of ^{239}Pu has been determined by immersing a sphere of ^{239}Pu of mass 120.1 gm in liquid nitrogen of a volume enough to stop all α-particles and measuring the rate of evaporation of the liquid. The evaporation rate corresponded to a power of 0.231 W. Calculate, to the nearest hundred years, the half-life of ^{239}Pu, given that the energy of its decay alpha-particles is 5.144 MeV. (Take into account the recoil energy of the product nucleus.) Given conversion factors:

$$1 \text{ MeV} = 1.60206 \times 10^{-13} \text{ joule} \,,$$

$$1 \text{ atomic mass unit} = 1.66 \times 10^{-24} \text{ gm} \,.$$

(SUNY, Buffalo)

Solution:

The decay takes place according to $^{239}Pu \rightarrow \alpha +^{235} U$.
The recoil energy of ^{235}U is

$$E_u = \frac{p_u^2}{2M_u} = \frac{p_\alpha^2}{2M_u} = \frac{2M_\alpha E_\alpha}{2M_u} = \frac{4}{235}E_\alpha \,.$$

The energy released per α-decay is

$$E = E_u + E_\alpha = \frac{239}{235}E_\alpha = 5.232 \text{ MeV} \,.$$

The decay rate is

$$\frac{dN}{dt} = \frac{0.231}{5.232 \times 1.60206 \times 10^{-13}} = 2.756 \times 10^{11} \ s^{-1} \ .$$

The number of ^{239}Pu is

$$N = \frac{120.1 \times 5.61 \times 10^{26}}{239 \times 939} = 3.002 \times 10^{23} \ .$$

The half-life is

$$T_{1/2} = \frac{\ln 2}{\lambda} = \frac{N \ln 2}{\frac{dN}{dt}} = \frac{3.002 \times 10^{23} \times \ln 2}{2.756 \times 10^{11}} = 7.55 \times 10^{11} \ s = 2.39 \times 10^4 \ yr \ .$$

2080

^{8}Li is an example of a β-delayed particle emitter. The ^{8}Li ground state has a half-life of 0.85 s and decays to the 2.9 MeV level in Be as shown in Fig. 2.24. The 2.9 MeV level then decays into 2 alpha-particles with a half-life of 10^{-22} s.

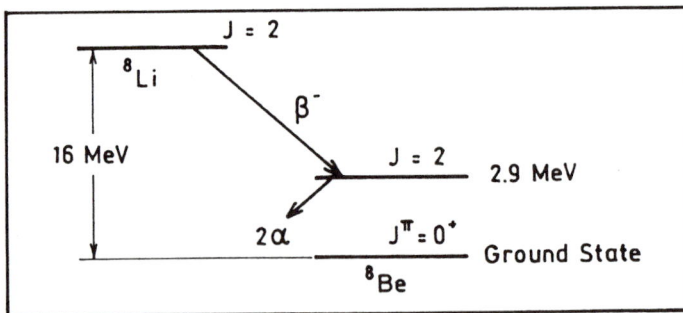

Fig. 2.24

(a) What is the parity of the 2.9 MeV level in ^{8}Be? Give your reasoning.
(b) Why is the half-life of the ^{8}Be 2.9 MeV level so much smaller than the half life of the ^{8}Li ground state?
(c) Where in energy, with respect to the ^{8}Be ground state, would you expect the threshold for ^{7}Li neutron capture? Why?

(*Wisconsin*)

Solution:

(a) The spin-parity of α-particle is $J^P = 0^+$. In $^8Be \to \alpha + \alpha$, as the decay final state is that of two identical bosons, the wave function is required to be exchange-symmetric. This means that the relative orbital quantum number l of the α-particles is even, and so the parity of the final state of the two α-particle system is

$$\pi_f = (+1)^2(-1)^l = +1 .$$

As the α-decay is a strong-interaction process, (extremely short half-life), parity is conserved. Hence the parity of the 2.9 MeV excited state of ^8Be is positive.

(b) The β-decay of the ^8Li ground state is a weak-interaction process. However, the α-decay of the 2.9 MeV excited state of ^8Be is a strong-interaction process with a low Coulomb barrier. The difference in the two interaction intensities leads to the vast difference in the lifetimes.

(c) The threshold energy for ^7Li neutron capture is higher than the ^8Be ground state by

$$M(^7Li) + m(n) - M(^8Be) = M(^7Li) + m(n) - M(^8Li)$$
$$+ M(^8Li) - M(^8Be) = S_n(^8Li) + 16 \text{ MeV} .$$

where $S_n(^8Li)$ is the energy of dissociation of ^8Li into ^7Li and a neutron. As

$$S_n(^8Li) = M(^7Li) + M_n(n) - M(^8Li) = 7.018223 + 1.00892 - 8.025018$$
$$= 0.002187 \text{ amu} = 2.0 \text{ MeV} ,$$

the threshold of neutron capture by ^7Li is about 18 MeV higher than the ground state of ^8Be. Note that as ^8Li is outside the stability curve against β-decay, the energy required for removal of a neutron from it in rather small.

2081

The following atomic masses have been determined (in amu):

(1) $\begin{matrix} ^7_3Li & 7.0182 \\ ^7_4Be & 7.0192 \end{matrix}$

(2) $\begin{array}{ll} {}^{13}_{6}C & 13.0076 \\ {}^{13}_{7}N & 13.0100 \end{array}$

(3) $\begin{array}{ll} {}^{19}_{9}F & 19.0045 \\ {}^{19}_{10}Ne & 19.0080 \end{array}$

(4) $\begin{array}{ll} {}^{34}_{15}P & 33.9983 \\ {}^{34}_{16}S & 33.9978 \end{array}$

(5) $\begin{array}{ll} {}^{35}_{16}S & 34.9791 \\ {}^{35}_{17}Cl & 34.9789 \end{array}$

Remembering that the mass of the electron is 0.00055 amu, indicate which nuclide of each pair is unstable, its mode(s) of decay, and the approximate energy released in the disintegration. Derive the conditions for stability which you used.

(Columbia)

Solution:

As for each pair of isobars the atomic numbers differ by one, only β-decay or orbital electron capture is possible between them.

Consider β-decay. Let M_x, M_y, m_e represent the masses of the original nucleus, the daughter nucleus, and the electron respectively. Then the energy release in the β-decay is $E_d(\beta^-) = [M_x(Z,A) - M_y(Z+1,A) - m_e]c^2$. Expressing this relation in amu and neglecting the variation of the binding energy of the electrons in different atoms and shells, we have

$$E_d(\beta^-) = [M_x(Z,A) - Zm_e - M_y(Z+1,A) + (Z+1)m_e - m_e]c^2$$
$$= [M_x(Z,A) - M_y(Z+1,A)]c^2,$$

where M indicates atomic mass. Thus β-decay can take place only if $M_x > M_y$. Similarly for β^+-decay, we have

$$E_d(\beta^+) = [M_x(Z,A) - M_y(Z-1,A) - 2m_e]c^2,$$

and so β^+-decay can take place only if $M_x - M_y > 2m_e = 0.0011$ amu. In the same way we have for orbital electron capture (usually from the K shell)

$$E_d(i) = [M_x(Z,A) - M_y(Z-1,A)]c^2 - W_i.$$

where W_i is the binding energy of an electron in the ith shell, ~ 10 eV or 1.1×10^{-8} amu for K-shell, and so we require $M_x - M_y > W_i/c^2$

Let $\Delta = M(Z+1, A) - M(Z, A)$.

Pair (1), $\Delta = 0.001$ amu < 0.0011 amu, $^7_4$Be is unstable against K-electron capture.

Pair (2), $\Delta = 0.0024$ amu > 0.0011 amu, $^{13}_7$N is unstable against β-decay and K-electron capture.

Pair (3), $\Delta = 0.0035$ amu > 0.0011 amu, $^{19}_{10}$Ne is unstable against β^+-decay and K-electron capture.

Pair (4), $\Delta = -0.0005$ amu, $^{34}_{15}$P is unstable against β^--decay.

Pair (5), $\Delta = -0.0002$ amu, $^{35}_{16}$S is unstable against β^--decay.

2082

^{34}Cl positron-decays to ^{34}S. Plot a spectrum of the number of positrons emitted with momentum p as a function of p. If the difference in the masses of the neutral atoms of ^{34}Cl and ^{34}S is 5.52 MeV/c^2, what is the maximum positron energy?

(Wisconsin)

Solution:

^{34}Cl decays according to

$$^{34}Cl \rightarrow {}^{34}S + e^+ + \nu \,.$$

The process is similar to β^--decay and the same theory applies. The number of decays per unit time that emit a positron of momentum between p and $p + dp$ is (**Problem 2077(b)**)

$$I(p)dp = \frac{g^2|M_{fi}|^2}{2\pi^3\hbar^7c^3}(E_m - E)^2p^2dp\,,$$

where E_m is the end-point (total) energy of the β^+-spectrum. Thus $I(p)$ is proportional to $(E_m - E)^2p^2$, as shown in Fig. 2.25. The maximum β^+-particle energy is

$$E_{\max \beta^+} = [M(^{34}Cl) - M(^{34}S) - 2m_e]c^2 = 5.52 \text{ MeV} - 1.022 \text{ MeV}$$

$$= 4.50 \text{ MeV}\,.$$

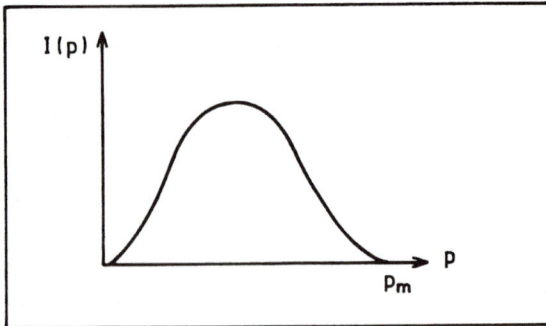

Fig. 2.25

2083

Both ^{161}Ho and ^{163}Ho decay by allowed electron capture to Dy isotopes, but the Q_{EC} values are about 850 keV and about 2.5 keV respectively. (Q_{EC} is the mass difference between the final ground state of nucleus plus atomic electrons and the initial ground state of nucleus plus atomic electrons.) The Dy orbital electron binding energies are listed in the table bellow. The capture rate for $3p_{1/2}$ electrons in ^{161}Ho is about 5% of the 3s capture rate. Calculate the $3p_{1/2}$ to 3s relative capture rate in ^{161}Ho. How much do the $3p_{1/2}$ and 3s capture rates change for both ^{161}Ho and ^{163}Ho if the Q_{EC} values remain the same, but the neutrino, instead of being massless, is assumed to have a mass of 50 eV?

Orbital	Binding Energy (keV)
1s	54
2s	9
$2p_{1/2}$	8.6
3s	2.0
$3p_{1/2}$	1.8

(Princeton)

Solution:

As ^{161}Ho and ^{163}Ho have the same nuclear charge Z, their orbital-electron wave functions are the same, their 3s and $3p_{1/2}$ waves differing only in phase. So the transition matrix elements for electron capture are also the same.

The decay constant is given by

$$\lambda \approx A|M_{if}|^2 \rho(E),$$

where M_{if} is the transition matrix element, $\rho(E)$ is the density of states, and A is a constant. For electron capture, the nucleus emits only a neutrino, and so the process is a two-body one. We have

$$\rho(E) \propto E_\nu^2 \approx (Q_{EC} - B)^2,$$

where B is the binding energy of an electron in s or p state. As

$$\frac{\lambda(3p_{1/2})}{\lambda(3s)} = \frac{|M(3p_{1/2})|^2 (Q_{EC} - B_p)^2}{|M(3s)|^2 (Q_{EC} - B_s)^2} = 0.05,$$

$$\frac{|M(3p_{1/2})^2}{|M(3s)|^2} = 0.05 \times \left(\frac{850 - 2.0}{850 - 1.8}\right)^2 = 0.04998.$$

Hence for ^{163}Ho,

$$\frac{\lambda(3p_{1/2})}{\lambda(3s)} = \frac{|M(3p_{1/2})|^2 (Q_{EC} - B_p)^2}{|M(3s)|^2 (Q_{EC} - B_s)^2}$$

$$= 0.04998 \times \left(\frac{2.5 - 1.8}{2.5 - 2.0}\right)^2 \approx 9.8\%.$$

If $m_\nu = 50$ eV, then, as $E_\nu^2 = p_\nu^2 + m_\nu^2$, the phase-space factor in $P(E)$ changes:

$$p_\nu^2 \frac{dp_\nu}{dE_\nu} = (E_\nu^2 - m_\nu^2)\frac{E_\nu}{p_\nu} = E_\nu\sqrt{E_\nu^2 - m_\nu^2} \approx E_\nu^2\left(1 - \frac{m_\nu^2}{2E_\nu^2}\right).$$

Hence the decay constant for every channel for ^{161}Ho and ^{163}Ho changes from λ_0 to λ:

$$\lambda \approx \lambda_0 \left(1 - \frac{1}{2}\frac{m_\nu^2}{E_\nu^2}\right),$$

or

$$\frac{\lambda_0 - \lambda}{\lambda_0} \approx \frac{1}{2}\frac{m_\nu^2}{E_\nu^2}.$$

Thus for ^{161}Ho, 3s state:

$$\frac{\lambda_0 - \lambda}{\lambda_0} = \frac{1}{2} \times \frac{50^2}{848^2 \times 10^6} = 1.74 \times 10^{-9} \,,$$

$3p_{1/2}$ state:

$$\frac{\lambda_0 - \lambda}{\lambda_0} = \frac{1}{2} \times \frac{50^2}{848.2^2 \times 10^6} = 1.74 \times 10^{-9} \,;$$

for ^{163}Ho, 3s state:

$$\frac{\lambda_0 - \lambda}{\lambda_0} = \frac{1}{2} \times \frac{50^2}{0.5 \times 10^6} = 5 \times 10^{-3} \,,$$

$3p_{1/2}$ state:

$$\frac{\lambda_0 - \lambda}{\lambda_0} = \frac{1}{2} \times \frac{50^2}{0.7^2 \times 10^6} = 2.25 \times 10^{-3} \,.$$

2084

An element of low atomic number Z can undergo allowed positron β-decay. Let p_0 be the maximum possible momentum of the positron, supposing $p_0 \ll mc$ (m =positron mass); and let Γ_β be the beta-decay rate. An alternative process is K-capture, the nucleus capturing a K-shell electron and undergoing the same nuclear transition with emission of a neutrino. Let Γ_K be the decay rate for this process. Compute the ratio Γ_K/Γ_β. You can treat the wave function of the K-shell electron as hydrogenic, and can ignore the electron binding energy.

(Princeton)

Solution:

The quantum perturbation theory gives the probability of a β^+-decay per unit time with decay energy E as

$$\omega = \frac{2\pi}{\hbar} \left| \int \psi_f^* H \psi_i d\tau \right|^2 \frac{dn}{dE} \,,$$

where ψ_i is the initial wave function, ψ_f is the final wave function and $\frac{dn}{dE}$ is the number of final states per unit interval of E. As the final state has three particles (nucleus, β^+ and ν), $\psi_f = u_f\phi_\beta\phi_\nu$ (assuming no interaction among the final particles or, if there is, the interaction is very weak), where u_f is the wave function of the final nucleus, ϕ_β, ϕ_ν are respectively the wave functions of the positron and neutrino.

In Fermi's theory of β-decay, H is taken to be a constant. Let it be g. Furthermore, the β^+-particle and neutrino are considered free particles and represented by plane waves:

$$\phi_\beta^* = \frac{1}{\sqrt{V}}e^{-i\mathbf{k}_\beta\cdot\mathbf{r}}, \qquad \phi_\nu^* = \frac{1}{\sqrt{V}}e^{-i\mathbf{k}_\nu\cdot\mathbf{r}},$$

where V is the volume of normalization, $\mathbf{k}_\beta$ and $\mathbf{k}_\nu$ are respectively the wave vectors of the β^+-particle and neutrino. Let

$$\int \psi_i u_f^* e^{-i(\mathbf{k}_\beta+\mathbf{k}_\nu)\cdot\mathbf{r}}d\tau = M_{fi}.$$

The final state is a three-particle state, and so dn is the product of the numbers of state of the final nucleus, the β^+-particle and neutrino. For β^+-decay, the number of states of the final nucleus is 1, while the number of states of β^+-particle with momentum between p and $p + dp$ is

$$dn_\beta = \frac{4\pi p^2 dp}{(2\pi\hbar)^3}V,$$

and that of the neutrino is

$$dn_\nu = \frac{4\pi p_\nu^2 dp_\nu}{(2\pi\hbar)^3}V.$$

Hence

$$\frac{dn}{dE} = \frac{dn_\beta dn_\nu}{dE} = \frac{p^2 p_\nu^2 dp dp_\nu}{4\pi^4\hbar^6 dE}V^2.$$

The sum of the β^+-particle and neutrino energies equals the decay energy E (neglecting nuclear recoil):

$$E_e + E_\nu \approx E,$$

and so for a given positron energy E_e, $dE_\nu = dE$. Then as the rest mass of neutrino is zero or very small, $E_\nu = cp_\nu$, and

$$p_\nu = (E - E_e)/c, \qquad dp_\nu = \frac{dE}{c}.$$

Therefore

$$\frac{dn}{dE} = \frac{(E - E_e)^2 p^2 dp}{4\pi^4 \hbar^6 c^3} V^2 .$$

On writing

$$\omega = \int I(p) dp ,$$

the above gives

$$I(p)dp = \frac{g^2 |M_{fi}|^2}{2\pi^3 \hbar^7 c^3} (E - E_e)^2 p^2 dp .$$

The β^+-decay rate Γ_β is

$$\Gamma_\beta = \int_0^{p_0} I(p)dp = B \int_0^{p_0} (E - E_e)^2 p^2 dp ,$$

where

$$B = \frac{g^2 |M_{fi}|^2}{2\pi^3 \hbar^7 c^3}$$

and p_0 is the maximum momentum of the positron, corresponding to a maximum kinetic energy $E_0 \approx E$. As $E_0 \ll m_e c^2$, and so $E_0 = \frac{p_0^2}{2m_e}$, $E_e \approx \frac{p^2}{2m_e}$, we have

$$\Gamma_\beta = B \int_0^{p_0} \frac{1}{(2m_e)^2} (p_0^4 + p^4 - 2p_0^2 p^2) p^2 dp$$

$$= \frac{B p_0^7}{4m_e^2} \left(\frac{1}{3} + \frac{1}{7} - \frac{2}{5} \right)$$

$$\approx 1.9 \times 10^{-2} \frac{B p_0^7}{m_e^2} .$$

In K-capture, the final state is a two-body system, and so monoenergetic neutrinos are emitted. Consider

$$\Gamma_K = \frac{2\pi}{\hbar} \left| \int \psi_f^* H \psi_i d\tau \right|^2 \frac{dn}{dE} .$$

The final state wave function ψ_f^* is the product of the daughter nucleus wave function u_f^* and the neutrino wave function ϕ_ν^*. The neutrino can be considered a free particle and its wave a plane wave

$$\phi_\nu^* = \frac{1}{\sqrt{V}} e^{-i\mathbf{k}_\nu \cdot \mathbf{r}} .$$

The initial wave function can be taken to be approximately the product of the wave functions of the parent nucleus and K-shell electron:

$$\phi_K = \frac{1}{\sqrt{\pi}} \left(\frac{Zm_ee^2}{\hbar^2}\right)^{3/2} e^{-Zm_ee^2r/\hbar^2}.$$

Then as

$$\left|\int \psi_f^* H\psi_i d\tau\right| = \frac{g}{\sqrt{V\pi}} \left(\frac{Zm_ee^2}{\hbar^2}\right)^{\frac{3}{2}} \left|\int u_f^* u_i e^{-i\mathbf{k}_\nu \cdot \mathbf{r}} e^{-\frac{Zm_ee^2}{\hbar^2}r} d\tau\right|$$

$$\approx \frac{g}{\sqrt{V\pi}} \left(\frac{Zm_ee^2}{\hbar^2}\right)^{3/2} |M_{fi}|,$$

$$\frac{dn}{dE} = \frac{4\pi V p_\nu^2 dp_\nu}{(2\pi\hbar)^3 dE} = \frac{4\pi V}{(2\pi\hbar)^3} \frac{E_\nu^2}{c^3},$$

taking $E_\nu \approx E$ and neglecting nuclear recoil, we have

$$\Gamma_K = \frac{m_e^3 g^2 |M_{fi}|^2}{\pi^2 \hbar^7 e^3} \left(\frac{Ze^2}{\hbar}\right)^3 E_\nu^2 = 2\pi m_e^3 B \left(\frac{Ze^2}{\hbar}\right)^3 E_\nu^2.$$

Ignoring the electron binding energy, we can take $E_\nu \approx E_0 + 2m_ec^2 \approx 2m_ec^2$, and hence

$$\frac{\Gamma_K}{\Gamma_\beta} = \frac{8\pi Z^3}{1.9 \times 10^{-2}} \left(\frac{e^2}{\hbar c}\right)^3 \left(\frac{m_ec}{p_0}\right)^7 = 5.1 \times 10^{-4} Z^3 \left(\frac{m_ec}{p_0}\right)^7.$$

Thus $\frac{\Gamma_k}{\Gamma_\beta} \propto \frac{1}{p_0^7}$. It increases rapidly with decreasing p_0.

2085

Tritium, the isotope ^{3}H, undergoes beta-decay with a half-life of 12.5 years. An enriched sample of hydrogen gas containing 0.1 gram of tritium produces 21 calories of heat per hour.

(a) For these data calculate the average energy of the β-particles emitted.

(b) What specific measurements on the beta spectrum (including the decay nucleus) indicate that there is an additional decay product and specifically that it is light and neutral.

(c) Give a critical, quantitative analysis of how a careful measurement of the beta spectrum of tritium can be used to determine (or put an upper limit on) the mass of the electron's neutrino.

(Columbia)

Solution:

(a) The decay constant is

$$\lambda = \frac{\ln 2}{T_{\frac{1}{2}}} = \frac{\ln 2}{12.5 \times 365 \times 24} = 6.33 \times 10^{-6} hr^{-1} .$$

Hence

$$-\frac{dN}{dt} = \lambda N = \frac{0.1 \times 6.023 \times 10^{23}}{3} \times 6.33 \times 10^{-6} = 1.27 \times 10^{17}$$

decay per hour and the average energy of the β-particles is

$$\bar{E} = \frac{21 \times 4.18}{1.27 \times 10^{17}} = 6.91 \times 10^{-16} \; J = 4.3 \text{ keV} .$$

(b) Both α- and β-decays represent transitions between two states of definite energies. However, the former is a two-body decay (daughter nucleus $+\alpha$-particle) and the conservation laws of energy and momentum require the α-particles to be emitted monoenergetic, whereas β-transition is a three-body decay (daughter nucleus + electron or position + neutrino) and so the electrons emitted have a continuous energy distribution with a definite maximum approximately equal to the transition energy. Thus the α-spectrum consists of a vertical line (or peak) while the β-spectrum is continuous from zero to a definite end-point energy. Thus a measurement of the β spectrum indicates the emission of a third, neutral particle. Conservation of energy indicates that it is very light.

(c) Pauli suggested that β-decay takes place according to

$$^A_Z X \rightarrow ^A_{Z+1} Y + \beta^- + \bar{\nu}_e .$$

As shown in Fig. 2.25, β^- has a continuous energy spectrum with a maximum energy E_m. When the kinetic energy of $\bar{\nu}_e$ trends to zero, the energy of β^- trends to E_m. Energy conservation requires

$$M(^A_Z X) = M(^A_{Z+1} Y) + \frac{E_m}{c^2} + m_\nu ,$$

or, for the process under consideration,

$$m_\nu = M(^3_1H) - M(^3_2He) - E_m/c^2 \,.$$

If E_m is precisely measured, the neutrino mass can be calculated. It has been found to be so small that only an upper limit can be given.

2086

(a) Describe briefly the energy spectra of alpha- and beta-particles emitted by radioactive nuclei. Emphasize the differences and qualitatively explain the reasons for them.

(b) Draw a schematic diagram of an instrument which can measure one of these spectra. Give numerical estimates of essential parameters and explain how they are chosen.

(UC, Berkeley)

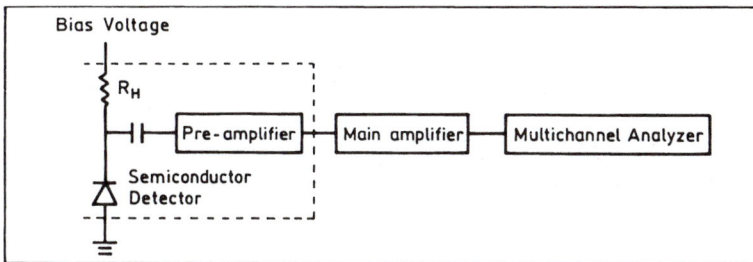

Fig. 2.26

Solution:

(a) α-particles from a radioactive nuclide are monoenergetic; the spectrum consists of vertical lines. β-particles have a continuous energy spectrum with a definite end-point energy. This is because emission of a β-particle is accompanied by a neutrino which takes away some decay energy.

(b) Figure 2.26 is a schematic sketch of a semiconductor α-spectrometer.

The energy of an α-particle emitted in α-decay is several MeV in most cases, so a thin-window, gold-silicon surface-barrier semiconductor detector is used which has an energy resolution of about 1 percent at room-temperature. As the α-particle energy is rather low, a thick, sensitive layer is not needed and a bias voltage from several tens to 100 V is sufficient. For good measurements the multichannel analyzer should have more than 1024 channels, using about 10 channels for the full width at half maximum of a peak.

2087

The two lowest states of scandium-42, $^{42}_{21}Sc_{21}$, are known to have spins 0^+ and 7^+. They respectively undergo positron-decay to the first 0^+ and 6^+ states of calcium-42, $^{42}_{20}Ca_{22}$, with the positron reduced half-lives $(ft)_{0^+} = 3.2 \times 10^3$ seconds, $(ft)_{7^+} = 1.6 \times 10^4$ seconds. No positron decay has been detected to the 0^+ state at 1.84 MeV. (See Fig. 2.27.)

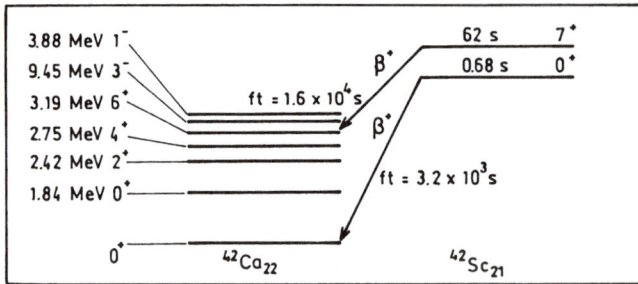

Fig. 2.27

(a) The two states of ^{42}Sc can be simply accounted for by assuming two valence nucleons with the configuration $(f_{7/2})^2$. Determine which of the indicated states of ^{42}Ca are compatible with this configuration. Briefly outline your reasoning. Assuming charge independence, assign isospin quantum numbers $|T, T_Z\rangle$ for all $(f_{7/2})^2$ states. Classify the nature of the two beta-transitions and explain your reasoning.

(b) With suitable wave functions for the $|J, M_J\rangle = |7,7\rangle$ state of scandium-42 and the $|6,6\rangle$ state of calcium-42, calculate the ratio $(ft)_{7^+}/(ft)_{0^+}$ expected for the two positron-decays.

For $j = l + \frac{1}{2}$:

$$\hat{S}_-|j,j\rangle = \frac{1}{(2j)^{1/2}}|j,j-1\rangle + \left(\frac{2j-1}{2j}\right)^{1/2}|j-1,j-1\rangle,$$

$$\hat{S}_z |j, j\rangle = \frac{1}{2} |j, j\rangle \,,$$

$$G_v = 1.4 \times 10^{-49} \text{ erg cm}^3 \,,$$

$$G_A = 1.6 \times 10^{-49} \text{ erg cm}^3 \,.$$

(Princeton)

Solution:

(a) For ^{42}S, $T_z = \frac{1}{2}(Z - N) = 0$. As the angular momenta of the two nucleons are 7/2 each and the isospins are 1/2 each, vector addition gives for the nuclear spin an integer from 0 to 7, and for the nuclear isospin 0 or 1. The generalized Pauli's principle requires the total wave function to be antisymmetric, and so $J + T = $ odd. Hence the states compatible with the configuration $(f_{7/2})^2$ are $J = 0^+, 2^+, 4^+, 6^+$ when $T = 1$, and $J = 1^+, 3^+, 5^+, 7^+$ when $T = 0$.

The transition $7^+ \to 6^+$ is a Gamow–Teller transition as for such transitions $\Delta J = 0$ or 1 ($J_i = 0$ to $J_f = 0$ is forbidden), $\Delta T = 0$ or 1, $\pi_i = \pi_f$.

The transition $0^+ \to 0^+$ is a Fermi transition as for such transitions $\Delta J = 0$, $\Delta T = 0$, $\pi_i = \pi_f$.

(b) The probability per unit time of β-transition is $\Gamma(\beta) \propto G_v^2 \langle M_F \rangle^2 + G_A^2 \langle M_{GT} \rangle^2$, where $\langle M_F \rangle^2$ and $\langle M_{GT} \rangle^2$ are the squares of the spin-averaged weak interaction matrix elements:

$$\langle M_F \rangle^2 = \frac{1}{2J_i + 1} \sum_{M_i, M_f} \langle J_f M_f T_f T_{fz} | 1 \cdot \sum_{k=1}^{A} t_{\pm}(k) | J_i M_i T_i T_{iz} \rangle^2$$

$$= \langle J_f M T_f T_{fz} | 1 \cdot \sum_{k=1}^{A} t_{\pm}(k) | J_i M T_i T_{iz} \rangle^2 \,,$$

$$\langle M_{GT} \rangle^2 = \frac{1}{2J_i + 1} \sum_{m, M_i, M_f} |\langle J_f M_f T_f T_{fz} | \sum_{k=1}^{A} \sigma_m(k) t_{\pm}(k) | J_i M_i T_i T_{iz} \rangle|^2 \,,$$

where m takes the values $+1, 0, -1$, for which

$$\sigma_{+1} = \sigma_x + i\sigma_y, \quad \sigma_0 = \sigma_z, \quad \sigma_{-1} = \sigma_x - i\sigma_y \,.$$

Then the half-life is

$$ft = \frac{K}{G_v^2 \langle M_F \rangle^2 + G_A^2 \langle M_{GT} \rangle^2} \,,$$

where $K = 2\pi^3 \hbar^7 \ln 2 / m^5 c^4$, a constant. Hence

$$\frac{ft(7^+ \rightarrow 6^+)}{ft(0^+ \rightarrow 0^+)} = \frac{G_v^2 \langle M_F \rangle_{0+}^2}{G_A^2 \langle M_{GT} \rangle_{7+}^2} \, .$$

Consider

$$\langle M_F \rangle = \langle JMTT_{fz} | 1 \cdot \sum_{k=1}^{A} t_{\pm}(k) | JMTT_{iz} \rangle = \langle JMTT_{fz} | T_{\pm} | JMTT_{iz} \rangle$$

$$= \sqrt{T(T+1) - T_{iz} T_{fz}} \, ,$$

replacing the sum of the z components of the isospins of the nucleons by the z-component of the total isospin. Taking $T = 1$, $T_{iz} = 0$, we have

$$\langle M_F \rangle^2 = 2 \, .$$

Consider

$$\langle M_{GT} \rangle^2 = \sum_m |\langle 6, 6, 1, -1 | \{ \sigma_m(1) t_{\pm}(1) + \sigma_m(2) t_{\pm}(2) \} | 7, 7, 1, 0 \rangle|^2 \, ,$$

where only the two nucleons outside full shells, which are identical, are taken into account. Then

$$\langle M_{GT} \rangle^2 = 4 \sum_m |\langle 6, 6, 1, -1 | \sigma_m(1) t_{\pm}(1) | 7, 7, 1, 0 \rangle|^2 \, .$$

Writing the wave functions as combinations of nucleon wave functions:

$$|7, 7\rangle = \left| \frac{7}{2}, \frac{7}{2}; \frac{7}{2}, \frac{7}{2} \right\rangle \, ,$$

$$|7, 6\rangle = \frac{1}{\sqrt{2}} \left(\left| \frac{7}{2}, \frac{6}{2}; \frac{7}{2}, \frac{7}{2} \right\rangle + \left| \frac{7}{2}, \frac{7}{2}; \frac{7}{2}, \frac{6}{2} \right\rangle \right) \, ,$$

$$|6, 6\rangle = \frac{1}{\sqrt{2}} \left(\left| \frac{7}{2}, \frac{6}{2}; \frac{7}{2}, \frac{7}{2} \right\rangle - \left| \frac{7}{2}, \frac{7}{2}; \frac{7}{2}, \frac{6}{2} \right\rangle \right) \, ,$$

we have

$$\langle M_{GT} \rangle^2 = 4 \left| \left\langle \frac{7}{2}, \frac{6}{2}; \frac{7}{2}, \frac{7}{2}, ; 1, -1 \left| \frac{\sigma_-(1) t_{\pm}(1)}{2} \right| \frac{7}{2}, \frac{7}{2}; \frac{7}{2}, \frac{7}{2}; 1, 0 \right\rangle \right|^2 = 2 \, .$$

Thus

$$\frac{(ft)_{7+}}{(ft)_{0+}} = \frac{G_v^2}{G_A^2} \approx \left(\frac{1.4}{1.6}\right)^2 \approx 0.77 .$$

2088

The still-undetected isotope copper-57 ($^{57}_{29}\text{Cu}_{28}$) is expected to decay by positron emission to nickel-57 ($^{57}_{28}\text{Ni}_{29}$).

(a) Suggest shell-model spin-parity assignments for the ground and first excited states of these nuclei.

(b) Estimate the positron end-point energy for decay from the ground state of copper-57 to the ground state of nickel-57. Estimate the half-life for this decay (order of magnitude).

(c) Discuss what one means by Fermi and by Gamow–Teller contributions to allowed β-decays, and indicate the corresponding spin-parity selection rules. For the above decay process, estimate the ratio Γ_F/Γ_{GT} of the two contributions to the decay rate. Does one expect appreciable β^+-decay from the copper-57 ground state to the first excited state of nickel-57? Explain.

(d) Nickel-58 occurs naturally. Briefly sketch an experimental arrangement for study of copper-57 positron-decay.

(Princeton)

Solution:

(a) ^{57}Cu and ^{57}Ni are mirror nuclei with the same energy-level structure of a single nucleon outside of double-full shells. The valence nucleon is proton for ^{57}Cu and neutron for ^{57}Ni, the two nuclei having the same features of ground and first excited states.

For the ground state, the last nucleon is in state $2p_{3/2}$ (Fig. 2.11), and so $J^{\pi} = (\frac{3}{2})^-$; for the first excited state, the nucleon is in state $1f_{5/2}$, and so $J^{\pi} = (\frac{5}{2})^- (E_1 = 0.76 \text{ MeV})$.

(b) As ^{57}Cu and ^{57}Ni are mirror nuclei, their mass difference is (**Problem 2067(c)**)

$$\Delta E = M(Z+1, A)c^2 - M(Z, A)c^2$$

$$= \frac{3e^2}{5R}[(Z+1)^2 - Z^2] - (m_n - M_H)c^2$$

$$= \frac{3c\hbar}{5R}\left(\frac{e^2}{c\hbar}\right)(2Z+1)-(m_n-M_H)c^2$$

$$= \frac{3\times 197\times(2\times 28+1)}{5\times 1.2\times 57^{1/3}\times 137}-0.78$$

$$= 9.87\ \text{MeV}.$$

Thus the ground state of ^{57}Cu is 9.87 MeV higher than that of ^{57}Ni. The positron end-point energy for decay from the ground state of ^{57}Cu to that of ^{57}Ni is

$$E_0 = \Delta E - 2m_e c^2 \approx 9.87 - 1.02 \approx 8.85\ \text{MeV}.$$

As the β^+-decay is from $(\frac{3}{2})^-$ to $(\frac{3}{2})^-$, $\Delta J = 0$, $\Delta\pi = +$, $\Delta T = 0$, $\Delta T_z = -1$, the decay is of a superallowed type. To simplify calculation take $F(Z, E) = 1$. Then (**Problem 2084**)

$$\lambda_\beta \approx \int_0^{p_0} I(p)dp \approx B\int_0^{E_0}(E_0-E)^2 E^2 dE$$

$$= BE_0^5\left(\frac{1}{3}+\frac{1}{5}-\frac{1}{2}\right) = \frac{1}{30}BE_0^5,$$

where

$$B = \frac{g^2|M_{fi}|^2}{2\pi^3 c^6 \hbar^7} = 3.36\times 10^{-3}\ \text{MeV}^{-5}\text{s}^{-1},$$

with $|M_{fi}|^2 \approx 1$, $g = 8.95\times 10^{-44}$ MeV $\cdot$ cm^3 (experimental value). Hence

$$\tau_{1/2} = \ln 2/\lambda = \frac{30\ln 2}{BE_0^5} = 0.114\ s.$$

(c) In β^+-decay between mirror nuclei ground states $\frac{3^-}{2}\to\frac{3^-}{2}$, as the nuclear structures of the initial and final states are similar, the transition is of a superallowed type. Such transitions can be classified into Fermi and Gamow–Teller types. For the Fermi type, the selection rules are $\Delta J = 0$, $\Delta\pi = +$, the emitted neutrino and electron have antiparallel spins. For the Gamow–Teller type, the selection rules are $\Delta J = 0, \pm 1$, $\Delta\pi = +$, the emitted neutrino and electron have parallel spins.

For transition $\frac{3^-}{2}\to\frac{3^-}{2}$ of the Fermi type,

$$|M_F|^2 = T(T+1) - T_{iz}T_{fz} = \frac{1}{2}\left(\frac{1}{2}+1\right)+\frac{1}{2}\times\frac{1}{2} = 1.$$

For transition $\frac{3^-}{2} \rightarrow \frac{3^-}{2}$ of the Gamow–Teller type,

$$|M_{GT}|^2 = \frac{J_f + 1}{J_f} = \frac{3/2 + 1}{3/2} = \frac{5}{3}.$$

The coupling constants for the two types are related roughly by $|g_{GT}| \approx 1.24|g_F|$. So the ratio of the transition probabilities is

$$\frac{\lambda_F}{\lambda_{GT}} = \frac{g_F^2 |M_F|^2}{g_{GT}^2 |M_{GT}|^2} = \frac{1}{1.24^2 \times 5/3} = 0.39.$$

The transition from ^{57}Cu to the first excited state of ^{57}Ni is a normal-allowed transition because $\Delta J = 1$, $\Delta \pi = +$. As the initial and final states are $2p_{3/2}$ and $1f_{5/2}$, and so the difference in nuclear structure is greater, the fT of this transition is larger than that of the superallowed one by 2 to 3 orders of magnitude. In addition, there is the space phase factor $\left(\frac{8.85-0.76}{8.85}\right)^5 = 0.64$. Hence the branching ratio is very small, rendering such a transition difficult to detect.

(d) When we bombard ^{58}Ni target with protons, the following reaction may occur:

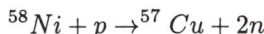

$$^{58}Ni + p \rightarrow ^{57}Cu + 2n$$

As the mass-excess $\Delta = (M - A)$ values (in MeV) are

$$\Delta(n) \approx 8.071, \quad \Delta(^1H) = 7.289,$$

$$\Delta(^{58}Ni) = -60.235, \quad \Delta(^{57}Cu) \approx -46.234.$$

We have

$$Q = \Delta(^{58}Ni) + \Delta(^1H) - \Delta(^{57}Cu) - 2\Delta(n)$$

$$= -60.235 + 7.289 + 46.234 - 2 \times 8.071 = -22.854 \text{ MeV}.$$

Hence the reaction is endoergic and protons of sufficient energy are needed. The neutrons can be used to monitor the formation of ^{57}Cu, and measuring the delay in β^+ emission relative to n emission provides a means to study β^+-decay of ^{57}Cu.

2089

Suppose a search for solar neutrinos is to be mounted using a large sample of lithium enriched in the isotope $^7_3$Li. Detection depends on production, separation and detection of the electron-capturing isotope $^7_4$Be with a half-life of 53 days. The low lying levels of these two nuclei are shown below in Fig. 2.28. The atomic mass of $^7_4$Be in its ground state lies 0.86 MeV above the atomic mass of $^7_3$Li in its ground state.

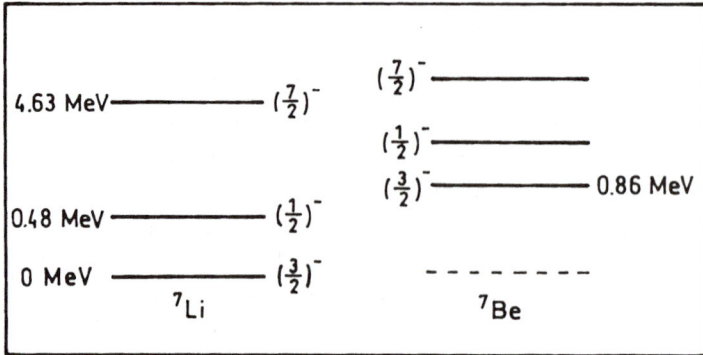

Fig. 2.28

(a) Discuss the electron-capture modes of the ground state of beryllium-7 by providing estimates for the branching ratios and relative decay probabilities (ft ratios).

(b) To calibrate this detector, a point source emitting 10^{17} monochromatic neutrinos/sec with energy 1.5 MeV is placed in the center of a one metric ton sphere of lithium-7. Estimate the total equilibrium disintegration rate of the beryllium-7, given

$$G_V = 1.42 \times 10^{-49} \text{ erg cm}^3 ,$$

$$G_A = 1.60 \times 10^{-49} \text{ erg cm}^3 ,$$

$$\rho_{\text{Li}} = 0.53 \text{ gm/cm}^3 .$$

(Princeton)

Solution:

(a) Two modes of electron capture are possible:

$$\left(\frac{3^-}{2}\right)^- \to \left(\frac{3}{2}\right)^- : \quad \Delta J = 0, \Delta P = + ,$$

which is a combination of F and G-T type transitions;

$$\left(\frac{3}{2}\right)^{-} \rightarrow \left(\frac{1}{2}\right)^{-} \ : \ \Delta J = 1, \Delta P = +,$$

which is a pure G-T type transition.

$^{7}_{3}\text{Li}$ and $^{7}_{4}\text{Be}$ are mirror nuclei with $T = \frac{1}{2}$, and $T_z = \frac{1}{2}$ and $-\frac{1}{2}$ respectively.

For the F-type transition $\left(\frac{3}{2}\right)^{-} \rightarrow \left(\frac{3}{2}\right)^{-}$ the initial and final wave functions are similar and so

$$\langle M_F \rangle^2 = T(T+1) - T_{zi}T_{zf} = \frac{1}{2}\cdot\frac{3}{2} + \frac{1}{2}\cdot\frac{1}{2} = \frac{3}{4} + \frac{1}{4} = 1 \,.$$

For the G-T-type transition $\frac{3}{2}^{-} \rightarrow \frac{3}{2}^{-}$, the single-particle model gives

$$\langle M_{G-T}\rangle^2 = \frac{J_f + 1}{J_f} = \frac{3/2 + 1}{3/2} = \frac{5}{3} \,.$$

For the G-T-type transition $\left(\frac{3}{2}\right)^{-} \rightarrow \left(\frac{1}{2}\right)^{-}$, the transition is form $l + \frac{1}{2}$ to $l - \frac{1}{2}$ with $l = 1$, and the single-particle model gives

$$\langle M_{G-T}\rangle^2 = \frac{4l}{2l+1} = \frac{4}{3} \,.$$

As $\lambda_K(M^2, W_\nu) = |M|^2 W_\nu^2$, where W_ν is the decay energy,

$$\frac{\lambda_K\left(\frac{3}{2}^{-} \rightarrow \frac{3}{2}^{-}\right)}{\lambda_K\left(\frac{3}{2}^{-} \rightarrow \frac{1}{2}^{-}\right)} = \frac{\langle M_{G-T}\rangle^2_{3/2} + \frac{G_V^2}{G_A^2}\langle M_F\rangle^2}{\langle M_{G-T}\rangle^2_{1/2}} \cdot \frac{W_{\nu_1}^2}{W_{\nu_2}^2}$$

$$= \frac{\frac{5}{3} + \left(\frac{1.42}{1.60}\right)^2}{\frac{4}{3}} \times \left(\frac{0.86}{0.86 - 0.48}\right)^2$$

$$= \frac{(5 + 0.79 \times 3) \times 0.86^2}{4 \times (0.86 - 0.48)^2} = 9.43 \,.$$

Hence the branching ratios are $B(\frac{3}{2}^{-} \rightarrow \frac{3}{2}^{-}) = \frac{9.43}{10.43} = 90.4\%$,

$$B\left(\frac{3^{-}}{2} \rightarrow \frac{1^{-}}{2}\right) = \frac{1}{10.43} = 9.6\% \,.$$

The fT ratio of the two transitions is

$$\frac{(fT)_{3/2^-}}{(fT)_{1/2^-}} = \frac{\langle M_{G-T}\rangle^2_{1/2}}{\langle M_{G-T}\rangle^2_{3/2} + \frac{G^2_V}{G^2_A}\langle M_F\rangle^2} = \frac{4}{3 \times 0.79 + 5} = 0.543\,.$$

(b) When irradiating ^{7}Li with neutrinos, ^{7}Li captures neutrino and becomes ^{7}Be. On the other hand, ^{7}Be undergoes decay to ^{7}Li. Let the number of ^{7}Be formed per unit time in the irradiation be ΔN_1. Consider a shell of ^{7}Li of radius r and thickness dr. It contains

$$\frac{4\pi r^2 \rho n dr}{A}$$

^{7}Li nuclei, where $n =$ Avogadro's number, $A =$ mass number of ^{7}Li. The neutrino flux at r is $\frac{I_0}{4\pi r^2}$. If $\sigma =$ cross section for electron-capture by ^{7}Li, $a =$ activity ratio of ^{7}Li for forming ^{7}Be, $R =$ radius of the sphere of ^{7}Li, the number of ^{7}Be nuclei produced per unit time is

$$\Delta N_1 = \int \frac{I_0}{4\pi r^2}\rho n \sigma a \cdot 4\pi r^2 dr/A = I_0 \rho n \sigma a R/A\,.$$

With $a = 0.925$, $\rho = 0.53$ g cm^{-3}, $A = 7$, $n = 6.023 \times 10^{23}$, $R = \left(\frac{3 \times 10^6}{4\pi\rho}\right)^{\frac{1}{3}} = 76.7$ cm, $I_0 = 10^{17}$ s^{-1}, $\sigma \approx 10^{-43}$ cm^2, we have

$$\Delta N_1 = \frac{10^{17} \times 0.53 \times 6.023 \times 10^{23} \times 10^{-43} \times 0.925 \times 76.7}{7}$$

$$= 3.2 \times 10^{-2}\ s^{-1}\,.$$

At equilibrium this is also the number of ^{7}Be that decay to ^{7}Li.

Hence the rate of disintegration of ^{7}Be at equilibrium is 3.2×10^{-2} s^{-1}. Note that the number of ^{7}Li produced in ^{7}Be decays is negligible compared with the total number present.

2090

It is believed that nucleons (N) interact directly through the weak interaction and that the latter violates parity conservation. One way to study the nature of the N-N weak interaction is by means of α-decay, as typified by the decays of the 3^+, $T = 1$ and 3^-, $T = 0$ states of ^{20}Ne (Fig. 2.29).

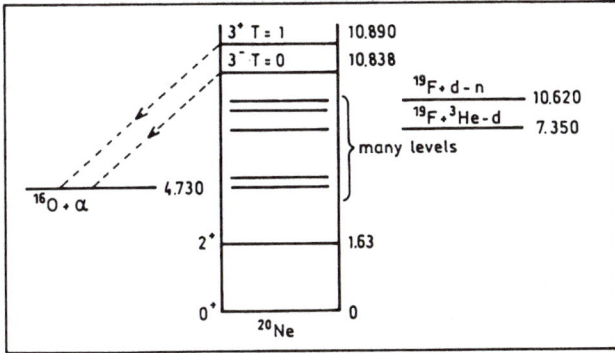

Fig. 2.29

In the following you will be asked to explain the principles of an experiment to measure the weak-interaction matrix element between these states, $\langle 3^+|H_{\text{weak}}|3^-\rangle$.

(a) The N-N weak interaction has isoscalar, isovector, and isotensor components (i.e., ranks 0,1, and 2 in isospin). Which components contribute to the matrix element $\langle 3^+|H_{\text{weak}}|3^-\rangle$?

(b) Explain the parity and isospin selection rules for α-decay. In particular, explain which of the two ^{20}Ne states would decay to the ground state of ^{16}O $+ \alpha$ if there were no parity-violating N-N interaction.

(c) Allowing for a parity-violating matrix element $\langle 3^+|H_{\text{weak}}|3^-\rangle$ of 1 eV, estimate the α width of the parity-forbidden transition, Γ_α (forbidden), in terms of the α width of the parity-allowed transition, Γ_α (allowed). Assume Γ_α (allowed) is small compared with the separation energy between the 3^+, 3^- states.

(d) The α width of the parity-allowed transition is Γ_α (allowed) $=$ 45 keV, which is not small compared with the separation energy. Do you expect the finite width of this state to modify your result of part (c) above? Discuss.

(e) The direct reaction ^{19}F(^{3}He,d)^{20}Ne* populates one of the excited states strongly. Which one do you expect this to be and why?

(f) There is also a $1^+/1^-$ parity doublet at ~ 11.23 MeV. Both states have $T = 1$.

(i) In this case which state is parity-forbidden to α-decay?

(ii) As in part(a), which isospin components of the weak N-N interaction contribute to the mixing matrix element? (Note that ^{20}Ne is self-conjugate) Which would be determined by a measurement of the parity-forbidden α width?

(Princeton)

Solution:

(a) As $T = 1$ for the 3^+ state and $T = 0$ for the 3^- state, only the isovector component with $\Delta T = 1$ contributes to $\langle 3^+|H_{\text{weak}}|3^-\rangle$.

(b) α-decay is a strong interaction for which isospin is conserved. Hence $\Delta T = 0$. As the isospin of α-particle is zero, the isospin of the daughter nucleus should equal that of the parent. As ^{16}O has $T = 0$, only the 3^-, $T = 0$ state can undergo α-decay to ^{16}O $+ \alpha$. As both the spins of ^{16}O and α are zero, and the total angular momentum does not change in α-decay, the final state orbital angular momentum is $l = 3$ and so the parity is $(-1)^3 = -1$. As it is the same as for the initial state, the transition is parity-allowed.

(c) Fermi's golden rule gives the first order transition probability per unit time as

$$\lambda = \frac{2\pi}{\hbar}|V_{fi}|^2\rho(E_f),$$

where V_{fi} is the transition matrix element and $\rho(E_f)$ is the final state density. Then the width of the parity-allowed transition $(3^-, T = 0$ to ^{16}O $+ \alpha)$ is

$$\Gamma_\alpha = \frac{2\pi}{\hbar}|V_{3^- \to {}^{16}O}|^2\rho(E_f).$$

The parity-forbidden transition $(3^+, T = 1$ to ^{16}O $+ \alpha)$ is a second order process, for which

$$\lambda = \frac{2\pi}{\hbar}\left|\sum_{n \neq i}\frac{V_{fn}V_{ni}}{E_i - E_n + i\varepsilon}\right|^2\rho(E_f),$$

where 2ε is the width of the intermediate state, and the summation is to include all intermediate states. In this case, the only intermediate state is that with 3^-, $T = 0$. Hence

$$\Gamma'_\alpha = \frac{2\pi}{\hbar}|V_{3^- \to {}^{16}O}|^2\frac{1}{(E_i - E_n)^2 + \varepsilon^2}|\langle 3^+|H_{\text{weak}}|3^-\rangle|^2\rho(E_f)$$

$$= \Gamma_\alpha\frac{|\langle 3^+|H_{\text{weak}}|3^-\rangle|^2}{(\Delta E)^2 + (\Gamma_\alpha/2)^2},$$

where ΔE is the energy spacing between the $3^+, 3^-$ states, Γ_α is the width of the parity-allowed transition. If $\Gamma_\alpha \ll \Delta E$, as when $\langle 3^+|H_{\text{weak}}|3^-\rangle = 1$ eV, $\Delta E = 0.052$ MeV $= 52 \times 10^3$ eV, we have

$$\Gamma'_\alpha \approx \frac{|\langle 3^+|H_{\text{weak}}|3^-\rangle|^2}{(\Delta E)^2}\Gamma_\alpha = \frac{\Gamma_\alpha}{52^2 \times 10^6} = 3.7 \times 10^{-10}\Gamma_\alpha .$$

(d) As $\Gamma_\alpha = 45$ keV, $(\Gamma_\alpha/2)^2$ cannot be ignored when compared with $(\Delta E)^2$. Hence

$$\Gamma'_\alpha = \frac{10^{-6}}{52^2 + \frac{45^6}{4}}\Gamma_\alpha = 3.1 \times 10^{-10}\Gamma_\alpha = 1.4 \times 10^{-5} \text{ eV} .$$

(e) Consider the reaction $^{19}F(^3He, d)^{20}Ne^*$. Let the spins of ^{19}F, ^3He, d, ^{20}Ne, and the captured proton be $\mathbf{J}_A$, $\mathbf{J}_a$, $\mathbf{J}_b$, $\mathbf{J}_B$, J_p, the orbital angular momenta of ^3He, d and the captured proton be $\mathbf{l}_a$, $\mathbf{l}_b$, $\mathbf{l}_p$, respectively. Then

$$\mathbf{J}_A + \mathbf{J}_a + \mathbf{l}_a = \mathbf{J}_B + \mathbf{J}_b + \mathbf{l}_b .$$

As

$$\mathbf{J}_A = \mathbf{J}_p + \mathbf{l}_b, \quad \mathbf{l}_a = \mathbf{l}_p + \mathbf{l}_b, \quad \mathbf{J}_A + \mathbf{s}_p + \mathbf{l}_p = \mathbf{J}_B ,$$

and $J_A = \frac{1}{2}$, $J_B = 3$, $J_b = 1$, $l_b = 0$, $s_p = \frac{1}{2}$, we have $J_p = \frac{1}{2}$, $l_p = 2, 3, 4$. Parity conservation requires $P(^{19}F)P(p)(-1)^{l_p} = P(^{20}Ne^*)$, $P(^{20}Ne^*) = (-1)^{l_p}$.

Experimentally l_p is found from the angular distribution to be $l_p = 2$. Then $P(^{20}Ne^*) = +$, and so the reaction should populate the 3^+ state of Ne^*, not the 3^- state.

(f) (i) The 1^+ state is parity-forbidden to α-decay. On the other hand, in the α-decay of the 1^- state, $l_f + J_\alpha + J_{^{16}O} = 1$, $P_f = P(\alpha)P(^{16}O)(-1)^{l_f} = -1$, so that its α-decay is parity-allowed

(ii) As ^{20}Ne is a self-conjugate nucleus, $T_3 = 0$ because $\langle 1, 0|1, 0; 1, 0\rangle = 0$. So only the components of $T = 0, 2$ can contribute. However in weak interaction, $|\Delta T| \leq 1$, and so only the component with $\Delta T = 0$ can contribute to the experiment result.

2091

Consider the following energy level structure (Fig. 2.30):

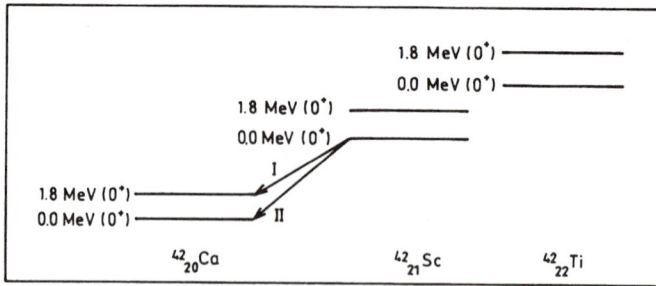

Fig. 2.30

The ground states form an isotriplet as do the excited states (all states have a spin-parity of 0^+). The ground state of $^{42}_{21}\text{Sc}$ can β-decay to the ground state of $^{42}_{20}\text{Ca}$ with a kinetic end-point energy of 5.4 MeV (transition II in Fig. 2.30).

(a) Using phase space considerations only, calculate the ratio of rates for transitions I and II.

(b) Suppose that the nuclear states were, in fact, pure (i.e. unmixed) eigenstates of isospin. Why would the fact that the Fermi matrix element is an isospin ladder operator forbid transition I from occurring?

(c) Consider isospin mixing due to electromagnetic interactions. In general

$$H_{EM} = H_0 + H_1 + H_2,$$

where the subscripts refer to the isospin tensor rank of each term. Write the branching ratio $\frac{\Gamma_I}{\Gamma_{II}}$ in terms of the reduced matrix elements of each part of H_{EM} which mixes the states.

(d) Using the results of parts (a) and (c), ignoring H_2, and given that $\frac{\Gamma_I}{\Gamma_{II}} = 6 \times 10^{-5}$, calculate the value of the reduced matrix element which mixes the ground and excited states of $^{42}_{20}\text{Ca}$.

(*Princeton*)

Solution:

(a) From phase space consideration only, for β-decay of $E_0 \gg m_e c^2$, $\Gamma \approx E_0^5$ (**Problem 2077**). Thus

$$\frac{\Gamma_I}{\Gamma_{II}} = \frac{(5.4 - 1.8)^5}{(5.4 - 0)^5} \approx 0.13 \,.$$

(b) For Fermi transitions within the same isospin multiplet, because the structures of the initial and final states are similar, the transition probability is large. Such transitions are generally said to be superallowed. For $0^+ \to 0^+ (T = 1)$, there is only the Fermi type transition, for which

$$\langle M_F \rangle^2 = \langle \alpha, T_f, T_{f3} | \sum_{K=1}^{A} t_{\pm}(K) | \alpha', T_i, T_{i3} \rangle^2$$

$$= \left(\delta_{\alpha\alpha'} \delta_{T_i T_f} \sqrt{T(T+1) - T_{i3} T_{f3}} \right)^2$$

$$= \begin{cases} T(T+1) - T_{i3} T_{f3}, & \text{if } \alpha = \alpha', \ T_f = T_i, \\ 0, & \text{otherwise}, \end{cases}$$

ignoring higher order corrections to the Fermi matrix element. Here α is any nuclear state quantum number other than isospin. From this we see that channel II is a transition within the same isospin multiplet, i.e., a superallowed one, channel I is a transition between different isospin multiplets, i.e., a Fermi-forbidden transition.

(c) We make use of the perturbation theory. Let the ground and excited states of ^{42}Ca be $|1\rangle$ and $|2\rangle$ respectively. Because of the effect of H_{EM}, the states become mixed. Let the mixed states be $|1\rangle'$ and $|2\rangle'$, noting that the mixing due to H_{EM} is very small. We have

$$H^0 |1\rangle = E_1 |1\rangle ,$$

$$H^0 |2\rangle = E_2 |2\rangle ,$$

where E_1 and E_2 are the energies of the two states ($E_1 \approx E_0$, $E_2 - E_1 = 1.8$ MeV).

Consider

$$H = H^0 + H_{EM} ,$$

where $H_{EM} = H_0 + H_1 + H_2$. As the index refers to isospin tensor rank, we write H_0, H_1 H_2 as $P_{0,0}$, $P_{1,0}$, $P_{2,0}$ and define

$$\langle J_1 m_1 | P_{\mu\nu} | J_2 m_2 \rangle = C^{J_1 m_1}_{\mu\nu J_2 m_2} \langle J_1 || P_{\mu\nu} || J_2 \rangle .$$

Then

$$H_{EM} = P_{0,0} + P_{1,0} + P_{2,0},$$

$$\langle 1|H_{EM}|2\rangle = \langle \alpha, 1, -1|(P_{0,0} + P_{1,0} + P_{2,0})|\alpha', 1, -1\rangle$$

$$= \left(\langle \alpha, 1||P_0||\alpha', 1\rangle - \sqrt{\frac{1}{2}}\langle \alpha, 1||P_1||\alpha', 1\rangle \right.$$

$$\left. + \sqrt{\frac{1}{10}}\langle \alpha, 1||P_2||\alpha', 1\rangle \right),$$

$$\langle 1|H_{EM}|1\rangle = \langle 2|H_{EM}|2\rangle = \langle \alpha, 1, -1|(P_{0,0} + P_{1,0} + P_{2,0})|\alpha, 1, -1\rangle$$

$$= \langle \alpha, 1||P_0||\alpha, 1\rangle - \sqrt{\frac{1}{2}}\langle \alpha, 1||P_1||\alpha, 1\rangle + \sqrt{\frac{1}{10}}\langle \alpha, 1||P_2||\alpha, 1\rangle.$$

In the above equations, α and α' denote the quantum numbers of $|1\rangle$ and $|2\rangle$ other than the isospin, and $\langle \alpha, 1||P||\alpha, 1\rangle$ denote the relevant part of the reduced matrix element. Thus

$$\frac{\Gamma_I}{\Gamma_{II}} = \frac{E_1^5|M_1|^2}{E_2^5|M_2|^2} = \frac{(5.4 - 1.8 - \langle 2|H_{EM}|2\rangle)^5}{(5.4 - \langle 1|H_{EM}|1\rangle)^5}\frac{\langle 1|H_{EM}|2\rangle^2}{(E_2 - E_1)^2}.$$

If energy level corrections can be ignored, then $\langle 1|H_{EM}|1\rangle \ll E_1, E_2$, and

$$\frac{\Gamma_I}{\Gamma_{II}} = \frac{E_{10}^5}{E_{20}^5(E_2 - E_1)^2}|\langle 1|H_{EM}|2\rangle|^2$$

$$= \frac{(5.4 - 1.8)^5}{5.4^5 \times 1.8^2}\left(\langle 1||P_0||2\rangle - \sqrt{\frac{1}{2}}\langle 1||P_1||2\rangle + \sqrt{\frac{1}{10}}\langle 1||P_2||2\rangle \right)^2.$$

If we ignore the contribution of H_2 and assume $\langle 1||P_0||2\rangle = 0$, then the isoscalar H does not mix the two isospin states and we have

$$\frac{\Gamma_I}{\Gamma_{II}} = \frac{E_{10}^5}{E_{20}^5(E_2 - E_1)^2}|\langle \alpha, 1||P_1||\alpha', 1\rangle|^2.$$

(d) In the simplified case above,

$$\frac{\Gamma_I}{\Gamma_{II}} = \frac{(5.4 - 1.8)^5}{5.4^5 \times 1.8^2}|\langle \alpha, 1||P_1||\alpha', 1\rangle|^2 = 6 \times 10^{-5}$$

gives

$$|\langle\alpha,1||P_1||\alpha',1\rangle|^2 = 24.6 \times 6 \times 10^{-5} = 1.48 \times 10^{-3}\ \text{MeV}^2\,,$$

or

$$|\langle\alpha,1||P_1||\alpha',1\rangle| = 38\ \text{keV}\,.$$

2092

"Unlike atomic spectroscopy, electric dipole (E1) transitions are not usually observed between the first few nuclear states".

(a) For light nuclei, give arguments that support this statement on the basis of the shell model. Indicate situations where exceptions might be expected.

(b) Make an order-of-magnitude "guesstimate" for the energy and radioactive lifetime of the lowest-energy electric dipole transition expected for $^{17}_9F_8$, outlining your choice of input parameters.

(c) Show that for nuclei containing an equal number of neutrons and protons $(N = Z)$, no electric dipole transitions are expected between two states with the same isospin T.

The following Clebch–Gordan coefficient may be of use:
Using notation $\langle J_1 J_2 M_1 M_2 | J_{TOT} M_{TOT}\rangle$, $\langle J100|J0\rangle = 0$.

(Princeton)

Solution:

(a) Based on single-particle energy levels given by shell model, we see that levels in the same shell generally have the same parity, especially the lowest-lying levels like $1s, 1p, 1d, 2s$ shells, etc. For light nuclei, γ-transition occurs mainly between different single-nucleon levels. In transitions between different energy levels of the same shell, parity does not change. On the other hand, electric dipole transition $E1$ follows selection rules $\Delta J = 0$ or 1, $\Delta P = -1$. Transitions that conserve parity cannot be electric dipole in nature. However if the ground and excited states are not in the same shell, parity may change in a transition. For example in the transition $1p_{3/2} \to 1s_{1/2}$, $\Delta J = 1$, $\Delta P = -1$. This is an electric dipole transition.

(b) In the single-particle model, the lowest-energy electric dipole transition $E1$ of ^{17}F is $2s_{1/2} \to 1p_{1/2}$. The transition probability per unit time can be estimated by (**Problem 2093** with $L = 1$)

$$\lambda \approx \frac{c}{4}\left(\frac{e^2}{\hbar c}\right)\left(\frac{E_\gamma}{\hbar c}\right)^3 \langle r \rangle^2 \,,$$

where E_γ is the transition energy and $\langle r \rangle \sim R = 1.4 \times 10^{-13} \, A^{1/3}$ cm. Thus

$$\lambda \approx \frac{3 \times 10^{10} \times (1.4 \times 10^{-13})^2}{4 \times 137 \times (197 \times 10^{-13})^3} \, A^{2/3} E_\gamma^3 = 1.4 \times 10^{14} A^{2/3} E_\gamma^3 \,,$$

where E_γ is in MeV. For ^{17}F we may take $E_\gamma \approx 5$ MeV, $A = 17$, and so

$$\lambda = 1.2 \times 10^{17} \, s \,,$$

or

$$\tau = \lambda^{-1} = 9 \times 10^{-18} \, s \,.$$

(c) For light or medium nuclei, the isospin is a good quantum number. A nucleus state can be written as $|JmTT_z\rangle$, where J, m refer to angular momentum, T, T_z refer to isospin. The electric multipole transition operator between two states is

$$O_E(L,E) = \sum_{K=1}^{A} \left[\frac{1}{2}(1 + \tau_z(K))e_p + \frac{1}{2}(1 - \tau_z(K))e_n\right] r^L(K)Y_{LM}(r(K))$$

$$= \sum_{K=1}^{A} S(L,M,K)\cdot 1 + \sum_{K=1}^{A} V(L,M,K)\tau_z(K)$$

with

$$S(L,M,K) = \frac{1}{2}(e_p + e_n)r^L(K)Y_{LM}(r(K)) \,,$$

$$V(L,M,K) = \frac{1}{2}(e_p - e_n)r^L(K)Y_{LM}(r(K)) \,,$$

where τ_z is the z component of the isospin matrix, for which $\tau_z\phi_n = -\phi_n$, $\tau_z\phi_p = +\phi_p$.

The first term is related to isospin scalar, the second term to isospin vector. An electric multipole transition from J, T, T_z to J', T', T_z' can be written as

$$B_E(L : J_i T_i T_z \rightarrow J_f T_f T_z) = \langle J_f T_f T_z | O_E(L) | J_i T_i T_z \rangle^2 / (2J_i + 1)$$

$$= \frac{1}{(2J_i + 1)(2T_f + 1)} [\delta_{T_i T_f} \langle J_f T_f | \sum_{K=1}^{A} S(L, K) \cdot 1 | J_i T_i \rangle$$

$$+ \langle T_i T_z 10 | T_f T_z \rangle \langle J_f T_f | \sum_{K=1}^{A} V(L, K) \tau_z(K) | J_i T_i \rangle]^2 .$$

From the above equation, we see that for electric multipole transitions between two states the isospin selection rule is $\Delta T \leq 1$. When $\Delta T = 0$, $\delta'_{TT} \neq 0$, there is an isospin scalar component; when $\Delta T = 1$, the scalar component is zero.

For electric dipole transition,

$$\sum_{K=1}^{A} S(L, K) \cdot 1 = \sum_{K=1}^{A} \frac{1}{2}(e_p + e_n) r(K) Y_{LM}(r(K))$$

$$= \frac{1}{2}(e_p + e_n) \sum_{K=1}^{A} r(K) Y_{LM}(r(K)),$$

r being nucleon coordinate relative to the center of mass of the nucleus.

For spherically or axially symmetric nuclei, as $\sum_{K=1}^{A} r Y_{LM}(r(K))$ is zero, the isospin scalar term makes no contribution to electric dipole transition. For the isospin vector term, when $T_i = T_f = T$,

$$\langle T_i T_z 10 | T_f T_z \rangle = \frac{T_z}{\sqrt{T(T + 1)}} .$$

Then for nuclei with $Z = N$, in transitions between two levels of $\Delta T = 0$, as $T_z = 0$,

$$\langle T_i T_z 10 | T_f T_z \rangle = 0 .$$

and so both the isospin scalar and vector terms make no contribution. Thus for self-conjugate nuclei, states with $T_i = T_f$ cannot undergo electric dipole transition.

2093

(a) Explain why electromagnetic E_λ radiation is emitted predominantly with the lowest allowed multipolarity L. Give an estimate for the ratios $E_1 : E_2 : E_3 : E_4 : E_5$ for the indicated transitions in ^{16}O (as shown in Fig. 2.31).

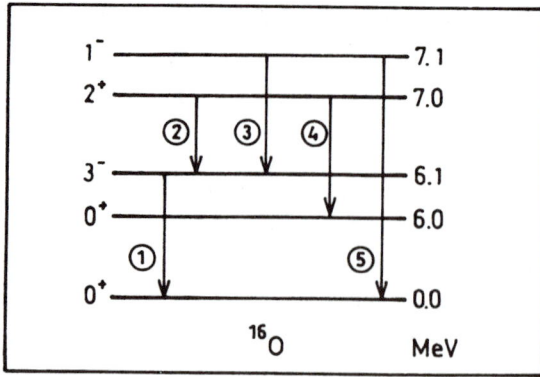

Fig. 2.31

(b) Estimate the lifetime of the 7.1 MeV state. Justify your approximations.

(c) List the possible decay modes of the 6.0 MeV state.

(Princeton)

Solution:

(a) In nuclear shell theory, γ-ray emission represents transition between nucleon energy states in a nucleus. For a proton moving in a central field radiation is emitted when it transits from a higher energy state to a lower one in the nucleus. If L is the degree of the electric multipole radiation, the transition probability per unit time is given by

$$\lambda_E(L) \approx \frac{2(L+1)}{L[(2L+1)!!]^2} \left(\frac{3}{L+3}\right)^2 \left(\frac{e^2}{\hbar}\right) k^{2L+1} \langle r^L \rangle^2 ,$$

where $k = \frac{w}{c} = \frac{E_\gamma}{\hbar c}$ is the wave number of the radiation, E_γ being the transition energy, and $\langle \gamma^L \rangle^2 \approx R^{2L}$, $R = 1.4 \times 10^{-13} A^{1/3}$ cm being the nuclear radius. Thus

$$\lambda_E(L) \approx \frac{2(L+1)}{L[(2L+1)!!]^2} \left(\frac{3}{L+3}\right)^2 \left(\frac{e^2}{R\hbar c}\right) \left(\frac{E_\gamma c}{\hbar c}\right) \left(\frac{E_\gamma R}{\hbar c}\right)^{2L}$$

$$= \frac{2(L+1)}{L[(2L+1)!!]^2} \left(\frac{3}{L+3}\right)^2 \frac{1}{137} \left(\frac{3 \times 10^{10} E_\gamma}{197 \times 10^{-13}}\right)$$

$$\times \left(\frac{E_\gamma \times 1.4 \times 10^{-13} A^{1/3}}{197 \times 10^{-13}} \right)^{2L}$$

$$= \frac{4.4(L+1)}{L[(2L+1)!!]^2} \left(\frac{3}{L+3} \right)^2 \left(\frac{E_\gamma}{197} \right)^{2L+1} (1.4A^{1/3})^{2L} \times 10^{21} \ s^{-1}$$

with E_γ in MeV. Consider ^{16}O. If $E_\gamma \sim 1$ MeV, we have

$$\frac{\lambda_E(L+1)}{\lambda_E(L)} \sim (kR)^2 = \left(\frac{E_\gamma R}{\hbar c} \right)^2 = \left(\frac{1.4 \times 10^{-13} \times 16^{1/3}}{197 \times 10^{-13}} \right)^2 \approx 3 \times 10^{-4} \ .$$

Hence $\lambda_E(L)$ decreases by a factor 10^{-4} as L increases by 1. This means that E_L radiation is emitted predominantly with the lowest allowed multipolarity L.

The tranistions of ^{16}O indicated in Fig. 2.31 can be summarized in the table below.

Transition	$\Delta \pi$	Δl	Type	L	E_γ (MeV)
E_1	yes	3	octopole	3	6.1
E_2	yes	1	dipole	1	0.9
E_3	no	2	quadrupole	2	1.0
E_4	no	2	quadrupole	2	1.0
E_5	yes	1	dipole	1	7.1

Thus we have

$$\lambda_{E_1} : \lambda_{E_2} : \lambda_{E_3} : \lambda_{E_4} : \lambda_{E_5} = \frac{4}{3(7!!)^2} \left(\frac{1}{2} \right)^2 \left(\frac{6.1}{197} \right)^7 (1.4A^{1/3})^6$$

$$: \frac{2}{(3!!)^2} \left(\frac{3}{4} \right)^2 \left(\frac{0.9}{197} \right)^3 (1.4A^{1/3})^2$$

$$: \frac{3}{2(5!!)^2} \left(\frac{3}{5} \right)^2 \left(\frac{1}{197} \right)^5 (1.4A^{1/3})^4$$

$$: \frac{3}{2(5!!)^2} \left(\frac{3}{5} \right)^2 \left(\frac{1}{197} \right)^5 (1.4A^{1/3})^4$$

$$: \frac{2}{(3!!)^2} \left(\frac{3}{4}\right)^2 \left(\frac{7.1}{197}\right)^3 (1.4 A^{1/3})^2$$

$$= 1.59 \times 10^{-12} : 1.48 \times 10^{-7} : 1.25 \times 10^{-12}$$

$$: 1.25 \times 10^{-12} : 7.28 \times 10^{-5}$$

$$= 2.18 \times 10^{-8} : 2.03 \times 10^{-3}$$

$$: 1.72 \times 10^{-8} : 1.72 \times 10^{-8} : 1$$

Thus the transition probability of E_5 is the largest, that of E_2 is the second, those of E_3, E_4 and E_1 are the smallest.

(b) The half-life of the 7.1 MeV level can be determined from λ_{E_5}:

$$\lambda_{E_5} = \frac{4.4 \times 2}{(3!!)^2} \left(\frac{3}{4}\right)^2 \left(\frac{7.1}{197}\right)^3 (1.4 \times 16^{1/3})^2 \times 10^{21} = 3.2 \times 10^{17} \ s^{-1} \,,$$

giving

$$T_{1/2}(7.1 \text{ MeV}) = \ln 2/\lambda_{E_5} = 2.2 \times 10^{-18} \ s \,.$$

Neglecting transitions to other levels is justified as their probabilities are much smaller, e.g.,

$$\lambda_{E_3} : \lambda_{E_5} = 1.7 \times 10^{-8} : 1 \,.$$

In addition, use of the single-particle model is reasonable as it assumes the nucleus to be spherically symmetric, the initial and final state wave functions to be constant inside the nucleus and zero outside which are plausible for ^{16}O.

(c) The γ-transition $0^+ \to 0^+$ from the 6.0 MeV states to the ground state of ^{16}O is forbidden. However, the nucleus can still go to the ground state by internal conversion.

2094

The γ-ray total nuclear cross section σ_{total} (excluding e^+e^- pair production) on neodymium 142 is given in Fig. 2.32

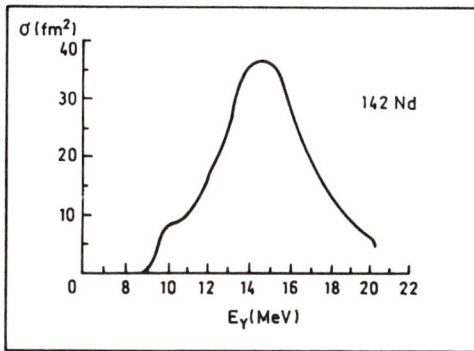

Fig. 2.32

(a) Which electric or magnetic multipole is expected to dominate the cross section and why?

(b) Considering the nucleus simply as two fluids of nucleons (protons and neutrons), explain qualitatively the origin of the resonance shown in the figure.

(c) Using a simple model of the nucleus as A particles bound in an harmonic oscillator potential, estimate the resonance energy as a function of A. Does this agree with the observed value in the figure for A = 142?

(d) Discuss the role of residual two-body interactions in modifying the estimate in (c).

(e) What are the physical processes responsible for the width of the resonance? Make rough estimates of the width due to different mechanisms.

(Princeton)

Solution:

(a) The excitation curves of reactions (γ, n) and (γ, p) show a broad resonance of several MeV width from $E_\gamma = 10$ to 20 MeV. This can be explained as follows. When the nuclear excitation energy increases, the density of states increases and the level widths become broader. When the level spacing and level width become comparable, separate levels join together, so that γ-rays of a wide range of energy can excite the nucleus, thus producing a broad resonance. If $E_\gamma \approx 15$ MeV, greater than the nucleon harmonic oscillator energy $\hbar \omega \approx 44/A^{1/3}$ MeV, dipole transition can occur. The single-particle model gives (**Problem 2093(a)**)

$$\frac{\Gamma(E2\,or\,M1)}{\Gamma(E1)} \approx (kR)^2 = \left(\frac{15 \times 1.4 \times 10^{-13} \times 142^{1/3}}{197 \times 10^{-13}}\right)^2 = 0.3\,.$$

Hence the nuclear cross section is due mainly to electric dipole absorption. We can also consider the collective absorption of the nucleus. We see that absorption of γ-rays causes the nucleus to deform and when the γ-ray energy equals the nuclear collective vibrational energy levels, resonant absorption can take place. As $E_\gamma \approx 15$ MeV, for ^{142}Nd nucleus, electric dipole, quadrupole, octopole vibrations are all possible. However as the energy is nearest to the electric dipole energy level, $E1$ resonant absorption predominates.

(b) Consider the protons and neutrons inside the nucleus as liquids that can seep into each other but cannot be compressed. Upon impact of the incoming photon, the protons and neutrons inside the nucleus tend to move to different sides, and their centers of mass become separated. Consequently, the potential energy of the nucleus increases, which generates restoring forces resulting in dipole vibration. Resonant absorption occurs if the photon frequency equals the resonant frequency of the harmonic oscillator.

(c) In a simple harmonic-oscillator model we consider a particle of mass $M = Am_N$, m_N being the nucleon mass, moving in a potential $V = \frac{1}{2}Kx^2$, where K, the force constant, is proportional to the nuclear cross-sectional area. The resonant frequency is $f \approx \sqrt{K/M}$. As $K \propto R^2 \propto A^{2/3}$, $M \propto A$, we have

$$f \propto A^{-1/6} \approx A^{-0.17}\,.$$

This agrees with the experimental result $E_\gamma \propto A^{-0.19}$ fairly well.

(d) The residual two-body force is non-centric. It can cause the nucleus to deform and so vibrate more easily. The disparity between the rough theoretical derivation and experimental results can be explained in terms of the residual force. In particular, for a much deformed nucleus double resonance peaks may occur. This has been observed experimentally.

(e) The broadening of the width of the giant resonance is due mainly to nuclear deformation and resonance under the action of the incident photons. First, the deformation and restoring force are related to many factors and so the hypothetical harmonic oscillator does not have a "good" quality (Q value is small), correspondingly the resonance width is broad. Second, the photon energy can pass on to other nucleons, forming a compound nucleus and redistribution of energy according to the degree of freedom. This may generate a broad resonance of width from several to 10 MeV. In addition there are other broadening effects like the Doppler effect of an order of magnitude of several keV. For a nucleus of $A = 142$, the broadening due to Doppler effect is

$$\Delta E_D \approx \frac{E_\gamma^2}{Mc^2} \approx \frac{15^2}{142 \times 940} = 1.7 \times 10^{-3} \ MeV = 1.7 \ \text{keV}.$$

2095

The total cross section for the absorption of γ-rays by ^{208}Pb (whose ground state has spin-parity $J^\pi = 0^+$) is shown in Fig. 2.33. The peak at 2.6 MeV corresponds to a $J^\pi = 3^-$ level in ^{208}Pb which γ-decays to a 1^- level at 1.2 MeV (see Fig. 2.34).

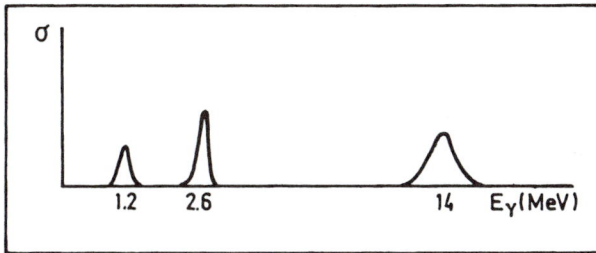

Fig. 2.33

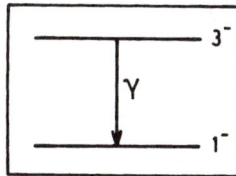

Fig. 2.34

(a) What are the possible electric and/or magnetic multipolarities of the γ-rays emitted in the transition between the 2.6 MeV and 1.2 MeV levels? Which one do you expect to dominate?

(b) The width of the 2.6 MeV level is less than 1 eV, whereas the width of the level seen at 14 MeV is 1 MeV. Can you suggest a plausible reason for this large difference? What experiment might be done to test your conjecture?

(Wisconsin)

Solution:

(a) In the transition $3^- \to 1^-$, the emitted photon can carry away an angular momentum $l = 4, 3, 2$. As there is no parity change, $l = 4, 2$. Hence the possible multipolarities of the transition are $E4$, $M3$ or $E2$. The electric quadrupole transition $E2$ is expected to dominate.

(b) The width of the 2.6 MeV level, which is less than 1 eV, is typical of an electromagnetic decay, whereas the 14 MeV obsorption peak is a giant dipole resonance (**Problem 2094**). As the resonance energy is high, the processes are mostly strong interactions with emission of nucleons, where the single-level widths are broader and many levels merge to form a broad, giant resonance. Thus the difference in decay mode leads to the large difference in level width.

Experimentally, only γ-rays should be found to be emitted from the 2.6 MeV level while nucleons should also be observed to be emitted from the 14 MeV level.

2096

Gamma-rays that are emitted from an excited nuclear state frequently have non-isotropic angular distribution with respect to the spin direction of the excited nucleus. Since generally the nuclear spins are not aligned, but their directions distributed at random, this anisotropy cannot be measured. However, for nuclides which undergo a cascade of γ-emissions (e.g., ^{60}Ni which is used for this problem-see Fig. 2.35), the direction of one of the cascading γ-rays can be used as a reference for the orientation of a specific nucleus. Thus, assuming a negligible half-life for the intermediate state, a measurement of the coincidence rate between the two γ-rays can give the angular correlation which may be used to determine the nuclear spins.

In the case of ^{60}Ni we find such a cascade, namely $J^p = 4^+ \to J^p = 2^+ \to J^p = 0^+$. The angular correlation function is of the form $W(\theta) \sim 1 + 0.1248 \cos^2 \theta + 0.0418 \cos^4 \theta$.

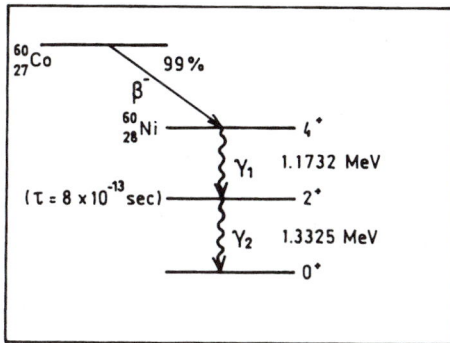

Fig. 2.35

(a) Of what types are the transitions?

(b) Why are the odd powers of $\cos\theta$ missing? Why is $\cos^4\theta$ the highest power?

(c) Draw a schematic diagram of an experimental setup showing how you would make the measurements. Identify all components. (Give block diagram.)

(d) Describe the γ-ray detectors.

(e) How do you determine the coefficients in the correlation function which would prove that ^{60}Ni undergoes the transition $4 \to 2 \to 0$?

(f) Accidental coincidences will occur between the two γ-ray detectors. How can you take account of them?

(g) How would a source of ^{22}Na be used to calibrate the detectors and electronics? (^{22}Na emits 0.511 MeV gammas from β^+ annihilation.)

(h) How would Compton scattering of γ-rays within the ^{60}Co source modify the measurements?

(*Chicago*)

Solution:

(a) Each of the two gamma-ray cascading emissions subtracts 2 from the angular momentum of the excited nucleus, but does not change the parity. Hence the two emissions are of electric-quadrupole $E2$ type.

(b) The angular correlation function for cascading emission can be written as

$$W(\theta) = \sum_{K=0}^{K_{\max}} A_{2K} P_{2K}(\cos\theta),$$

where $0 \leq K_{\max} \leq \min(J_b, L_1, L_2)$,

$$A_{2K} = F_{2K}(L_1, J_a, J_b) F_{2K}(L_2, J_c, J_b),$$

L_1, L_2 being the angular momenta of the two γ-rays, J_a, J_b, J_c being respectively the initial, intermediate and final nuclear spins, $P_{2K}(\cos\theta)$ are Legendre polynomials.

Since $W(\theta)$ depends on $P_{2K}(\cos\theta)$ only, it consists of even powers of $\cos\theta$. For the $4^+ \to 2^+ \to 0^+$ transition of ^{60}Ni, $K_{\max}$ is 2. Hence the highest power of $\cos\theta$ in $P_4(\cos\theta)$ is 4, and so is in $W(\theta)$.

(c) Figure 2.36 shows a block diagram of the experimental apparatus to measure the angular correlation of the γ-rays. With probe 1 fixed, rotate probe 2 in the plane of the source and probe 1 about the source to change the angle θ between the two probes, while keeping the distance between the probes constant. A fast-slow-coincidence method may be used to reduce spurious coincidences and multiscattering.

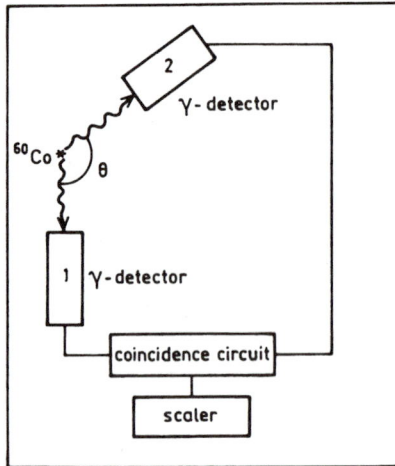

Fig. 2.36

(d) A γ-ray detector usually consists of a scintillator, a photomultiplier, and a signal-amplifying high-voltage circuit for the photomultiplier. When the scintillator absorbs a γ-ray, it fluoresces. The fluorescent photons hit the cathode of the photomultiplier, causing emission of primary photoelectrons, which are multiplied under the high voltage, giving a signal on the anode. The signal is then amplified and processed.

(e) The coincidence counting rate $W(\theta)$ is measured for various θ. Fitting the experimental data to the angular correlation function we can deduce the coefficients.

(f) We can link a delay line to one of the γ-detectors. If the delay time is long compared to the lifetime of the intermediate state the signals from the two detectors can be considered independent, and the coincidence counting rate accidental. This may then be used to correct the observed data.

(g) The two γ-photons of 0.511 MeV produced in the annihilation of β^+ from ^{22}Na are emitted at the same time and in opposite directions. They can be used as a basis for adjusting the relative time delay between the two detectors to compensate for any inherent delays of the probes and electronic circuits to get the best result.

(h) The Compton scattering of γ-rays in the ^{60}Co source will increase the irregularity of the γ-emission and reduce its anisotropy, thereby reducing the deduced coefficients in the angular correlation function.

2097

A nucleus of mass M is initially in an excited state whose energy is ΔE above the ground state of the nucleus. The nucleus emits a gamma-ray of energy $h\nu$ and makes a transition to its ground state.

Explain why the gamma-ray $h\nu$ is not equal to the energy level difference ΔE and determine the fractional change $\frac{h\nu - \Delta E}{\Delta E}$. (You may assume $\Delta E < Mc^2$)

(Wisconsin)

Solution:

The nucleus will recoil when it emits a γ-ray because of the conservation of momentum. It will thereby acquire some recoil energy from the excitation energy and make $h\nu$ less than ΔE.

Let the total energy of the nucleus be E and its recoil momentum be p. The conservation of energy and of momentum give

$$p = p_\gamma, \qquad E + E_\gamma = Mc^2 + \Delta E.$$

As

$$E_\gamma = P_\gamma c = h\nu, \qquad E = \sqrt{p^2 c^2 + M^2 c^4},$$

we have

$$E_\gamma = \frac{1}{2Mc^2} \cdot \frac{(\Delta E)^2 + 2Mc^2 \Delta E}{\left(1 + \frac{\Delta E}{Mc^2}\right)} \approx \Delta E - \frac{(\Delta E)^2}{2Mc^2},$$

or

$$\frac{h\nu - \Delta E}{\Delta E} = -\frac{\Delta E}{2Mc^2}.$$

2098

A (hypothetical) particle of rest mass m has an excited state of excitation energy ΔE, which can be reached by γ-ray absorption. It is assumed that $\Delta E/c^2$ is not small compared to m.

Find the resonant γ-ray energy, E_γ, to excite the particle which is initially at rest.

(*Wisconsin*)

Solution:

Denote the particle by A. The reactions is $\gamma + A \to A^*$. Let E_γ and p_γ be the energy and momentum of the γ-ray, p be the momentum of A, initially at rest, after it absorbs the γ-ray. Conservation of energy requires

$$E_\gamma + mc^2 = \sqrt{\left(m + \frac{\Delta E}{c^2}\right)^2 c^4 + p^2 c^2}.$$

Momentum conservation requires

$$p = p_\gamma,$$

or

$$pc = p_\gamma c = E_\gamma.$$

Its substitution in the energy equation gives

$$E_\gamma = \Delta E + \frac{(\Delta E)^2}{2mc^2}.$$

Thus the required γ-ray energy is higher than ΔE by $\frac{\Delta E^2}{2mc^2}$, which provides for the recoil energy of the particle.

2099

(a) Use the equivalence principle and special relativity to calculate, to first order in y, the frequency shift of a photon which falls straight down through a distance y at the surface of the earth. (Be sure to specify the sign.)

(b) It is possible to measure this frequency shift in the laboratory using the Mössbauer effect.

Describe such an experiment — specifically:

What is the Mössbauer effect and why is it useful here?

What energy would you require the photons to have?

How would you generate such photons?

How would you measure such a small frequency shift?

Estimate the number of photons you would need to detect in order to have a meaningful measurement.

(Columbia)

Solution:

(a) Let the original frequency of the photon be ν_0, and the frequency it has after falling a distance y in the earth's gravitational field be ν. Then the equivalent masses of the photon are respectively $h\nu_0/c^2$ and $h\nu/c^2$. Suppose the earth has mass M and radius R. Conservation of energy requires

$$h\nu_0 - G\frac{M \cdot \frac{h\nu_0}{c^2}}{R+y} = h\nu - G\frac{M \cdot \frac{h\nu}{c^2}}{R},$$

where G is the gravitational constant, or, to first order in y,

$$\frac{\nu - \nu_0}{\nu_0} = \frac{GM}{c^2}\left(\frac{1}{R} - \frac{1}{R+y}\right) \approx \frac{gy}{c^2} = 1.09 \times 10^{-16}y,$$

where g is the acceleration due to gravity and y is in meters. For example, taking $y = 20 \, m$ we have

$$\frac{\nu - \nu_0}{\nu_0} = 2.2 \times 10^{-15}.$$

(b) In principle, photons emitted by a nucleus should have energy E_γ equal to the excitation energy E_0 of the nucleus. However, on account of the recoil of the nucleus which takes away some energy, $E_\gamma < E_0$, or more precisely (**Problem 2097**),

$$E_\gamma = E_0 - \frac{E_0^2}{2Mc^2},$$

where M is the mass of the nucleus. Likewise, when the nucleus absorbs a photon by resonant absorption the latter must have energy (**Problem 2098**)

$$E_\gamma = E_0 + \frac{E_0^2}{2Mc^2}.$$

As $\frac{E_0^2}{2Mc^2}$ is usually larger than the natural width of the excited state, γ-rays emitted by a nucleus cannot be absorbed by resonant absorption by the same kind of nucleus.

However, when both the γ source and the absorber are fixed in crystals, the whole crystal recoils in either process, $M \to \infty$, $\frac{E_0^2}{2Mc^2} \to 0$. Resonant absorption can now occur for absorber nuclei which are the same as the source nuclei. This is known as the Mössbauer effect. It allows accurate measurement of γ-ray energy, the precision being limited only by the natural width of the level.

To measure the frequency shift $\frac{\Delta\nu}{\nu_0} = 2.2 \times 10^{-15}$, the γ source used must have a level of natural width Γ/E_γ less than $\Delta\nu/\nu_0$. A possible choice is ^{67}Zn which has $E_\gamma = 93$ keV, $\Gamma/E_\gamma = 5.0 \times 10^{-16}$. Crystals of ^{67}Zn are used both as source and absorber. At $y = 0$, both are kept fixed in the same horizontal plane and the resonant aborption curve is measured. Then move the source crystal to 20 m above the absorber. The frequency of the photons arriving at the fixed absorber is $\nu_0 + \Delta\nu$ and resonant absorption does not occur. If the absorber is given a downward velocity of v such that by the Doppler effect the photons have frequency ν_0 as seen by the absorber, resonant absorption can take place. As

$$\nu_0 = (\nu_0 + \Delta\nu)\left(1 - \frac{v}{c}\right) \approx \nu_0 + \Delta\nu - \nu_0\left(\frac{v}{c}\right),$$

$$v \approx c\left(\frac{\Delta\nu}{\nu_0}\right) = 3 \times 10^{10} \times 2.2 \times 10^{-15}$$

$$= 6.6 \times 10^{-5} \ cm \ s^{-1},$$

which is the velocity required for the absorber.

Because the natural width for γ-emission of ^{67}Zn is much smaller than $\Delta\nu/\nu_0$, there is no need for a high counting rate. A statistical error of 5% at the spectrum peak is sufficient for establishing the frequency shift, corresponding to a photon count of 400.

2100

A parent isotope has a half-life $\tau_{1/2} = 10^4$ yr $= 3.15 \times 10^{11}$ s. It decays through a series of radioactive daughters to a final stable isotope. Among the daughters the greatest half-life is 20 yr. Others are less than a year. At $t = 0$ one has 10^{20} parent nuclei but no daughters.

(a) At $t = 0$ what is the activity (decays/sec) of the parent isotope?

(b) How long does it take for the population of the 20 yr isotope to reach approximately 97% of its equilibrium value?

(c) At $t = 10^4$ yr how many nuclei of the 20 yr isotope are present? Assume that none of the decays leading to the 20 yr isotope is branched.

(d) The 20 yr isotope has two competing decay modes: α, 99.5%; β, 0.5%. At $t = 10^4$ yr, what is the activity of the isotope which results from the β-decay?

(e) Among the radioactive daughters, could any reach their equilibrium populations much more quickly (or much more slowly) than the 20 yr isotope?

(Wisconsin)

Solution:

(a) The decay constant of the parent isotope is

$$\lambda_1 = \frac{\ln 2}{\tau_{1/2}} = 6.93 \times 10^{-5} \ yr^{-1} = 2.2 \times 10^{-12} \ s^{-1} .$$

When $t = 0$, the activity of the parent isotope is

$$A_1(0) = \lambda_1 N_1(t = 0) = \frac{2.2 \times 10^{-12} \times 10^{20}}{3.7 \times 10^7} = 5.95 \text{ millicurie} .$$

(b) Suppose the 20 yr isotope is the nth-generation daughter in a radioactive series. Then its population is a function of time:

$$N_n(t) = N_1(0)(h_1 e^{-\lambda_1 t} + h_2 e^{-\lambda_2 t} + \cdots + h_n e^{-\lambda_n t}),$$

where

$$h_1 = \frac{\lambda_1 \lambda_2 \cdots \lambda_{n-1}}{(\lambda_2 - \lambda_1)(\lambda_3 - \lambda_1) \cdots (\lambda_n - \lambda_1)},$$

$$h_2 = \frac{\lambda_1 \lambda_2 \cdots \lambda_{n-1}}{(\lambda_1 - \lambda_2)(\lambda_3 - \lambda_2) \cdots (\lambda_n - \lambda_2)},$$

$$\vdots$$

$$h_n = \frac{\lambda_1 \lambda_2 \cdots \lambda_{n-1}}{(\lambda_1 - \lambda_n)(\lambda_2 - \lambda_n) \cdots (\lambda_{n-1} - \lambda_n)},$$

where $N_1(0)$ is the number of the parent nuclei at $t = 0$, λ_i is the decay constant of the ith-generation daughter. For secular equilibrium we require $\lambda_1 \ll \lambda_j$, $j = 2, 3, \ldots, n, \ldots$. As the nth daugther has the largest half-life of 10^{20} yr, we also have $\lambda_n \ll \lambda_j$, $j = 2, 3, \ldots, (j \neq n)$, $\lambda_n = \ln 2/\tau_{1/2} = 3.466 \times 10^{-2}$ yr^{-1}. Thus

$$h_1 \approx \frac{\lambda_1}{\lambda_n}, \qquad h_n \approx -\frac{\lambda_1}{\lambda_n}.$$

After a sufficiently long time the system will reach an equilibrium at which $\lambda_n N_n^e(t) = \lambda_1 N_1^e(t)$, the superscript e denoting equilibrium values, or

$$N_n^e(t) = \frac{\lambda_1}{\lambda_n} N_1^e(t) = \frac{\lambda_1}{\lambda_n} N_1(0) e^{-\lambda_1 t}.$$

At time t before equilibrium is reached we have

$$N_n(t) \approx N_1(0) \left(\frac{\lambda_1}{\lambda_n} e^{-\lambda_1 t} - \frac{\lambda_1}{\lambda_n} e^{-\lambda_n t} \right).$$

When $N_n(t) = 0.97 N_n^e(t)$, or

$$0.97 \frac{\lambda_1}{\lambda_n} N_1(0) e^{-\lambda_1 t} \approx N_1(0) \left(\frac{\lambda_1}{\lambda_n} e^{-\lambda_1 t} - \frac{\lambda_1}{\lambda_n} e^{-\lambda_n t} \right),$$

the time is $t = t_0$ given by

$$t_0 = \frac{\ln 0.03}{\lambda_1 - \lambda_n} \approx 101 \ yr.$$

Hence after about 101 years the population of the 20 yr isotope will reach 97% of its equilibrium value.

(c) At $t = 10^4$ yr, the system can be considered as in equilibrium. Hence the population of the 20 yr isotope at that time is

$$N_n(10^4) = \frac{\lambda_1}{\lambda_n} N_1(0)e^{-\lambda_1 t} = 10^{17}.$$

(d) After the system has reached equilibrium, all the isotopes will have the same activity. At $t = 10^4$ years, the activity of the parent isotope is

$$A_1(10^4) = \lambda_1 N(0)e^{-\lambda_1 t} = 6.93 \times 10^{-5} \times 10^{20} \times \exp(-6.93 \times 10^{-5} \times 10^4)$$

$$= 3.47 \times 10^{15} \ yr^{-1} = 3.0 \ mc.$$

The activity of the β-decay product of the 20 yr isotope is

$$A_\beta = 3 \times 0.05 = 0.15 \ mc.$$

(e) The daughter nuclei ahead of the 20 yr isotope will reach their equilibrium populations more quickly than the 20 yr isotope, while the daughter nuclei after the 20 yr isotope will reach their equilibrium populations approximately as fast as the 20 yr isotope.

2101

A gold foil 0.02 cm thick is irradiated by a beam of thermal neutrons with a flux of 10^{12} neutrons/cm^2/s. The nuclide ^{198}Au with a half-life of 2.7 days is produced by the reaction ^{197}Au$(n, \gamma)^{198}$Au. The density of gold is 19.3 gm/cm^3 and the cross section for the above reaction is 97.8×10^{-24} cm^2. ^{197}Au is 100% naturally abundant.

(a) If the foil is irradiated for 5 minutes, what is the ^{198}Au activity of the foil in decays/cm^2/s?

(b) What is the maximum amount of ^{198}Au/cm^2 that can be produced in the foil?

(c) How long must the foil be irradiated if it is to have 2/3 of its maximum activity?

(Columbia)

Solution:

(a) Initially the number of ^{197}Au nuclei per unit area of foil is

$$N_1(0) = \frac{0.02 \times 19.3}{197} \times 6.023 \times 10^{23} = 1.18 \times 10^{21} \ cm^{-2}.$$

Let the numbers of ^{197}Au and ^{198}Au nuclei at time t be N_1, N_2 respectively, σ be the cross section of the (n, γ) reaction, I be flux of the incident neutron beam, and λ be the decay constant of ^{198}Au. Then

$$\frac{dN_1}{dt} = -\sigma I N_1 \,,$$

$$\frac{dN_2}{dt} = \sigma I N_1 - \lambda N_2 \,.$$

Integrating we have

$$N_1 = N_1(0)e^{-\sigma I t} \,,$$

$$N_2 = \frac{\sigma I}{\lambda - \sigma I} N_1(0)(\bar{e}^{\sigma I t} - e^{-\lambda t}) \,.$$

As

$$\lambda = \frac{\ln 2}{2.7 \times 24 \times 3600} = 2.97 \times 10^{-6} \ s^{-1} \,,$$

$$\sigma I = 9.78 \times 10^{-23} \times 10^{12} = 9.78 \times 10^{-11} \ s^{-1} \ll \lambda \,,$$

at $t = 5 \ min = 300 \ s$ the activity of ^{198}Au is

$$A(300s) = \lambda N_2(t) = \frac{\lambda \sigma I N_1(0)}{\lambda - \sigma I}(e^{-\sigma I t} - e^{-\lambda t}) \approx \sigma I N_1(0)(1 - e^{-\lambda t})$$

$$= 9.78 \times 10^{-11} \times 1.18 \times 10^{21} \times [1 - \exp(-2.97 \times 10^{-6} \times 300)]$$

$$= 1.03 \times 10^8 \ cm^{-2} \ s^{-1} \,.$$

(b) After equilibrium is attained, the activity of a nuclide, and hence the number of its nuclei, remain constant. This is the maximum amount of ^{198}Au that can be produced. As

$$\frac{dN_2}{dt} = 0 \,,$$

we have

$$\lambda N_2 = \sigma I N_1 \approx \sigma I N_1(0)$$

giving

$$N_2 = \frac{\sigma I}{\lambda} N_1(0) = \frac{9.78 \times 10^{-11}}{2.97 \times 10^{-6}} \times 1.18 \times 10^{21}$$

$$= 3.89 \times 10^{16} \text{ cm}^{-2}.$$

(c) As

$$A = \frac{2}{3} A_{max} \approx \sigma I N_1(0)(1 - e^{-\lambda t}),$$

$$t = -\frac{1}{\lambda} \ln \left(1 - \frac{2}{3} \frac{A_{max}}{\sigma I N_1(0)} \right) = -\frac{1}{\lambda} \ln \left(1 - \frac{2}{3} \right) = 3.70 \times 10^5 \text{ s} = 4.28 \text{ day}.$$

2102

In the fission of ^{235}U, 4.5% of the fission lead to ^{133}Sb. This isotope is unstable and is the parent of a chain of β-emitters ending in stable ^{133}Cs:

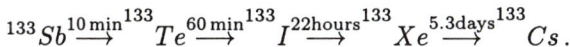

$$^{133}Sb \xrightarrow{10 \text{ min}} {}^{133}Te \xrightarrow{60 \text{ min}} {}^{133}I \xrightarrow{22 \text{hours}} {}^{133}Xe \xrightarrow{5.3 \text{days}} {}^{133}Cs.$$

(a) A sample of 1 gram of uranium is irradiated in a pile for 60 minutes. During this time it is exposed to a uniform flux of 10^{11} neutrons/cm^2 sec. Calculate the number of atoms of Sb, Te, and I present upon removal from the pile. Note that uranium consists of 99.3% ^{238}U and 0.7% ^{235}U, and the neutron fission cross section of ^{235}U is 500 barns. (You may neglect the shadowing of one part of the sample by another.)

(b) Twelve hours after removal from the pile the iodine present is removed by chemical separation. How many atoms of iodine would be obtained if the separation process was 75% efficient?

(Columbia)

Solution:

(a) The number of Sb atoms produced in the pile per second is

$$C = N_0 \cdot f \cdot \sigma \cdot 4.5\%$$

$$= \frac{1 \times 0.007}{235} \times 6.023 \times 10^{23} \times 10^{11} \times 500 \times 10^{-24} \times 0.045$$

$$= 4.04 \times 10^7 \text{ s}^{-1}.$$

Let the numbers of atoms of Sb, Te, I present upon removal from the pile be N_1, N_2, N_3 and their decay constants be $\lambda_1, \lambda_2, \lambda_3$ respectively. Then $\lambda_1 = \frac{\ln 2}{600} = 1.16 \times 10^{-3}\ s^{-1}$, $\lambda_2 = 1.93 \times 10^{-4}\ s^{-1}$, $\lambda_3 = 8.75 \times 10^{-6}\ s^{-1}$, and $\frac{dN_1}{dt} = C - \lambda_1 N_1$, with $N_1 = 0$ at $t = 0$, giving for $T = 3600\ s$,

$$N_1(T) = \frac{C}{\lambda_1}(1 - e^{-\lambda_1 T}) = 3.43 \times 10^{10}\,,$$

$\frac{dN_2}{dt} = \lambda_1 N_1 - \lambda_2 N_2$, with $N_2 = 0$, at $t = 0$, giving

$$N_2(T) = \frac{C}{\lambda_2}\left(1 + \frac{\lambda_2}{\lambda_1 - \lambda_2}e^{-\lambda_1 T} - \frac{\lambda_1}{\lambda_1 - \lambda_2}e^{-\lambda_2 T}\right) = 8.38 \times 10^{10}\,,$$

$\frac{dN_3}{dt} = \lambda_2 N_2 - \lambda_3 N_3$, with $N_3 = 0$, at $t = 0$, giving

$$N_3(T) = \frac{C}{\lambda_3}\left[1 - \frac{\lambda_2\lambda_3 e^{-\lambda_1 T}}{(\lambda_1 - \lambda_2)(\lambda_1 - \lambda_3)} - \frac{\lambda_3\lambda_1 e^{-\lambda_2 T}}{(\lambda_2 - \lambda_3)(\lambda_2 - \lambda_1)}\right]$$
$$+ \frac{C}{\lambda_3}\left[\frac{\lambda_2\lambda_3}{(\lambda_1 - \lambda_2)(\lambda_1 - \lambda_3)} - \frac{\lambda_1\lambda_3}{(\lambda_1 - \lambda_2)(\lambda_2 - \lambda_3)} - 1\right]e^{-\lambda_3 T}$$
$$= \frac{C}{\lambda_3}\left[1 - \frac{\lambda_2\lambda_3 e^{-\lambda_1 T}}{(\lambda_1 - \lambda_2)(\lambda_1 - \lambda_3)} - \frac{\lambda_3\lambda_1 e^{-\lambda_2 T}}{(\lambda_2 - \lambda_3)(\lambda_2 - \lambda_1)}\right.$$
$$\left. - \frac{\lambda_1\lambda_2 e^{-\lambda_3 T}}{(\lambda_3 - \lambda_1)(\lambda_3 - \lambda_2)}\right] \approx \frac{C}{\lambda_3}(1 - e^{-\lambda_3 T})$$
$$= \frac{C}{\lambda_3}(1 - 0.969) = 2.77 \times 10^{10}\,.$$

(b) After the sample is removed from the pile, no more Sb is produced, but the number of Sb atoms will decrease with time. Also, at the initial time $t = T$, N_1, N_2, N_3 are not zero. We now have

$$N_1(t) = N_1(T)e^{-\lambda_1 t}\,,$$

$$N_2(t) = \frac{\lambda_1}{\lambda_2 - \lambda_1}N_1(T)e^{-\lambda_1 t} + \left[N_2(T) + \frac{\lambda_1 N_1(T)}{\lambda_1 - \lambda_2}e^{-\lambda_2 t}\right]\,,$$

$$N_3(t) = \frac{\lambda_1\lambda_2 N_1(T)}{(\lambda_2 - \lambda_1)(\lambda_3 - \lambda_1)}e^{-\lambda_1 t} + \frac{\lambda_2}{\lambda_3 - \lambda_2}\left[N_2(T) + \frac{\lambda_1 N_1(T)}{\lambda_1 - \lambda_2}\right]e^{-\lambda_2 t}$$
$$+ \left[N_3(T) + \frac{\lambda_2}{\lambda_2 - \lambda_3}N_2(T) + \frac{\lambda_1\lambda_2 N_1(T)}{(\lambda_1 - \lambda_3)(\lambda_2 - \lambda_3)}\right]e^{-\lambda_3 t}\,.$$

For $t = 12$ hours, as $t \gg \tau_1, \tau_2$,

$$N_3(12 \text{ hours}) \approx \left[N_3(T) + \frac{\lambda_2}{\lambda_2 - \lambda_3} N_2(T) + \frac{\lambda_1 \lambda_2 N_1(T)}{(\lambda_1 - \lambda_3)(\lambda_2 - \lambda_3)} \right] e^{-\lambda_3 t}$$

$$= 10^{10} \times [2.77 + 8.80 + 3.62]$$

$$\times \exp(-8.75 \times 10^{-6} \times 12 \times 3600)$$

$$= 1.04 \times 10^{11}.$$

The number of atoms of I isotope obtained is

$$N = 0.75 \times N_3 = 7.81 \times 10^{10}.$$

2103

A foil of ^{7}Li of mass 0.05 gram is irradiated with thermal neutrons (capture cross section 37 milllibars) and forms ^{8}Li, which decays by β^--decay with a half-life of 0.85 sec. Find the equilibrium activity (number of β-decays per second) when the foil is exposed to a steady neutron flux of 3×10^{12} neutrons/sec·cm^2.

(Columbia)

Solution:

Let the ^{7}Li population be $N_1(t)$, the ^{8}Li population be $N_2(t)$. Initially

$$N_1(0) = \frac{0.05}{7} \times 6.023 \times 10^{23} = 4.3 \times 10^{21}, \quad N_2(0) = 0.$$

During the neutron irradiation, $N_1(t)$ changes according to

$$\frac{dN_1}{dt} = -\sigma \phi N_1,$$

where σ is the neutron capture cross section and ϕ is the neutron flux, or

$$N_1(t) = N_1(0)e^{-\sigma \phi t}.$$

$N_2(t)$ changes according to

$$\frac{dN_2}{dt} = -\frac{dN_1}{dt} - \lambda N_2(t) = N_1(0)\sigma \phi e^{-\sigma \phi t} - \lambda N_2(t),$$

where λ is the β-decay constant of ^{8}Li. Integration gives

$$N_2(t) = \frac{\sigma\phi}{\lambda - \sigma\phi}(e^{-\sigma\phi t} - e^{-\lambda t})N_1(0).$$

At equilibrium, $\frac{dN_2}{dt} = 0$, which gives the time t it takes to reach equilibrium:

$$t = \frac{1}{\lambda - \sigma\phi}\ln\left(\frac{\lambda}{\sigma\phi}\right).$$

As $\lambda = \frac{\ln 2}{0.85} = 0.816\ s^{-1}$, $\sigma\phi = 3.7 \times 10^{-26} \times 3 \times 10^{12} = 1.11 \times 10^{-13}\ s^{-1}$,

$$t \approx \frac{1}{\lambda}\ln\left(\frac{\lambda}{\sigma\phi}\right) = 3.63\ s.$$

The equilibrium activity is

$$A = \lambda N_2 \approx \frac{\lambda\sigma\phi N_1(0)}{\lambda - \sigma\phi} \approx \sigma\phi N_1(0) = 4.77 \times 10^8\ Bq = 12.9\ mc.$$

2104

In a neutron-activation experiment, a flux of 10^8 neutrons/cm^2·sec is incident normally on a foil of area 1 cm^2, density 10^{22} atoms/cm^3, and thickness 10^{-2} cm (Fig. 2.37). The target nuclei have a total cross section for neutron capture of 1 barn (10^{-24} cm^2), and the capture leads uniquely to a nuclear state which β-decays with a lifetime of 10^4 sec. At the end of 100 sec of neutron irradiation, at what rate will the foil be emitting β-rays?

(Wisconsin)

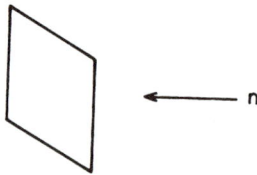

Fig. 2.37

Solution:

Let the number of target nuclei be $N(t)$, and that of the unstable nuclei resulting from neutron irradiation be $N_\beta(t)$. As the thickness of the target is 10^{-2} cm, it can be considered thin so that

$$\frac{dN(t)}{dt} = -\sigma\phi N(t),$$

where ϕ is the neutron flux, σ is the total neutron capture cross section of the target nuclei. Integration gives $N(t) = N(0)e^{-\sigma\phi t}$. As $\sigma\phi = 10^{-24} \times 10^8 = 10^{-16}$ s^{-1}, $\sigma\phi t = 10^{-14} \ll 1$ and we can take $N(t) \approx N(0)$, then

$$\frac{dN}{dt} \approx -\sigma\phi N(0),$$

indicating that the rate of production is approximately constant.

Consider the unstable nuclide. We have

$$\frac{dN_\beta(t)}{dt} \approx \sigma\phi N(0) - \lambda N_\beta(t),$$

where λ is the β-decay constant. Integrating we have

$$N_\beta(t) = \frac{\sigma\phi N(0)}{\lambda}(1 - e^{-\lambda t}),$$

and so

$$A = N_\beta(t)\lambda = \sigma\phi N(0)(1 - e^{-\lambda t}).$$

At $t = 100$ s, the activity of the foil is

$$A = 10^{-16} \times 10^{22} \times 1 \times 10^{-2} \times (1 - e^{-10^{-2}}) = 99.5 \ s^{-1}$$

as

$$\lambda = \frac{1}{10^4} = 10^{-4} \ s.$$

2105

Radioactive dating is done using the isotope
(a) ^{238}U.
(b) ^{12}C.
(c) ^{14}C.

(CCT)

Solution:

^{14}C. The radioactive isotope ^{14}C maintains a small but fixed proportion in the carbon of the atomsphere as it is continually produced by bombardment of cosmic rays. A living entity, by exchanging carbon with the atmosphere, also maintains the same isotopic proportion of ^{14}C. After it dies, the exchange ceases and the isotopic proportion attenuates, thus providing a means of dating the time of death. ^{12}C is stable and cannot be used for this purpose. ^{238}U has a half-life of 4.5×10^9 years, too long for dating.

2106

^{14}C decays with a half-life of about 5500 years.

(a) What would you guess to be the nature of the decay, and what are the final products? Very briefly explain.

(b) If no more ^{14}C enters biological systems after their death, estimate the age of the remains of a tree whose radioactivity (decays/sec) of the type given in (a) is 1/3 of that of a comparable but relatively young tree.

(Wisconsin)

Solution:

(a) ^{14}C is a nuclide with excess neutrons, and so it will β^--decay to ^{14}N according to

$$^{14}C \rightarrow^{14} N + e^- + \bar{\nu}_e \,.$$

(b) The number of ^{14}C of a biological system attenuates with time after death according to $N(t) = N(0)e^{-\lambda t}$, which gives the activity of ^{14}C as

$$A(t) = \lambda N(t) = A(0)e^{-\lambda t} \,.$$

Thus the age of the dead tree is

$$t = \frac{1}{\lambda} \ln \frac{A(0)}{A(t)} = \frac{\tau_{1/2}}{\ln 2} \ln \frac{A(0)}{A(t)}$$

$$= \frac{5500}{\ln 2} \ln \left(\frac{3}{1}\right) = 8717 \text{ years} \,.$$

2107

Plutonium (^{238}Pu, $Z = 94$) has been used as power source in space flights. ^{238}Pu has an α-decay half-life of 90 years (2.7×10^9 sec).

(a) What are the Z and N of the nucleus which remains after α-decay?

(b) Why is ^{238}Pu more likely to emit α's than deuterons as radiation?

(c) Each of the α-particles is emitted with 5.5 MeV. What is the power released if there are 238 gms of ^{238}Pu (6×10^{23} atoms)? (Use any units you wish but specify.)

(d) If the power source in (c) produces 8 times the minimum required to run a piece of apparatus, for what period will the source produce sufficient power for that function.

(Wisconsin)

Solution:

(a) The daughter nucleus has $N = 142$, $Z = 92$.

(b) This is because the binding energy of α-particle is higher than that of deuteron and so more energy will be released in an α-decay. For ^{238}Pu,

$$^{238}_{94}Pu \rightarrow {}^{234}_{92}U + \alpha, \quad Q = 46.186 - 38.168 - 2.645 \approx 5.4 \text{ MeV},$$

$$^{238}_{94}Pu \rightarrow {}^{236}_{93}Np + d, \quad Q = 46.186 - 43.437 - 13.136 \approx -10.4 \text{ MeV}.$$

Deuteron-decay is not possible as $Q < 0$.

(c) Because of the recoil of ^{234}U, the decay energy per ^{238}Pu is

$$E_d = E_\alpha + E_U = \frac{p_\alpha^2}{2m_\alpha} + \frac{p_\alpha^2}{2m_U} = E_\alpha \left(1 + \frac{m_\alpha}{m_U}\right) = 5.5 \left(\frac{238}{234}\right) = 5.6 \text{ MeV}.$$

As the half-life of ^{238}Pu is $T_{1/2} = 90 \ yr = 2.7 \times 10^9 \ s$, the decay constant is

$$\lambda = \ln 2 / T_{1/2} = 2.57 \times 10^{-10} \ s^{-1}.$$

For 238 g of ^{238}Pu, the energy released per second at the beginning is

$$\frac{dE}{dt} = E_d \frac{dN}{dt} = E_d \lambda N_0 = 5.6 \times 2.57 \times 10^{-10} \times 6 \times 10^{23} = 8.6 \times 10^{14} \text{ MeV/s}.$$

(d) As the amount of ^{238}Pu nuclei attenuates, so does the power output:

$$W(t) = W(0)e^{-\lambda t}.$$

When $W(t_0) = W(0)/8$,

$$t_0 = \ln 8 / \lambda = 3 \ln 2 / \lambda = 3T_{1/2} = 270 \ yr.$$

Thus the apparatus can run normally for 270 years.

6. NUCLEAR REACTIONS (2108–2120)

2108

Typical nuclear excitation energies are about 10^{-2}, 10^1, 10^3, 10^5 MeV.

(Columbia)

Solution:

10^1 MeV.

2109

The following are atomic masses in units of u ($1\ u = 932$ MeV/c^2).

Electron	0.000549	$^{152}_{62}Sm$	151.919756
Neutron	1.008665	$^{152}_{63}Eu$	151.921749
$^1_1$H	1.007825	$^{152}_{64}Gd$	151.919794

(a) What is the Q-value of the reaction ^{152}Eu(n,p)?

(b) What types of weak-interaction decay can occur for ^{152}Eu?

(c) What is the maximum energy of the particles emitted in each of the processes given in (b)?

(Wisconsin)

Solution:

(a) The reaction $^{152}Eu + n \rightarrow {}^{152}Sm + p$ has Q-value

$$Q = [m(^{152}Eu) + m(n) - m(^{152}Sm) - m(p)]c^2$$

$$= [M(^{152}Eu) + m(n) - M(^{152}Sm) - M(^1H)]c^2$$

$$= 0.002833\ u = 2.64\ \text{MeV},$$

where m denotes nuclear masses, M denotes atomic masses. The effects of the binding energy of the orbiting electrons have been neglected in the calculation.

(b) The possible weak-interaction decays for ^{152}Eu are β-decays and electron capture:

$$\beta^-\text{-decay} :^{152}Eu \rightarrow^{152} Gd + e^- + \bar{\nu}_e \,,$$

$$\beta^+\text{-decay} :^{152}Eu \rightarrow^{152} Sm + e^+ + \nu_e \,,$$

$$\text{orbital electron capture} :^{152}Eu + e^- \rightarrow^{152} Sm + \nu_e \,.$$

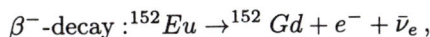

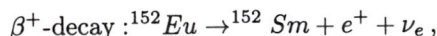

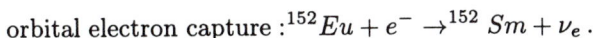

Consider the respective Q-values:

$$\beta^-\text{-decay} : \; E_d(\beta^-) = [M(^{152}Eu) - M(^{152}Gd)]c^2 = 1.822 \text{ MeV} > 0 \,,$$

energetically possible.

$$\beta^+\text{-decay} : \; E_d(\beta^+) = [M(^{152}Eu) - M(^{152}Sm) - 2m(e)]c^2$$
$$= 0.831 \text{ MeV} > 0 \,,$$

energetically possible.

Orbital electron capture:

$$E_d(EC) = [M(^{152}Eu) - M(^{152}Sm)]c^2 - W_j = 1.858 \text{ MeV} - W_j \,,$$

where W_j is the electron binding energy in atomic orbits, the subscript j indicating the shell K, L, M, etc., of the electron. Generally $W_j \ll 1$ MeV, and orbital electron capture is also energetically possible for ^{152}Eu.

(c) As the mass of electron is much smaller than that of the daughter nucleus, the latter's recoil can be neglected. Then the maximum energies of the particles emitted in the processes given in (b) are just the decay energies. Thus

for β^--decay, the maximum energy of electron is 1.822 MeV,

for β^+-decay, the maximum energy of positron is 0.831 MeV.

For orbital electron capture, the neutrinos are monoenergetic, their energies depending on the binding energies of the electron shells from which they are captured. For example, for K capture, $W_k \approx 50$ keV, $E_\nu \approx 1.8$ MeV.

2110

(a) Consider the nuclear reaction

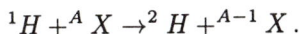

$$^1H +^A X \rightarrow^2 H +^{A-1} X \,.$$

For which of the following target nuclei $^A X$ do you expect the reaction to be the strongest, and why?

$$^A X = {}^{39}Ca, \ {}^{40}Ca, \ {}^{41}Ca.$$

(b) Use whatever general information you have about nuclei to estimate the temperature necessary in a fusion reactor to support the reaction

$$^2 H + {}^2 H \to {}^3 He + n.$$

(Wisconsin)

Solution:

(a) The reaction is strongest with a target of ^{41}Ca. In the reaction the proton combines with a neutron in ^{41}Ca to form a deuteron. The isotope ^{41}Ca has an excess neutron outside of a double-full shell, which means that the binding energy of the last neutron is lower than those of ^{40}Ca, ^{39}Ca, and so it is easier to pick up.

(b) To facilitate the reaction $^2 H + {}^2 H \to^3 H + n$, the two deuterons must be able to overcome the Coulomb barrier $V(r) = \frac{1}{4\pi\varepsilon_0}\frac{e^2}{r}$, where r is the distance between the deuterons. Take the radius of deuteron as 2 fm. Then $r_{\min} = 4 \times 10^{-15}$ m, and $V_{\max} = \frac{1}{4\pi\varepsilon_0}\frac{e^2}{r_{\min}}$. The temperature required is

$$T \gtrsim \frac{V_{\max}}{k} = \frac{1}{4\pi\varepsilon_0}\frac{e^2}{r_{\min}}\frac{1}{k} = \left(\frac{1}{4\pi\varepsilon_0}\frac{e^2}{\hbar c}\right)\left(\frac{\hbar c}{r_{\min}}\right)\frac{1}{k}$$

$$= \frac{1}{137} \times \left(\frac{197 \times 10^{-15}}{4 \times 10^{-15}}\right)\frac{1}{8.6 \times 10^{-11}} = 4 \times 10^9 \ K.$$

In the above k is Boltzmann's constant. Thus the temperature must be higher than 4×10^9 K for the fusion reaction $^2 H + {}^2 H \to {}^3 He + n$ to occur.

2111

(a) Describe one possible experiment to determine the positions (excitation energies) of the excited states (energy levels) of a nucleus such as ^{13}C. State the target, reaction process, and detector used.

(b) In the proposed experiment, what type of observation relates to the angular momentum of the excited state?

(Wisconsin)

Solution:

(a) Bombard a target of ^{12}C with deuterons and detect the energy spectrum of the protons emitted in the reaction ^{12}C(d,p)^{13}C with a gold-silicon surface-barrier semiconductor detector. This, combined with the known energy of the incident deuterons, then gives the energy levels of the excited states of ^{13}C. One can also use a Ge detector to measure the energy of the γ-rays emitted in the de-excitation of $^{13}C^*$ and deduce the excited energy levels.

(b) From the known spin-parity of ^{12}C and the measured angular distribution of the reaction product p we can deduced the spin-parity of the resultant nucleus ^{13}C.

2112

Given the atomic mass excess $(M - A)$ in keV:

$$^1 n = 8071 \text{ keV}, \ ^1 H = 7289 \text{ keV}, \ ^7 Li = 14907 \text{ keV}, \ ^7 Be = 15769 \text{ keV},$$

and for an electron $m_0 c^2 = 511$ keV.

(a) Under what circumstances will the reaction ^{7}Li(p,n)^{7}Be occur?

(b) What will be the laboratory energy of the neutrons at threshold for neutron emission?

(Wisconsin)

Solution:

(a) In $^7 Li + p \rightarrow^7 Be + n + Q$ the reaction energy Q is

$$Q = \Delta M(^7 Li) + \Delta M(^1 H) - \Delta M(^7 Be) - \Delta M(n)$$

$$= 14907 + 7289 - 15769 - 8071 = -1644 \text{ keV}.$$

This means that in the center-of-mass system, the total kinetic energy of ^{7}Li and p must reach 1644 keV for the reaction to occur. Let E, P be the total energy and momentum of the proton in the laboratory system. We require

$$(E + m_{Li} c^2)^2 - P^2 c^2 = (|Q| + m_{Li} c^2 + m_p c^2)^2.$$

As $E^2 = m_p^2 c^4 + P^2 c^2$, $E \approx T + m_p c^2$, $|Q| \ll m_{Li}$, m_p, we have $2(E - m_p c^2)m_{Li}c^2 \approx 2|Q|(m_{Li} + m_p)c^2$, or

$$T = \frac{m_p + m_{Li}}{m_{Li}} \times |Q| \approx \frac{1+7}{7} \times 1644 = 1879 \text{ keV}.$$

Thus the kinetic energy T of the incident proton must be higher than 1879 keV.

(b) The velocity of the center of mass in the laboratory is

$$V_c = \frac{m_p}{m_p + m_{Li}} V_p.$$

As at threshold the neutron is produced at rest in the center-of-mass system, its velocity the laboratory is V_c. Its laboratory kinetic energy is therefore

$$\frac{1}{2}m_n V_c^2 = \frac{1}{2}\frac{m_n m_p^2}{(m_p + m_{Li})^2} \cdot \frac{2T}{m_p} = \frac{m_n m_p T}{(m_p + m_{Li})^2} \approx \frac{T}{64} = 29.4 \text{ keV}.$$

2113

The nucleus ^{8}Be is unstable with respect to dissociation into two α-particles, but experiments on nuclear reactions characterize the two lowest unstable levels as

$J = 0$, even parity, $\sim$95 keV above the dissociation level,
$J = 2$, even parity, $\sim$3 MeV above the dissociation level.

Consider how the existence of these levels influence the scattering of α-particles from helium gas. Specifically:

(a) Write the wave function for the elastic scattering in its partial wave expansion for $r \to \infty$.

(b) Describe qualitatively how the relevant phase shifts vary as functions of energy in the proximity of each level.

(c) Describe how the variation affects the angular distribution of α-particles.

(Chicago)

Solution:

(a) The wave function for elastic scattering of α-particle (He^{++}) by a helium nucleus involves two additive phase shifts arising from Coulomb interaction (δ_l) and nuclear forces (η_l). To account for the identity of the two (spinless) particles, the spatial wave function must be symmetric with an even value of l. Its partial wave at $r \to \infty$ is

$$\sum_{l=0}^{\infty} \frac{1 + (-1)^l}{2}(2l+1)i^l e^{i(\delta_l + \eta_l)} \frac{1}{kr}$$

$$\times \sin\left[kr - \frac{l\pi}{2} - \gamma \ln(2kr) + \delta_l + \eta_l\right] P_l(\cos\theta),$$

where k is the wave number in the center-of-mass system and $\gamma = (2e)^2/\hbar v$, v being the relative velocity of the α-particles.

(b) The attractive nuclear forces cause each η_l to rise from zero as the center-of-mass energy increases to moderately high values. Specifically each η_l rises rather rapidly, by nearly π radians at each resonance, as the energy approaches and then surpasses any unstable level of a definite l of the compound nucleus, e.g., near 95 keV for $l = 0$ and near 3 MeV for $l = 2$ in the case of ^{8}Be.

However, the effect of nuclear forces remains generally negligible at energies lower than the Coulomb barrier, or whenever the combination of Coulomb repulsion and centrifugal forces reduces the amplitude of the relevant partial wave at values of r within the range of nuclear forces. Thus η_l remains ~ 0 (or $\sim n\pi$) except when very near a resonance, where η_l, rises by π anyhow. Taking $R \sim 1.5$ fm as the radius of each He^{++} nucleus, the height of the Coulomb barrier when two such nuclei touch each other is $B \sim (2e)^2/2R \sim 2$ MeV. Therefore the width of the $l = 0$ resonance at 95 keV is greatly suppressed by the Coulomb barrier, while the $l = 2$ resonance remains broad.

(c) To show the effect of nuclear forces on the angular distribution one may rewrite the partial wave expansion as

$$\sum_{l=0}^{\infty} \frac{1 + (-1)^l}{2}(2l+1)i^l e^{i\delta_l} \frac{1}{kr} \left\{ \sin\left(kr - \frac{l\pi}{2} - \gamma\ln(2kr) + \delta_l\right) \right.$$

$$\left. + \left(\frac{e^{2i\eta_l} - 1}{2i}\right) \exp\left[i\left(kr - \frac{l\pi}{2} - \gamma\ln(2kr) + \delta_l\right)\right] \right\} P_l(\cos\theta).$$

Here the first term inside the brackets represents the Coulomb scattering wave function unaffected by nuclear forces. The contribution of this term can be summed over l to give

$$\exp i\{kr \cos\theta - \gamma \ln[kr(1 - \cos\theta)] + \delta_0\} - \gamma(kr)^{-1} \exp i\{kr \cos\theta$$

$$- \gamma \ln(kr) + \delta_0\} \cdot \frac{1}{\sqrt{2}} \left[\frac{e^{-i\gamma \ln(1-\cos\theta)}}{1 - \cos\theta} + \frac{e^{-i\gamma \ln(1+\cos\theta)}}{1 + \cos\theta} \right].$$

The second term represents the scattering wave due to nuclear forces, which interferes with the Coulomb scattering wave in each direction. However, it is extremely small for η_l very close to $n\pi$, as for energies below the Coulomb barrier. Accordingly, detection of such interference may signal the occurence of a resonance at some lower energy.

An experiment in 1956 showed no significant interference from nuclear scattering below 300 keV center-of-mass energy, at which energy it was found $\eta_0 = (178 \pm 1)$ degrees.

2114

A 3-MV Van de Graaff generator is equipped to accelerate protons, deuterons, doubly ionized ^{3}He particles, and alpha-particles.

(a) What are the maximum energies of the various particles available from this machine?

(b) List the reactions by which the isotope ^{15}O can be prepared with this equipment.

(c) List at least six reactions in which ^{15}N is the compound nucleus.

Z					^{17}F	^{18}F	^{19}F	^{20}F	^{21}F	
8			^{14}O	^{15}O	^{16}O	^{17}O	^{18}O	^{19}O		
7			^{12}N	^{13}N	^{14}N	^{15}N	^{16}N	^{17}N		
6		^{10}C	^{11}C	^{12}C	^{13}C	^{14}C	^{15}C			
5		^{9}B	^{10}B	^{11}B	^{12}B					
	4	5	6	7	8	9	10	11	12	N

= Stable

Fig. 2.38

(d) Describe two types of reaction experiment which can be carried out with this accelerator to determine energy levels in ^{15}N. Derive any equations needed. (Assume all masses are known. Figure 2.38 shows the isotopes of light nuclei.)

(Columbia)

Solution:

(a) The available maximum energies of the various particles are: 3 MeV for proton, 3 MeV for deuteron, 6 MeV for doubly ionized ^{3}He, 6 MeV for α-particle.

(b) Based energy consideration, the reactions that can produce ^{15}O are

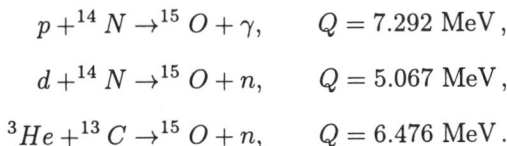

$$p + ^{14}N \rightarrow ^{15}O + \gamma, \qquad Q = 7.292 \text{ MeV},$$

$$d + ^{14}N \rightarrow ^{15}O + n, \qquad Q = 5.067 \text{ MeV},$$

$$^3He + ^{13}C \rightarrow ^{15}O + n, \qquad Q = 6.476 \text{ MeV}.$$

^{15}O cannot be produced with α-particles because of their high binding energy and small mass, which result in $Q = -8.35$ MeV.

(c) The reactions in which ^{15}N is the compound nucleus are

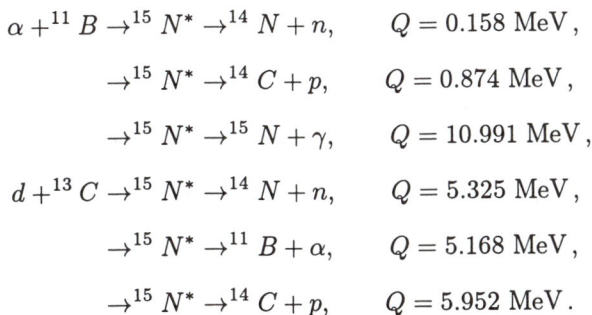

$$\alpha + ^{11}B \rightarrow ^{15}N^* \rightarrow ^{14}N + n, \qquad Q = 0.158 \text{ MeV},$$

$$\rightarrow ^{15}N^* \rightarrow ^{14}C + p, \qquad Q = 0.874 \text{ MeV},$$

$$\rightarrow ^{15}N^* \rightarrow ^{15}N + \gamma, \qquad Q = 10.991 \text{ MeV},$$

$$d + ^{13}C \rightarrow ^{15}N^* \rightarrow ^{14}N + n, \qquad Q = 5.325 \text{ MeV},$$

$$\rightarrow ^{15}N^* \rightarrow ^{11}B + \alpha, \qquad Q = 5.168 \text{ MeV},$$

$$\rightarrow ^{15}N^* \rightarrow ^{14}C + p, \qquad Q = 5.952 \text{ MeV}.$$

(d) (1) For the reaction $\alpha + ^{11}B \rightarrow ^{15}N^* \rightarrow ^{15}N + \gamma$, measure the γ-ray yield curve as a function of the energy E_α of the incoming α-particles. A resonance peak corresponds to an energy level of the compound nucleus $^{15}N^*$, which can be calculated:

$$E^* = \frac{11}{15}E_\alpha + m(^4He)c^2 + m(^{11}B)c^2 - m(^{15}N)c^2.$$

(2) With incoming particles of known energy, measuring the energy spectrums of the produced particles enables one to determine the energy levels of $^{15}N^*$. For instance, the reaction

$$^3He + {}^{14}N \rightarrow {}^{15}N + d$$

has $Q = 4.558$ MeV for ground state ^{15}N. If the incoming ^3He has energy E_0, the outgoing deuteron has energy E' and angle of emission θ, the excitation energy E^* is given by

$$E^* = Q - Q',$$

where

$$Q' = \left[1 + \frac{m(d)}{m(^{15}N)}\right] E' - \left[1 - \frac{m(^3He)}{m(^{15}N)}\right] E_0 - \frac{2\sqrt{m(^3He)m(d)E_0 E'}}{m(^{15}N)} \cos\theta$$

$$= \left(1 + \frac{2}{15}\right) E' - \left(1 - \frac{3}{15}\right) E_0 - 2\frac{\sqrt{3 \times 2E_0 E'}}{15} \cos\theta$$

$$= \frac{1}{15}(17E' - 12E_0 - 2\sqrt{6E_0 E'} \cos\theta).$$

2115

When Li^6 (whose ground state has $J = 1$, even parity) is bombarded by deuterons, the reaction rate in the reaction $Li^6 + d \rightarrow \alpha + \alpha$ shows a resonance peak at E (deuteron)= 0.6 MeV. The angular distribution of the α-particle produced shows a $(1 + A\cos^2\theta)$ dependence where θ is the emission angle relative to the direction of incidence of the deuterons. The ground state of the deuteron consists of a proton and a neutron in $^3S_1$ configuration. The masses of the relevant nuclides are

$$m_d = 2.0147 \text{ amu}, \ m_\alpha = 4.003 \text{ amu},$$

$$m_{Li} = 6.0170 \text{ amu}, \ m_{Be} = 8.0079 \text{ amu},$$

where 1 amu $= 938.2$ MeV.

From this information alone, determine the energy, angular momentum, and parity of the excited level in the compound nucleus. What partial wave deuterons (s,p,d, etc.) are effective in producing this excited level? (explain)

(Columbia)

Solution:

The excitation energy of the compound nucleus $^8Be^*$ in the reaction $d +^6 Li \to^8 Be^*$ is

$$E(^8Be^*) = [m(^2H) + m(^6Li) - m(^8Be)] + E_d \frac{m(^6Li)}{m(^6Li) + m(^2H)}$$

$$= (2.0147 + 6.0170 - 8.0079) \times 938.2 + 0.6 \times \frac{6}{8} = 22.779 \text{ MeV}.$$

In the decay $^8Be^* \to \alpha + \alpha$, as J^π of α is 0^+, the symmetry of the total wave function of the final state requires that l_f, the relative orbital angular momentum of the two α-particles, be even and the decay, being a strong interaction, conserve parity, the parity of $^8Be^*$ is $\pi(^8Be^*) = (-1)^{l_f}(+1)^2 = +1$.

As the angular distribution of the final state α-particles is not spherically symmetric but corresponds to $l_f = 2$, we have

$$J^\pi(^8Be^*) = 2^+.$$

Then the total angular momentum of the initial state $d +^6 Li$ is also $J_i = 2$. As $\mathbf{J}_i = \mathbf{J}_d + \mathbf{J}_{Li} + \mathbf{l}_i = 1 + 1 + \mathbf{l}_i$ and as

$$1 + 1 = \begin{cases} 0 \\ 1, \\ 2 \end{cases} \text{the possible values of } l_i \text{ are } 0,1,2,3,4.$$

However, the ground state parities of ^6Li and d are both positive, l_i must be even. As the angular distribution of the final state is not isotropic, $l_i \neq 0$ and the possible values of l_i are 2,4. So d-waves produce the main effect.

2116

Fast neutrons impinge on a 10-cm thick sample containing 10^{21} ^{53}Cr atoms/cm^3. One-tenth of one percent of the neutrons are captured into a spin-parity $J^\pi = 0^+$ excited state in ^{54}Cr. What is the neutron capture cross section for this state? The excited ^{54}Cr sometimes γ-decays as shown in Fig. 2.39. What is the most likely J^π for the excited state at 9.2 MeV? What are the multipolarities of the γ-rays?

(Wisconsin)

Solution:

Let the number of neutrons impinging on the sample be n and the neutron capture cross section for forming the 0^+ state be σ. Then $10 \times 10^{21} n \sigma = 10^{-3} n$, or

$$\sigma = 10^{-25} \text{ cm}^2 = 0.1 \ b \,.$$

Let the spin-parity of the 9.2 MeV level be J^p. As ^{54}Cr only occasionally γ-decays, the transitions are probably not of the $E1$ type, but correspond to the next lowest order. Consider $0^+ \to J^p$. If $\Delta J = 2$, the electric multipole field has parity $(-1)^{\Delta J} = +$, i.e. $J^p = 2^+$, and the transition is of the $E2$ type. The transitions γ_2, γ_3 are also between 0^+ and 2^+ states, so they are probably of the $E2$ type too. For $\gamma_4 : 2^+ \to 2^+$, we have $\Delta L = 1, 2, 3$ or 4. For no parity change between the initial and final states, γ_4 must be $E2$, $E4$ or $M1$, $M3$. Hence most probably $\gamma_4 = E2$, or $M1$, or both.

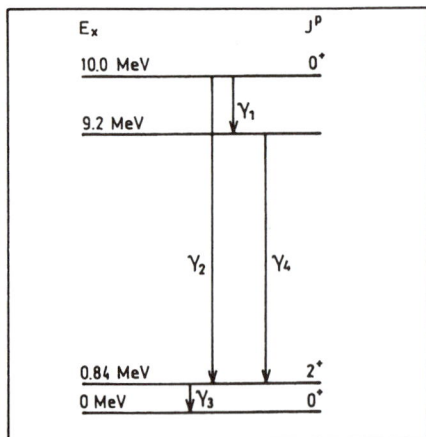

Fig. 2.39

2117

The surface of a detector is coated with a thin layer of a naturally fissioning heavy nuclei. The detector area is 2 cm^2 and the mean life of the fissioning isotope is $\frac{1}{3} \times 10^9$ years (1 $yr = 3 \times 10^7$ sec.). Twenty fissions are detected per second. The detector is then placed in a uniform neutron flux of 10^{11} neutrons/cm^2/sec. The number of fissions detected in the neutron flux is 120 per second. What is the cross section for neutron-induced fission?

(*Wisconsin*)

Solution:

Let the number of the heavy nuclei be N. Then the number of natural fissions taking place per second is

$$\frac{dN}{dt} = -\lambda N \approx -\lambda N_0 \,,$$

where $N_0 = N|_{t=0}$, as $\lambda = \frac{1}{\frac{1}{3} \times 10^9 \times 3 \times 10^7} = 10^{-16} \ll 1$.

The number of induced fissions per second is $\sigma N \phi \approx \sigma N_0 \phi$, where ϕ is the neutron flux, σ is the cross section for neutron-induced fission. As

$$\frac{\sigma N_0 \phi + \lambda N_0}{\lambda N_0} = \frac{120}{20} \,,$$

or

$$\frac{\sigma \phi}{\lambda} = \frac{100}{20} = 5 \,,$$

we have

$$\sigma = \frac{5\lambda}{\phi} = \frac{5 \times 10^{-16}}{10^{11}} = 5 \times 10^{-27} \ \text{cm}^2 = 5 \ \text{mb} \,.$$

2118

(a) How do you expect the neutron elastic scattering cross section to depend on energy for very low energy neutrons?

(b) Assuming nonresonant scattering, estimate the thermal neutron elastic cross section for ^{3}He.

(c) Use the information in the partial level scheme for $A = 4$ shown in Fig. 2.40 to estimate the thermal neutron absorption cross section for ^{3}He. Resonant scattering may be important here.

(*Princeton*)

Solution:

(a) For thermal neutrons of very low energies, the elastic scattering cross section of light nuclei does not depend on the neutron energy, but is constant

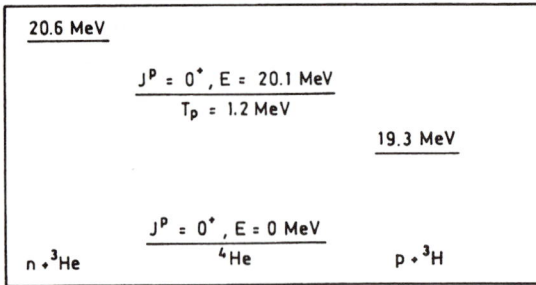

Fig. 2.40

for a large range of energy. But for heavier nuclei, resonant scattering can occur in some cases at very low neutron energies. For instance, resonant scattering with ^{157}Gd occurs at $E_n = 0.044$ eV.

(b) The thermal neutron nonresonant scattering cross section for nuclei is about $4\pi R_0^2$, where R_0 is the channel radius, which is equal to the sum of the radii of the incoming particle and the target nucleus. Taking the nuclear radius as

$$R \approx 1.5 \times 10^{-13} \, A^{1/3} \, ,$$

the elastic scattering cross section of ^{3}He for thermal neutron is

$$\sigma = 4\pi R_0^2 \approx 4\pi[1.5 \times 10^{-13}(3^{1/3} + 1)]^2 = 1.7 \times 10^{-24} \text{ cm}^2 = 1.7 \, b \, .$$

(c) The Breit–Wigner formula

$$\sigma_{nb} = \pi \lambdabar^2 \frac{\Gamma_n \Gamma_b}{(E' - E_0)^2 + \Gamma^2/4}$$

can be used to calculate the neutron capture cross section for ^{3}He in the neighborhood of a single resonance. Here $\lambdabar$ is the reduced wavelength of the incident particle, E' is the energy and E_0 is the energy at resonance peak of the compound nucleus $A = 4$, Γ_n and Γ_b are the partial widths of the resonant state for absorption of neutron and for emission of b respectively, and Γ is the total level width.

For laboratory thermal neutrons, $E_n \approx 0.025\ eV$,

$$\lambdabar = \frac{\hbar}{\sqrt{2\mu E_n}} = \frac{\hbar}{\sqrt{\frac{2m_n m_{He}}{m_n + m_{He}} E_n}} = \frac{\hbar c}{\sqrt{\frac{3}{2} E_n m_n c^2}}$$

$$= \frac{197 \times 10^{-13}}{\sqrt{\frac{3}{2} \times 2.5 \times 10^{-8} \times 940}} = 3.3 \times 10^{-9}\ cm.$$

As both the first excited and ground states of ^{4}He have 0^+, $\Gamma_\gamma = 0$, and the only outgoing channel is for the excited state of ^{4}He to emit a proton. The total width is $\Gamma = \Gamma_n + \Gamma_p$. With $\Gamma_n \approx 150\ eV$, $\Gamma \approx \Gamma_p = 1.2$ MeV, $E' = 20.6$ MeV, $E = 20.1$ MeV, we obtain

$$\sigma = \pi \lambdabar^2 \frac{\Gamma_n \Gamma_p}{(E' - E_0)^2 + \Gamma^2/4} = 1 \times 10^{-20}\ cm^2 = 1 \times 10^4\ b.$$

2119

Typical cross section for low energy neutron-nucleus scattering is 10^{-16}, 10^{-24}, 10^{-32}, 10^{-40} cm^2.

(Columbia)

Solution:

10^{-24} cm^2. The radius of the sphere of action of nuclear forces is $\sim 10^{-12} - 10^{-13}$ cm, and a typical scattering cross-section can be expected to be of the same order of magnitude as its cross-sectional area.

2120

In experiments on the reaction ^{21}Ne(d, ^{3}He)^{20}F with 26 MeV deuterons, many states in ^{20}F are excited. The angular distributions are characteristic of the direct reaction mechanism and therefore are easily sorted into those for which the angular momentum of the transferred proton is $l_p = 0$ or 1 or 2.

The lowest energy levels of ^{21}Ne and the known negative-parity states of ^{20}F below 4 MeV are as shown in Fig. 2.41 (the many positive-parity excited states of ^{20}F are omitted).

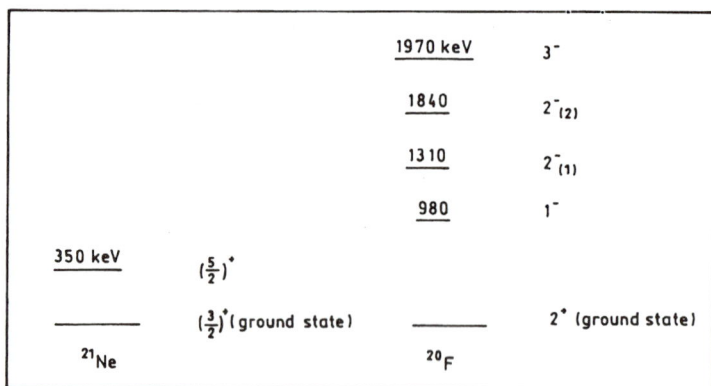

		1970 keV	3^-
		1840	$2^-_{(2)}$
		1310	$2^-_{(1)}$
		980	1^-
350 keV	$(\frac{5}{2})^*$		
———	$(\frac{3}{2})^*$ (ground state)	———	2^* (ground state)
^{21}Ne		^{20}F	

Fig. 2.41

The relative $l_p = 1$ strengths $S(J^\pi)$ observed in the (d, ^{3}He) reaction are approximately

$$S(1^-) = 0.84,$$

$$S(2^-_1) = 0.78,$$

$$S(2^-_2) = 0.79,$$

$$S(3^-) = 0.00.$$

(a) If the ^{21}Ne target and a ^{20}F state both have (1s-0d) configuration, they both have positive parity and therefore some $l_p = 0$ or $l_p = 2$ transitions are expected. On the other hand, the final states of ^{20}F with negative parity are excited with $l_p = 1$. Explain.

(b) In order to explain the observed negative-parity states in ^{20}F, one can try a coupling model of a hole weakly coupled to states of ^{21}Ne. With this model of a ^{21}Ne nucleus with an appropriate missing proton and level diagrams as given above, show how one can account for the negative-parity states in ^{20}F.

(c) In the limit of weak coupling; i.e., with no residual interaction between the hole and the particles, what would be the (relative) energies of the 4 negative-parity states?

(d) What would be the effect if now a weak particle-hole interaction were turned on? Do the appropriate centroids of the reported energies of the 1^-, 2^-, 2^-, 3^- states conform to this new situation?

(e) The weak coupling model and the theory of direct reactions lead to specific predictions about the relative cross sections (strengths) for the various final states. Compare these predictions with the observed S-factors given above. Show how the latter can be used to obtain better agreement with the prediction in part (d).

(Princeton)

Solution:

(a) The reactions are strong interactions, in which parity is conserved. So the parity change from initial to final state must equal the parity of the proton that is emitted as part of ^{3}He:

$$P(^{21}Ne) = P(^{20}F)P(p) = P(^{20}F)(-1)^{l_p} .$$

When both ^{20}F and ^{21}Ne have even parity, $(-1)^{l_p} = 1$ and so $l_p = 0, 2 \cdots$. As conservation of the total angular momentum requires that l_p be 0, 1, 2, we have $l_p = 0, 2$. Similarly, for the negative-parity states of ^{20}F, the angular momentum that the proton takes away can only be $1, 3 \cdots$. In particular for 1^- and 2^- states of ^{20}F, $l_p = 1$.

(b) In the weak coupling model, ^{20}F can be considered as consisting of ^{21}Ne and a proton hole (p^-). J^P of ^{20}F is then determined by a neutron in $1d_{3/2}$, $1d_{5/2}$, or $2s_{1/2}$ and a proton hole in $1p_{1/2}$, $1p_{3/2}$ or $2s_{1/2}$, etc., outside of full shells (Fig. 2.16). For example, the 1^- state of ^{20}F can be denoted as

$$|1M\rangle = |1p_{1/2}^{-1}, 1d_{3/2}; 1, M\rangle$$

$$= \sum_{m_1, m_2} \left\langle \frac{1}{2}, \frac{3}{2}, m_1, m_2 \middle| 1, M \right\rangle \psi_{1/2m} \psi_{3/2m} .$$

where $^1p_{1/2}^{-1}$ means a proton hole in $1p_{1/2}$ state, $1d_{3/2}$ means a neutron in $1d_{3/2}$ state. In the same way, the 2^- can be denoted as

$$|1p_{1/2}^{-1}, 1d_{3/2}; 2, M\rangle \quad \text{and} \quad |1p_{1/2}^{-1}, 1d_{5/2}; 2, M\rangle ,$$

the 3^- state can be denoted as

$$|1p_{1/2}^{-1}, 1d_{5/2}; 3, M\rangle .$$

(c) We have $H = H_p + H_h + V_{ph}$, where H_p and H_h are respectively the Hamiltonian of the nuclear center and the hole, and V_{ph} is the potential due to the interaction of the hole and the nuclear center. In the limit of weak coupling,

$$V_{ph} = 0,$$

$$H_p \psi(a_1, j_1, m_1) = E_{a_1, j_1, m_1} \psi(a_1, j_1, m_1),$$

$$H_h \phi(a_2, j_2, m_2) = E_{a_2, j_2, m_2} \phi(a_2, j_2, m_2).$$

Then for the four negative-parity states we have

$$3^- : E_{3^-} = E_p(1d_{5/2}) + E_h(1p_{1/2}),$$

$$2_1^- : E_{2_1^-} = E_p(1d_{5/2}) + E_h(1p_{1/2}),$$

$$2_2^- : E_{2_2^-} = E_p(1d_{3/2}) + E_h(1p_{1/2}),$$

$$1^- : E_{1^-} = E_p(1d_{3/2}) + E_h(1p_{1/2}).$$

Thus $E_{3^-} = E_{2_1^-}$, $E_{2_2^-} = E_{1^-}$, as shown in Fig. 2.42, with values

$$E_{3^-} = E_{2_1^-} = 1230 \text{ keV}, \qquad E_{2_2^-} = E_{1^-} = 890 \text{ keV}.$$

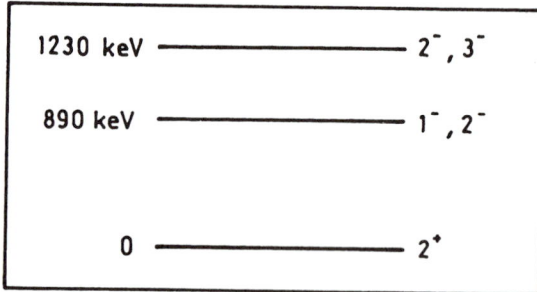

1230 keV	———————	$2^-, 3^-$
890 keV	———————	$1^-, 2^-$
0	———————	2^+

Fig. 2.42

(d) If $V_{ph} \neq 0$, i.e., coupling exists, then

$$E_{3^-} = H_p(1d_{5/2}) + H_h(1p_{1/2}) + \langle 1p_{1/2}^{-1}, 1d_{5/2}, 3|V_{ph}|1p_{1/2}^{-1}, 1d_{5/2}, 3 \rangle,$$

$$E_{1^-} = H_p(1d_{3/2}) + H_h(1p_{1/2}) + \langle 1p_{1/2}^{-1}, 1d_{3/2}, 1|V_{ph}|1p_{1/2}^{-1}, 1d_{3/2}, 1 \rangle.$$

As

$$\langle 1p_{1/2}^{-1}, 1d_{5/2}, 3^- | V_{ph} | 1p_{1/2}^{-1}, 1d_{5/2}, 3^- \rangle \approx 0.7 \text{ MeV},$$

$$\langle 1p_{1/2}^{-1}, 1d_{3/2}, 1^- | V_{ph} | 1p_{1/2}^{-1}, 1d_{3/2}, 1^- \rangle \approx 0.1 \text{ MeV},$$

$$\langle 1p_{1/2}^{-1}, 1d_{5/2}, 2^- | V_{ph} | 1p_{1/2}^{-1}, 1d_{5/2}, 2^- \rangle = 0.45 \text{ MeV},$$

$$\langle 1p_{1/2}^{-1}, 1d_{3/2}, 2^- | V_{ph} | 1p_{1/2}^{-1}, 1d_{3/2}, 2^- \rangle = 0.25 \text{ MeV},$$

$$\langle 1p_{1/2}^{-1}, 1d_{5/2}, 2^- | V_{ph} | 1p_{1/2}^{-1}, 1d_{3/2}, 2^- \rangle$$

$$= \langle 1p_{1/2}^{-1}, 1d_{3/2}, 2^- | V_{ph} | 1p_{1/2}^{-1}, 1d_{5/2}, 2^- \rangle$$

$$= 0.3 \text{ MeV}.$$

the above gives

$$E'_{3^-} = 0.9 + 0.35 + 0.7 = 1.95 \text{ MeV}$$

$$E'_{1^-} = 0.9 + 0.1 = 1.0 \text{ MeV}.$$

$E'_{2_1^-}$ and $E'_{2_2^-}$ are the eigenvalues of the matrix

$$\begin{pmatrix} \langle 1p_{1/2}^{-1}; 1d_{5/2}, 2^- | H | 1p_{1/2}^{-1}, 1d_{5/2}, 2^- \rangle & \langle 1p_{1/2}^{-1}; 1d_{5/2}, 2^- | H | 1p_{1/2}^{-1}, 1d_{3/2}, 2^- \rangle \\ \langle 1p_{1/2}^{-1}; 1d_{3/2}, 2^- | H | 1p_{1/2}^{-1}, 1d_{5/2}, 2^- \rangle & \langle 1p_{1/2}^{-1}; 1d_{3/2}, 2^- | H | 1p_{1/2}^{-1}, 1d_{3/2}, 2^- \rangle \end{pmatrix}.$$

The secular equation

$$\begin{pmatrix} \lambda - 1.95 & -0.3 \\ -0.3 & \lambda - 1.1 \end{pmatrix} = 0$$

gives $E'_{2_1^-} = \lambda_1 = 1.80 \text{ MeV}$, $E'_{2_2^-} = \lambda_2 = 1.26 \text{ MeV}$.

The energy levels are shown in Fig. 2.43.

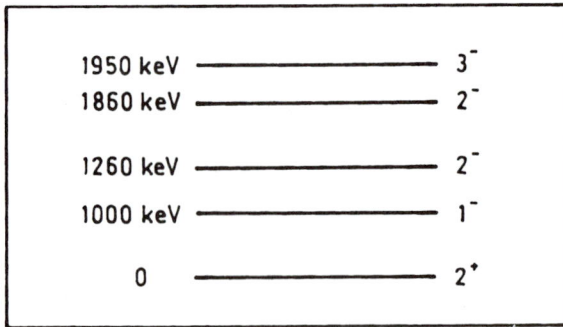

Fig. 2.43

(e) The relative strengths of the various final states as given by different theories are compared in the table below:

	Nilson model	PHF	Shell model	Experimental
$S(1^-)$	0.70	0.76	0.59	0.84
$S(2_1^-)$	0.93	0.20	0.72	0.78
$S(2_2^-)$	0.28	0.20	0.23	0.79
$S(3^-)$			0.002	0.00

It is noted in particular that for $S(2_2^-)$, the theoretical values are much smaller than the experimental values.

PART III

PARTICLES PHYSICS

1. INTERACTIONS AND SYMMETRIES (3001–3037)

3001

The interactions between elementary particles are commonly classified in order of decreasing strength as strong, electromagnetic, weak and gravitational.

(a) Explain, as precisely and quantitatively as possible, what is meant by 'strength' in this context, and how the relative strengths of these interactions are compared.

(b) For each of the first three classes state what conservation laws apply to the interaction. Justify your answers by reference to experimental evidence.

(Columbia)

Solution:

(a) The interactions can be classified according to the value of a characteristic dimensionless constant related through a coupling constant to the interaction cross section and interaction time. The stronger the interaction, the larger is the interaction cross section and the shorter is the interaction time.

Strong interaction: Range of interaction $\sim 10^{-13}$ cm. For example, the interaction potential between two nuclei has the from

$$V(r) = \frac{g_h}{r} \exp\left(-\frac{r}{R}\right),$$

where $R \approx \hbar/m_\pi c$ is the Compton wavelength of pion. Note the exponential function indicates a short interaction length. The dimensionless constant

$$g_h^2/\hbar c \approx 1 \sim 10$$

gives the interaction strength.

Electromagnetic interaction: The potential for two particles of charge e at distance r apart has the form

$$V_e(r) = e^2/r.$$

The dimensionless constant characteristic of interaction strength is the fine structure constant

$$\alpha = e^2/\hbar c \approx 1/137.$$

Weak interaction: Also a short-range interaction, its strength is represented by the Fermi coupling constant for β-decay

$$G_F = 1.4 \times 10^{-49} \text{ erg cm}^3 .$$

The potential of weak interaction has the form

$$V_w(r) = \frac{g_w}{r} \exp\left(-\frac{r}{R_w}\right),$$

where it is generally accepted that $R_w \approx 10^{-16}$ cm. The dimensionless constant characteristic of its strength is

$$g_w^2/\hbar c = G_F m_p^2 c/\hbar^3 \approx 10^{-5} .$$

Gravitational interaction: For example the interaction potential between two protons has the form

$$G m_p^2/r .$$

The dimensionless constant is

$$G m_p^2/\hbar c \approx 6 \times 10^{-39} .$$

As the constants are dimensionless they can be used to compare the interaction strengths directly. For example, the ratio of the strengths of gravitational and electromagnetic forces between two protons is

$$G m_p^2/e^2 \approx 10^{-36} .$$

Because of its much smaller strength, the gravitational force can usually be neglected in particle physics. The characteristics of the four interactions are listed in Table 3.1.

Table 3.1

Interaction	Characteristic constant	Strength	Range of interaction	Typical cross section	Typical lifetime
Strong	$\frac{g_h^2}{\hbar c}$	$1 \sim 10$	10^{-13} cm	10^{-26} cm^2	10^{-23} s
Electromagnetic	$\frac{e^2}{\hbar c}$	$\frac{1}{137}$	∞	10^{-29} cm^2	10^{-16} s
Weak	$\frac{g_w^2}{\hbar c} = \frac{G_F m_p^2 c}{\hbar^3}$	10^{-5}	10^{-16} cm	10^{-38} cm^2	10^{-10} s
Gravitational	$\frac{G m_p^2}{\hbar c}$	10^{-39}	∞		

Table 3.2

Quantity	E	J	P	Q	B	$L_e(L_\mu)$	I	I_3	S	P	C	T	CP	G
Strong	Y	Y	Y	Y	Y	Y	Y	Y	Y	Y	Y	Y	Y	Y
Electromagnetic	Y	Y	Y	Y	Y	Y	N	Y	Y	Y	Y	Y	Y	N
Weak	Y	Y	Y	Y	Y	Y	N	N	N	N	N	N	N	N

(b) The conservation laws valid for strong, electromagnetic, and weak interactions are listed in Table 3.2, where y =conserved, N =not conserved.

The quantities listed are all conserved in strong interaction. This agrees well with experiment. For example nucleon-nucleus and pion-nucleus scattering cross sections calculated using isospin coupling method based on strong forces agree well with observations.

In electromagnetic interaction I is not conserved, e.g. $\Delta I = 1$ in electromagnetic decay of Σ^0 ($\Sigma^0 \to \Lambda^0 + \gamma$).

In weak interaction I, I_3, S, P, C, T, PC are not conserved, e.g. 2π-decay of K_L^0. The process $K_L^0 \to \pi^+\pi^-$ violates PC conservation. As PCT is conserved, time-reversal invariance is also violated. All these agree with experiment.

3002

The electrostatic force between the earth and the moon can be ignored

(a) because it is much smaller than the gravitational force.
(b) because the bodies are electrically neutral.
(c) because of the tidal effect.

(CCT)

Solution:

For electrostatic interaction the bodies should be electrically charged. As the earth and the moon are both electrically neutral, they do not have electrostatic interaction. Thus answer is (b).

3003

(a) Explain the meaning of the terms: boson, fermion, hadron, lepton, baryon,

(b) Give one example of a particle for each of the above.

(c) Which of the above name is, and which is not, applicable to the photon?

(Wisconsin)

Solution:

(a) Fermion: All particles of half-integer spins.

Boson: All particles of integer spins.

Hardron: Particles which are subject to strong interaction are called hadrons.

Lepton: Particles which are not subject to strong interaction but to weak interaction are called leptons.

Baryon: Hadrons of half-integer spins are called baryons.

(b) Boson: π meson;

Fermion: proton;

Hardron: proton;

Lepton: neutrino;

Baryon: proton;

(c) The name boson is applicable to photon, but not the other names.

3004

Why does the proton have a parity while the muon does not? Because

(a) parity is not conserved in electromagnetism.

(b) the proton is better known.

(c) parity is defined from reactions relative to each other. Therefore, it is meaningful for the proton but not for the muon.

(CCT)

Solution:

The answer is (c).

3005

What is the G-parity operator and why was it introduced in particle physics? What are the eigenvalues of the G-operator for pions of different charges, and for a state of n pions?

What are the G values for ρ, ω, ϕ, and η mesons?

(*Buffalo*)

Solution:

The G-operator can be defined as $G = Ce^{i\pi I_2}$ where I_2 is the second component of isospin I, and C is the charge conjugation operator.

As the C-operator has eigenvalues only for the photon and neutral mesons and their systems, it is useful to be able to extend the operation to include charged particles as well. The G-parity is so defined that charged particles can also be eigenstates of G-parity. Since strong interaction is invariant under both isospin rotation and charge conjugation, G-parity is conserved in strong interaction, which indicates a certain symmetry in strong interaction. This can be used as a selection rule for certain charged systems.

For an isospin multiplet containing a neutral particle, the eigenvalue of G-operator is

$$G = C(-1)^I,$$

where C is the C eigenvalue of the neutral particle, I is the isospin. For π meson, $C(\pi^0) = +1$, $I = 1$, so $G = -1$; for a system of n π-mesons, $G(n\pi) = (-1)^n$. Similarly for

$$\rho: \quad C(\rho^0) = -1, \quad I(\rho) = 1, \quad G(\rho) = +1;$$

$$\omega: \quad C(\omega^0) = -1, \quad I(\omega^0) = 0, \quad G(\omega) = -1;$$

$$\phi: \quad C(\phi) = -1, \quad I(\phi) = 0, \quad G(\phi) = -1;$$

$$\eta^0: \quad C(\eta^0) = +1, \quad I(\eta^0) = 0, \quad C(\eta^0) = +1.$$

ρ, ω, ϕ decay by strong interaction. As G-parity is conserved in strong interaction, their G-parities can also be deduced from the decays. Thus as

$$\rho^0 \to \pi^+\pi^-, \quad G(\rho) = (-1)^2 = 1;$$

$$\omega \to 3\pi, \quad G(\omega) = (-1)^3 = -1;$$

$$\phi \to 3\pi, \quad G(\phi) = (-1)^3 = -1.$$

Note as η^0 decays by electromagnetic interaction, in which G-parity is not conserved, its G-parity cannot be deduced from the decay.

3006

Following is a list of conservation laws (or symmetries) for interactions between particles. For each indicate by S,E,W those classes of interactions — strong, electromagnetic, weak — for which no violation of the symmetry or conservation law has been observed. For any one of these conservation laws, indicate an experiment which established a violation.

(a) I-spin conservation

(b) I_3 conservation

(c) strangeness conservation

(d) invariance under CP

(*Wisconsin*)

Solution:

(a) I-spin conservation — S.

(b) I_3 conservation — S, E.

(c) Strangeness conservation — S, E.

(d) CP invariance — S, E, and W generally. CP violation in weak interaction is found only in K_L decay. Isospin nonconservation can be observed in the electromagnetic decay $\Sigma^0 \to \Lambda^0 + \gamma$. I_3 nonconservation can be observed in the weak decay $\pi^- \to \mu^- + \bar{\nu}_\mu$.

Strangeness nonconservation is found in the weak decay of strange particles. For example, in $\Lambda^0 \to \pi^- + p$, $S = -1$ for the initial state, $S = 0$ for the final state, and so $\Delta S = -1$.

The only observed case of CP violation is the K_L^0 decay, in which the 3π and 2π decay modes have the ratio

$$\eta = \frac{B(K_L^0 \to \pi^+\pi^-)}{B(K_L^0 \to \text{all charged particles})} \approx 2 \times 10^{-3}.$$

It shows that CP conservation is violated in K_L^0 decay, but only to a very small extent.

3007

A state containing only one strange particle

(a) can decay into a state of zero strangeness.

(b) can be created strongly from a state of zero strangeness.

(c) cannot exist.

(CCT)

Solution:

Strange particles are produced in strong interaction but decay in weak interaction, and the strangeness number is conserved in strong interaction but not in weak interaction. Hence the answer is (a).

3008

A particle and its antiparticle

(a) must have the same mass.

(b) must be different from each other.

(c) can always annihilate into two photons.

(CCT)

Solution:

Symmetry requires that a particle and its antiparticle must have the same mass. Hence the answer is (a).

3009

Discuss briefly four of the following:

(1) J/ψ particle.

(2) Neutral K meson system, including regeneration of K_s.

(3) The two types of neutrino.

(4) Neutron electric dipole moment.

(5) Associated production.

(6) Fermi theory of beta-decay.

(7) Abnormal magnetic moment of the muon.

(Columbia)

Solution:

(1) J/ψ *particle*. In 1974, C. C. Ting, B. Richter and others, using different methods discovered a heavy meson of mass $M = 3.1 \text{ GeV}/c^2$. Its

lifetime was 3 ~ 4 orders of magnitude larger than mesons of similar masses, which makes it unique in particle physics. Named J/ψ particle, it was later shown to be the bound state of a new kind of quark, called the charm quark, and its antiquark. The J/ψ particle decays into charmless particles via the OZI rule or into a lepton pair via electromagnetic interaction, and thus has a long lifetime. Some of its quantum numbers are

$$m(J/\psi) = (3096.9 \pm 0.1) \text{ MeV}/c^2, \qquad \Gamma = (63 \pm 9)\text{keV},$$

$$I^G(J^P)C = 0^-(1^-) - .$$

All of its decay channels have been fully studied. J/ψ particle and other charmed mesons and baryons make up the family of charmed particles, which adds significantly to the content of particle physics.

(2) *Neutral K mesons* Detailed discussions are given in **Problems 3056–3058**.

(3) *Two kinds of neutrino.* Experiments have shown that there are two types of neutrino: one (ν_e) is associated with electron (as in β-decay), the other (ν_μ) with muon (as in $\pi \to \mu$ decay). Also a neutrino and its antineutrino are different particles.

The scattering of high energy neutrinos can lead to the following reactions:

$$\nu_e + n \to p + e^-, \qquad \bar{\nu}_e + p \to n + e^+,$$

$$\nu_\mu + n \to p + \mu^-, \qquad \bar{\nu}_\mu + p \to n + \mu^+.$$

Suppose a neutrino beam from a certain source is scattered and it contains $\nu_\mu(\bar{\nu}_\mu)$. If $\nu_e(\bar{\nu}_e)$ and $\nu_\mu(\bar{\nu}_\mu)$ are the same, approximately the same numbers of $e^\mp$ and $\mu^\mp$ should be observed experimentally. If they are not the same, the reactions producing $e^\mp$ are forbidden and no electrons should be observed. An experiment carried out in 1962 used a proton beam of energy > 20 GeV to bombard a target of protons to produce energetic pions and kaons. Most of the secondary particles were emitted in a cone of very small opening angle and decayed with neutrinos among the final products. A massive shielding block was used which absorbed all the particles except the neutrinos. The resulting neutrino beam (98–99% ν_μ, 1–2% ν_e) was used to bombard protons to produce muons or electrons. Experimentally, 51 muon events, but not one confirmed electron event, were observed. This proved that $\nu_e(\bar{\nu}_e)$ and $\nu_\mu(\bar{\nu}_\mu)$ are different particles.

That ν and $\bar{\nu}$ are different can be proved by measuring the reaction cross section for neutrinos in ^{37}Cl. Consider the electron capture process

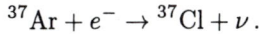

$$^{37}\text{Ar} + e^- \rightarrow {}^{37}\text{Cl} + \nu \,.$$

The reverse process can also occur:

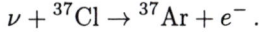

$$\nu + {}^{37}\text{Cl} \rightarrow {}^{37}\text{Ar} + e^- \,.$$

If $\bar{\nu}$ and ν are the same, so can the process below:

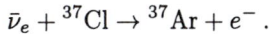

$$\bar{\nu}_e + {}^{37}\text{Cl} \rightarrow {}^{37}\text{Ar} + e^- \,.$$

In an experiment by R. Davis and coworkers, 4000 liters of CCl_4 were placed next to a nuclear reactor, where $\bar{\nu}$ were generated. Absorption of the antineutrinos by ^{37}Cl produced ^{37}Ar gas, which was separated from CCl_4 and whose rate of K-capture radioactivity was measured. The measured cross section was far less than the theoretical value $\sigma \approx 10^{-43}$ cm^2 expected if ν_e and $\bar{\nu}_e$ were the same. This showed that $\bar{\nu}$ is different from ν.

(4) *Electric dipole moment of neutron*

Measurement of the electric dipole moment of the neutron had been of considerable interest for a long time as it offered a means of directly examining time reversal invariance. One method for this purpose is described in Fig. 3.1, which makes use of nuclear magnetic resonance and electrostatic

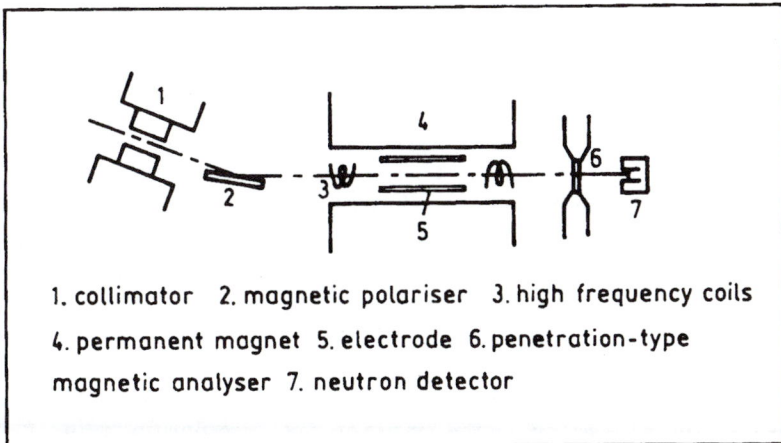

1. collimator 2. magnetic polariser 3. high frequency coils
4. permanent magnet 5. electrode 6. penetration-type
magnetic analyser 7. neutron detector

Fig. 3.1

deflection. It gave $P_n = eD$, where $D = (-1 \pm 4) \times 10^{-21}$ cm is the effective length of the dipole moment and e is the electron charge. Later, an experiment with cold neutrons gave $D = (0.4 \pm 1.1) \times 10^{-24}$ cm. This means that, within the experimental errors, no electric dipole moment was observed for the neutron.

(5) *Associated production*

Many new particles were discovered in cosmic rays around 1950 in two main categories — mesons and baryons. One peculiar characteristics of these particles was that they were produced in strong interaction (interaction time $\sim 10^{-23}$ s) but decayed in weak interaction ($\tau \sim 10^{-10} \sim 10^{-8}$ s). Also, they were usually produced in pairs. This latter phenomenon is called associated production and the particles are called strange particles. To account for the "strange" behavior a new additive quantum number called strangeness was assigned to all hadrons and the photon. The strangeness number S is zero for γ and the "ordinary" particles and is a small, positive or negative, integer for the strange particles K, Λ, Σ etc. A particle and its antiparticle have opposite strangeness numbers. S is conserved for strong and electromagnetic interactions but not for weak interaction. Thus in production by strong interaction from ordinary particles, two or more strange particles must be produced together to conserve S. This accounts for the associated production. In the decay of a strange particle into ordinary particles it must proceed by weak interaction as S is not conserved. The basic reason for the strange behavior of these particles is that they contain strange quarks or antiquarks.

(6) The Fermi theory of β-decay

Fermi put forward a theory of β-decay in 1934, which is analogous to the theory of electromagnetic transition. The basic idea is that just as γ-ray is emitted from an atom or nucleus in an electromagnetic transition, an electron and a neutrino are produced in the decay process. Then the energy spectrum of emitted electrons can be derived in a simple way to be

$$\left[\frac{dI(p_e)}{p_e^2 F dp_e}\right]^{1/2} = C|M_{ij}|^2 (E_0 - E_e),$$

where $dI(p_e)$ is the probability of emitting an electron of momentum between p_e and $p_e + dp_e$, E_e is the kinetic energy corresponding to p_e, E_0 is the maximum kinetic energy of the electrons, C is a constant, M_{ij} is the matrix element for weak interaction transition, $F(Z, E_e)$ is a factor which takes account of the effect of the Coulomb field of the nucleus on the emission of the electron. The theory, which explains well the phenomenon of β-decay, had been used for weak interaction processes until nonconservation of parity in weak interaction was discovered, when it was replaced by a revised version still close to the original form. Thus the Fermi theory may be considered the fundamental theory for describing weak interaction processes.

(7) *Abnormal magnetic moment of muon*

According to Dirac's theory, a singly-charged exact Dirac particle of spin J and mass m has a magnetic moment given by

$$\mu = \frac{J}{me} = g\frac{J}{2mc},$$

where $g = -2$ for muon. However muon is not an exact Dirac particle, nor its g-factor exactly -2. It is said to have an abnormal magnetic moment, whose value can be calculated using quantum electrodynamics (QED) in accordance with the Feynman diagrams shown in Fig. 3.2. Let $\alpha = \frac{|g|-2}{2}$. QED gives

$$\alpha_{\mu}^{\text{th}} = \alpha/(2\pi) + 0.76578(\alpha/\pi)^2 + 2.55(\alpha/\pi)^3 + \cdots$$

$$= (116592.1 \pm 1.0) \times 10^{-8},$$

in excellent agreement with the experimental value

$$\alpha_{\mu}^{\text{exp}} = (116592.2 \pm 0.9) \times 10^{-8}.$$

This has been hailed as the most brilliant achievement of QED.

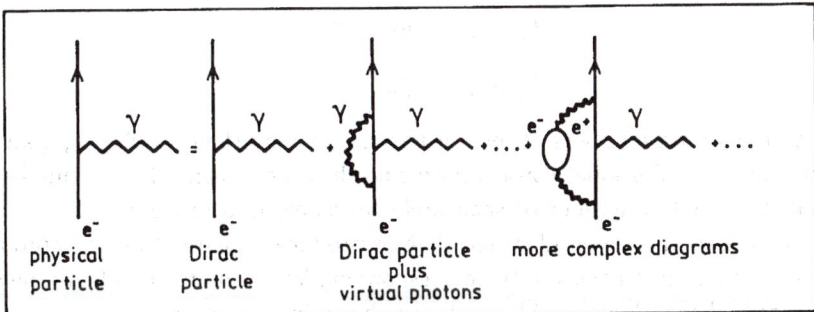

Fig. 3.2

3010

The lifetime of the muon is 10^9, 10^2, 10^{-2}, 10^{-6} second.

(Columbia)

Solution:

10^{-6} s (more precisely $\tau_\mu = 2.2 \times 10^{-6}$ s).

3011

List all of the known leptons. How does μ^+ decay? Considering this decay and the fact that $\nu_\mu + n \to e^- + p$ is found to be forbidden, discuss possible lepton quantum number assignments that satisfy additive quantum number conservation laws. How could ν_μ produce a new charged "heavy lepton"?

(Wisconsin)

Solution:

Up to now 10 kinds of leptons have been found. These are e^-, ν_e, μ^-, ν_μ, τ^- and their antiparticles e^+, $\bar{\nu}_e$, μ^+, $\bar{\nu}_\mu$, τ^+. ν_τ and $\bar{\nu}_\tau$ have been predicted theoretically, but not yet directly observed.

μ^+ decays according to $\mu^+ \to e^+ + \nu_e + \bar{\nu}_\mu$. It follows that $\bar{\nu}_e + \mu^+ \to e^+ + \bar{\nu}_\mu$. On the other hand the reaction $\nu_\mu + n \to e^- + p$ is forbidden. From these two reactions we see that for allowed reactions involving leptons, if there is a lepton in the initial state there must be a corresponding lepton in the final state. Accordingly we can define an electron lepton number L_e and a muon lepton number L_μ such that

$$L_e = 1 \quad \text{for} \quad e^-, \nu_e,$$

$$L_\mu = 1 \quad \text{for} \quad \mu^-, \nu_\mu,$$

with the lepton numbers of the antiparticles having the opposite sign, and introduce an additional conservation rule that the electron lepton number and the μ lepton number be separately conserved in a reaction.

It follows from a similar rule that to produce a charged heavy lepton, the reaction must conserve the corresponding lepton number. Then a new charged "heavy lepton" A^+ can be produced in a reaction

$$\nu_\mu + n \to A^+ + \nu_A + \mu^- + X,$$

where ν_A is the neutrino corresponding to A^+, X is a baryon. For example, $A^+ = \tau^+$, $\nu_A = \nu_\tau$.

3012

Give a non-trivial (rate greater than 5%) decay mode for each particle in the following list. If you include neutrinos in the final state, be sure to specify their type.

$$n \to, \pi^+ \to, \rho^0 \to, K^0 \to, \Lambda^0 \to, \Delta^{++} \to, \mu^- \to, \phi \to, \Omega^- \to, J/\Psi \to .$$

(*Wisconsin*)

Solution:

$n \to pe^-\bar{\nu}_e$; $\pi^+ \to \mu^+\nu_\mu$; $\rho^0 \to \pi^+\pi^-$; $K^0 \to \pi^+\pi^-$, $\pi^0\pi^0$, $\pi^0\pi^0\pi^0$, $\pi^+\pi^-\pi^0$, $\pi^\pm\mu^\mp\nu_\mu$, $\pi^0\mu^\pm e^\mp\nu_e$; $\Lambda^0 \to p\pi^-$, $n\pi^0$; $\Delta^{++} \to p\pi^+$; $\mu^- \to e^-\bar{\nu}_e\nu_\mu$; $\phi \to K^+K^-$, $K^0_L K^0_S$, $\pi^+\pi^-\pi^0$; $\Omega^- \to \Lambda K^-$, $\Xi^0\pi^-$, $\Xi^-\pi^0$; $J/\psi \to e^+e^-$, $\mu^+\mu^-$, hadrons.

3013

Consider the following high-energy reactions or particle decays:

(1) $\pi^- + p \to \pi^0 + n$
(2) $\pi^0 \to \gamma + \gamma + \gamma$
(3) $\pi^0 \to \gamma + \gamma$
(4) $\pi^+ \to \mu^+ + \nu_\mu$
(5) $\pi^+ \to \mu^+ + \bar{\nu}_\mu$
(6) $p + \bar{p} \to \Lambda^0 + \Lambda^0$
(7) $p + \bar{p} \to \gamma$.

Indicate for each case:

(a) allowed or forbidden,
(b) reason if forbidden,
(c) type of interaction if allowed (i.e., strong, weak, electromagnetic, etc.)

(*Wisconsin*)

Solution:

(1) $\pi^- + p \to \pi^0 + n$: All quantum numbers conserved, allowed by strong interaction.

(2) $\pi^0 \to \gamma + \gamma + \gamma : C(\pi^0) = +1$, $C(3\gamma) = (-1)^3 \neq C(\pi^0)$, forbidden as C-parity is not conserved.

(3) $\pi^0 \to \gamma + \gamma$: electromagnetic decay allowed.

(4) $\pi^+ \to \mu^+ + \nu_\mu$: weak decay allowed.

(5) $\pi^+ \to \mu^+ + \bar{\nu}_\mu$: left-hand side $L_\mu = 0$, right-hand side $L_\mu = -2$, forbidden as μ-lepton number is not conserved.

(6)

$$p + \quad \bar{p} \to \quad \Lambda^0 + \quad \Lambda^0$$

B	1	-1	1	1	$\Delta B = +2$
S	0	0	-1	-1	$\Delta S = -2$

it is forbidden as baryon number is not conserved.

(7) $p + \bar{p} \to \gamma$ is forbidden, for the angular momentum and parity cannot both be conserved. Also the momentum and energy cannot both be conserved, for

$$W^2(p, \bar{p}) = (E_p + E_{\bar{p}})^2 - (\mathbf{p}_p + \mathbf{p}_{\bar{p}})^2 = m_p^2 + m_{\bar{p}}^2 + 2(E_p E_{\bar{p}} - \mathbf{p}_p \cdot \mathbf{p}_{\bar{p}}) \geq 2m_p^2 > 0, \text{ as } E^2 = p^2 + m^2, E_p E_{\bar{p}} \geq p_p p_{\bar{p}} \geq \mathbf{p}_p \cdot \mathbf{p}_{\bar{p}}, W^2(\gamma) = E_\gamma^2 - p_\gamma^2 = E_\gamma^2 - E_\gamma^2 = 0, \text{ and so } W(p, \bar{p}) \neq W^2(\gamma).$$

3014

For each of the following decays state a conservation law that forbids it:

$$n \to p + e^-$$

$$n \to \pi^+ + e^-$$

$$n \to p + \pi^-$$

$$n \to p + \gamma$$

(Wisconsin)

Solution:

$n \to p + e^-$: conservation of angular momentum and conservation of lepton number are both violated.

$n \to \pi^+ + e^-$: conservation of baryon number and conservation lepton number are both violated.

$n \to p + \pi^-$: conservation of energy is violated.

$n \to p + \gamma$: conservation of electric charge is violated.

3015

What conservation laws, invariance principles, or other mechanisms account for the suppressing or forbidding of the following processes?

(1) $p + n \to p + \Lambda^0$

(2) $K^+ \to \pi^+ + \pi^- + \pi^+ + \pi^- + \pi^+ + \pi^0$

(3) $\bar{K}^0 \to \pi^- + e^+ + \nu_e$

(4) $\Lambda^0 \to K^0 + \pi^0$

(5) $\pi^+ \to e^+ + \nu_e$ (relative to $\pi^+ \to \mu^+ + \nu_\mu$)

(6) $K_L^0 \to e^+ + e^-$

(7) $K^- \to \pi^0 + e^-$

(8) $\pi^0 \to \gamma + \gamma + \gamma$

(9) $K_L^0 \to \pi^+ + \pi^-$

(10) $K^+ \to \pi^+ + \pi^+ + \pi^0$

(Wisconsin)

Solution:

(1) Conservation of strangeness number and conservation of isospin are violated.

(2) Conservation of energy is violated.

(3) $\Delta S = 1$, $\Delta Q = 0$, the rule that if $|\Delta S| = 1$ in weak interaction, ΔS must be equals to ΔQ is violated

(4) Conservation of baryon number is violated.

(5) The process go through weak interaction and the ratio of rates is **(Problem 3040)**

$$\frac{\Gamma(\pi^+ \to e^+ + \nu_e)}{\Gamma(\pi^+ \to \mu^+ + \nu_\mu)} = \left(\frac{m_e}{m_\mu}\right)^2 \left(\frac{m_\pi^2 - m_e^2}{m_\pi^2 - m_\mu^2}\right)^2 = 1.2 \times 10^{-4}.$$

Hence the $\pi \to e\nu$ mode is quite negligible.

(6) $\Delta S = -1$, $\Delta Q = 0$, same reason as for (3).

(7) Conservation of lepton number is violated.

(8) Conservation of C-parity is violated.

(9) CP parity conservation is violated.

(10) Conservation of electric charge is violated.

3016

Which of the following reactions violate a conservation law? Where there is a violation, state the law that is violated.

$$\mu^+ \to e^+ + \gamma$$
$$e^- \to \nu_e + \gamma$$
$$p + p \to p + \Sigma^+ + K^-$$
$$p \to e^+ + \nu_e$$
$$p \to e^+ + n + \nu_e$$
$$n \to p + e^- + \bar{\nu}_e$$
$$\pi^+ \to \mu^+ + \nu_\mu$$

(*Buffalo*)

Solution:

$\mu^+ \to e^+ + \gamma$ is forbidden because it violates the conservation of lepton number, which must hold for any interaction.

$e^- \to \nu_e + \gamma$, $p + p \to p + \Sigma^+ + K^-$ are forbidden because they violate electric charge conservation.

$p \to e^+ + \nu_e$ is forbidden because it violates baryon number conservation.

$p \to e^+ + n + \nu_e$ is forbidden because it violates energy conservation.

$n \to p + e^- + \bar{\nu}_e$, $\pi^+ \to \mu^+ + \nu_\mu$ are allowed.

3017

(a) Explain why the following reactions are not observed, even if the kinetic energy of the first proton is several BeV:

(1) $p + p \to K^+ + \Sigma^+$

(2) $p + n \to \Lambda^0 + \Sigma^+$

(3) $p + n \to \Xi^0 + p$

(4) $p + n \to \Xi^- + K^+ + \Sigma^+$

(b) Explain why the following decay processes are not observed:

(1) $\Xi^0 \to \Sigma^0 + \Lambda^0$

(2) $\Sigma^+ \to \Lambda^0 + K^+$

(3) $\Xi^- \to n + \pi^-$
(4) $\Lambda^0 \to K^+ + K^-$
(5) $\Xi^0 \to p + \pi^-$

(*Columbia*)

Solution:

(a) The reactions involve only strongly interacting particles and should obey all the conservation laws. If some are violated then the process is forbidden and not observed. Some of the relevant data and quantum numbers are given in Table 3.3.

(1) $p + p \to K^+ + \Sigma^+$, the baryon number, the isospin and its third component are not conserved.

(2) $p + n \to \Lambda^0 + \Sigma^+$, the strangeness number ($\Delta S = -2$) and the third component of isospin are not conserved.

(3) $p + n \to \Xi^0 + p$, for the same reasons as for (2).

(4) $p + n \to \Xi^- + K^+ + \Sigma^+$, for the same reasons as for (2).

(b) All the decays are nonleptonic weak decays of strange particles, where the change of strangeness number S, isospin I and its third component I_3 should obey the rules $|\Delta S| = 1$, $|\Delta I_3| = 1/2$, $|\Delta I| = 1/2$.

(1) $\Xi^0 \to \Sigma^0 + \Lambda^0$, the energy and the baryon number are not conserved.

(2) $\Sigma^+ \to \Lambda^0 + K^+$, the energy is not conserved.

(3) $\Xi^- \to n + \pi^-$, $|\Delta S| = 2 > 1$, $|\Delta I_3| = 1 > 1/2$.

(4) $\Lambda^0 \to K^+ + K^-$, the baryon number is not conserved.

(5) $\Xi^0 \to p + \pi^-$, $|\Delta S| = 2 > 1$, $|\Delta I_3| = 1 > 1/2$.

Table 3.3

Particle	Lifetime(s)	Mass(MeV/c²)	Spin J	Strangeness number S	Isospin I
$\pi^\pm$	2.55×10^{-8}	139.58	0	0	1
K	1.23×10^{-8}	493.98	0	± 1	1/2
p	stable	938.21	1/2	0	1/2
n	1.0×10^3	939.51	1/2	0	1/2
Λ^0	2.52×10^{-10}	1115.5	1/2	-1	0
Σ^+	0.81×10^{-10}	1189.5	1/2	-1	1
Σ^0	$< 10^{-14}$	1192.2	1/2	-1	1
Ξ^-	1.7×10^{-10}	1321	1/2	-2	1/2
Ξ^0	2.9×10^{-10}	1315	1/2	-2	1/2

3018

Listed below are a number of decay processes.

(a) Which do not occur in nature? For each of these specify the conservation law which forbids its occurrence.

(b) Order the remaining decays in order of increasing lifetime. For each case name the interaction responsible for the decay and give an order-of-magnitude estimate of the lifetime. Give a brief explanation for your answer.

$$p \to e^+ + \pi^0$$
$$\Omega^- \to \Xi^0 + K^-$$
$$\rho^0 \to \pi^+ + \pi^-$$
$$\pi^0 \to \gamma + \gamma$$
$$D^0 \to K^- + \pi^+$$
$$\Xi^- \to \Lambda^0 + \pi^-$$
$$\mu^- \to e^- + \bar{\nu}_e + \nu_\mu$$

Table 3.4

particle	mass (MeV/c^2)	J	B	L	I	S	G
γ	0	1	0	0	0	0	0
ν_e	0	1/2	0	1	0	0	0
ν_μ	0	1/2	0	1	0	0	0
e^-	0.5	1/2	0	1	0	0	0
μ^-	106	1/2	0	1	0	0	0
π^0	135	0	0	0	1	0	0
κ^-	494	0	0	0	1/2	-1	0
ρ^0	770	1	0	0	1	0	0
p	938	1/2	1	0	1/2	0	0
Λ^0	1116	1/2	1	0	0	-1	0
Ξ^-	1321	1/2	1	0	1/2	-2	0
Ω^-	1672	3/2	1	0	0	-3	0
D^0	1865	0	0	0	1/2	0	1

(Columbia)

Solution:

(a) $p \to e^+ + \pi^0$, forbidden as the lepton number and the baryon number are not conserved.

$\Omega^- \to \Xi^0 + K^-$, forbidden because the energy is not conserved as $m_\Omega < (m_\Xi + m_K)$.

(b) The allowed decays are arranged below in increasing order of life-time:

$\rho^0 \to \pi^+ + \pi^-$, lifetime $\approx 10^{-24}$ s, strong decay,

$\pi^0 \to \gamma + \gamma$, lifetime $\approx 10^{-16}$ s, electromagnetic decay,

$D^0 \to K^- + \pi^+$, lifetime $\approx 10^{-13}$ s, weak decay,

$\Xi^- \to \Lambda^0 + \pi^-$, lifetime $\approx 10^{-10}$ s, weak decay,

$\mu^- \to e^- + \bar{\nu}_e + \nu_\mu$, lifetime $\approx 10^{-6}$ s, weak decay.

The first two decays are typical of strong and electromagnetic decays, the third and fourth are weak decays in which the strangeness number and the charm number are changed, while the last is the weak decay of a non-strange particle.

3019

An experiment is performed to search for evidence of the reaction $pp \to HK^+K^+$.

(a) What are the values of electric charge, strangeness and baryon number of the particle H? How many quarks must H contain?

(b) A theoretical calculation for the mass of this state H yields a predicted value of $m_H = 2150$ MeV.

What is the minimum value of incident-beam proton momentum necessary to produce this state? (Assume that the target protons are at rest)

(c) If the mass prediction is correct, what can you say about the possible decay modes of H? Consider both strong and weak decays.

(*Princeton*)

Solution:

(a) As K^+ has $S = 1$, $B = 0$, H is expected to have electric charge $Q = 0$, strangeness number $S = -2$, baryon number $B = 2$. To satisfy these requirements, H must contain at least six quarks (uu dd ss).

(b) At minimum incident energy, the particles are produced at rest in the center-of-mass frame. As $(\Sigma E)^2 - (\Sigma \mathbf{p})^2$ is invariant, we have

$$(E_0 + m_p)^2 - p_0^2 = (m_H + 2m_K)^2,$$

giving

$$E_0 = \frac{(m_H + 2m_K)^2 - 2m_p^2}{2m_p}$$

$$= \frac{(2.15 + 2 \times 0.494)^2 - 2 \times 0.938^2}{2 \times 0.938} = 4.311 \text{ GeV},$$

and hence the minimum incident momentum

$$p_0 = \sqrt{E_0^2 - m_p^2} = 4.208 \text{ GeV}/e.$$

(c) As for strong decays, $\Delta S = 0$, $\Delta B = 0$, the possible channels are $H \rightarrow \Lambda^0 \Lambda^0$, $\Lambda^0 \Sigma^0$, $\Xi^- p$, $\Xi^0 n$.

However they all violate the conservation of energy and are forbidden. Consider possible weak decays. The possible decays are nonleptonic decays $H \rightarrow \Lambda + n$, $\Sigma^0 + n$, $\Sigma^- + p$, and semi-leptonic decays

$$H \rightarrow \Lambda + p + e^- + \bar{\nu}, \quad \Sigma^0 + p + e^- + \bar{\nu}.$$

3020

Having 4.5 GeV free energy, what is the most massive isotope one could theoretically produce from nothing?

(a) ^{2}D.
(b) ^{3}He.
(c) ^{3}T.

(*CCT*)

Solution:

With a free energy of 4.5 GeV, one could create baryons with energy below 2.25 GeV (To conserve baryon number, the same number of baryons and antibaryons must be produced together. Thus only half the energy is available for baryon creation). Of the three particles only ^{2}D has rest energy below this. Hence the answer is (a).

3021

(i) The decay $K \rightarrow \pi\gamma$ is absolutely forbidden by a certain conservation law, which is believed to hold exactly. Which conservation law is this?

(ii) There are no known mesons of electric charge two. Can you give a simple explanation of this?

(iii) Explain how the parity of pion can be measured by observation of the polarizations of the photons in $\pi^0 \to \gamma\gamma$.

(iv) To a very high accuracy, the cross section for e^-p scattering equals the cross section for e^+p scattering. Is this equality a consequence of a conservation law? If so, which one? If not, explain the observed equality. To what extent (if any at all) do you expect this equality to be violated?

(v) It has recently been observed that in inclusive Λ production (Fig. 3.3), for example $\pi p \to \Lambda+$anything, the Λ is produced with a surprisingly high polarization. Do you believe this polarization is

(a) along (or opposite to) the direction of the incident beam,

(b) along (or opposite to) the direction of motion of the outgoing Λ, or

(c) perpendicular to both?

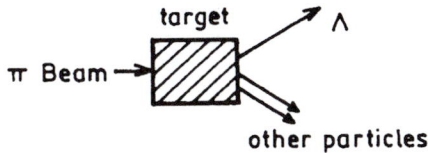

Fig. 3.3

(Princeton)

Solution:

(i) The decay is forbidden by the conservation of strangeness number, which holds exactly in electromagnetic interaction.

(ii) According to the prevailing theory, a meson consists of a quark and an antiquark. The absolute value of a quark's charge is not more than $2/3$. So it is impossible for the charge of a meson consisting of two quarks to be equal to 2.

(iii) Let the wave vectors of the two photons be $\mathbf{k}_1$, $\mathbf{k}_2$, the directions of the polarization of their electric fields be $\mathbf{e}_1$, $\mathbf{e}_2$, and let $\mathbf{k} = \mathbf{k}_1 - \mathbf{k}_2$. Since the spin of π^0 is 0, the possible forms of the decay amplitude are $A\mathbf{e}_1 \cdot \mathbf{e}_2$ and $B\mathbf{k} \cdot (\mathbf{e}_1 \times \mathbf{e}_2)$, which, under space inversion, respectively does not and does change sign. Thus the former form has even parity, and the latter, odd parity. These two cases stand for the two different relative polarizations of the photons. The former describes mainly parallel polarizations, while the latter describes mainly perpendicular polarizations between the two photons. It is difficult to measure the polarization of high energy photons ($E \sim 70$ MeV) directly. But in π^0 decays, in a fraction α^2 of the cases the two photons convert directly to two electron-positron pairs. In such cases the relative polarization of the photons can be determined by measuring the angle between the two electron-positron pairs. The experimental results tend to favor the perpendicular polarization. Since parity is conserved in electromagnetic interaction, the parity of π^0 is odd.

(iv) No. To first order accuracy, the probability of electromagnetic interaction is not related to the sign of the charge of the incident particle. Only when higher order corrections are considered will the effect of the sign of the charge come in. As the strength of each higher order of electromagnetic interaction decreases by a factor α^2, this equality is violated by a fraction $\alpha^2 \approx 5.3 \times 10^{-5}$.

(v) The polarization $\boldsymbol{\sigma}$ of Λ is perpendicular to the plane of interaction. As parity is conserved in strong interaction, $\boldsymbol{\sigma}$ is perpendicular to the plane of production, i.e.,

$$\boldsymbol{\sigma} \propto \mathbf{p}_\pi \times \mathbf{p}_\Lambda$$

3022

Recently a stir was caused by the reported discovery of the decay $\mu^+ \to e^+ + \gamma$ at a branching ratio of $\sim 10^{-9}$.

(a) What general principle is believed to be responsible for the suppression of this decay?

(b) The apparatus consists of a stopping μ^+ beam and two NaI crystals, which respond to the total energy of the positrons or gamma rays. How would you arrange the crystals relative to the stopping target and beam, and what signal in the crystals would indicate that an event is such a μ decay?

(c) The background events are the decays $\mu^+ \to e^+ + \nu_e + \bar{\nu}_\mu + \gamma$ with the neutrinos undetected. Describe qualitatively how one would distinguish events of this type from the $\mu^+ \to e^+ + \gamma$ events of interest.

(Wisconsin)

Solution:

(a) This decay is suppressed by the separate conservation of electron-lepton number and μ-lepton number,

(b) $\mu^+ \to e^+ + \gamma$ is a two-body decay. When the muon decays at rest into e^+ and γ, we have $E_e \approx E_\gamma = \frac{m_\mu c^2}{2}$. As e^+ and γ are emitted in opposite directions the two crystals should be placed face to face. Also, to minimize the effect of any directly incident mesons they should be placed perpendicular to the μ beam (see Fig. 3.4). The coincidence of e^+ and γ signals gives the μ decay events, including the background events given in (c).

(c) $\mu^+ \to e^+ + \gamma$ is a two-body decay and $\mu^+ \to e^+ + \nu_e + \bar{\nu}_\mu + \gamma$ is a four-body decay. In the former e^+ and γ are monoenergetic, while in the latter e^+ and γ have continuous energies up to a maximum. We can separate them by the amplitudes of the signals from the two crystals. For $\mu^+ \to e^+ + \gamma$, $(E_e + E_\gamma) = m_\mu$, while for $\mu^+ \to e^+ + \nu_e + \bar{\nu}_\mu + \gamma$, $(E_e + E_\gamma) < m_\mu$.

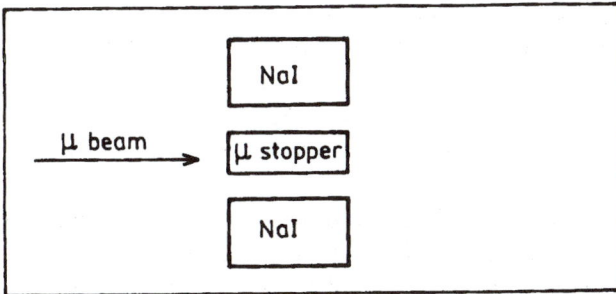

Fig. 3.4

3023

Describe the properties of the various types of pion and discuss in detail the experiments which have been carried out to determine their spin, parity, and isospin.

(Buffalo)

Solution:

There are three kinds of pion: $\pi^0, \pi^+\pi^-$, with π^+ being the antiparticle of π^- and π^0 its own antiparticle, forming an isospin triplet of $I = 1$. Their main properties are listed in Table 3.5.

Table 3.5

Particle	Mass(MeV)	Spin	Parity	C-Parity	Isospin	I_3	G
π^+	139.6	0	−		1	1	−1
π^0	135	0	−	+	1	0	−1
π^-	139.6	0	−		1	−1	−1

To determine the spin of π^+, we apply the principle of detailed balance to the reversible reaction $\pi^+ + d \rightleftarrows p + p$, where the forward reaction and its inverse have the same transition matrix element. Thus

$$\frac{d\sigma}{d\Omega}(pp \to d\pi^+) = \frac{d\sigma}{d\Omega}(d\pi^+ \to pp) \times 2\frac{p_\pi^2(2J_\pi + 1)(2J_d + 1)}{p_p^2(2J_p + 1)^2},$$

where p_π, p_p are the momenta of π and p, respectively, in the center-of-mass frame. Experimental cross sections give $2J_\pi + 1 = 1.00 \pm 0.01$, or $J_\pi = 0$.

The spin of π^- can be determined directly from the hyperfine structure of the π-mesic atom spectrum. Also the symmetry of particle and antiparticle requires π^+ and π^- to have the same spin. So the spin of π^- is also 0.

The spin of π^0 can be determined by studying the decay $\pi^0 \to 2\gamma$. First we shall see that a particle of spin 1 cannot decay into 2 γ's. Consider the decay in the center-of-mass frame of the 2 γ's, letting their momenta be $\mathbf{k}$ and $-\mathbf{k}$, their polarization vectors be ε_1 and ε_2 respectively. Because the spin of the initial state is 1, the final state must have a vector form. As a real photon has only transverse polarization, only the following vectors can be constructed from $\mathbf{k}$, ε_1, ε_2:

$$\varepsilon_1 \times \varepsilon_2, \quad (\varepsilon_1 \cdot \varepsilon_2)\mathbf{k}, \quad (\varepsilon \times \varepsilon_2 \cdot \mathbf{k})\mathbf{k}.$$

All the three vector forms change sign when the 2 γ's are exchanged. However the 2γ system is a system of two bosons which is exchange-symmetric and so none of three forms can be the wave function of the system. Hence the spin of π^0 cannot be 1. On the other hand, consider the reaction $\pi^- + p \to \pi^0 + n$ using low energy (s-wave) π^-. The reaction is forbidden for $J_{\pi^0} \geq 2$. Experimentally, the cross section for the charge-exchange reaction is very large. The above proves that $J_{\pi^0} = 0$.

The parity of π^- can be determined from the reaction $\pi^- + d \to n + n$, employing low energy (s-wave) π^-. It is well known that $J_d^P = 1^+$, so $P(\pi^-) = P^2(n)(-1)^l$, l being the orbital angular momentum of the relative motion of the two neutrons. Since an n–n system is a Fermion system and so is exchange antisymmetric, $l = 1$, $J = 1$, giving $P(\pi^-) = -1$.

The parity of π^+ can be determined by studying the cross section for the reaction $\pi^+ + d \to p + p$ as a function of energy of the incident low energy (s-wave) π^+. This gives $P(\pi^+) = -1$.

The parity of π^0 can be determined by measuring the polarization of the decay $\pi^0 \to 2\gamma$. As $J(\pi^0) = 0$, and the 2γ system in the final state is exchange symmetric, possible forms of the decay amplitude are

$$\varepsilon_1 \cdot \varepsilon_2, \text{ corresponding to } P(\pi^0) = +1,$$

$$\mathbf{k} \cdot (\varepsilon_1 \times \varepsilon_2), \text{ corresponding to } P(\pi^0) = -1,$$

where $\mathbf{k}$ is the momentum of a γ in the π^0 rest frame. The two forms respectively represent the case of dominantly parallel polarizations and the case of dominantly perpendicular polarizations of the two photons. Consider then the production of electron-positron pairs by the 2 γ's:

$$\pi^0 \to \gamma + \gamma$$
$$\quad \big\lfloor \quad \big\lfloor_{\!\!\to}\ e^+ + e^-$$
$$\quad \big\lfloor_{\!\!\to}\ e^+ + e^-$$

An electron-positron pair is created in the plane of the electric vector of the γ ray. As the experimental results show that the planes of the two pairs are mainly perpendicular to each other, the parity of π^0 is -1.

The isospin of π can be deduced by studying strong interaction processes such as

$$n + p \to d + \pi^0, \qquad p + p \to d + \pi^+.$$

Consider the latter reaction. The isospin of the initial state $(p+p)$ is $|1, 1\rangle$, the isospin of the final state is also $|1, 1\rangle$. As isospin is conserved, the transition to the final state $(d + \pi^+)$ has a probability of 100%. Whereas, in the former reaction, the isospin of the initial state is $\frac{1}{\sqrt{2}}(|1,0\rangle - |0,0\rangle)$, of which only the state $|1,0\rangle$ can transit to the $(d + \pi^0)$ system of isospin $|1,0\rangle$. Hence the probability for the transition from $(n + p)$ to $(d + \pi^0)$ is only 50%. In other words, if $I(\pi) = 1$, we would have

$$\sigma(pp \to d\pi^+) = 2\sigma(pn \to d\pi^0).$$

As this agrees with experiment, $I(\pi) = 1$.

3024

The electrically neutral baryon Σ^0 (1915) (of mass 1915 MeV/c^2) has isospin $I = 1$, $I_3 = 0$. Call Γ_{K^-p}, $\Gamma_{\bar{K}^0n}$, Γ_{π^-p}, $\Gamma_{\pi^+\pi^-}$ respectively the rates for the decays $\Sigma^0(1915) \to K^-p$, $\Sigma^0(1915) \to \bar{K}^0n$, $\Sigma^0(1915) \to \pi^-p$, $\Sigma^0(1915) \to \pi^+\pi^-$. Find the ratios

$$\frac{\Gamma_{\bar{K}^0n}}{\Gamma_{K^-p}}, \quad \frac{\Gamma_{\pi^-p}}{\Gamma_{K^-p}}, \quad \frac{\Gamma_{\pi^+\pi^-}}{\Gamma_{K^-p}}.$$

(The masses of the nucleons, K^-, and π^- mesons are such that all these decays are kinetically possible. You can disregard the small mass splitting within an isospin multiplet.)

(Chicago)

Solution:

n, p form an isospin doublet, π^+, π^0, π^- form an isospin triplet, and K^+, K^0 form an isospin doublet. K^- and $\bar{K}^0$, the antiparticles of K^+ and K^0 respectively, also form an isospin doublet. Write the isospin state of $\Sigma^0(1915)$ as $|1,0\rangle$, those of p and n as $|1/2, 1/2\rangle$ and $|1/2, -1/2\rangle$, and those of $\bar{K}^0$ and K^- as $|1/2, 1/2\rangle$ and $|1/2, -1/2\rangle$, respectively. As

$$\Psi(\bar{K}^0n) = \left|\frac{1}{2}, \frac{1}{2}\right\rangle \left|\frac{1}{2}, -\frac{1}{2}\right\rangle = \sqrt{\frac{1}{2}}(|1,0\rangle + |0,0\rangle),$$

$$\Psi(\bar{K}^-p) = \left|\frac{1}{2}, -\frac{1}{2}\right\rangle \left|\frac{1}{2}, \frac{1}{2}\right\rangle = \sqrt{\frac{1}{2}}(|1,0\rangle - |0,0\rangle),$$

$\Sigma^0(1915) \to \bar{K}^0 n$ and $\Sigma^0(1915) \to K^- p$ are both strong decays, the partial widths are

$$\Gamma_{\bar{K}^0 n} \propto |\langle \Psi(\Sigma^0)|H|\Psi(\bar{K}^0 n)\rangle|^2 = \left(\frac{a_1}{\sqrt{2}}\right)^2 = \frac{a_1^2}{2},$$

$$\Gamma_{K^- p} \propto |\langle \Psi(\Sigma^0)|H|\Psi(K^- p)\rangle|^2 = \left(\frac{a_1}{\sqrt{2}}\right)^2 = \frac{a_1^2}{2},$$

where $a_1 = \langle 1|H|1 \rangle$. Note $\langle 1|H|0 \rangle = 0$ and, as strong interaction is charge independent, a_1 only depends on I but not on I_3. Hence

$$\frac{\Gamma_{\bar{K}^0 n}}{\Gamma_{K^- p}} = 1.$$

$\Sigma^0(1915) \to p\pi^-$ is a weak decay ($\Delta I_3 = -\frac{1}{2} \neq 0$) and so

$$\frac{\Gamma_{\pi^- p}}{\Gamma_{K^- p}} \ll 1$$

(actually $\sim 10^{-10}$).

In the $\Sigma^0(1915) \to \pi^+ \pi^-$ mode baryon number is not conserved, and so the reaction is forbidden. Thus

$$\Gamma_{\pi^+ \pi^-} = 0,$$

or

$$\frac{\Gamma_{\pi^+ \pi^-}}{\Gamma_{K^- p}} = 0.$$

3025

Which of the following reactions are allowed? If forbidden, state the reason.

(a) $\pi^- + p \to K^- + \Sigma^+$
(b) $d + d \to {}^4He + \pi^0$
(c) $K^- + p \to \Xi^- + K^+$

What is the ratio of reaction cross sections $\sigma(p+p \to \pi^+ + d)/\sigma(n+p \to \pi^0 + d)$ at the same center-of-mass energy?

(Chicago)

Solution:

(a) Forbidden as $\Delta I_3 = (-1/2) + (+1) - (-1) - 1/2 = 1 \neq 0$, $\Delta S = (-1) + (-1) - 0 - 0 = -2 \neq 0$.

(b) Forbidden as $I(d) = I(^4He) = 0$, $I(\pi^0) = 1$, $\Delta I = 1 \neq 0$

(c) Allowed by strong interaction as Q, I, I_3, and S are all conserved.

The difference in cross section between $pp \rightarrow \pi^+ d$ and $np \rightarrow \pi^0 d$ relates to isospin only. Using the coupling presentation for isospins and noting the orthogonality of the isospin wave functions, we have

$$|pp\rangle = \left|\frac{1}{2}, \frac{1}{2}\right\rangle \left|\frac{1}{2}, \frac{1}{2}\right\rangle = |1, 1\rangle ,$$

$$|\pi^+ d\rangle = |1, 1\rangle |0, 0\rangle = |1, 1\rangle ,$$

$$|np\rangle = \left|\frac{1}{2}, -\frac{1}{2}\right\rangle \left|\frac{1}{2}, \frac{1}{2}\right\rangle = \frac{1}{\sqrt{2}}|1, 0\rangle - \frac{1}{\sqrt{2}}|0, 0\rangle ,$$

$$|\pi^0 d\rangle = |1, 0\rangle |0, 0\rangle = |1, 0\rangle .$$

Hence the matrix element of $pp \rightarrow \pi^+ d$ is

$$\langle \pi^+ d|\hat{H}|pp\rangle \propto \langle 1, 1|\hat{H}|1, 1\rangle = \langle 1|\hat{H}|1\rangle = a_1 .$$

Similarly, the matrix element of $np \rightarrow \pi^0 d$ is

$$\langle \pi^0 d|\hat{H}|np\rangle \propto \frac{1}{\sqrt{2}}\langle 1, 0|\hat{H}|1, 0\rangle - \frac{1}{\sqrt{2}}\langle 1, 0|\hat{H}|0, 0\rangle$$

$$\propto \frac{1}{\sqrt{2}}\langle 1, 0|\hat{H}|1, 0\rangle = \frac{1}{\sqrt{2}}\langle 1|\hat{H}|1\rangle = \frac{a_1}{\sqrt{2}} ,$$

as $\langle 1, 0|\hat{H}|0, 0\rangle = 0$ and strong interaction is independent of I_3. Therefore,

$$\frac{\sigma(pp \rightarrow \pi^+ d)}{\sigma(np \rightarrow \pi^0 d)} = \frac{|\langle \pi^+ d|\hat{H}|pp\rangle|^2}{|\langle \pi^0 d|\hat{H}|np\rangle|^2} = \frac{a_1^2}{\frac{1}{2}a_1^2} = 2 .$$

3026

Given two angular momenta $\mathbf{J}_1$ and $\mathbf{J}_2$ (for example $\mathbf{L}$ and $\mathbf{S}$) and the corresponding wave functions.

(a) Compute the Clebsch–Gordan coefficients for the states with $\mathbf{J} = \mathbf{j}_1 + \mathbf{j}_2$, $M = m_1 + m_2$, where $j_1 = 1$ and $j_2 = 1/2$, $J = 3/2$, $M = 1/2$, for the various possible m_1 and m_2 values.

(b) Consider the reactions

(1) $\pi^+ p \to \pi^+ p$,

(2) $\pi^- p \to \pi^- p$,

(3) $\pi^- p \to \pi^0 n$.

These reactions, which conserve isospin, can occur in the isospin $I = 3/2$ state (Δ resonance) or $I = 1/2$ state (N^* resonance). Calculate the ratio of these cross sections $\sigma_1 : \sigma_2 : \sigma_3$ for an energy corresponding to a Δ resonance and to an N^* resonance. At a resonance energy you can neglect the effect due to the other isospin state. Note that the pion is an isospin $I_\pi = 1$ state and the nucleon an isospin $I_n = 1/2$ state.

(UC, Berkeley)

Solution:

(a) First consider

$$\left| \frac{3}{2}, \frac{3}{2} \right\rangle = |1, 1\rangle \left| \frac{1}{2}, \frac{1}{2} \right\rangle .$$

Applying the operator

$$L_- = J_x - iJ_y = (j_{1x} - ij_{1y}) + (j_{2x} - ij_{2y}) \equiv L_-^{(1)} + L_-^{(2)}$$

to the above:

$$L_- \left| \frac{3}{2}, \frac{3}{2} \right\rangle = L_-^{(1)} |1, 1\rangle \left| \frac{1}{2}, \frac{1}{2} \right\rangle + L_-^{(2)} |1, 1\rangle \left| \frac{1}{2}, \frac{1}{2} \right\rangle ,$$

as

$$L_- |J, M\rangle = \sqrt{J(J+1) - M(M-1)} |J, M-1\rangle ,$$

we have

$$\sqrt{3} \left| \frac{3}{2}, \frac{1}{2} \right\rangle = \sqrt{2} |1, 0\rangle \left| \frac{1}{2}, \frac{1}{2} \right\rangle + |1, 1\rangle \left| \frac{1}{2}, -\frac{1}{2} \right\rangle ,$$

or

$$\left| \frac{3}{2}, \frac{1}{2} \right\rangle = \sqrt{\frac{2}{3}} |1, 0\rangle \left| \frac{1}{2}, \frac{1}{2} \right\rangle + \sqrt{\frac{1}{3}} |1, 1\rangle \left| \frac{1}{2}, -\frac{1}{2} \right\rangle .$$

(b) We couple each initial pair in the isospin space:

$$|\pi^+ p\rangle = |1, 1\rangle \left|\frac{1}{2}, \frac{1}{2}\right\rangle = \left|\frac{3}{2}, \frac{3}{2}\right\rangle,$$

$$|\pi^- p\rangle = |1, -1\rangle \left|\frac{1}{2}, \frac{1}{2}\right\rangle = \sqrt{\frac{2}{3}}\left|\frac{1}{2}, -\frac{1}{2}\right\rangle + \sqrt{\frac{1}{3}}\left|\frac{3}{2}, -\frac{1}{2}\right\rangle,$$

$$|\pi^0 n\rangle = |1, 0\rangle \left|\frac{1}{2}, -\frac{1}{2}\right\rangle = \sqrt{\frac{2}{3}}\left|\frac{3}{2}, -\frac{1}{2}\right\rangle - \sqrt{\frac{1}{3}}\left|\frac{1}{2}, -\frac{1}{2}\right\rangle.$$

Because of charge independence in strong interaction, we can write

$$\left\langle \frac{3}{2}, m_j \left| \hat{H} \right| \frac{3}{2}, m_i \right\rangle = a_1,$$

$$\left\langle \frac{1}{2}, m_j \left| \hat{H} \right| \frac{1}{2}, m_i \right\rangle = a_2,$$

independent of the value of m. Furthermore the orthogonality of the wave functions requires

$$\left\langle \frac{1}{2} \left| \hat{H} \right| \frac{3}{2} \right\rangle = 0.$$

Hence the transition cross sections are

$$\sigma_1(\pi^+ p \to \pi^+ p) \propto \left| \left\langle \frac{3}{2}, \frac{3}{2} \left| \hat{H} \right| \frac{3}{2}, \frac{3}{2} \right\rangle \right|^2 = |a_1|^2,$$

$$\sigma_2(\pi^- p \to \pi^- p) \propto \left| \left(\sqrt{\frac{2}{3}} \left\langle \frac{1}{2}, -\frac{1}{2} \right| + \sqrt{\frac{1}{3}} \left\langle \frac{3}{2}, -\frac{1}{2} \right| \right) \right.$$

$$\left. \hat{H} \left(\sqrt{\frac{2}{3}} \left| \frac{1}{2}, -\frac{1}{2} \right\rangle + \sqrt{\frac{1}{3}} \left| \frac{3}{2}, -\frac{1}{2} \right\rangle \right) \right|^2$$

$$= \left| \frac{2}{3} a_2 + \frac{1}{3} a_1 \right|^2,$$

$$\sigma_3(\pi^- p \to \pi^0 n) \propto \left| \left(\sqrt{\frac{2}{3}} \left\langle \frac{1}{2}, -\frac{1}{2} \right| + \sqrt{\frac{1}{3}} \left\langle \frac{3}{2}, -\frac{1}{2} \right| \right) \right.$$

$$\left. \hat{H} \left(\sqrt{\frac{2}{3}} \left| \frac{3}{2}, -\frac{1}{2} \right\rangle - \sqrt{\frac{1}{3}} \left| \frac{1}{2}, -\frac{1}{2} \right\rangle \right) \right|^2$$

$$= \left| -\frac{\sqrt{2}}{3} a_2 + \frac{\sqrt{2}}{3} a_1 \right|^2 ,$$

When Δ resonance takes place, $|a_1| \gg |a_2|$, and the effect of a_2 can be neglected. Hence

$$\sigma_1 \propto |a_1|^2 ,$$

$$\sigma_2 \propto \frac{1}{9} |a_1|^2 ,$$

$$\sigma_3 \propto \frac{2}{9} |a_1|^2 ,$$

and $\sigma_1 : \sigma_2 : \sigma_3 = 9 : 1 : 2$.

When N^* resonance occurs, $|a_1| \ll |a_2|$, and we have

$$\sigma_1 \approx 0 ,$$

$$\sigma_2 \propto \frac{4}{9} |a_2|^2 ,$$

$$\sigma_3 \propto \frac{2}{9} |a_2|^2 ,$$

$$\sigma_1 : \sigma_2 : \sigma_3 = 0 : 2 : 1 .$$

3027

Estimate the ratios of decay rates given below, stating clearly the selection rules ("fundamental" or phenomenological) which are operating. Also state whether each decay (regardless of the ratio) is strong, electromagnetic or weak. If at all possible, express your answer in terms of the fundamental constants G, α, θ_c, m_K, etc. Assume that the strong interactions have unit strength (i.e. , unit dimensionless coupling constant).

(a) $\dfrac{K^+ \rightarrow \pi^+ \pi^0}{K_s^0 \rightarrow \pi^+ \pi^-}$

(b) $\dfrac{\rho^0 \rightarrow \pi^0 \pi^0}{\rho^0 \rightarrow \pi^+ \pi^-}$

(c) $\dfrac{K_L^0 \rightarrow \mu^+ \mu^-}{K_L^0 \rightarrow \pi^0 \pi^0}$

(d) $\dfrac{K^+ \rightarrow \pi^+ \pi^+ e^- \nu}{K^- \rightarrow \pi^+ \pi^- e^- \nu}$

(e) $\dfrac{\Omega^- \rightarrow \Sigma^- \pi^0}{\Omega^- \rightarrow \Xi^0 \pi^-}$

(f) $\dfrac{\eta^0 \rightarrow \pi^+ \pi^-}{\eta^0 \rightarrow \pi^+ \pi^- \pi^0}$

(g) $\dfrac{\Lambda^0 \rightarrow K^- \pi^+}{\Lambda^0 \rightarrow p \pi^-}$

(h) $\dfrac{\theta^0 \rightarrow \pi^+ \pi^- \pi^0}{\omega^0 \rightarrow \pi^+ \pi^- \pi^0}$

(i) $\dfrac{\Sigma^- \rightarrow \Lambda^0 \pi^-}{\Sigma^- \rightarrow n \pi^-}$

(j) $\dfrac{\pi^- \rightarrow e^- \nu}{K^+ \rightarrow \mu^+ \nu}$

(Princeton)

Solution:

(a) Consider $K^+ \rightarrow \pi^+ \pi^0$. For nonleptonic weak decays $\Delta I = 1/2$. As $I(K) = 1/2$, the isospin of the 2π system must be 0 or 1. The generalized Pauli's principle requires the total wave function of the 2π system to be symmetric. As the spin of K is 0, conservation of the total angular momentum requires $J(2\pi) = J(K) = 0$. Then as the spin of π is 0, $l(2\pi) = 0$. Thus the spatial and spin parts of the wave function of the 2π system are both symmetric, so the isospin wave function must also be symmetric. It follows that the isospin of the 2π system has two possible values, 0 or 2. Hence $I(\pi^+ \pi^0) = 0$. However, $I_3(\pi^+ \pi^0) = 1 + 0 = 1$. As the rule $I_3 \le I$ is violated, the decay is forbidden. On the other hand, $K_s^0 \rightarrow \pi^+ \pi^-$ is allowed as it satisfies the rule $\Delta I = 1/2$.

Therefore,

$$\frac{K^+ \to \pi^+\pi^-}{K_s^0 \to \pi^+\pi^-} \ll 1.$$

Note the ratio of the probability amplitudes for $\Delta I = 1/2, 3/2$ in K-decay, A_0 and A_2, can be deduced from

$$\frac{\Gamma(K^+ \to \pi^+\pi^0)}{\Gamma(K_s^0 \to \pi^+\pi^-)} = \frac{3}{4}\left(\frac{A_2}{A_0}\right)^2 \approx 1.5 \times 10^{-3},$$

giving

$$\frac{A_2}{A_0} \approx 4.5\%.$$

(b) Consider the decay modes $\rho^0 \to \pi^+\pi^-$, $\pi^0\pi^0$. $\rho^0 \to \pi^+\pi^-$ is an allowed strong decay, while for $\rho^0 \to \pi^0\pi^0$, the C-parities are $C(\rho^0) = -1$, $C(\pi^0\pi^0) = 1$, and the decay is forbidden by conservation of C-parity. Hence

$$\frac{\rho^0 \to \pi^0\pi^0}{\rho^0 \to \pi^+\pi^-} \approx 0.$$

(c) As K_L^0 is not the eigenstate of CP, $K_L^0 \to \pi^0\pi^0$ has a nonzero branching ratio, which is approximately 9.4×10^{-4}. The decay $K_L^0 \to \mu^+\mu^-$, being a second order weak decay, has a probability even less than that of $K_L^0 \to \pi^0\pi^0$. It is actually a flavor-changing neutral weak current decay. Thus

$$1 \gg \frac{K_L^0 \to \mu^+\mu^-}{K_L^0 \to \pi^0\pi^0} \approx 0.$$

Experimentally, the ratio $\approx 10^{-8}/10^{-3} = 10^{-5}$.

(d) $K^+ \to \pi^+\pi^+e^-\bar{\nu}$ is a semileptonic weak decay and so ΔQ should be equal to ΔS, where ΔQ is the change of hadronic charge. As $\Delta S = 1$, $\Delta Q = -1$, it is forbidden. But as $K^- \to \pi^+\pi^-e^-\bar{\nu}$ is an allowed decay,

$$\frac{K^+ \to \pi^+\pi^+e^-\bar{\nu}}{K^- \to \pi^+\pi^-e^-\bar{\nu}} = 0.$$

(e) In $\Omega^- \to \Sigma^-\pi^0$, $\Delta S = 2$. Thus it is forbidden. As $\Omega^- \to \Xi^0\pi^-$ is allowed by weak interaction,

$$\frac{\Omega^- \to \Sigma^-\pi^0}{\Omega^- \to \Xi^0\pi^-} = 0.$$

(f) Consider $\eta^0 \to \pi^+\pi^-$. η^0 has $J^P = 0^-$ and decays electromagnetically ($\Gamma = 0.83$ keV). As J^P of $\pi^\pm$ is 0^-, a $\pi^+\pi^-$ system can only form states

$0^+, 1^-, 2^+$. Since parity is conserved in electromagnetic decay, this decay mode is forbidden. On the other hand, $\eta^0 \to \pi^+\pi^-\pi^0$ is an electromagnetic decay with all the required conservation rules holding. Hence

$$\frac{\eta^0 \to \pi^+\pi^-}{\eta^0 \to \pi^+\pi^-\pi^0} = 0.$$

(g) $\Lambda^0 \to K^-\pi^+$ is a nonleptonic decay mode. As $\Delta I_3 = 1/2$, $\Delta S = 0$, it is forbidden. $\Lambda^0 \to p\pi^-$ is also a nonleptonic weak decay satisfying $|\Delta S| = 1$, $|\Delta I| = 1/2$, $|\Delta I_3| = 1/2$ and is allowed. Hence

$$\frac{\Lambda^0 \to K^-\pi^+}{\Lambda^0 \to p\pi^-} = 0.$$

(h) Consider $\theta^0 \to \pi^+\pi^-\pi^0$. θ^0 has strong decays ($\Gamma = 180$ MeV) and $I^G J^{PC} = 0^+2^{++}$. As $G(\pi^+\pi^-\pi^0) = (-1)^3 = -1$, $G(\theta^0) = +1$, G-parity is not conserved and the decay mode is forbidden. Consider $\omega^0 \to \pi^+\pi^-\pi^0$. As $I^G J^{PC}$ of ω^0 is 0^-1^{--}, it is allowed. Hence

$$\frac{\theta^0 \to \pi^+\pi^-\pi^0}{\omega^0 \to \pi^+\pi^-\pi^0} = 0.$$

(i) Consider $\Sigma^- \to \Lambda^0\pi^-$. As $\Delta S = 0$, it is forbidden. $\Sigma^- \to n\pi^-$ is an allowed nonleptonic weak decay. Hence

$$\frac{\Sigma^- \to \Lambda^0\pi^-}{\Sigma^- \to n\pi^-} = 0.$$

(j) $\pi^- \to e^-\bar{\nu}$ and $K^+ \to \mu^+\nu$ are both semileptonic two-body decays. For the former, $\Delta S = 0$ and the coupling constant is $G\cos\theta_c$, for the latter $\Delta S = 1$ and the coupling constant is $G\sin\theta_c$, where θ_c is the Cabbibo angle. By coupling of axial vectors we have

$$\omega'(\varphi \to l\nu) = \frac{f_\varphi^2 m_l^2 (m_\varphi^2 - m_l^2)^2}{4\pi m_\varphi^3},$$

where f_φ is the coupling constant. Hence

$$\frac{\pi^- \to e^-\bar{\nu}}{K^+ \to \mu^+\nu} = \frac{f_\pi^2 m_e^2 (m_\pi^2 - m_e^2)^2 m_K^3}{f_K^2 m_\mu^2 (m_K^2 - m_\mu^2)^2 m_\pi^3}$$

$$= \frac{m_K^3 m_e^2 (m_\pi^2 - m_e^2)^2}{m_\pi^3 m_\mu^2 (m_K^2 - m_\mu^2)^2} \cot^2\theta_c$$

$$= 1.35 \times 10^{-4},$$

using $\theta_c = 13.1^0$ as deduced from experiment.

3028

The Σ^* is an unstable hyperon with mass $m = 1385$ MeV and decay width $\Gamma = 35$ MeV, with a branching ratio into the channel $\Sigma^{*+} \to \pi^+ \Lambda$ of 88%. It is produced in the reaction $K^- p \to \pi^- \Sigma^{*+}$, but the reaction $K^+ p \to \pi^+ \Sigma^{*+}$ does not occur.

(a) What is the strangeness of the Σ^*? Explain on the basis of the reactions given.

(b) Is the decay of the Σ^* strong or weak? Explain.

(c) What is the isospin of the Σ^*? Explain using the information above.

(Wisconsin)

Solution:

(a) As Σ^{*+} is produced in the strong interaction $K^- p \to \pi \Sigma^{*+}$, which conserves strangeness number, the strangeness number of Σ^{*+} is equal to that of K^-, namely, -1. As $S(K^+) = +1$, the reaction $K^+ p \to \pi^+ \Sigma^{*+}$ violates the conservation of strangeness number and is forbidden.

(b) The partial width of the decay $\Sigma^{*+} \to \Lambda \pi^+$ is

$$\Gamma_{\Lambda\pi} = 88\% \times 35 = 30.8 \text{ MeV},$$

corresponding to a lifetime

$$\tau_{\Lambda\pi} \approx \frac{\hbar}{\Gamma_{\Lambda\pi}} = \frac{6.62 \times 10^{-22}}{30.8} = 2.15 \times 10^{-23} \text{ s}.$$

As its order of magnitude is typical of the strong interaction time, the decay is a strong decay.

(c) Isospin is conserved in strong interaction. The strong decay $\Sigma^{*+} \to \Lambda \pi^+$ shows that, as $I(\Lambda) = I_3(\Lambda) = 0$,

$$I(\Sigma^*) = I(\pi) = 1.$$

3029

A particle X has two decay modes with partial decay rates $\gamma_1 (\text{sec}^{-1})$ and $\gamma_2 (\text{sec}^{-1})$.

(a) What is the inherent uncertainty in the mass of X?

(b) One of the decay modes of X is the strong interaction decay

$$X \to \pi^+ + \pi^+ .$$

What can you conclude about the isotopic spin of X?

(Wisconsin)

Solution:

(a) The total decay rate of particle X is

$$\lambda = \gamma_1 + \gamma_2 .$$

So the mean lifetime of the particle is

$$\tau = \frac{1}{\lambda} = \frac{1}{\gamma_1 + \gamma_2} .$$

The inherent uncertainty in the mass of the particle, Γ, is given by the uncertainty principle $\Gamma\tau \sim \hbar$. Hence

$$\Gamma \sim \frac{\hbar}{\tau} = \hbar(\gamma_1 + \gamma_2) .$$

(b) As $X \to \pi^+\pi^+$ is a strong decay, isospin is conserved. π^+ has $I = 1$ and $I_3 = +1$. Thus final state has $I = 2$ and so the isospin of X is 2.

3030

Suppose that π^- has spin 0 and negative intrinsic parity. If it is captured by a deuterium nucleus from a p orbit in the reaction

$$\pi^- + d \to n + n ,$$

show that the two neutrons must be in a singlet state. The deuteron's spin-parity is 1^+.

(Wisconsin)

Solution:

The parity of the initial state $\pi^- d$ is

$$P_i = P(\pi^-)P(d)(-1)^l = (-1) \times (+1) \times (-1)^1 = +1 .$$

As the reaction is by strong interaction, parity is conserved, and so the parity of the final state is $+1$.

As the intrinsic parity of the neutron is $+1$, the parity of the final state nn is $P_f = (+1)^2(-1)^l = P_i = (-1)^1(-1)(+1)$, where l is the orbital momentum quantum number of the relative motion of the two neutrons in the final state. Thus $l = 0, 2, 4, \ldots$. However, the total wave function of the final state, which consists of two identical fermions, has to be exchange-antisymmetric. Now as l is even, i.e., the orbital wave function is exchange-symmetric, the spin wave function has to be exchange-antisymmetric. Hence the two neutrons must be in a singlet spin state.

3031

A negatively charged π-meson (a pseudoscalar particle: zero spin and odd parity) is initially bound in the lowest-energy Coulomb wave function around a deuteron. It is captured by the deuteron (a proton and neutron in $^3S_1$ state), which is converted into a pair of neutrons:

$$\pi^- + d \to n + n.$$

(a) What is the orbital angular momentum of the neutron pair?

(b) What is their total spin angular momentum?

(c) What is the probability for finding both neutron spins directed opposite the spin of the deuteron?

(d) If the deuteron's spin is initially 100% polarized in the $\mathbf{k}$ direction, what is the angular dependence of the neutron emission probability (per unit solid angle) for a neutron whose spin is opposite to that of the initial deuteron? (See Fig. 3.5) You may find some of the first few (not normalized) spherical harmonics useful:

$$Y_0^0 = 1,$$

$$Y_1^{\pm 1} = \mp \sin \theta e^{\pm i\phi},$$

$$Y_1^0 = \cos \theta,$$

$$Y_2^{\pm 1} = \mp \sin 2\theta e^{\pm i\phi}.$$

(CUSPEA)

Solution:

(a) As $J^P(d) = 1^+$, $J^P(\pi^-) = 0^-$, $J^P(n) = \frac{1}{2}^+$, angular momentum conservation demands $J = 1$, parity conservation demands $(+1)^2(-1)^L$

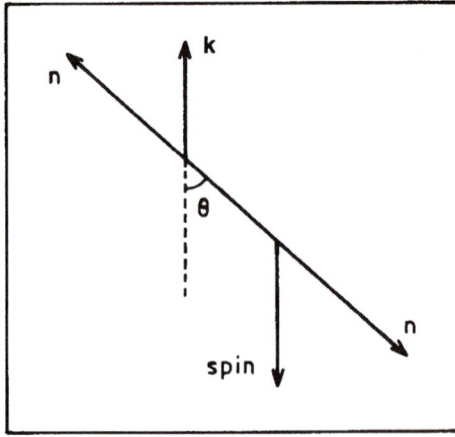

Fig. 3.5

$$= (-1)(+1)(-1)^0, \text{ or } (-1)^L = -1 \text{ for the final state. As neutrons are}$$
fermions the total wave function of the final state is antisymmetric. Thus
$(-1)^L(-1)^{S+1} = -1$, and $L + S$ is an even number. For a two-neutron
system $S = 0, 1$. If $S = 0$, then $L = 0, 2, 4, \ldots$. But this would mean
$(-1)^L = +1$, which is not true. If $S = 1$, the $L = 1, 3, 5, \ldots$, which satisfies
$(-1)^L = -1$. Now if $L \geq 3$, then J cannot be 1. Hence the neutron pair
has $L = 1$.

(b) The total spin angular momentum is $S = 1$.

(c) If the neutrons have spins opposite to the deuteron spin, $S_z = -\frac{1}{2} -$
$\frac{1}{2} = -1$. Then $J_z = L_z + S_z = L_z - 1$. As $L = 1$, $L_z = 0, \pm 1$. In either
case, $|\langle 1, L_z - 1|1, 1\rangle|^2 = 0$, i.e. the proability for such a case is zero.

(d) The wave function for the neutron-neutron system is

$$\Psi = |1, 1\rangle = C_1 Y_1^1 \chi_{10} + C_2 Y_1^0 \chi_{11},$$

where C_1, C_2 are constants such that $|C_1|^2 = |C_2|^2 = 1/2$, and

$$\chi_{10} = \frac{1}{\sqrt{2}}(\uparrow\downarrow + \downarrow\uparrow), \qquad \chi_{11} = (\uparrow\uparrow).$$

From the symmetry of the above wave function and the normalization condition, we get

$$\frac{dP}{d\Omega} = |C_1|^2 (Y_1^1 \chi_{10})^* (Y_1^1 \chi_{10})$$

$$= \frac{1}{2} (Y_1^1)^* Y_1^1$$

$$= \frac{3}{8\pi} \sin^2 \theta .$$

3032

(a) The η^0-particle can be produced by s-waves in the reaction

$$\pi^- + p \to \eta^0 + n .$$

(Note no corresponding process $\pi^- + p \to \eta^- + p$ is observed)

(b) In the η^0 decay the following modes are observed, with the probabilities as indicated:

$$\eta^0 \to 2\gamma (38\% \text{ of total})$$

$$\to 3\pi (30\% \text{ of total})$$

$$\to 2\pi (< 0.15\% \text{ of total}) .$$

(c) The rest mass of the η^0 is 548.8 MeV.

Describe experiments/measurements from which the above facts (a) (b) (c) may have been ascertained. On the basis of these facts show, as precisely as possible, how the spin, isospin, and charge of the η^0 can be inferred.

(*Columbia*)

Solution:

An experiment for this purpose should consist of a π^- beam with variable momentum, a hydrogen target, and a detector system with good spatial and energy resolutions for detecting γ-rays and charged particles. The π^- momentum is varied to obtain more 2γ and 3π events. The threshold energy E_0 of the reaction is given by

$$(E_0 + m_p)^2 - P_0^2 = (m_\eta + m_n)^2 .$$

where P_0 is the threshold momentum of the incident π^-, or

$$E_0 = \frac{(m_\eta + m_n)^2 - m_p^2 - m_\pi^2}{2m_p}$$

$$= \frac{(0.5488 + 0.94)^2 - 0.938^2 - 0.14^2}{2 \times 0.938}$$

$$= 0.702 \text{ GeV} = 702 \text{ MeV},$$

giving

$$P_0 = \sqrt{E_0^2 - m_\pi^2} \approx 0.688 \text{ GeV}/c = 688 \text{ MeV}/c.$$

Thus η^0 can be produced only if the π^- momentum is equal to or larger than 688 MeV/c.

Suppose the center of mass of the $\pi^- p$ system moves with velocity $\beta_c c$ and let $\gamma_c = (1 - \beta_c^2)^{-\frac{1}{2}}$. Indicate quantities in the center-of-mass system (cms) by a bar. Lorentz transformation gives

$$\bar{P}_0 = \gamma_c(P_0 + \beta_c E_0).$$

As $\bar{P}_0 = \bar{P}_p = m_p \gamma_c \beta_c$, we have

$$\beta_c = \frac{P_0}{m_p + E_0} = \frac{688}{702 + 938} = 0.420,$$

$$\gamma_c = 1.10,$$

and hence

$$\bar{P}_0 = \gamma_c(P_0 - \beta_c E_0) = 433 \text{ Mev}/c.$$

The de Broglie wavelength of the incident π^- meson in cms is

$$\lambda = \frac{\hbar c}{\bar{P}_0 C} = \frac{197 \times 10^{-13}}{433} = 0.45 \times 10^{-13} \text{ cm}.$$

As the radius of proton $\approx 0.5 \times 10^{-13}$ cm, s-waves play the key role in the $\pi^- p$ interaction.

Among the final products, we can measure the invariant-mass spectrum of 2γ's. If we find an invariant mass peak at 548.8 MeV, or for 6γ events, 3 pairs of γ's with invariant mass peaking at m_π^0, or the total invariant mass of 6 γ's peaking at 548.8 MeV, we can conclude that η^0 particles have been created. One can also search for $\pi^+ \pi^- \pi^0$ events. All these show the

occurrence of

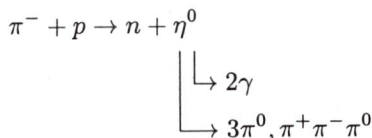

$$\pi^- + p \to n + \eta^0$$

$$\llcorner \to 2\gamma$$

$$\llcorner \to 3\pi^0, \pi^+\pi^-\pi^0$$

If the reaction $\pi^- + p \to p + \eta^-$ did occur, one would expect η^- to decay via the process

$$\eta^- \to \pi^+\pi^-\pi^- .$$

Experimentally no $\pi^+\pi^-\pi^-$ events have been observed.

The quantum numbers of η^0 can be deduced as follows.

Spin: As η^0 can be produced using s-waves, conservation of angular momentum requires the spin of η^0 to be either 0 or 1. However since a vector meson of spin 1 cannot decay into 2 γ's, $J(\eta^0) = 0$.

Parity: The branching ratios suggest η^0 can decay via electromagnetic interaction into 2 γ's, via strong interaction into 3 π's, but the branching ratio of 2π-decay is very small. From the 3π-decay we find

$$P(\eta^0) = P^3(\pi)(-1)^{l+l'} ,$$

where l and l' are respectively the orbital angular momentum of a 2π system and the relative orbital angular momentum of the third π relative to the 2π system. As $J(\eta^0) = 0$, conservation of total angular momentum requires $l' = -l$ and so

$$P(\eta^0) = (-1)^3 = -1 .$$

Isospin: Because η^- is not observed, η^0 forms an isospin singlet. Hence $I(\eta^0) = 0$.

Charge: Conservation of charge shows $Q(\eta^0) = 0$. In addition, from the 2γ-decay channel we can further infer that $C(\eta^0) = +1$.

To summarize, the quantum numbers of η^0 are $I(\eta^0) = 0$, $Q(\eta^0) = 0$, $J^{PC}(\eta^0) = 0^{-+}$. Like π and K mesons, η^0 is a pseudoscalar meson, and it forms an isospin singlet.

3033

A beam of K^+ or K^- mesons enters from the left a bubble chamber to which a uniform magentic field of $B \approx 12$ kGs is applied perpendicular to the observation window.

(a) Label with symbols (π^+, π^-, p, etc.) all the products of the decay of the K^+ in the bubble chamber pictures in Fig. 3.6 and give the complete reaction equation for K^+ applicable to each picture.

(b) In Fig. 3.7 the K^- particles come to rest in the bubble chamber. Label with symbols all tracks of particles associated with the K^- particle and identify any neutral particle by a dashed-line "track". Give the complete reaction equation for the K^- interaction applicable to each picture.

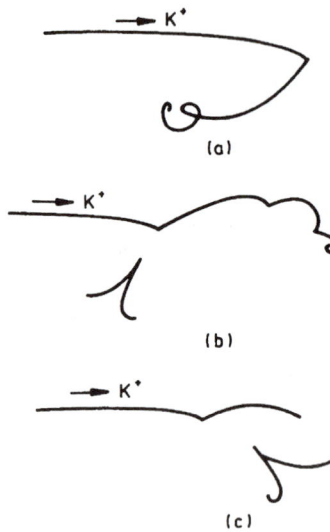

Fig. 3.6

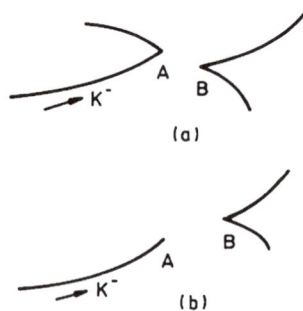

Fig. 3.7

(c) Assuming that tracks in Fig. 3.7(a) and Fig. 3.7(b) above all lie in the plane of the drawing determine the expressions for the lifetime of the neutral particle and its mass.

(*Chicago*)

Solution:

(a) The modes and branching ratios of K^+ decay are as follows:

$$K^+ \rightarrow \mu^+ \nu_\mu \quad 63.50\%,$$
$$\pi^+ \pi^0 \quad 21.16\%,$$
$$\pi^+ \pi^+ \pi^- \quad 5.59\%,$$
$$\pi^+ \pi^0 \pi^0 \quad 1.73\%,$$
$$\mu^+ \nu_\mu \pi^0 \quad 3.20\%,$$
$$e^+ \nu_e \pi^0 \quad 4.82\%.$$

The products from decays of K^+ consist of three kinds of positively charged particle π^+, μ^+, e^+, one kind of negatively charged particle π^-, plus some neutral particles π^0, ν_μ, ν_e. Where π^+ is produced, there should be four linearly connected tracks of positively charged particles arising from $K^+ \rightarrow \pi^+ \rightarrow \mu^+ \rightarrow e^+$. Where μ^+ or e^+ is produced there should be three or two linearly connected tracks of positively charged particles in the picture arising from $K^+ \rightarrow \mu^+ \rightarrow e^+$ or $K^+ \rightarrow e^+$, respectively. Where π^0 is produced, because of the decay $\pi^0 \rightarrow 2\gamma (\tau \approx 10^{-16} \text{ s})$ and the subsequent electron-positron pair production of the γ-rays, we can see the e^+, e^- tracks starting out as a fork.

Analysing Fig. 3.6(a) we have Fig. 3.8. The decay of K^+ could produce either $\mu^+ \nu$ or $\mu^+ \gamma \pi^0$. As the probability is much larger for the former we assume that it was what actually happened. Then the sequence of events is as follows:

$$K^+ \rightarrow \mu^+ + \nu_\mu$$
$$\downarrow$$
$$e^+ \bar{\nu}_\mu \nu_e$$
$$\mathrel{\rightarrow} e^+ + e^- \rightarrow \gamma_1 + \gamma_2$$

Note the sudden termination of the e^+ track, which is due to the annihilation of the positron with an electron of the chamber producing two oppositely directed γ-rays.

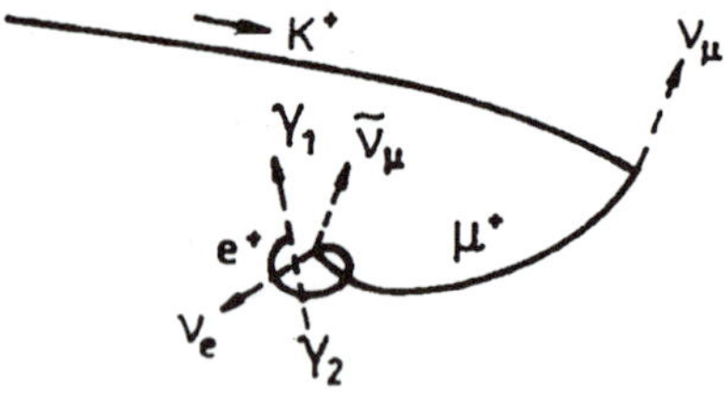

Fig. 3.8

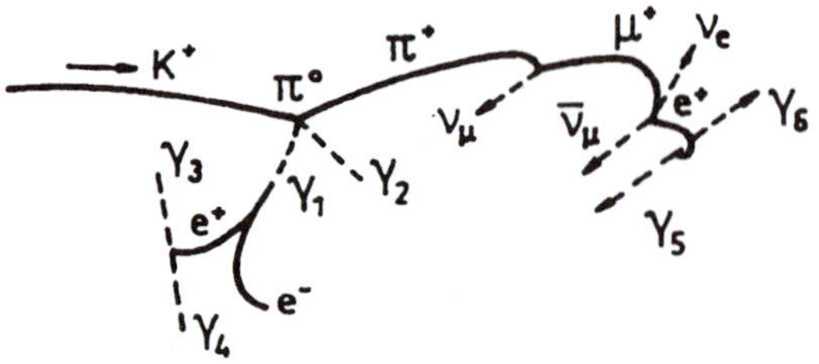

Fig. 3.9

Analysing Fig. 3.6(b) we have Fig. 3.9. The sequence of events is as follows:

$$K^+ \to \pi^0 + \pi^+$$
$$\pi^+ \longrightarrow \mu^+ + \nu_\mu$$
$$\mu^+ \longrightarrow e^+ + \nu_e + \bar{\nu}_\mu\,, \quad e^+ + e^- \to \gamma_5 + \gamma_6$$
$$\pi^0 \longrightarrow \gamma_1 + \gamma_2$$
$$\gamma_1 \longrightarrow e^+ + e^-, \quad e^+ + e^- \to \gamma_3 + \gamma_4$$

or

$$K^+ \to \pi^0 + \pi^0 + \pi^+$$
$$\pi^+ \longrightarrow \mu^+ + \nu_\mu$$
$$\pi^0 \longrightarrow \gamma + \gamma$$
$$\pi^0 \longrightarrow \gamma + \gamma$$

with the subsequent μ^+ decay and pair production of the γ-rays.

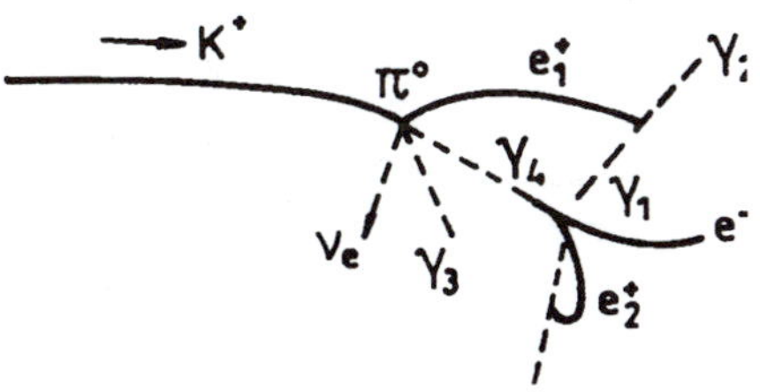

Fig. 3.10

Note that because of its short lifetime, π^0 decays almost immediately as it is produced. From Fig. 3.6(c) we have Fig. 3.10.

The sequence of events is as follows:

$$K^+ \to \nu_e + e_1^+ + \pi^0$$

$$\quad\quad\quad\quad\quad \longmapsto \gamma_3 + \gamma_4$$

$$\quad\quad\quad\quad\quad\quad\quad\quad\quad \longmapsto e_2^+ + e^-, \quad e_2^+ + e^- \to \gamma_5 + \gamma_6$$

$$\quad\quad\quad \longmapsto e_1^+ + e^- \to \gamma_1 + \gamma_2$$

Figure 3.7(a) is interpreted as follows:

$$K^- + n \to \Lambda^0 + \pi^-$$

$$\quad\quad\quad\quad\quad \longmapsto p + \pi^-$$

The tracks are labelled in Fig. 3.11 below:

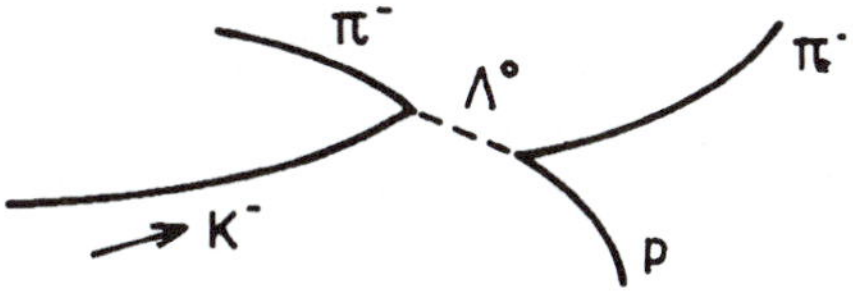

Fig. 3.11

Figure 3.7(b), is interpreted as

$$K^- + p \to \Lambda^0 + \pi^0$$

$$\quad\quad\quad\quad\quad \longmapsto \gamma + \gamma$$

$$\quad\quad\quad \longmapsto p + \pi^-$$

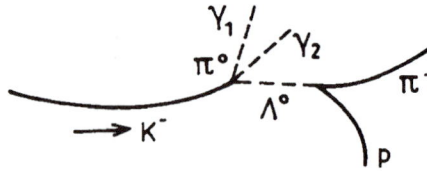

Fig. 3.12

Figure 3.12 shows the tracks with labels. Note that Λ^0 has a lifetime $\sim 10^{-10}$ s, sufficient to travel an appreciable distance in the chamber.

(c) To determine the mass and lifetime of the neutral particle Λ^0, we measure the length of the track of the neutral particle and the angles it makes with the tracks of p and π^-, θ_p and θ_π, and the radii of curvature, R_p and R_π, of the tracks of p and π^-. Force considerations give the momentum of a particle of charge e moving perpendicular to a magnetic field of flux density B as

$$P = eBR,$$

where R is the radius of curvature of its track. With e in C, B in T, R in m, we have

$$P = eBRc \left(\frac{\text{joule}}{c}\right) = \left(\frac{1.6 \times 10^{-19} \times 3 \times 10^8}{1.6 \times 10^{-19} \times 10^9}\right) BR \left(\frac{\text{GeV}}{c}\right)$$

$$= 0.3BR \left(\frac{\text{GeV}}{c}\right).$$

The momenta P_p, P_π of p and π^- from Λ^0 decay can then be determined from the radii of curvature of their tracks.

As $(\Sigma E)^2 - (\Sigma P)^2$ is invariant, we have

$$m_\Lambda^2 = (E_p + E_\pi)^2 - (\mathbf{P}_p + \mathbf{P}_\pi)^2,$$

where m_Λ is the rest mass of Λ^0.

As

$$E_p^2 = P_p^2 + m_p^2,$$

$$E_\pi^2 = P_\pi^2 + m_\pi^2,$$

we have

$$m_\Lambda = \sqrt{m_p^2 + m_\pi^2 + 2E_pE_\pi - 2P_\pi P_p \cos(\theta_p + \theta_\pi)}.$$

The energy and momentum of the Λ^0 particle are given by

$$E_\Lambda = E_p + E_\pi,$$

$$P_\Lambda = P_p \cos\theta_p + P_\pi \cos\theta_\pi.$$

If the path length of Λ is l, its laboratory lifetime is $\tau = \frac{l}{\beta c}$, and its proper lifetime is

$$\tau_0 = \frac{l}{\gamma\beta\tau} = \frac{lm_\Lambda}{P_\Lambda} = l(P_p\cos\theta_p + P_\pi\cos\theta_\pi)^{-1}$$
$$\times [m_p^2 + m_\pi^2 + 2E_pE_\pi - 2P_\pi P_p \cos(\theta_p + \theta_\pi)]^{1/2}.$$

3034

The invariant-mass spectrum of Λ^0 and π^+ in the reaction $K^- + p \rightarrow \Lambda^0 + \pi^+ + \pi^-$ shows a peak at 1385 MeV with a full width of 50 MeV. It is called Y_1^*. The $\Lambda^0\pi^-$ invariant-mass spectrum from the same reaction (but different events) shows a similar peak.

(a) From these data determine the strangeness, hypercharge and isospin of Y_1^*.

(b) Evidence indicates that the product $\Lambda^0 + \pi^+$ from a Y_1^* is in a relative p state of angular momentum. What spin assignments J are possible for the Y_1^*? What is its intrinsic parity? (Hint: the intrinsic parity of Λ^0 is $+$ and that of π^+ is $-$)

(c) What (if any) other strong decay modes do you expect for Y_1^*?

(*Columbia*)

Solution:

(a) The resonance state Y_1^* with full width $\Gamma = 50$ MeV has a lifetime $\tau = \hbar/\Gamma = 6.6 \times 10^{-22}/50 = 1.3 \times 10^{-23}$ s. The time scale means that Y_1^* decays via strong interaction, and so the strangeness number S, hypercharge

Y, isospin I and its z-component I_3 are conserved. Hence

$$S(Y_1^*) = S(\Lambda^0) + S(\pi^+) = -1 + 0 = -1,$$

$$Y(Y_1^*) = Y(\Lambda^0) + Y(\pi^+) = 0 + 0 = 0,$$

$$I(Y_1^*) = I(\Lambda^0) + I(\pi^+) = 0 + 1 = 1,$$

$$I_3(Y_1^*) = I_3(\Lambda^0) + I_3(\pi^+) = 0 + 1 = 1.$$

Y_1^* is actually an isospin triplet, its three states being Y_1^{*+}, Y_1^{*0}, and Y_1^{*-}. The resonance peak of $\Lambda^0 \pi^-$ corresponds to Y_1^{*-}.

(b) Λ^0 has spin $J_\Lambda = 1/2$, π^+ has spin $J_\pi = 0$. The relative motion is a p state, so $l = 1$. Then $J_{Y_1^*} = 1/2 + 1$, the possible values being $1/2$ and $3/2$. The intrinsic parity of Y_1^* is $P(Y_1^*) = P(\pi)P(\Lambda)(-1)^l = (-1)(1)(-1) = 1$.

(c) Another possible strong decay channel is

$$Y_1^* \to \Sigma \pi.$$

As the intrinsic parity of Σ is $(+1)$, that of π, (-1), the particles emitted are in a relative p state

3035

Consider the hyperon nonleptonic weak decays:

$$\Lambda^0 \to p\pi^-$$

$$\Lambda^0 \to n\pi^0$$

$$\Sigma^- \to n\pi^-$$

$$\Sigma^+ \to p\pi^0$$

$$\Sigma^+ \to n\pi^+$$

$$\Xi^- \to \Lambda^0 \pi^-$$

$$\Xi^0 \to \Lambda^0 \pi^0$$

On assuming that these $\Delta S = 1$ weak decays satisfy the $\Delta I = 1/2$ rule, use relevant tables to find the values of x, y, z, as defined below:

$$x = \frac{A(\Lambda^0 \to p\pi^-)}{A(\Lambda^0 \to n\pi^0)},$$

$$y = \frac{A(\Sigma^+ \to \pi^+ n) - A(\Sigma^- \to \pi^- n)}{A(\Sigma^+ \to \pi^0 p)},$$

$$z = \frac{A(\Xi^0 \to \Lambda^0 \pi^0)}{A(\Xi^- \to \Lambda^0 \pi^-)},$$

where A denotes the transition amplitude.

(Columbia)

Solution:

As nonleptonic decays of hyperon require $\Delta I = 1/2$, we can introduce an "imaginary particle" a having $I = \frac{1}{2}$, $I_3 = -\frac{1}{2}$, and combine the hyperon with a in isospin compling:

$$|\Lambda^0, a\rangle = |0,0\rangle \left|\frac{1}{2}, -\frac{1}{2}\right\rangle = \left|\frac{1}{2}, -\frac{1}{2}\right\rangle,$$

$$|\Sigma^-, a\rangle = |1,-1\rangle \left|\frac{1}{2}, -\frac{1}{2}\right\rangle = \left|\frac{3}{2}, -\frac{3}{2}\right\rangle,$$

$$|\Sigma^+, a\rangle = |1,1\rangle \left|\frac{1}{2}, -\frac{1}{2}\right\rangle = \sqrt{\frac{1}{3}}\left|\frac{3}{2}, \frac{1}{2}\right\rangle + \sqrt{\frac{2}{3}}\left|\frac{1}{2}, \frac{1}{2}\right\rangle,$$

$$|\Xi^0, a\rangle = \left|\frac{1}{2}, \frac{1}{2}\right\rangle \left|\frac{1}{2}, -\frac{1}{2}\right\rangle = \sqrt{\frac{1}{2}}|1,0\rangle + \sqrt{\frac{1}{2}}|0,0\rangle,$$

$$|\Xi^-, a\rangle = \left|\frac{1}{2}, -\frac{1}{2}\right\rangle \left|\frac{1}{2}, -\frac{1}{2}\right\rangle = |1,-1\rangle.$$

Similarly, we find the isospin wave functions for the final states:

$$|\pi^-, p\rangle = |1,-1\rangle \left|\frac{1}{2}, \frac{1}{2}\right\rangle = \sqrt{\frac{1}{3}}\left|\frac{3}{2}, -\frac{1}{2}\right\rangle - \sqrt{\frac{2}{3}}\left|\frac{1}{2}, -\frac{1}{2}\right\rangle,$$

$$|\pi^0, p\rangle = |1,0\rangle \left|\frac{1}{2}, \frac{1}{2}\right\rangle = \sqrt{\frac{2}{3}}\left|\frac{3}{2}, \frac{1}{2}\right\rangle - \sqrt{\frac{1}{3}}\left|\frac{1}{2}, \frac{1}{2}\right\rangle,$$

$$|\pi^+, n\rangle = |1, 1\rangle \left|\frac{1}{2}, -\frac{1}{2}\right\rangle = \sqrt{\frac{1}{3}}\left|\frac{3}{2}, \frac{1}{2}\right\rangle + \sqrt{\frac{2}{3}}\left|\frac{1}{2}, \frac{1}{2}\right\rangle,$$

$$|\pi^0, n\rangle = |1, 0\rangle \left|\frac{1}{2}, -\frac{1}{2}\right\rangle = \sqrt{\frac{2}{3}}\left|\frac{3}{2}, -\frac{1}{2}\right\rangle + \sqrt{\frac{1}{3}}\left|\frac{1}{2}, -\frac{1}{2}\right\rangle,$$

$$|\pi^-, n\rangle = |1, -1\rangle \left|\frac{1}{2}, -\frac{1}{2}\right\rangle = \left|\frac{3}{2}, -\frac{3}{2}\right\rangle,$$

$$|\Lambda^0, \pi^0\rangle = |0, 0\rangle|1, 0\rangle = |1, 0\rangle,$$

$$|\Lambda^0, \pi^-\rangle = |0, 0\rangle|1, -1\rangle = |1, -1\rangle.$$

The coefficients have been obtained from Clebsch–Gordan tables. The transition amplitudes are thus

$$A_1(\Lambda^0 \to n\pi^0) = \sqrt{\frac{1}{3}}M_{1/2}$$

$$A_2(\Lambda^0 \to p\pi^-) = -\sqrt{\frac{2}{3}}M_{1/2},$$

with

$$M_{1/2} = \left\langle \frac{1}{2}\left|H_w\right|\frac{1}{2}\right\rangle.$$

Hence

$$x = \frac{A_2}{A_1} = -\sqrt{2}.$$

Similarly,

$$A_3(\Sigma^- \to \pi^- n) = M_{3/2},$$

$$A_4(\Sigma^+ \to \pi^0 p) = \sqrt{\frac{1}{3}}\sqrt{\frac{2}{3}}M_{3/2} - \sqrt{\frac{2}{3}}\sqrt{\frac{1}{3}}M_{1/2} = \frac{\sqrt{2}}{3}(M_{3/2} - M_{1/2}),$$

$$A_5(\Sigma^+ \to \pi^+ n) = \sqrt{\frac{1}{3}}\sqrt{\frac{1}{3}}M_{3/2} + \sqrt{\frac{2}{3}}\sqrt{\frac{2}{3}}M_{1/2} = \frac{1}{3}(M_{3/2} + 2M_{1/2}),$$

with

$$M_{3/2} = \left\langle \frac{3}{2}\left|H_\omega\right|\frac{3}{2}\right\rangle.$$

Hence

$$y = \frac{A_5 - A_3}{A_4} = \frac{M_{3/2} + 2M_{1/2} - 3M_{3/2}}{\sqrt{2}(M_{3/2} - M_{1/2})} = -\sqrt{2}.$$

Also,

$$A_6(\Xi^0 \to \Lambda^0 \pi^0) = \sqrt{\frac{1}{2}} M_1,$$

$$A_7(\Xi^- \to \Lambda^0 \pi^-) = M_1$$

with

$$M_1 = \langle 1 | H_\omega | 1 \rangle.$$

Hence

$$z = \frac{A_6}{A_7} = \frac{1}{\sqrt{2}}.$$

3036

(a) The principle of detailed balance rests on the validity of time reversal invariance and serves to relate the cross section for a given reaction $a + b \to c + d$ to the cross section for the inverse reaction $c + d \to a + b$. Let $\sigma_I(W)$ be cross section for

$$\gamma + p \to \pi^+ + n$$

at total center-of-mass energy W, where one integrates over scattering angle, sums over final spins, and averages over initial spins. Let $\sigma_{II}(W)$ be the similarly defined cross section, at the same center-of-mass energy, for

$$\pi^+ + n \to \gamma + p.$$

Let μ be the pion mass, m the nucleon mass (neglect the small difference between the n and p masses). Given $\sigma_I(W)$, what does detailed balance predict for $\sigma_{II}(W)$?

(b) For reaction II, what is the threshold value W_{thresh} and how does $\sigma_{II}(W)$ vary with W just above threshold?

(Princeton)

Solution:

(a) For simplicity denote the state (a,b) by α and the state (c,d) by β. Let $\sigma_{\alpha\beta}$ be the cross section of the process

$$a + b \to c + d$$

and $\sigma_{\beta\alpha}$ be the cross section of the inverse process

$$c + d \to a + b.$$

If T invariance holds true, then when the forward and inverse reactions have the same energy W in the center-of-mass frame, $\sigma_{\alpha\beta}$ and $\sigma_{\beta\alpha}$ are related by

$$\frac{\sigma_{\alpha\beta}}{\sigma_{\beta\alpha}} = \frac{P_\beta^2(2I_c + 1)(2I_d + 1)}{P_\alpha^2(2I_a + 1)(2I_b + 1)},$$

which is the principle of detailed balance. Here P_α is the relative momentum of the incident channel of the reaction $a + b \to c + d$, P_β is the relative momentum of the incident channel of the inverse reaction, I_a, I_b, I_c, I_d are respectively the spins of a, b, c, d.

For the reaction $\gamma + p \to \pi^+ + n$, in the center-of-mass frame of the incident channel let the momentum of the γ be P_γ, the energy of the proton be E_p. Then $W = E_\gamma + E_p$. As the γ has zero rest mass,

$$E_\gamma^2 - P_\gamma^2 = 0,$$

or

$$(W - E_p)^2 - P_\gamma^2 = 0.$$

With $P_\gamma = P_p$, $E_p^2 - P_p^2 = m^2$,

$$E_p = \frac{W^2 + m^2}{2W}.$$

Hence the relative momentum is

$$P_\alpha^2 = P_\gamma^2 = E_p^2 - m^2 = \frac{W^2 - m^2}{2W}.$$

For the inverse reaction $\pi^+ + n \to \gamma + p$, in the center-of-mass frame let the energy of π^+ be E_π, its momentum be P_π, and the energy of the neutron be E_n, then as $W = E_\pi + E_n$,

$$(W - E_n)^2 - E_\pi^2 = 0.$$

With $P_\pi = P_n$, $E_n^2 = P_\pi^2 + m^2$, $E_\pi^2 = P_\pi^2 + \mu^2$, we have

$$E_n = \frac{W^2 + m^2 - \mu^2}{2W},$$

and hence

$$P_\beta^2 = P_\pi^2 = E_n^2 - m^2 = \frac{(W^2 + m^2 - \mu^2)^2 - 4W^2m^2}{4W^2}.$$

We have $I_\gamma = 1$, $I_p = 1/2$, $I_n = 1/2$, $I_\pi = 0$. However as photon has only left and right circular polarizations, $2I_\gamma + 1$ should be replaced by 2. Hence

$$\frac{\sigma_I(W)}{\sigma_{II}(W)} = \frac{P_\beta^2(2I_\pi + 1)(2I_n + 1)}{P_\alpha^2(2I_\gamma + 1)(2I_p + 1)} = \frac{P_\beta^2}{2P_\alpha^2},$$

or

$$\sigma_{II}(W) = \frac{(W^2 - m^2)^2}{(W^2 + m^2 - \mu^2)^2 - 4W^2m^2}\sigma_I(W).$$

(b) At threshold all the final particles are produced at rest in the center-of-mass frame. The energy of the center of mass is $W^{th*} = m + \mu$. In the laboratory let the energy of the photon be E_γ. As the proton is at rest, at the threshold

$$(E_\gamma + m)^2 - P_\gamma^2 = (m + \mu)^2,$$

or, since $E_\gamma = P_\gamma$,

$$E_\gamma^{th} = \mu\left(1 + \frac{\mu}{2m}\right) = 150 \text{ MeV}.$$

When $E_\gamma > E_\gamma^{th}$, $\sigma(\gamma + p \to \pi^+ + n)$ increases rapidly with increasing E_γ. When $E_\gamma = 340$ MeV, a wide resonance peak appears, corresponding to an invariant mass

$$E^* = \sqrt{(E_\gamma + m_p)^2 - P_\gamma^2} = \sqrt{2m_p E_\gamma + m_p^2} = 1232 \text{ MeV}.$$

It is called the Δ particle. The width $\Gamma = 115$ MeV and $\sigma \approx 280$ μb at the peak.

3037

The following questions require rough, qualitative, or magnitude answers.

(a) How large is the cross section for $e^+e^- \to \mu^+\mu^-$ at a center-of-mass energy of 20 GeV? How does it depend on energy?

(b) How large is the neutrino-nucleon total cross section for incident neutrinos of 100 GeV (in the nucleon rest frame)? How does it depend on energy? At what energy is this energy dependence expected to change, according to the Weinberg–Salam theory?

(c) How long is the lifetime of the muon? Of the tau lepton? If a new lepton is discovered ten times heavier than tau, how long-lived is it expected to be, assuming it decays by the same mechanism as the muon and tau?

(d) How large is the nucleon-nucleon total cross section at accelerator energies?

(e) In pion-nucleon elastic scattering, a large peak is observed in the forward direction (scattering through small angles). A smaller but quite distinct peak is observed in the backward direction (scattering through approximately 180° in the center-of-mass frame). Can you explain the backward peak? A similar backward peak is observed in K^+p elastic scattering; but in K^-p scattering it is absent. Can you explain this?

(Princeton)

Solution:

(a) The energy dependence of the cross section for $e^+e^- \to \mu^+\mu^-$ can be estimated by the following method. At high energies $s^{\frac{1}{2}} \gg m_e, m_\mu$, where $s = E_{cm}^2$, and we can take $m_e \approx m_\mu \approx 0$. As there are two vertexes in the lowest order electromagnetic interaction, we have

$$\sigma = f(s)\alpha^2.$$

where α is the fine structure constant $\frac{e^2}{\hbar c} = \frac{1}{137}$. Dimensionally $\sigma = [M]^{-2}$, $s = [M]^2$, $\alpha = [0]$, and so

$$f(s) \approx \frac{1}{s},$$

or

$$\sigma \approx \frac{\alpha^2}{s}.$$

A calculation using quantum electrodynamics without taking account of radiation correction gives

$$\sigma \approx \frac{4\pi\alpha^2}{3s}.$$

At $E_{cm} = 20$ GeV,

$$\sigma = \frac{4\pi\alpha^2}{3 \times 20^2} = 5.6 \times 10^{-7} \text{ GeV}^{-2} = 2.2 \times 10^{-34} \text{ cm}^2 = 220 \text{ pb},$$

as 1 MeV$^{-1} = 197 \times 10^{-13}$ cm.

(b) We can estimate the neutrino-nucleon total cross section in a similar manner. In the high energy range $s^{\frac{1}{2}} \gg m_p$, ν and p react by weak interaction, and

$$\sigma \approx G_F^2 f(s).$$

Again using dimensional analysis, we have $G_F = [M]^{-2}$, $s = [M]^2$, $\sigma = [M]^{-2}$, and so $f(s) = [M]^2$, or

$$f(s) \approx s,$$

i.e.,

$$\sigma \approx G_F^2 s.$$

Let the energy of the neutrino in the neutron's rest frame be E_ν. Then

$$s = (E_\nu + m_p)^2 - p_\nu^2 = m_p^2 + 2m_p E_\nu \approx 2m_p E_\nu,$$

or

$$\sigma \approx G_F^2 s \approx G_F^2 m_p E_\nu.$$

For weak interaction (**Problem 3001**)

$$G_F m_p^2 = 10^{-5}.$$

With $m_p \approx 1$ GeV, at $E_\nu = 100$ GeV.

$$\sigma \approx 10^{-10} E_\nu \text{ GeV}^{-2}$$

$$= 10^{-10} \times 10^2 \times 10^{-6} \text{ MeV}^{-2} = 10^{-14} \times (197 \times 10^{-13})^2 \text{ cm}^2$$

$$= 4 \times 10^{-36} \text{ cm}^2.$$

Experimentally, $\sigma \approx 0.6 \times 10^{-38}$ cm^2. According to the Weinberg–Salam theory, σ changes greatly in the neighborhood of $s \approx m_W^2$, where m_W is the mass of the intermediate vector boson W, 82 GeV.

(c) μ has lifetime $\tau_\mu \approx 2.2 \times 10^{-6}$ s and τ has lifetime $\tau_\tau \approx 2.86 \times 10^{-13}$ s.

Label the new lepton by H. Then $m_H = 10 m_\tau$. On assuming that it decays by the same mechanism as muon and tau, its lifetime would be

$$\tau_H = \left(\frac{m_\tau}{m_H}\right)^5 \tau_\tau \approx 10^{-5}\tau_\tau = 2.86 \times 10^{-18} \text{ s}.$$

(d) Nucleons interact by strong interaction. In the energy range of presentday accelerators the interaction cross section between nucleons is

$$\sigma_{NN} \approx \pi R_N^2 \,,$$

R_N being the radius of the nucleon. With $R_N \approx 10^{-13}$ cm,

$$\sigma_{NN} \approx 3 \times 10^{-26} \text{ cm}^2 = 30 \text{ mb} \,.$$

Experimentally, $\sigma_{pp} \approx 30 \sim 50$ mb for $E_p = 2 \sim 10 \times 10^3$ GeV,

$$\sigma_{np} \approx 30 \sim 50 \text{ mb for } E_p = 5 \sim 10 \times 10^2 \text{ GeV} \,.$$

(e) Analogous to the physical picture of electromagnetic interaction, the interaction between hadrons can be considered as proceeding by exchanging virtual hadrons. Any hadron can be the exchanged particle and can be created by other hadrons, so all hadrons are equal. It is generally accepted that strong interaction arises from the exchange of a single particle, the effect of multiparticle exchange being considered negligible. This is the single-particle exchange model.

Figure 3.13(a) shows a t channel, where $t = -(P_{\pi^+} - P_{\pi^+{'}})^2$ is the square of the 4-momentum transfer of π^+ with respect to $\pi^{+'}$. Figure 3.13(b) shows

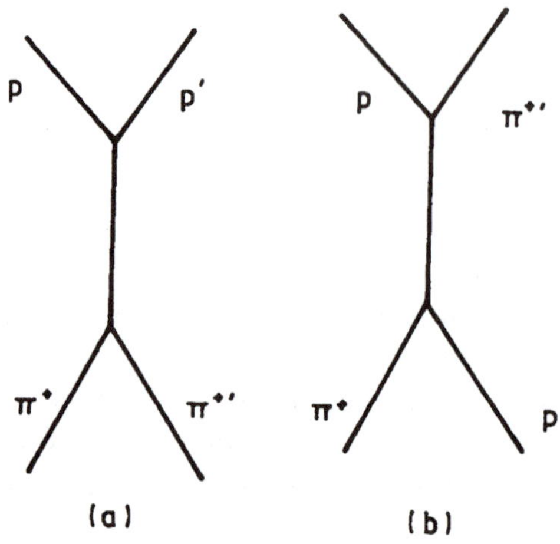

Fig. 3.13

a u channel, where $u = -(P_{\pi^+} - P_{p'})^2$ is the square of the 4-momentum transfer of π^+ with respect to p'. Let θ be the angle of the incident π^+ with respect to the emergent π^+. When $\theta = 0$, $|t|$ is very small; when $\theta = 180^0$, $|u|$ is very small. The former corresponds to the π^+ being scattered forwards and the latter corresponds to the π^+ being scattered backwards. As quantum numbers are conserved at each vertex, for the t channel the virtual exchange particle is a meson, for the u channel it is a baryon. This means that there is a backward peak for baryon-exchange scattering. Generally speaking, the amplitude for meson exchange is larger. Hence the forward peak is larger. For example, in $\pi^+ p$ scattering there is a u channel for exchanging n, and so there is a backward peak. In $K^+ p$ scattering, a virtual baryon ($S = -1, B = 1$) or Λ^0 is exchanged. But in $K^- p$ scattering, if there is a baryon exchanged, it must have $S = 1, B = 1$. Since there is no such a baryon, $K^- p$ scattering does not have a backward peak.

2. WEAK AND ELECTROWEAK INTERACTIONS, GRAND UNIFICATION THEORIES(3038–3071)

3038

Consider the leptonic decays:

$$\mu^+ \to e^+ \nu \bar{\nu} \quad \text{and} \quad \tau^+ \to e^+ \nu \bar{\nu}$$

which are both believed to proceed via the same interaction.

(a) If the μ^+ mean life is 2.2×10^{-6} s, estimate the τ^+ mean life given that the experimental branching ratio for $\tau^+ \to e^+ \nu \bar{\nu}$ is 16%

Note that:

$$m_\mu = 106 \text{ MeV}/c^2,$$

$$m_\tau = 1784 \text{ MeV}/c^2,$$

$$m_e = 0.5 \text{ MeV}/c^2,$$

$$m_\nu = 0 \text{ MeV}/c^2,$$

(b) If the τ^+ is produced in a colliding beam accelerator (like PEP), $e^+e^- \rightarrow \tau^+\tau^-$ at $E_{em} = 29$ GeV (e^+ and e^- have equal and opposite momenta), find the mean distance (in the laboratory) the τ^+ will travel before decay.

(UC, Berkeley)

Solution:

(a) The theory of weak interaction gives the decay probabilities per unit time as

$$\lambda_\mu = \tau_\mu^{-1} = \frac{G_\mu^2 m_\mu^5}{192\pi^3}, \qquad \lambda_\tau = \frac{G_\tau^2 m_\tau^5}{192\pi^3}.$$

As the same weak interaction constant applies, $G_\mu = G_\tau$ and

$$\lambda_\tau / \lambda_\mu = m_\tau^5 / m_\mu^5.$$

If λ is the total decay probability per unit time of τ^+, the branching ratio is $R = \lambda_\tau(\tau^+ \rightarrow e^+\nu\bar{\nu})/\lambda$.

Hence $\tau = \lambda^{-1} = R/\lambda_\tau(\tau^+ \rightarrow e^+\nu\bar{\nu}) = R\left(\frac{m_\mu}{m_\tau}\right)^5 \tau_\mu = 16\% \times \left(\frac{106}{1784}\right)^5 \times 2.2 \times 10^{-6} = 2.6 \times 10^{-13}$ s.

(b) In the center-of-mass system, τ^+ and τ^- have the same energy. Thus

$$E_\tau = E_{cm}/2 = 14.5 \text{ GeV}.$$

As the collision is between two particles of equal and opposite momenta, the center-of-mass frame coincides with the laboratory frame. Hence the laboratory Lorentz factor of τ is

$$\gamma = E_\tau/m_\tau = 14.5 \times 10^3/1784 = 8.13,$$

giving

$$\beta = \sqrt{1 - \gamma^{-2}} = \sqrt{1 - 8.13^{-2}} = 0.992.$$

Hence the mean flight length in the laboratory is

$$L = \beta c\gamma\tau = 0.992 \times 3 \times 10^{10} \times 8.13 \times 2.6 \times 10^{-13} = 6.29 \times 10^{-2} \text{ cm}.$$

3039

Assume that the same basic weak interaction is responsible for the beta decay processes $n \rightarrow pe^-\bar{\nu}$ and $\Sigma^- \rightarrow \Lambda e^-\bar{\nu}$, and that the matrix elements describing these decays are the same. Estimate the decay rate of the process $\Sigma^- \to \Lambda e^- \bar{\nu}$ given the lifetime of a free neutron is about 10^3 seconds.

Given:

$$m_n = 939.57 \text{ MeV}/c^2, \qquad m_\Sigma = 1197.35 \text{ MeV}/c^2,$$

$$m_p = 938.28 \text{ MeV}/c^2, \qquad m_\Lambda = 1116.058 \text{ MeV}/c^2,$$

$$m_e = 0.51 \text{ MeV}/c^2, \qquad m_\nu = 0.$$

(UC, Berkeley)

Solution:

β-decay theory gives the transition probability per unit time as $W = 2\pi G^2 |M|^2 dN/dE_0$ and the total decay rate as $\lambda \propto E_0^5$, where E_0 is the maximum energy of the decay neutrino. For two decay processes of the same transition matrix element and the same coupling constant we have

$$\frac{\lambda_1}{\lambda_2} = \left(\frac{E_{01}}{E_{02}} \right)^5.$$

Hence

$$\lambda(\Sigma^- \to \Lambda e \bar{\nu}) = \left[\frac{E_0(\Sigma^- \to \Lambda e^- \bar{\nu})}{E_0(n \to pe^- \bar{\nu})} \right]^5 \lambda_n$$

$$= \left(\frac{m_\Sigma - m_\Lambda - m_e}{m_n - m_p - m_e} \right)^5 \frac{1}{\tau_n}$$

$$= \left(\frac{1197.35 - 1116.058 - 0.51}{939.57 - 938.28 - 0.51} \right)^5 \times 10^{-3}$$

$$= 1.19 \times 10^7 \ s^{-1}.$$

3040

Although the weak interaction coupling is thought to be universal, different weak processes occur at vastly different rates for kinematics reasons.

(a) Assume a universal V-A interaction, compute (or estimate) the ratio of rates:

$$\gamma = \frac{\Gamma(\pi^- \to \mu^- \bar{\nu})}{\Gamma(\pi^- \to e^- \bar{\nu})} \, .$$

Be as quantitative as you can.

(b) How would this ratio change (if the universal weak interaction coupling were scalar? Pseudoscalar?

(c) What would you expect (with V-A) for

$$\gamma' = \frac{\Gamma(\Lambda \to p\mu^- \bar{\nu})}{\Gamma(\Lambda \to pe^- \bar{\nu})} \, .$$

Here a qualitative answer will do.

Data:

$$J^P(\pi^-) = 0^-; \qquad M_\Lambda = 1190 \text{ MeV}/c^2;$$

$$M_\mu = 105 \text{ MeV}/c^2; \qquad M_e = 0.5 \text{ MeV}/c^2; \qquad M_p = 938 \text{ MeV}/c^2 \, .$$

(*Princeton*)

Solution:

(a) The weak interaction reaction rate is given by

$$\Gamma = 2\pi G^2 |M|^2 \frac{dN}{dE_0} \, ,$$

where $\frac{dN}{dE_0}$ is the number of the final states per unit energy interval, M is the transition matrix element, G is the weak interaction coupling constant.

Consider the two decay modes of π^-:

$$\pi^- \to \mu^- \bar{\nu}_\mu , \qquad \pi^- \to e^- \bar{\nu} \, .$$

Each can be considered as the interaction of four fermions through an intermediate nucleon-antinucleon state as shown in Fig. 3.14:

$$\pi^- \xrightarrow{\text{Strong-interaction}} \bar{p} + n \xrightarrow{\text{Weak-interaction}} e^- + \bar{\nu}_e \text{ or } \mu^- + \bar{\nu}_\mu \, .$$

From a consideration of parities and angular momenta, and basing on the $V-A$ theory, we can take the coupling to be of the axial vector (A) type.

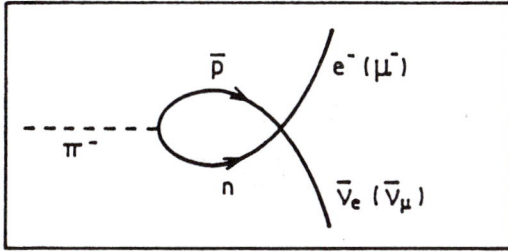

Fig. 3.14

For A coupling, $M^2 \approx 1 - \beta$, where β is the velocity of the charged lepton. The phase space factor is

$$\frac{dN}{dE_0} = Cp^2 \frac{dp}{dE_0},$$

where C is a constant, p is the momentum of the charged lepton in the rest frame of the pion. The total energy of the system is

$$E_0 = m_\pi = p + \sqrt{p^2 + m^2},$$

where m is the rest mass of the charged lepton, and the neutrino is assumed to have zero rest mass. Differentiating we have

$$\frac{dp}{dE_0} = \frac{E_0 - p}{E_0}.$$

From

$$m_\pi = p + \sqrt{p^2 + m^2}$$

we have

$$p = \frac{m_\pi^2 - m^2}{2m_\pi}.$$

Combining the above gives

$$\frac{dp}{dE_0} = \frac{m_\pi^2 + m^2}{2m_\pi^2}.$$

We also have

$$\beta = \frac{p}{\sqrt{p^2 + m^2}} = \frac{p}{m_\pi - p},$$

and so

$$1 - \beta = \frac{2m^2}{m_\pi^2 + m^2}.$$

Thus the decay rate is proportional to

$$(1 - \beta)p^2 \frac{dp}{dE_0} = \frac{1}{4}\left(\frac{m}{m_\pi}\right)^2 \left(\frac{m_\pi^2 - m^2}{m_\pi}\right)^2.$$

Hence the ratio is

$$\gamma = \frac{\Gamma(\pi^- \to \mu^- \bar\nu_\mu)}{\Gamma(\pi^- \to e^- \bar\nu_e)} = \frac{m_\mu^2(m_\pi^2 - m_\mu^2)^2}{m_e^2(m_\pi^2 - m_e^2)^2} = 8.13 \times 10^3.$$

(b) For scalar coupling, $M^2 \approx 1 - \beta$ also and the ratio R would not change.

For pseudoscalar coupling, $M^2 \approx 1 + \beta$, and the decay rate would be proportional to

$$(1 + \beta)p^2 \frac{dp}{dE} = \frac{1}{4}\left(\frac{m_\pi^2 - m^2}{m_\pi}\right)^2.$$

Then

$$\gamma = \frac{\Gamma(\pi^- \to \mu^- \bar\nu_\mu)}{\Gamma(\pi^- \to e^- \bar\nu_e)} = \frac{(m_\pi^2 - m_\mu^2)^2}{(m_\pi^2 - m_e^2)^2} = 0.18.$$

These may be compared with the experimental result

$$\gamma^{\exp} = 8.1 \times 10^3.$$

(c) For the semileptonic decay of Λ^0 the ratio

$$\gamma' = \frac{\Gamma(\Lambda \to p\mu^- \bar\nu_\mu)}{\Gamma(\Lambda \to pe^- \bar\nu_e)}$$

can be estimated in the same way. We have

$$\gamma'_{th} = 0.164 \qquad \gamma'_{\exp} = 0.187 \pm 0.042,$$

which means Λ decay can be described in terms of the V-A coupling theory.

3041

List the general properties of neutrinos and antineutrinos. What was the physical motivation for the original postulate of the existence of the neutrino? How was the neutrino first directly detected?

(Wisconsin)

Solution:

Table 3.6 lists some quantum numbers of neutrino and antineutrino.

Table 3.6

	Charge	Spin	Helicity	Lepton number
neutrino	0	1/2	−1	+1
antineutrino	0	1/2	+1	−1

Both neutrino and antineutrino are leptons and are subject to weak interaction only. Three kinds of neutrinos and their antiparticles are at present believed to exist in nature. These are electron-neutrino, muon-neutrino, τ-neutrino, and their antiparticles. (ν_τ and $\bar{\nu}_\tau$ have not been detected experimentally).

Originally, in order to explain the conflict between the continuous energy spectrum of electrons emitted in β-decays and the discrete nuclear energy levels, Pauli postulated in 1933 the emission in β-decay also of a light neutral particle called neutrino. As it is neutral the neutrino cannot be detected, but it takes away a part of the energy of the transition. As it is a three-body decay the electron has continuous energy up to a definite cutoff given by the transition energy.

As neutrinos take part in weak interaction only, their direct detection is very difficult. The first experimental detection was carried out by Reines and Cowan during 1953–1959, who used $\bar{\nu}$ from a nuclear reactor to bombard protons. From the neutron decay $n \rightarrow p + e^- + \bar{\nu}$ we expect $\bar{\nu} + p \rightarrow n + e^+$ to occur. Thus if a neutron and a positron are detected simultaneously the existence of $\bar{\nu}$ is proved. It took the workers six years to get a positive result.

3042

(a) How many neutrino types are known to exist? What is the spin of a neutrino?

(b) What properties of neutrinos are conserved in scattering processes? What is the difference between a neutrino and an antineutrino? Illustrate this by filling in the missing particle:

$$\nu_\mu + e^- \rightarrow \mu^- + ?\,.$$

(c) Assume the neutrino mass is exactly zero. Does the neutrino have a magnetic moment? Along what direction(s) does the neutrino spin point? Along what direction(s) does the antineutrino spin point?

(d) What is the velocity of a $3°K$ neutrino in the universe if the neutrino mass is 0.1 eV?

(Wisconsin)

Solution:

(a) Two kinds of neutrino have been found so far. These are electron-neutrinos and muon-neutrinos and their antiparticles. Theory predicts the existence of a third kind of neutrino, τ-neutrino and its antiparticle. The neutrino spin is 1/2.

(b) In a scattering process, the lepton number of each kind of neutrino is conserved. The difference between a neutrino and the corresponding antineutrino is that they have opposite lepton numbers. Furthermore if the neutrino mass is zero, the helicities of neutrino and antineutrino are opposite. The unknown particle in the reaction is ν_e:

$$\nu_\mu + e^- \rightarrow \mu^- + \nu_e\,.$$

(c) If the neutrino masses are strictly zero, they have no magnetic moment. The neutrino spin points along a direction opposite to its motion, while the antineutrino spin does the reverse.

(d) The average kinetic energy of a neutrino in a gas of temperature T is $E_k = 3kT/2$, where k is Boltzmann's constant. The velocity of the neutrino is then

$$\beta = \sqrt{2E_k/m} = \sqrt{3kT/m} = \sqrt{3 \times 8.62 \times 10^{-5} \times 3/0.1} = 0.088\,,$$

corresponding to 2.6×10^7 m/s.

3043

(a) Describe the experiments that prove
(1) there are two kinds of neutrino,
(2) the interaction cross section is very small.

(b) Write down the reactions in which an energetic neutrino may produce a single pion with

(1) a proton, and with

(2) a neutron.

(c) Define helicity and what are its values for neutrino and antineutrino.

(d) Can the following modes of μ^+ decay proceed naturally? Why?

(1) $\mu^+ \to e^+ + \gamma$,

(2) $\mu^+ \to e^+ + e^- + e^+$.

(SUNY Buffalo)

Solution:

(a) (1) For two-neutrino experiment see **Problem 3009(3)**.

(2) The first observation of the interaction of free neutrinos was made by Reines and Cowan during 1953–1959, who employed $\bar{\nu}_e$ from a nuclear reactor, which have a broad spectrum centered around 1 MeV, as projectiles and cadmium chloride ($CdCl_2$) and water as target to initiate the reaction

$$\bar{\nu}_e + p \to n + e^+ \,.$$

The e^+ produced in this reaction rapidly comes to rest due to ionization loss and forms a positronium which annihilates to give two γ-rays, each of energy 0.511 MeV. The time scale for this process is of the order 10^{-9} s. The neutron produced, after it has been moderated in the water, is captured by cadmium, which then radiates a γ-ray of ~ 9.1 MeV after a delay of several μs. A liquid scintillation counter which detects both rays gives two differential pulses with a time differential of about 10^{-5} s. The 200-litre target was sandwiched between two layers of liquid scintillator, viewed by banks of photomultipliers. The experiment gave $\sigma_\nu \sim 10^{-44}$ cm^2, consistent with theoretical expectation. Compared with the cross section σ_h of a hardon, $10^{-24\sim-26}$ cm^2, σ_ν is very small indeed.

(b) (1) $\nu_\mu + p \to \mu^- + p + \pi^+$.

(2)
$$\nu_\mu + n \to \mu^- + n + \pi^+$$
$$\to \mu^- + p + \pi^0 \,.$$

(c) The helicity of a particle is defined as $H = \frac{\mathbf{P} \cdot \boldsymbol{\sigma}}{|\mathbf{P}||\boldsymbol{\sigma}|}$, where $\mathbf{P}$ and $\boldsymbol{\sigma}$ are the momentum and spin of the particle. The neutrino has $H = -1$ and is said to be left-handed, the antineutrino has $H = +1$ and is right-handed.

(d) $\mu^+ \to e^+ + \gamma$, $\mu^+ \to e^+ + e^- + e^+$.

Neither decay mode can proceed because they violate the conservation of electron-lepton number and of muon-lepton number.

3044

A sensitive way to measure the mass of the electron neutrino is to measure

(a) the angular distribution in electron-neutrino scattering.

(b) the electron energy spectrum in beta-decay.

(c) the neutrino flux from the sun.

(CCT)

Solution:

In the Kurie plot of a β spectrum, the shape at the tail end depends on the neutrino mass. So the answer is (b).

3045

How many of one million 1-GeV neutrinos interact when traversing the earth? ($\sigma = 0.7 \times 10^{-38}$ cm^2/n, where n means a nucleon, $R = 6000$ km, $\rho \approx 5$ g/cm^2, $\langle A \rangle = 20$)

(a) all.

(b) ≈ 25.

(c) none.

(CCT)

Solution:

Each nucleon can be represented by an area σ. The number of nucleons encountered by a neutrino traversing earth is then

$$N_n = \frac{2R\sigma\rho N_A}{\langle A \rangle} \langle A \rangle \,,$$

where N_A =Avogadro's number. The total number of encounters (collisions) is

$$N = N_\nu N_n = 2R\sigma\rho N_A N_\nu$$
$$= 2 \times 6 \times 10^8 \times 0.7 \times 10^{-38} \times 5 \times 6.02 \times 10^{23} \times 10^6 = 25.2$$

So the answer is (b).

3046

The cross section rises linearly with E_ν. How long must a detector ($\rho \approx$ 5 g/cm^3, $\langle A \rangle = 20$) be so that 1 out of 10^6 neutrinos with $E_\nu = 1000$ GeV interacts?

(a) 6 km.
(b) 480 m.
(c) 5 m.

(CCT)

Solution:

Write $L = 2R$ in **Problem 3045**, then $N \propto L\sigma$. As $\sigma' = 1000\sigma$, we have

$$\frac{1}{25.2} = \frac{10^3 L}{2R},$$

or

$$L = \frac{2 \times 6000}{25.2 \times 10^3} = 0.476 \ km.$$

Hence the anser is (b).

3047

An experiment in a gold mine in South Dakota has been carried out to detect solar neutrinos using the reaction

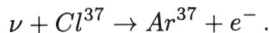

$$\nu + Cl^{37} \rightarrow Ar^{37} + e^-.$$

The detector contains approximately 4×10^5 liters of tetrachlorethylene (CCl_4). Estimate how many atoms of Ar^{37} would be produced per day. List your assumptions. How can you improve the experiment?

(Columbia)

Solution:

The threshold for the reaction $\nu + Cl^{37} \rightarrow Ar^{37} + e^-$ is $(M_{Ar} - M_{Cl})c^2 = 0.000874 \times 937.9 = 0.82$ MeV, so only neutrinos of $E_\nu > 0.82$ MeV can be detected. On the assumption that the density ρ of CCl_4 is near that of water, the number of Cl nuclei per unit volume is

$$n = \frac{4\rho N_0}{A} = (4/172) \times 6.02 \times 10^{23} = 1.4 \times 10^{22} \ \text{cm}^{-3},$$

where $A = 172$ is the molecular weight of CCl_4.

In general the interaction cross section of neutrino with matter is a function of E_ν. Suppose $\bar{\sigma} \approx 10^{-42}$ cm^2/Cl. The flux of solar neutrinos on the earth's surface depends on the model assumed for the sun. Suppose the flux with $E_\nu > 0.82$ MeV is $F = 10^9$ cm^{-2} s^{-1}. Then the number of neutrinos detected per day is $N_\nu = nV\bar{\sigma}Ft = 1.4 \times 10^{22} \times 4 \times 10^5 \times 10^3 \times 10^{-42} \times 10^9 \times 24 \times 3600 = 4.8 \times 10^2$.

However only neutrinos with energies $E_\nu > 0.82$ MeV can be detected in this experiment, whereas solar neutrinos produced in the main process in the sun $p + p \to {}^2H + e^+ + \nu_e$ have maximum energy 0.42 MeV. Most solar neutrinos will not be detected in this way. On the other hand, if Ga or In are used as the detection medium, it would be possible to detect neutrinos of lower energies.

3048

It has been suggested that the universe is filled with heavy neutrinos ν_H (mass m_H) which decay into a lighter neutrino ν_L (mass m_L) and a photon, $\nu_H \to \nu_L + \gamma$, with a lifetime similar to the age of the universe. The ν_H were produced at high temperatures in the early days, but have since cooled and, in fact, they are now so cold that the decay takes place with the ν_H essentially at rest.

(a) Show that the photons produced are monoenergetic and find their energy.

(b) Evaluate your expression for the photon energy in the limit $m_L \ll m_H$. If the heavy neutrinos have a mass of 50 eV as has been suggested by recent terrestrial experiments, and $m_L \ll m_H$, in what spectral regime should one look for these photons?

(c) Suppose the lifetime of the heavy neutrinos were short compared to the age of the universe, but that they were still "cold" (in the above sense) at the time of decay. How would this change your answer to part (b) (qualitatively)?

(Columbia)

Solution:

(a) As it is a two-body decay, conservation of energy and conservation of momentum determine uniquely the energy of each decay particle. Thus the photons are monoenergetic. The heavy neutrinos can be considered as decaying at rest. Thus

$$m_H = E_L + E_\gamma, \qquad P_L = P_\gamma.$$

As $E_\gamma = P_\gamma$, $E_L^2 = P_L^2 + m_L^2$, these give

$$E_\gamma = \frac{1}{2m_H}(m_H^2 - m_L^2).$$

(b) In the limit $m_H \gg m_L$, $E_\gamma \approx \frac{1}{2}m_H$. If $m_H = 50$ eV, $E_\gamma = 25$ eV. The photons emitted have wavelength

$$\lambda = \frac{h}{P_\gamma} = \frac{2\pi\hbar c}{P_\gamma c} = \frac{2\pi \times 197 \times 10^{-13}}{25 \times 10^{-6}} = 495 \times 10^{-8} \text{ cm} = 495 \text{ Å}.$$

This is in the regime of ultraviolet light. Thus one would have to look at extraterrestrial ultraviolet light for the detection of such photons.

(c) If the lifetime of the heavy neutrinos is far smaller than that of the universe, they would have almost all decayed into the lighter neutrinos. This would make their direct detection practically impossible.

3049

The particle decay sequence

$$\pi^+ \to \mu^+ + \nu_\mu, \qquad \mu^+ \to e^+ + \nu_e + \bar{\nu}_\mu$$

shows evidence of parity nonconservation.

(a) What observable quantity is measured to show this effect? Sketch or give a formula for the distribution of this observable.

(b) Does the process show that both decay processes violate parity conservation, or only one? Explain why.

(Wisconsin)

Solution:

(a) Suppose the pions decay in flight. We can study the forward muons μ^+ which stop and decay inside a carbon absorber. The angular distribution of the e^+ produced in the μ^+ decay can determine if parity is conserved. Relative to the initial direction of μ^+ the e^+ have angular distribution $dN/d\Omega = 1 - \frac{1}{3}\cos\theta$, which changes under space reflection $\theta \to \pi - \theta$. Hence parity is not conserved.

(b) Both the decay processes violate parity conservation since both proceed via weak interaction.

3050

Consider the following decay scheme:

$$\pi^+ \to \mu^+ + \nu_1$$
$$\;\;\;\;\longmapsto e^+ + \nu_2 + \bar{\nu}_3$$

(a) If the pion has momentum p, what is the value of the minimum (and maximum) momentum of the muon? Express the answer in terms of m_μ, m_π and p ($m_{\nu_1} = m_{\nu_2} = m_{\bar{\nu}_3} = 0$) and assume $p \gg m_\mu, m_\pi$.

(b) If the neutrino in π decay has negative helicity, what is the helicity of the muon for this decay?

(c) Given that ν_2 and $\bar{\nu}_3$ have negative and positive helicities respectively, what is the helicity of the positron?

(d) What conserved quantum number indicates that ν_1 and $\bar{\nu}_3(\nu_2)$ are associated with the muon (electron) respectively?

(e) The pion decays to an electron: $\pi^+ \to e^+ + \nu_e$. Even though the kinematics for the electron and muon decay modes are similar, the rate of muon decay is 10^4 times the rate of electron decay. Explain.

(*Princeton*)

Solution:

(a) Let γ be the Lorentz factor of π^+. Then $\beta\gamma = \frac{p}{m_\pi}$,

$$\gamma = \frac{\sqrt{p^2 + m_\pi^2}}{m_\pi} \approx \left(1 + \frac{m_\pi^2}{2p^2}\right)\frac{p}{m_\pi}.$$

In the rest system of π^+, $p_\mu^* = p_\nu^* = E_\nu^*$, $m_\pi = E_\mu^* + E_\nu^* = E_\mu^* + p_\mu^*$, giving

$$p_\mu^* = p_\nu^* = \frac{m_\pi^2 - m_\mu^2}{2m_\pi},$$

$$E_\mu^* = \sqrt{p_\mu^{*2} + m_\mu^2} = \frac{m_\pi^2 + m_\mu^2}{2m_\pi}.$$

Transforming to the laboratory system we have

$$p_\mu \cos\theta = \gamma p_\mu^* \cos\theta^* + \gamma\beta E_\mu^*.$$

In the direction of p ($\theta = 0$), p_μ has extreme values

$$(p_\mu)_{\max} \approx p\left(1 + \frac{m_\pi^2}{2p^2}\right)\frac{m_\pi^2 - m_\mu^2}{2m_\pi^2} + p\frac{m_\pi^2 - m_\mu^2}{2m_\pi^2}$$

$$= p + \frac{m_\pi^2 - m_\mu^2}{4p}, \qquad (\theta^* = 0)$$

$$(p_\mu)_{\min} \approx -p\left(1 + \frac{m_\pi^2}{2p^2}\right)\frac{m_\pi^2 - m_\mu^2}{2m_\pi^2} + p\frac{m_\pi^2 + m_\mu^2}{2m_\pi^2}$$

$$= \left(\frac{m_\mu^2}{m_\pi^2}\right)p - \frac{m_\pi^2 - m_\mu^2}{4p}. \qquad (\theta^* = \pi)$$

(b) If the neutrino in π^+ decay has negative helicity, from the fact that π^+ has zero spin and the conservation of total angular momentum and of momentum, we can conclude that μ^+ must have negative helicity in the rest system of π^+.

(c) Knowing that $\bar\nu_3$ and ν_2 respectively have positive and negative helicities, one still cannot decide on the helicity of e^+. If we moderate the decay muons and study their decay at rest, a peak is found at 53 MeV in the energy spectrum of the decay electrons. This means the electron and $\nu_2\bar\nu_3$ move in opposite directions. If the polarization direction of μ^+ does not change in the moderation process, the angular distribution of e^+ relative to p_μ,

$$\frac{dN_{e^+}}{d\Omega} \approx 1 - \frac{\alpha}{3}\cos\theta,$$

where $\alpha \approx 1$, shows that e^+ has a maximum probability of being emitted in a direction opposite to p_μ ($\theta = \pi$). Hence the helicity of e^+ is positive.

The longitudinal polarization of the electron suggests that parity is not conserved in π and μ decays.

(d) The separate conservation of the electron- and muon-lepton numbers indicates that ν_1 and $\bar{\nu}_3$ are associated with muon and that ν_2 is associated with electron since the electron-lepton numbers of ν_1, ν_2 and $\bar{\nu}_3$ are 0,1,0, and their muon-lepton numbers are 1,0, -1 respectively.

(e) See **Problem 3040**.

3051

A beam of unpolarized electrons

(a) can be described by a wave function that is an equal superposition of spin-up and spin-down wave functions.

(b) cannot be described by a wave function.

(c) neither of the above.

(*CCT*)

Solution:

The answer is (a).

3052

Let $\mathbf{s}, \mathbf{p}$ be the spin and linear momentum vectors of an elementary particle respectively.

(a) Write down the transformations of $\mathbf{s}, \mathbf{p}$ under the parity operator $\hat{P}$ and the time reversal operator $\hat{T}$.

(b) In view of the answers to part (a), suggest a way to look for time reversal violation in the decay $\Lambda \to N + \pi$. Are any experimental details or assumptions crucial to this suggestion?

(*Wisconsin*)

Solution:

(a) Under the operation of the parity operator, $\mathbf{s}$ and $\mathbf{p}$ are transformed according to

$$\hat{P}\mathbf{s}\hat{P}^{-1} = \mathbf{s}, \qquad \hat{P}\mathbf{p}\hat{P}^{-1} = -\mathbf{p}.$$

Under the time-reversal operator $\hat{T}$, **s** and **p** are transformed according to

$$\hat{T}\mathbf{s}\hat{T}^{-1} = -\mathbf{s}, \qquad \hat{T}\mathbf{p}\hat{T}^{-1} = -\mathbf{p}.$$

(b) Consider the angular correlation in the decay of polarized Λ particles. Define

$$Q = \mathbf{s}_\Lambda \cdot (\mathbf{p}_N \times \mathbf{p}_\pi),$$

where $\mathbf{s}_\Lambda$ is the spin of the Λ particle, $\mathbf{p}_N$ and $\mathbf{p}_\pi$ are the linear momenta of the nucleon and the pion respectively. Time reversal operation gives

$$\hat{T}Q\hat{T}^{-1} = \hat{T}\mathbf{s}_\Lambda\hat{T}^{-1} \cdot (\hat{T}\mathbf{p}_N\hat{T}^{-1} \times \hat{T}\mathbf{p}_\pi\hat{T}^{-1}) = -\mathbf{s}_\Lambda \cdot [(-\mathbf{p}_N) \times (-\mathbf{p}_\pi)] = -Q,$$

or

$$\bar{Q} = \langle\alpha|Q|\alpha\rangle = \langle\alpha|\hat{T}^{-1}\hat{T}Q\hat{T}^{-1}\hat{T}|\alpha\rangle = -\langle\alpha_T|Q|\alpha_T\rangle.$$

If time reversal invariance holds true, $|\alpha_T\rangle$ and $|\alpha\rangle$ would describe the same state and so

$$\bar{Q} = \langle\alpha|Q|\alpha\rangle = -\langle\alpha_T|Q|\alpha_T\rangle = -\bar{Q},$$

or

$$\bar{Q} = 0.$$

To detect possible time reversal violation, use experimental setup as in Fig. 3.15. The pion and nucleon detectors are placed perpendicular to each other with their plane perpendicular to the Λ-particle spin. Measure the number of Λ decay events $N(\uparrow)$. Now reverse the polarization of the Λ-particles and under the same conditions measure the Λ decay events $N(\downarrow)$.

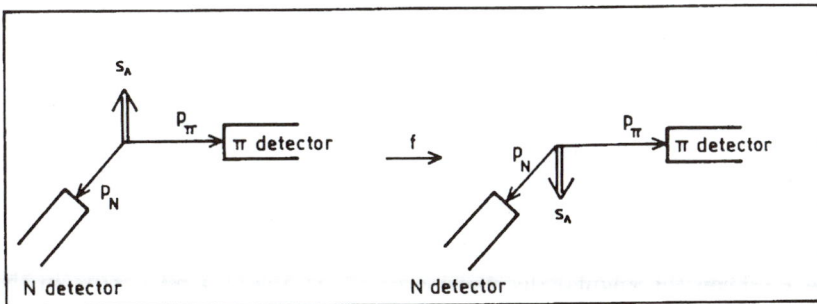

Fig. 3.15

A result $N(\uparrow) \neq N(\downarrow)$ would indicate time reversal violation in the decay $\Lambda \to \pi + N$.

This experiment requires all the Λ-particles to be strictly polarized.

3053

Consider the decay $\Lambda^0 \to p + \pi^-$. Describe a test for parity conservation in this decay. What circumstances may prevent this test from being useful?

(*Wisconsin*)

Solution:

$\Lambda^0 \to p + \pi^-$ is a nonleptonic decay. It is known Λ^0 and p both have spin 1/2 and positive parity, and π^- has spin 0 and negative parity.

As the total angular momentum is conserved, the final state may have relative orbital angular momentum 0 or 1. If

$l = 0$, the final-state parity is $P(p)P(\pi^-)(-1)^0 = -1$; if
$l = 1$, the final-state parity is $P(p)P(\pi^-)(-1)^1 = +1$.

Thus if parity is conserved in Λ^0 decay, $l = 0$ is forbidden. If parity is not conserved in Λ^0 decay, both the l values are allowed and the final-state proton wave function can be written as

$$\Psi = \Psi_s + \Psi_p = a_s Y_{0,0} \left| \frac{1}{2}, \frac{1}{2} \right\rangle + a_p \left(\sqrt{\frac{2}{3}} Y_{1,1} \left| \frac{1}{2}, -\frac{1}{2} \right\rangle - \sqrt{\frac{1}{3}} Y_{1,0} \left| \frac{1}{2}, \frac{1}{2} \right\rangle \right),$$

where $\left| \frac{1}{2}, \frac{1}{2} \right\rangle$ and $\left| \frac{1}{2}, -\frac{1}{2} \right\rangle$ are respectively the spin wave functions of the proton for $m = \pm \frac{1}{2}$, a_s and a_p are the amplitudes of the s and p waves.

Substitution of $Y_{1,1}$, $Y_{1,0}$, $Y_{0,0}$ gives

$$\Psi^* \Psi \propto |a_s - a_p \cos \theta|^2 + |a_p|^2 \sin^2 \theta$$

$$= |a_s|^2 + |a_p|^2 - 2\text{Re}(a_s a_p^*) \cos \theta \propto 1 + \alpha \cos \theta,$$

where $\alpha = 2\text{Re}(a_a a_p^*)/(|a_s|^2 + |a_p|^2)$.

If the Λ^0-particles are polarized, the angular distribution of p or π^- will be of the form $1 + \alpha \cos \theta$, (in the rest frame of Λ^0, p and π^- move in opposite directions). If Λ^0 are not fully polarized, let the polarizability be P. Then the angular distribution of π^- or p is $(1 + \alpha P \cos \theta)$. In the above θ is the angle between the direction of π^- or p and the polarization direction of Λ^0.

Measurement of the angular distribution can be carried out using the polarized Λ^0 arising from the associated production

$$\pi^- + p \rightarrow \Lambda^0 + K^0 .$$

Parity conservation in the associated production, which is a strong interaction, requires the Λ^0-particles to be transversally polarized with the spin direction perpendicular to the reaction plane. Experimentally if the momentum of the incident π^- is slightly larger than 1 GeV/c, the polarizability of Λ^0 is about 0.7. Take the plane of production of Λ^0, which is the plane containing the directions of the incident π^- and the produced Λ^0 (K^0 must also be in this plane to satisfy momentum conservation) and measure the counting rate disparity of the π^- (or p) emitted in Λ^0 decay between the spaces above and below this plane ($\theta = 0$ to $\pi/2$ and $\theta = \pi/2$ to π). A disparity would show that parity is not conserved in Λ^0 decay. An experiment by Eister in 1957 using incident π^- of momenta $910 \sim 1300$ MeV/c resulted in $P = 0.7$. Note in the above process, the asymmetry in the emission of π^- originates from the polarization of Λ^0. If the Λ^0 particles has $P = 0$ the experiment could not be used to test parity conservation.

3054

The Λ and p particles have spin 1/2, the π has spin 0.

(a) Suppose the Λ is polarized in the z direction and decays at rest, $\Lambda \rightarrow p + \pi^-$. What is the most general allowed angular distribution of π^-? What further restriction would be imposed by parity invariance?

(b) By the way, how does one produce polarized Λ's?

(Princeton)

Solution:

(a) The initial spin state of Λ-particle is $|\frac{1}{2}, \frac{1}{2}\rangle$. Conservation of angular momentum requires the final-state πp system orbital angular momentum quantum number to be $l = 0$ or 1 (**Problem 3053**).

If $l = 0$, the final-state wave function is $\Psi_s = a_s Y_{00} |\frac{1}{2}, \frac{1}{2}\rangle$, where a_s is the s-wave amplitude in the decay, $|\frac{1}{2}, \frac{1}{2}\rangle$ is the proton spin state, Y_{00} is the orbital angular motion wave function.

If $l = 1$, the final-state wave function is

$$\Psi_p = a_p \left(\sqrt{\frac{2}{3}} Y_{11} \left| \frac{1}{2}, -\frac{1}{2} \right\rangle - \sqrt{\frac{1}{3}} Y_{10} \left| \frac{1}{2}, \frac{1}{2} \right\rangle \right),$$

where a_p is the p-wave amplitude in the decay, $\sqrt{\frac{2}{3}}, -\sqrt{\frac{1}{3}}$ are Clebsch–Gordan coefficients.

With $Y_{00} = \frac{1}{\sqrt{4\pi}}$, $Y_{10} = \sqrt{\frac{3}{4\pi}} \cos\theta$, $Y_{11} = \sqrt{\frac{3}{8\pi}} e^{i\varphi} \sin\theta$, we have

$$\Psi_s = \frac{a_s}{\sqrt{4\pi}} \left| \frac{1}{2}, \frac{1}{2} \right\rangle,$$

$$\Psi_p = -\frac{a_p}{\sqrt{4\pi}} \left(e^{i\varphi} \sin\theta \left| \frac{1}{2}, -\frac{1}{2} \right\rangle + \cos\theta \left| \frac{1}{2}, \frac{1}{2} \right\rangle \right).$$

and the finalstate total wave function

$$\Psi = \frac{1}{\sqrt{4\pi}} \left((a_s - a_p \cos\theta) \left| \frac{1}{2}, \frac{1}{2} \right\rangle - a_p e^{i\varphi} \sin\theta \left| \frac{1}{2}, -\frac{1}{2} \right\rangle \right).$$

The probability distribution is then

$$\Psi^* \Psi \propto |a_s - a_p \cos\theta|^2 + |a_p \sin\theta|^2 = |a_s|^2 + |a_p|^2 - 2\mathrm{Re}(a_s a_p^*) \cos\theta.$$

Hence the pion angular distribution has the form

$$I(\theta) = C(1 + \alpha \cos\theta),$$

where α, C are constants.

The particles Λ, p, π have parities $+, +, -$ respectively. If parity is conserved in the decay, $l = 0$ is forbidden, i.e. $a_s = 0$, and the angular distribution of the pion is limited by the space-reflection symmetry to be symmetric above and below the decay plane. If observed otherwise, parity is not conserved.

(b) Polarized Λ^0-particles can be created by bombarding a proton target with pions:

$$\pi^- + p \to \Lambda^0 + K^0.$$

The Λ^0-particles are produced polarized perpendicular to the plane of production.

3055

(a) As is well known, parity is violated in the decay $\Lambda \to p + \pi^-$. This is reflected, for example, in the following fact. If the Λ-particle is fully polarized along, say, the z-axis, then the angular distribution of the proton obeys

$$\frac{d\Gamma}{d\Omega} = A(1 + \lambda \cos \theta).$$

Given the parameter λ, what is the longitudinal polarization of the proton if the Λ is unpolarized?

(b) For strangeness-changing hyperon decays in general, e.g., $\Lambda \to p\pi^-$, $\Lambda \to n\pi^0$, $\Sigma^+ \to n\pi^+$, $\Sigma^+ \to p\pi^0$, $\Sigma^- \to n\pi^-$, $K^+ \to \pi^+\pi^0$, $K_s^0 \to \pi^+\pi^-$, $K_s^0 \to \pi^0\pi^0$, etc., there is ample evidence for the approximate validity of the so-called $\Delta I = \frac{1}{2}$ rule (the transition Hamiltonian acts like a member of an isotopic spin doublet). What does the $\Delta I = \frac{1}{2}$ rule predict for the relative rates of $K^+ \to \pi^+\pi^0$, $K_s^0 \to \pi^+\pi^-$, $K_s^0 \to \pi^0\pi^0$?

(*Princeton*)

Solution:

(a) Parity is violated in Λ^0 decay and the decay process is described with s and p waves of amplitudes a_s and a_p (**Problem 3053**). According to the theory on decay helicity, a hyperon of spin 1/2 decaying at rest and emitting a proton along the direction $\Omega = (\theta, \phi)$ has decay amplitude

$$f_{\lambda M}(\theta, \phi) = (2\pi)^{-\frac{1}{2}} \mathcal{D}_{M\lambda'}^{1/2}(\phi, \theta, 0) a_{\lambda'},$$

where M and λ' are respectively the spin projection of Λ^0 and the proton helicity. We use a_+ and a_- to represent the decay amplitudes of the two different helicities. Conservation of parity would require $a_+ = -a_-$. The total decay rate is

$$W = |a_+|^2 + |a_-|^2.$$

The angular distribution of a particle produced in the decay at rest of a Λ^0 hyperon polarized along the z-axis is

$$\frac{dP}{d\Omega} = \frac{1}{W} \sum_{\lambda'} |f_{\lambda',1/2}(\theta,\phi)|^2$$

$$= (2\pi W)^{-1} \sum_{\lambda'} |a_{\lambda'}|^2 [d^{1/2}_{1/2,\lambda'}(\theta)]^2$$

$$= (2\pi W)^{-1} \left(|a_+|^2 \cos^2\frac{\theta}{2} + |a_-|^2 \sin^2\frac{\theta}{2} \right)$$

$$= A(1 + \lambda \cos\theta),$$

where $d^{1/2}_{1/2,\lambda'}(\theta) = D^{1/2}_{M\lambda'}(\phi,\theta,0)$, $A = \frac{1}{4\pi}$, $\lambda = \frac{|a_+|^2-|a_-|^2}{|a_+|^2+|a_-|^2}$. Note $\lambda = 0$ if parity is conserved.

The expectation value of the helicity of the protons from the decay of unpolarized Λ^0 particles is

$$P = (2W)^{-1} \sum_M \int \left(\frac{1}{2}|f_{1/2,M}|^2 - \frac{1}{2}|f_{-1/2,M}|^2 \right) d\Omega$$

$$= (2W)^{-1} \sum_M \int \sum_{\lambda'} \lambda' |f_{\lambda',M}|^2 d\Omega$$

$$= (2W)^{-1} \sum_M \sum_{\lambda'} \lambda' |a_{\lambda'}|^2 (2\pi)^{-1} \int |d^{1/2}_{M\lambda'}(\theta)|^2 d\Omega$$

$$= W^{-1} \sum_{\lambda'} \lambda' |a_{\lambda'}|^2,$$

where we have used

$$\sum_{M'} (d^J_{MM'}(\theta))^2 = \sum_{M'} d^J_{MM'}(-\theta) d^J_{M'M}(\theta) = d^J_{MM}(\theta).$$

Hence

$$P = \frac{1}{2}\frac{|a_+|^2 - |a_-|^2}{|a_+|^2 + |a_-|^2} = \frac{1}{2}\lambda.$$

(b) In the decays

$$K^+ \to \pi^+\pi^0,$$

$$K^0_s \to \pi^+\pi^-,$$

$$K^0_s \to \pi^0\pi^0,$$

the final states consist of two bosons and so the total wave functions should be symmetric. As the spin of K is zero, the final-state angular momentum is zero. Then as pions have spin zero, $l = 0$ for the final state, i.e., the space wave function is symmetric. Hence the symmetry of the total wave function requires the final-state isospin wave function to be symmetric, i.e., $I = 0, 2$ as pions have isospin 1. Weak decays require $\Delta I = \frac{1}{2}$. As K has isospin $\frac{1}{2}$, the two-π system must have $I = 0, 1$. Therefore, $I = 0$.

For $K^+ \to \pi^+\pi^0$, the final state has $I_3 = 0 + 1 = 1$. As $I = 0$ or 2 and $I \geq I_3$, we require $I = 2$ for the final state. This violates the $\Delta I = 1/2$ rule and so the process is forbidden. Experimentally we find

$$\sigma(K_S^0 \to \pi^+\pi^-)/\sigma(K^+ \to \pi^+\pi^0) \approx 455 \gg 1 \,.$$

On the other hand, in $K_S^0 \to \pi^+\pi^-$ or $\pi^0\pi^0$, as K^0, π^+, π^0, π^- have $I_3 = -\frac{1}{2}, 1, 0, -1$ respectively the final state has $I_3 = 0$, $I = 0$ or 2. The symmetry of the wave function requires $I = 0$. Hence the $\Delta I = \frac{1}{2}$ rule is satisfied and the final spin state is $|I, I_3\rangle = |0, 0\rangle$. Expanding the spin wave function we have

$$|I, I_3\rangle = |0, 0\rangle = \sqrt{\frac{1}{3}}(|1, 1; 1, -1\rangle + |1, -1; 1, 1\rangle - |1, 0, 1, 0\rangle)$$

$$= \sqrt{\frac{1}{3}}(|\pi^+\pi^-\rangle + |\pi^-\pi^+\rangle - |\pi^0\pi^0\rangle) \,.$$

Therefore

$$\frac{K_s^0 \to \pi^+\pi^-}{K_s^0 \to \pi^0\pi^0} = 2 \,.$$

3056

(a) Describe the CP violation experiment in K^0 decay and explain why this experiment is particularly appropriate.

(b) Find the ratio of K_S (K short) to K_L (K long) in a beam of 10 GeV/c neutral kaons at a distance of 20 meters from where the beam is produced.

$$(\tau_{K_L} = 5 \times 10^{-8} \text{ sec}, \quad \tau_{K_s} = 0.86 \times 10^{-10} \text{ sec})$$

(SUNY, Buffalo)

Solution:

(a) J. W. Cronin *et al.* observed in 1964 that a very few K^0 mesons decayed into 2 π's after a flight path of 5.7 feet from production. As the K_S^0 lifetime is short almost all K_S^0 should have decayed within centimeters from production. Hence the kaon beam at 5.7 feet from production should consist purely of K_L^0. If CP is conserved, K_L^0 should decay into 3 π's. The observation of 2π decay means that CP conservation is violated in K_L^0 decay. CP violation may be studied using K^0 decay because K^0 beam is a mixture of K_1^0 with $\eta_{CP} = 1$ and K_2^0 with $\eta_{CP} = -1$, which have different CP eigenvalues manifesting as 3π and 2π decay modes of different lifetimes. The branching ratio

$$R = \frac{K_L^0 \to \pi^+\pi^-}{K_L^0 \to all} \approx 2 \times 10^{-3}$$

quantizes the CP violation. K_L^0 corresponds to K_2^0 which has $\eta_{CP} = -1$ and should only decay into 3 π's. Experimentally it was found that K_L^0 also decays to 2 π's, i.e., $R \neq 0$. As $\eta_{CP}(\pi^+\pi^+) = +1$, CP violation occurs in K_L^0 decay.

(b) Take $M_{K^0} \approx 0.5$ GeV/c^2. Then $P_{K^0} \approx 10$ GeV/c gives

$$\beta\gamma \approx P_{K^0}/M_{K^0} = 20 \,.$$

When K^0 are generated, the intensities of the long-lived K_L^0 and the short-lived K_S^0 are equal:

$$I_{L0} = I_{S0} \,.$$

After 20 meters of flight,

$$I_L = I_{L0}e^{-t/\gamma\tau_L} = I_{L0}e^{-20/\beta\gamma c\tau_L} \,,$$

$$I_S = I_{S0}e^{-t/\gamma\tau_S} = I_{L0}e^{-20/\beta\gamma c\tau_S} \,,$$

and so

$$I_S/I_L = e^{-\frac{20}{\beta\gamma c}(\frac{1}{\tau_S} - \frac{1}{\tau_L})} \approx e^{-38.7} \approx 1.6 \times 10^{-17} \,.$$

Hence after 20 meters, the 2π decays are due entirely to K_L^0.

3057

The neutral K-meson states $|K^0\rangle$ and $|\bar{K}^0\rangle$ can be expressed in terms of states $|K_L\rangle$, $|K_S\rangle$:

$$|K^0\rangle = \frac{1}{\sqrt{2}}(|K_L\rangle + |K_S\rangle)\,,$$

$$|\bar{K}^0\rangle = \frac{1}{\sqrt{2}}(|K_L\rangle - |K_S\rangle)\,.$$

$|K_L\rangle$ and $|K_S\rangle$ are states with definite lifetimes $\tau_L \equiv \frac{1}{\gamma_L}$ and $\tau_S \equiv \frac{1}{\gamma_S}$, and distinct rest energies $m_L c^2 \neq m_S c^2$. At time $t = 0$, a meson is produced in the state $|\psi(t = 0)\rangle = |K^0\rangle$. Let the probability of finding the system in state $|K^0\rangle$ at time t be $P_0(t)$ and that of finding the system in state $|\bar{K}^0\rangle$ at time t be $\bar{P}_0(t)$. Find an expression for $P_0(t) - \bar{P}_0(t)$ in terms of γ_L, γ_S, $m_L c^2$ and $m_S c^2$. (Neglect CP violation)

(*Columbia*)

Solution:

We have at time t

$$|\Psi(t)\rangle = e^{-iHt}|\psi(0)\rangle = e^{-iHt}|K^0\rangle$$

$$= e^{-iHt}\frac{1}{\sqrt{2}}(|K_L\rangle + |K_S\rangle)$$

$$= \frac{1}{\sqrt{2}}[e^{-im_L t - \gamma_L t/2}|K_L\rangle + e^{-im_S t - \gamma_S t/2}|K_S\rangle]\,,$$

where the factors $\exp(-\gamma_L t/2)$, $\exp(-\gamma_S t/2)$ take account of the attenuation of the wave functions (particle number $\propto \bar{\Psi}\Psi$). Thus

$$|\Psi(t)\rangle = \frac{1}{\sqrt{2}}\left\{e^{-im_L t - \gamma_L t/2}\frac{1}{\sqrt{2}}(|K^0\rangle + |\bar{K}^0\rangle)\right.$$

$$\left. + e^{-im_S t - \gamma_S t/2}\frac{1}{\sqrt{2}}(|K^0\rangle - |\bar{K}^0\rangle)\right\}$$

$$= \frac{1}{2}\{[e^{-im_L t - \gamma_L t/2} + e^{-im_S t - \gamma_S t/2}]|K^0\rangle$$

$$+ [e^{-im_L t - \gamma_L t/2} - e^{-im_S t - \gamma_S t/2}]|\bar{K}^0\rangle\}\,,$$

and hence

$$\langle\psi(t)|\psi(t)\rangle = P_0(t) + \bar{P}_0(t)\,,$$

where

$$P_0(t) = \frac{1}{4}\{e^{-\gamma_L t} + e^{-\gamma_S t} + 2e^{-(\gamma_L + \gamma_S)t/2}\cos[(m_L - m_S)t]\},$$

$$\bar{P}_0(t) = \frac{1}{4}\{e^{-\gamma_L t} + e^{-\gamma_S t} - 2e^{-(\gamma_L + \gamma_S)t/2}\cos[(m_L - m_S)t]\}.$$

Thus we have

$$P_0(t) - \bar{P}_0(t) = e^{-(\gamma_L + \gamma_S)t/2}\cos[(m_L - m_S)t].$$

3058

(a) Explain how the dominance of one of the following four reactions can be used to produce a neutral kaon beam that is "pure" (i.e., uncontaminated by the presence of its antiparticle).

$$\pi^- p \to (\Lambda^0 \text{ or } K^0)(K^0 \text{ or } \bar{K}^0).$$

(b) A pure neutral kaon beam is prepared in this way. At time $t = 0$, what is the value of the charge asymmetry factor δ giving the number of $e^+ \pi^- \nu$ decays relative to the number of $e^- \pi^+ \nu$ decays as

$$\delta = \frac{N(e^+ \pi^- \nu) - N(e^- \pi^+ \bar{\nu})}{N(e^+ \pi^- \nu) + N(e^- \pi^+ \bar{\nu})}.$$

(c) In the approximation that CP is conserved, calculate the behavior of the charge asymmetry factor δ as a function of proper time. Explain how the observation of the time dependence of δ can be used to extract the mass difference Δm between the short-lived neutral kaon K_S^0 and the long-lived K_L^0.

(d) Now show the effect of a small nonconservation of CP on the proper time dependence of δ.

(Princeton)

Solution:

(a) The reaction $\pi^- p \to \Lambda^0 K^0$ obeys all the conservation laws, including $\Delta S = 0$, $\Delta I_z = 0$, for it to go by strong interaction. K^0 cannot be replaced by $\bar{K}^0$ without violating the rule $\Delta I_z = 0$. Hence it can be used to create a pure K^0 beam.

(b) When $t = 0$, the beam consists of only K^0. Decays through weak interaction obey selection rules

$$|\Delta S| = 1, \qquad |\Delta I| = |\Delta I_3| = \frac{1}{2}.$$

Then as $K^0 \to \pi^- e^+ \nu$ is allowed and $K^0 \to \pi^+ e^- \bar{\nu}$ is forbidden,

$$\delta(t = 0) = \frac{N(e^+ \pi^- \nu) - N(e^- \pi^+ \bar{\nu})}{N(e^+ \pi^- \nu) + N(e^- \pi^+ \bar{\nu})} = 1.$$

(c) At time $t = 0$,

$$|K_L^0(0)\rangle = \frac{1}{\sqrt{2}} |K^0(0)\rangle,$$

$$|K_S^0(0)\rangle = \frac{1}{\sqrt{2}} |K^0(0)\rangle.$$

At time t

$$|K_L^0(t)\rangle = \frac{1}{\sqrt{2}} |K^0(0)\rangle e^{-(im_L t + \Gamma_L t/2)},$$

$$|K_S^0(t)\rangle = \frac{1}{\sqrt{2}} |K^0(0)\rangle e^{-(im_S t + \Gamma_S t/2)}.$$

Hence

$$K^0(t)\rangle = \frac{1}{\sqrt{2}} (|K_S^0(t)\rangle + |K^0(t)\rangle)$$

$$= \frac{1}{2} |K^0(0)\rangle [e^{-(im_S t + \Gamma_S t/2)} + e^{-(im_L t + \Gamma_L t/2)}],$$

$$\bar{K}^0(t)\rangle = \frac{1}{2} |K^0(0)\rangle [e^{-(im_S t + \Gamma_S t/2)} - e^{-(im_L t + \Gamma_L t/2)}].$$

Note that the term $\Gamma t/2$ in the exponents accounts for the attenuation of K_S^0 and K_L^0 due to decay.

If the decay probabilities $N(K^0 \to \pi^- e^+ \nu) = N(\bar{K}^0 \to \pi^+ e^- \bar{\nu})$, then

$\delta(t)$

$$= \frac{|e^{-(im_S + \Gamma_S/2)t} + e^{-(im_L + \Gamma_L/2)t}|^2 - |e^{-(im_S + \Gamma_S/2)t} - e^{-(im_L + \Gamma_L/2)t}|^2}{|e^{-(im_S + \Gamma_S/2)t} + e^{-(im_L + \Gamma_L/2)t}|^2 + |e^{-(im_S + \Gamma_S/2)t} - e^{-(im_L + \Gamma_L/2)t}|^2}$$

$$= \frac{2e^{-(\Gamma_L + \Gamma_S)t/2} \cos(\Delta m t)}{e^{-\Gamma_L t} + e^{-\Gamma_S t}} .$$

Thus from the oscillation curve of $\delta(t)$, $\Delta m \equiv |m_L - m_S|$ can be deduced.

(d) If there is a small nonconservation of CP, let it be a small fraction ε. Then

$$|K^0(t)\rangle = \frac{1}{\sqrt{2}}[(1 + \varepsilon)|K_S^0(t)\rangle + (1 - \varepsilon)|K_L^0(t)\rangle]$$

$$= \frac{1}{2}|K^0(0)\rangle\{(e^{-(im_S t + \Gamma_S t/2)} + e^{-(im_L t + \Gamma_L t/2)})\}$$

$$+ \varepsilon(e^{-(im_S t + \Gamma_S t/2)} - e^{-(im_L t + \Gamma_L t/2)}),$$

$$|\bar{K}^0(t)\rangle = \frac{1}{2}|K^0(0)\rangle\{(e^{-(im_S t + \Gamma_S t/2)} - e^{-(im_L t + \Gamma_L t/2)})$$

$$+ \varepsilon(e^{-(im_S t + \Gamma_S t/2)} + e^{-(im_L t + \Gamma_L t/2)})\},$$

and so

$$\delta(t) = \frac{\langle K^0(t)|K^0(t)\rangle - \langle \bar{K}^0(t)|\bar{K}^0(t)\rangle}{\langle K^0(t)|K^0(t)\rangle + \langle \bar{K}^0(t)|\bar{K}^0(t)\rangle}$$

$$\approx \frac{2e^{-(\Gamma_L + \Gamma_S)t/2} \cos(\Delta m t)}{e^{-\Gamma_L t} + e^{-\Gamma_S t}} + \mathrm{Re}(\varepsilon) .$$

3059

In the Weinberg–Salam model, weak interactions are mediated by three heavy vector bosons, W^+, W^- and Z^0, with masses given by

$$M_W^2 = (\pi\alpha/\sqrt{2})G \sin^2\theta ,$$

$$M_Z^2 = M_W^2 / \cos^2\theta ,$$

where α is the fine structure constant, θ is the "weak mixing angle" or the "Weinberg angle", and G is the Fermi constant. The interaction Lagrangian between electrons, positrons, electron-neutrinos and W's, Z^0 is

$$L_{\text{INT}} = \frac{\sqrt{\pi\alpha}}{\sin\theta} \left\{ \frac{1}{\sqrt{2}} W_+^\mu \bar{\nu}\gamma_\mu(1-\gamma_5)e + \frac{1}{\sqrt{2}} W_-^\mu \bar{e}\gamma_\mu(1-\gamma_5)\nu \right.$$

$$\left. + \frac{1}{2\cos\theta} Z^\mu [\bar{\nu}\gamma_\mu(1-\gamma_5)\nu - \bar{e}\gamma_\mu(1-\gamma_5)e + 4\sin^2\theta\,\bar{e}\gamma_\mu e] \right\},$$

where ν and e are Dirac fields. Consider the elastic scattering of electron-antineutrinos off electrons

$$\bar{\nu}e^- \to \bar{\nu}e^-.$$

(a) Draw the lowest order Feynman diagram(s) for this process. Label each line.

(b) If the energies of the electron and antineutrino are small compared to M_W, the interaction between them can be represented by a four-fermion effective Lagrangian. Write down a correct effective Lagrangian, and put it into the form

$$L_{\text{eff}} = \frac{G}{\sqrt{2}} [\bar{\nu}\gamma^\mu(1-\gamma_5)\nu][\bar{e}\gamma_\mu(A-B\gamma_5)e],$$

where A and B are definite functions of θ.

NOTE: if ψ_1 and ψ_2 are anticommuting Dirac fields, then

$$[\bar{\psi}_1\gamma^\mu(1-\gamma_5)\psi_2][\bar{\psi}_2\gamma_\mu(1-\gamma_5)\psi_1] = [\bar{\psi}_1\gamma^\mu(1-\gamma_5)\psi_1][\bar{\psi}_2\gamma_\mu(1-\gamma_5)\psi_2].$$

(c) What experiments could be used to determine A and B?

(*Princeton*)

Solution:

(a) Elastic $\bar{\nu}e$ scattering can take place by exchanging W^- or Z^0. The respective lowest order Feynman diagrams are shown in Fig. 3.16.

(b) From the given Lagrangian, we can write down the Lagrangians for the two diagrams. For Fig. 3.16(a):

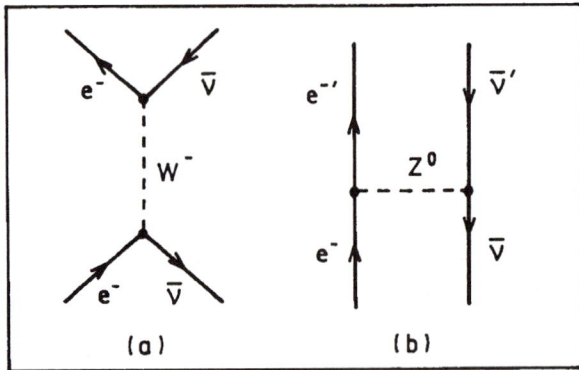

Fig. 3.16

$$L(e\bar{\nu}W) = \left(\sqrt{\frac{\pi\alpha}{2}}\frac{1}{\sin\theta}\right)^2 [\bar{\nu}\gamma^\mu(1-\gamma_5)e \cdot \frac{g^{\mu\nu} - (k^\mu k^\nu/M_W^2)}{M_W^2 - k^2}\bar{e}\gamma_\mu(1-\gamma_5)\nu]$$

$$= \frac{\pi\alpha}{2\sin^2\theta}[\bar{\nu}\gamma^\mu(1-\gamma_5)e \cdot \frac{g^{\mu\nu} - (k^\mu k^\nu/M_W^2)}{M_W^2 - k^2}\bar{e}\gamma_\mu(1-\gamma_5)\nu].$$

At low energies, $M_W^2 \gg k^2$ and the above equation can be simplified to

$$L(e\bar{\nu}W) = \frac{\pi\alpha}{2\sin^2\theta M_W^2}[\bar{\nu}\gamma^\mu(1-\gamma_5)e][\bar{e}\gamma_\mu(1-\gamma_5)\nu].$$

$\bar{\nu}$ and e being Dirac fields, we have

$$[\bar{\nu}\gamma^\mu(1-\gamma_5)e][\bar{e}\gamma_\mu(1-\gamma_5)\nu] = [\bar{\nu}\gamma^\mu(1-\gamma_5)\nu][\bar{e}\gamma_\mu(1-\gamma_5)e],$$

and the Lagrangian

$$L(e\bar{\nu}w) = \frac{G}{\sqrt{2}}[\bar{\nu}\gamma^\mu(1-\gamma_5)\nu][\bar{e}\gamma_\mu(1-\gamma_5)e],$$

as $G = \frac{\pi\alpha}{\sqrt{2}\sin^2\theta M_W^2}$.

For Fig. 3.16(b), the effective Lagrangian is

$$L(e\bar{\nu}Z^0) = \frac{\pi\alpha^2}{\sin^2\theta \cdot 4\cos^2\theta}$$

$$\times \left[\bar{\nu}\gamma^\mu(1-\gamma_5)\nu \cdot \frac{g^{\mu\nu} - (k^\mu k^\nu/M_Z^2)}{M_Z^2 - k^2} \cdot \bar{e}\gamma_\mu(g_V - g_A\gamma_5)e\right],$$

where $g_V = -1 + 4\sin^2\theta$, $g_A = -1$. If $M_Z^2 \gg k^2$ this can be simplified to a form for direct interaction of four Fermions:

$$L(e\bar{\nu}Z^0) = \frac{\pi\alpha^2}{4\sin^2\theta\cos^2\theta M_Z^2}[\bar{\nu}\gamma^\mu(1-\gamma_5)\nu][\bar{e}\gamma_\mu(g_V - g_A\gamma_5)e]$$

$$= \frac{G}{2\sqrt{2}}[\bar{\nu}\gamma^\mu(1-\gamma_5)\nu][\bar{e}\gamma_\mu(g_V - g_A\gamma_5)e]\,,$$

as

$$M_Z^2 = \frac{M_W^2}{\cos^2\theta} = \frac{\pi\alpha^2}{\sqrt{2}G\sin^2\theta\cos^2\theta}\,.$$

The total effective Lagrangian is the sum of the two diagrams:

$$L_{\text{eff}} = L(e\bar{\nu}W) + L(e\bar{\nu}Z^0)$$

$$= \frac{G}{\sqrt{2}}[\bar{\nu}\gamma^\mu(1-\gamma_5)\nu][\bar{e}\gamma_\mu(1-\gamma_5)e]$$

$$+ \frac{G}{2\sqrt{2}}[\bar{\nu}\gamma^\mu(1-\gamma_5)\nu][\bar{e}\gamma_\mu(g_V - g_A\gamma_5)e]$$

$$= \frac{G}{\sqrt{2}}[\bar{\nu}\gamma^\mu(1-\gamma_5)\nu]\left[\bar{e}\gamma_\mu\left(1 + \frac{g_V}{2} - \gamma_5 - \frac{g_A}{2}\gamma_5\right)e\right]$$

$$= \frac{G}{\sqrt{2}}[\bar{\nu}\gamma^\mu(1-\gamma_5)\nu][\bar{e}\gamma_\mu(A - B\gamma_5)e]\,,$$

where $A = 1 + g_V/2$, $B = 1 + g_A/2$.

(c) Many experiments have been carried out to measure A and B, with the best results coming from neutrino scatterings such as $\nu_\mu e^-$, $\bar{\nu}_\mu e$ scatterings. Also experiments on the asymmetry of l-charge in $e^+e^- \to l^+l^-$ can give g_V and g_A, and hence A and B.

Note the $p\bar{p}$ colliding beams of CERN have been used to measure the masses of W and Z directly, yielding

$$M_W = (80.8 \pm 2.7)\ \text{GeV}\,,$$

$$M_{Z^0} = (92.9 \pm 1.6)\ \text{GeV}\,,$$

and $\sin^2\theta = 0.224$.

3060

One of the important tests of the modern theory of weak interactions involves the elastic scattering of a μ-type neutrino off an electron:

$$\nu_\mu + e^- \rightarrow \nu_\mu + e^- \ .$$

For low energies this may be described by the effective interaction Hamiltonian density

$$H_{\text{eff}} = \frac{G_F}{\sqrt{2}} \bar{\psi}_\nu \gamma^\alpha (1 + \gamma_5) \psi_\nu \bar{\psi}_e \{g_V \gamma_\alpha + g_A \gamma_\alpha \gamma_5\} \psi_e \ ,$$

where G_F is the Fermi constant and g_V, g_A are dimensionless parameters. Let $\sigma(E)$ be the total cross section for this process, where E is the total center-of-mass energy, and take $E \gg m_e$. Suppose the target electron is unpolarized.

(a) On purely dimensional grounds, determine how $\sigma(E)$ depends on the energy E.

(b) Let $\frac{\partial \sigma}{\partial E}|_{0^\circ}$ be the differential cross section in the center-of-mass frame for forward scattering. Compute this in detail in terms of E, G_F, g_V, g_A.

(c) Discuss in a few words (and perhaps with a Feynman diagram) how this process is thought to arise from interaction of a vector boson with neutral "currents".

(Princeton)

Solution:

(a) Given $E \gg m_e$, we can take $m_e \approx 0$ and write the first order weak interaction cross section as $\sigma(E) \approx G_F^2 E^k$, where k is a constant to be determined. In our units, $\hbar = 1$, $c = 1$, $\hbar c = 1$. Then $[E] = M$. As $[\hbar c] = [ML] = 1$, $[\sigma] = [L^2] = M^{-2}$. Also, $[G_F] = \left[\frac{(\hbar c)^3}{(Mc^2)^2} \right] = M^{-2}$. Hence $k = -2 + 4 = 2$ and so

$$\sigma(E) \approx G_F^2 E^2 \ .$$

(b) The lowest order Faynman diagram for $\nu_\mu e \rightarrow \nu_\mu e$ is shown in Fig. 3.17. In the center-of-mass frame, taking $m_\nu = 0$, $m_e \approx 0$ we have

$$p_1 = p_3 = k = (p, \mathbf{p}) \ ,$$

$$p_2 = p_4 = p \approx (p, -\mathbf{p}) \ ,$$

$$\frac{d\sigma}{d\Omega} = \frac{|F|^2}{64\pi^2 S}$$

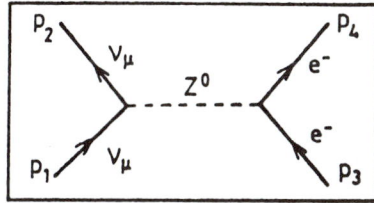

Fig. 3.17

with $S = E^2$. The square of the scattering amplitude based on H_{eff} is

$$|F|^2 = \frac{G_F^2}{2} \text{Tr} \left[\not{k} \gamma^\alpha (1 + \gamma_5) \not{k} \gamma^\beta (1 + \gamma_5) \right]$$

$$\times \frac{1}{2} \text{Tr} \left[\not{p} (g_V \gamma_\alpha + g_A \gamma_\alpha \gamma_5) \not{p} (g_V \gamma_\beta + g_A \gamma_\beta \gamma_5) \right],$$

use having been made of the relation $\sum_{P_s} \bar{u}u = \not{p} + m$, where $\not{p} = \gamma_\mu p^\mu$. Note the factor $\frac{1}{2}$ arises from averaging over the spins of the interacting electrons, whereas the neutrinos are all left-handed and need not be averaged. Consider

$$\text{Tr} \left[\not{k} \gamma^\alpha (1 + \gamma_5) \not{k} \gamma^\beta (1 + \gamma_5) \right] = 2 \text{Tr} \left[\not{k} \gamma^\alpha \not{k} \gamma^\beta (1 + \gamma_5) \right]$$

$$= 8 (k^\alpha k^\beta - k^2 g^{\alpha\beta} + k^\beta k^\alpha + i \varepsilon^{\alpha\beta\gamma\delta} k_\gamma k_\delta).$$

The last term in the brackets is zero because its sign changes when the indices γ, δ are interchanged. Also for a neutrino, $k^2 = 0$. Hence the above expression can be simplified:

$$\text{Tr} \left(\not{k} \gamma^\alpha (1 + \gamma_5) \not{k} \gamma^\beta (1 + \gamma_5) \right] = 16 k^\alpha k^\beta.$$

The second trace can be similarly simplified:

$$\frac{1}{2} \text{Tr} \left[\not{p} (g_V \gamma_\alpha + g_A \gamma_\alpha \gamma_5) \not{p} (g_V \gamma_\beta + g_A \gamma_\beta \gamma_5) \right]$$

$$= \frac{1}{2} \text{Tr} \left[g_V^2 \not{p} \gamma_\alpha \not{p} \gamma_\beta + 2 g_V g_A \not{p} \gamma_\alpha \not{p} \gamma_\beta \gamma_5 + g_A^2 \not{p} \gamma_\alpha \not{p} \gamma_\beta \right]$$

$$= 4 (g_A^2 + g_V^2)^2 p_\alpha p_\beta.$$

Then as

$$k^\alpha p_\alpha^- k^\beta p_\beta = (p_1 \cdot p_2)(p_3 \cdot p_4) = (p^2 + \mathbf{p} \cdot \mathbf{p})^2 = \left[2\left(\frac{E}{2}\right)^2\right]^2 = \left(\frac{S}{2}\right)^2 .$$

we have

$$|F|^2 = \frac{G_F^2}{2} \cdot 16 \times 4 \times \left(\frac{S}{2}\right)^2 (g_A^2 + g_V^2)^2$$

$$= 8G_F^2 S^2 (g_A^2 + g_V^2)^2 ,$$

and

$$\sigma = \int d\sigma = \int \frac{G_F^2}{8\pi^2} S(g_A^2 + g_V^2)^2 d\Omega = \frac{G_F^2 E^2}{2\pi}(g_A^2 + g_V^2)^2 .$$

Differentiating we have

$$\frac{d\sigma}{dE} = \frac{G_F^2 E}{\pi}(g_A^2 + g_V^2)^2 .$$

So the reaction cross section is isotropic in the center-of-mass frame, and the total cross section is proportional to E^2.

(c) The interaction is thought to take place by exchanging a neutral intermediate boson Z^0 as shown in Fig. 3.17 and is therefore called a neutral weak current interaction. Other such interactions are, for example,

$$\nu_\mu + N \to \nu_\mu + N , \qquad \nu_e + \mu \to \nu_e + \mu ,$$

where N is a nucleon.

3061

The Z-boson, mediator of the weak interaction, is eagerly anticipated and expected to weigh in at $M_Z \geq 80$ GeV.

(a) Given that the weak and electromagnetic interactions have roughly the same intrinsic strength (as in unified gauge theories) and that charged and neutral currents are of roughly comparable strength, show that this is a reasonable mass value (to a factor of 5).

(b) Estimate the width of Z^0 and its lifetime.

(c) Could you use Z^0 production in e^+e^- annihilation to experimentally determine the branching ratio of the Z^0 into neutrinos? If so, list explicitly what to measure and how to use it.

(Princeton)

Solution:

(a) The mediators of weak interactions are the massive intermediate vector bosons $W^\pm$ and Z^0. The weak coupling constant g_W can be related to the Fermi constant G_F in beta decays by

$$\frac{g_W^2}{8M_W^2} = \frac{G_F}{\sqrt{2}}.$$

In the Weinberg-Salam model, Z^0, which mediates neutrino and electron, has coupling constant g_Z related to the electromagnetic coupling constant g_e through

$$g_Z = \frac{g_e}{\sin\theta_W \cos\theta_W},$$

while g_W can be given as

$$g_W = \frac{g_e}{\sin\theta_W},$$

where θ_W is the weak mixing angle, called the Weinberg angle,

$$g_e = \sqrt{4\pi\alpha},$$

α being the fine structure constant. The model also gives

$$M_W = M_Z \cos\theta_W.$$

Thus

$$M_Z = \frac{M_W}{\cos\theta_W} = \frac{1}{\sin 2\theta_W}\left(\frac{4\pi\alpha}{\sqrt{2}G_F}\right)^{\frac{1}{2}}.$$

The Fermi constant G_F can be deduced from the observed muon mass and lifetime to be 1.166×10^{-5} GeV^{-2}. This gives

$$M_Z = \frac{74.6}{\sin 2\theta_W} \text{ GeV}.$$

For $M_Z \geq 80$ GeV, $\theta_W \leq 34.4^0$.

At the lower limit of 80 GeV, $\theta_W = 34.4°$ and the coupling constants are for electromagnetic interaction:

$$g_e = g_W \sin \theta_W = 0.6 g_W \,,$$

for neutral current interaction:

$$g_Z = \frac{g_W}{\cos \theta_W} = 1.2 g_W \,,$$

for charged current interaction:

$$g_W \,.$$

So the three interactions have strengths of the same order of magnitude if $M_Z \approx 80$ GeV.

(b) The coupling of Z^0 and a fermion can be written in the general form

$$L_{\text{int}}^Z = -\frac{g_W}{4 \cos \theta_W} \bar{f} \gamma^\mu (g_V - g_A \gamma_5) f Z_\mu \,,$$

where the values of g_V and g_A are for

$$\nu_e, \nu_\mu, \cdots \quad g_V = 1, \qquad g_A = 1 \,;$$

$$e, \mu, \cdots \quad g_V = -1 + 4 \sin^2 \theta_W, \qquad g_A = -1 \,;$$

$$u, c, \cdots \quad g_V = 1 - \frac{8}{3} \sin^2 \theta_W, \qquad g_A = 1 \,;$$

$$d, s, \cdots \quad g_V = -1 + \frac{4}{3} \sin^2 \theta_W, \qquad g_A = -1 \,.$$

Consider a general decay process

$$Z^0(P) \to f(p) + \bar{f}(q) \,.$$

The amplitude T is

$$T = -\frac{i g_W}{4 \cos \theta_W} \varepsilon_\mu^\nu(p) \bar{u}_\sigma(p) \gamma^\mu (g_V - g_A \gamma_5) \nu_\rho(q) \,.$$

Summing over the fermion spins, quark decay channels and quark colors, and averageing over the three polarization directions of Z^0, we have

$$\sum |T|^2 = \frac{4n}{3} \left(\frac{g_W}{4 \cos \theta_W} \right)^2 \left(-g_{\mu\nu} + \frac{p_\mu p_\nu}{M_Z^2} \right)$$
$$\times [(g_V^2 + g_A^2)(p^\mu q^\nu + p^\nu q^\mu - g^{\mu\nu} p \cdot g) - (g_V^2 - g_A^2)m^2 g^{\mu\nu}]$$
$$= \frac{4n}{3} \left(\frac{g_W}{4 \cos \theta_W} \right)^2 \left\{ (g_V^2 + g_A^2) \left[p \cdot q \right. \right.$$
$$\left. \left. + \frac{2}{M_Z^2}(P \cdot p)(P \cdot q) \right] + 3(g_V^2 - g_A^2)m^2 \right\},$$

where m is the fermion mass, n is the color number. In the rest system of Z^0, we have

$$E = M_Z, \qquad \mathbf{p} = 0,$$

$$E_p = E_q = \frac{1}{2} M_Z, \qquad p = (E_p, \mathbf{p}), \quad q = (E_q, -\mathbf{p}),$$

$$|\mathbf{p}| = |\mathbf{q}| = \frac{1}{2}(M_Z^2 - 4m^2)^{1/2},$$

and hence $p \cdot q = \left(\frac{M_Z}{2} \right)^2 + \frac{1}{4}(M_Z^2 - 4m^2) = \frac{1}{2} M_Z^2 - m^2$, $(P \cdot p)(P \cdot q) = \left(M_Z \cdot \frac{M_Z}{2} \right)^2 = \frac{M_Z^4}{4}$.

Substitution gives

$$\sum |T|^2 = \frac{4n}{3} \left(\frac{g_W}{4 \cos \theta_W} \right)^2 [(g_V^2 + g_A^2)M_Z^2 + 2(g_V^2 - 2g_A^2)m^2].$$

From the formula for the probability of two-body decay of a system at rest

$$d\Gamma = \frac{1}{32\pi^2} \frac{|\mathbf{p}|}{M_Z^2} \sum |T|^2 d\Omega,$$

and neglecting the fermion mas m, we obtain

$$\Gamma(Z^0 \rightarrow f\bar{f}) = \frac{nG_F M_Z^3}{24\sqrt{2}\pi}(g_V^2 + g_A^2).$$

Note that in the above we have used

$$|\mathbf{p}| \approx \frac{M_Z}{2} , \quad \int d\Omega = 4\pi , \quad \left(\frac{g_W}{4\cos\theta_W}\right)^2 = \frac{G_F}{2\sqrt{2}} M_Z^2 .$$

Putting in the values of g_V, g_A, and n (contribution of color) we find

$$\Gamma(Z^0 \to \nu_e \bar{\nu}_e) = \Gamma(Z^0 \to \nu_\mu \bar{\nu}_\mu) = \frac{G_F M_Z^3}{12\sqrt{2}\pi} ,$$

$$\Gamma(Z^0 \to e^+ e^-) = \Gamma(Z^0 \to \mu^+ \mu^-) = \frac{G_F M_Z^3}{12\sqrt{2}\pi}(1 - 4\sin^2\theta_W + 8\sin^4\theta_W) ,$$

$$\Gamma(Z^0 \to u\bar{u}) = \Gamma(Z^0 \to c\bar{c}) = \frac{G_F M_Z^3}{4\sqrt{2}\pi}\left(1 - \frac{8}{3}\sin^2\theta_W + \frac{32}{9}\sin^4\theta_W\right) ,$$

$$\Gamma(Z^0 \to d\bar{d}) = \Gamma(Z^0 \to s\bar{s}) = \frac{G_F M_Z^3}{4\sqrt{2}\pi}\left(1 - \frac{4}{3}\sin^2\theta_W + \frac{8}{9}\sin^4\theta_W\right) .$$

The sum of these branching widths gives the total width of Z^0:

$$\Gamma_Z = \frac{G_F M_Z^3}{12\sqrt{2}\pi} \cdot 8N\left(1 - 2\sin^2\theta_W + \frac{8}{3}\sin^4\theta_W\right) ,$$

where N is the number of generations of fermions, which is currently thought to be 3. The lifetime of Z^0 is $\tau = \Gamma_Z^{-1}$.

(c) Using the result of (b) and taking into accout the contribution of the quark colors, we have

$$\Gamma_{\nu\nu} : \Gamma_{\mu\mu} : \Gamma_{uu} : \Gamma_{dd} = 1 : \left(1 - 4\sin^2\theta_W + \frac{8}{3}\sin^4\theta_W\right)$$

$$: 3\left(1 - \frac{8}{3}\sin^2\theta_W + \frac{32}{9}\sin^4\theta_W\right)$$

$$: 3\left(1 - \frac{4}{3}\sin^2\theta_W + \frac{8}{9}\sin^4\theta_W\right)$$

$$\approx 1 : 0.5 : 1.8 : 2.3 ,$$

employing the currently accepted value $\sin^2\theta_W = 0.2196$.

If we adopt the currently accepted 3 generations of leptons and quarks, then

$$B_{\mu\mu} = \frac{\Gamma_{\mu\mu}}{\Gamma_z} = \frac{1 - 4\sin^2\theta_W + \frac{8}{3}\sin^4\theta_W}{8 \times 3(1 - 2\sin^2\theta_W + \frac{8}{3}\sin^4\theta_W)} \approx 3\% .$$

Similarly, for

$$\Sigma\Gamma_{\nu\nu} = \Gamma_{\nu_e} + \Gamma_{\nu_\mu} + \Gamma_{\mu_\tau} = \frac{G_F M_Z^3}{4\sqrt{2}\pi},$$

$$B_{\nu\nu} = \frac{\sum\Gamma_{\nu\nu}}{\Gamma_z} = 18\%.$$

According to the standard model, the numbers of generations of leptons and quarks correspond; so do those of ν_i and l_i. If we measure Γ_Z, we can deduce the number of generations N. Then by measuring $B_{\mu\mu}$ we can get $\Gamma_{\mu\mu}$.

Using the number of generations N and $\Gamma_{\mu\mu}$ we can obtain $\Gamma_{\nu\nu} \approx 2\Gamma_{\mu\mu}$, $\sum\Gamma_{\nu\nu} = 2N\Gamma_{\mu\mu}$.

In the production of Z^0 in e^+e^- annihilation, we can measure Γ_Z directly. Because the energy dispersion of the electron beam may be larger than Γ_Z, we should also measure $\Gamma_{\mu\mu}$ and Γ_h by measuring the numbers of muon pairs and hadrons in the resonance region, for as

$$A_h = \int_{\text{resonance region}} \sigma_h dE \approx \frac{6\pi^2}{M_Z^2}\frac{\Gamma_h\Gamma_{ee}}{\Gamma_Z} = \frac{6\pi^2}{M_Z^2}\frac{\Gamma_h\Gamma_{\mu\mu}}{\Gamma_Z},$$

$$A_\mu = \int_{\text{resonance region}} \sigma_{\mu\mu} dE \approx \frac{6\pi^2}{M_Z^2}\frac{\Gamma_h\Gamma_{\mu\mu}}{\Gamma_Z},$$

we have

$$A_\mu/A_h = \Gamma_{\mu\mu}/\Gamma_h.$$

Now for

$$N = 3, \qquad \Gamma_{\mu\mu} : \Gamma_h \approx 0.041;$$

$$N = 4, \qquad \Gamma_{\mu\mu} : \Gamma_h \approx 0.030;$$

$$N = 5, \qquad \Gamma_{\mu\mu} : \Gamma_h \approx 0.024.$$

From the observed A_μ and A_h we can get N, which then gives

$$B_{\nu\nu} = \sum\Gamma_{\nu\nu}/\Gamma_Z = 2N\Gamma_{\mu\mu}/\Gamma_Z, \quad \Gamma_Z = 3NT_{\mu\mu} + \Gamma_h.$$

3062

Experiments which scatter electrons off protons are used to investigate the charge structure of the proton on the assumption that the electromagnetic interaction of the electron is well understood. We consider an analogous process to study the charge structure of the neutral kaon, namely,

$K^0 + e \rightarrow K^0 + e$. (Call this amplitude A)

(a) Neglecting CP violation, express the amplitudes for the following processes in terms of A:

$K_L^0 + e \rightarrow K_L^0 + e$, (Scattering, call this A_s)

$K_L^0 + e \rightarrow K_S^0 + e$. (Regeneration, call this A_R)

(b) Consider the regeneration experiment

$K_L^0 + e \rightarrow K_S^0 + e$,

in which a kaon beam is incident on an electron target. At a very high energy E_K, what is the energy dependence of the differential cross section in the forward direction? (Forward means the scattering angle is zero, $\mathbf{p}_{KL} = \mathbf{p}_{KS}$). That is, how does $\left(\frac{d\sigma}{d\Omega}\right)_{0^0}$ vary with E_K? Define what you mean by very high energy.

(Princeton)

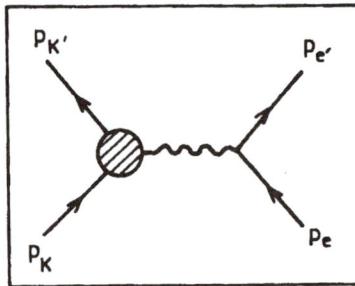

Fig. 3.18

Solution:

Consider the Feynman diagram Fig. 3.18, where p_K, $p_{K'}$, p_e, $p_{e'}$ are the initial and final momenta of K^0 and e with masses M and m, respectively. The S-matrix elements are:

$$S_{fi} = \delta_{fi} - i(2\pi)^4 \delta(p_K + p_e - p_{K'} - p_{e'}) \frac{t_{fi}}{(2\pi)^6} \sqrt{\frac{m^2}{4E_K E_{K'} E_e E_{e'}}}$$

where t_{fi} is the invariant amplitude

$$t_{fi} = ie^2(2\pi)^3 \sqrt{4E_K E_{K'}} \bar{u}(p_{e'})\gamma^\mu u(p_e)\frac{1}{q^2}\langle K^0 p_{K'}|j_\mu(0)|K^0 p_K\rangle \approx A,$$

j_μ being the current operator.

(a) We have

$$|K_S^0\rangle = \frac{1}{\sqrt{2}}(|K^0\rangle + |\bar{K}^0\rangle), \quad |K_L^0\rangle = \frac{1}{\sqrt{2}}(|K^0\rangle - |\bar{K}^0\rangle).$$

If CP violation is neglected, K_L^0, K_S^0, and K^0 have the same mass. Then

$$\langle K_L^0 p_{K'}|j_\mu(0)|K_L^0 p_K\rangle = \frac{1}{2}\{\langle K^0 p_{K'}|j_\mu(0)|K^0 p_K\rangle + \langle \bar{K}^0 p_{K'}|j_\mu(0)|\bar{K}^0 p_K\rangle\}.$$

As

$$\langle \bar{K}^0 p_{K'}|j_\mu(0)|\bar{K}^0 p_K\rangle = \langle \bar{K}^0 p_{K'}|C^{-1}Cj_\mu(0)C^{-1}C|\bar{K}^0 p_K\rangle$$
$$= -\langle K^0 p_{K'}|j_\mu(0)|K^0 p_K\rangle,$$

$A_S = 0$. Similarly we have $A_R = A$.

(b) Averaging over the spins of the initial electrons and summing over the final electrons we get the differential cross section

$$d\sigma = \frac{1}{2v_r}\frac{m^2}{4E_K E_{K'} E_e E_{e'}}(2\pi)^4\delta(p_e + p_K - p_{e'} - p_{K'})\sum_{\text{spin}}|t_{fi}|^2\frac{d\mathbf{p}_e d\mathbf{p}_{K'}}{(2\pi)^6}.$$

Integration over $\mathbf{p}_{e'}$ and $E_{K'}$ gives

$$\frac{d\sigma}{d\Omega'} = \frac{m}{32\pi^2}\frac{p_{K'}}{p_K}\frac{\sum_{\text{spin}}|t_{fi}|^2}{m + E_K - (p_K E_{K'}/p_{K'})\cos\theta'},$$

where θ' is the angle $\mathbf{p}_{K'}$ makes with $\mathbf{p}_K$. Momentum conservations requires

$$p_{e'} + p_{K'} - p_e - p_K = 0,$$

giving

$$m + \sqrt{M_L^2 + \mathbf{p}_K^2} = \sqrt{m^2 + \mathbf{p}_e^2} + \sqrt{M_S^2 + \mathbf{p}_{K'}^2},$$

where M_L, M_S are the masses of K_L^0 and K_S^0 respectively and m is the electron mass. Consider

$$E_L = \sqrt{M_L^2 + \mathbf{p}_K^2} = \sqrt{(M_S + \Delta M)^2 + \mathbf{p}_K^2}$$

with $\Delta M = M_L - M_S$. If $E_L^2 \gg M_S \Delta M$, or $E_L \gg \Delta M$, K_L^0 is said to have high energy. At this time the momentum equation becomes

$$m + \sqrt{M_S^2 + \mathbf{p}_K^2} = \sqrt{m^2 + \mathbf{p}_e^2} + \sqrt{M_S^2 + \mathbf{p}_{K'}^2},$$

which represents an elastic scattering process.

For forward scattering, $p_K = p_{K'}$, $p_e = 0$, and

$$\left. \frac{d\sigma}{d\Omega} \right|_0 = \frac{1}{32\pi} \sum_{\text{spin}} |t_{fi}|^2.$$

Now

$$(2\pi)^3 \sqrt{4E_K E_{K'}} \langle K^0 p_{K'} | j_\mu(0) | K^0 p_K \rangle = (p_K + p_{K'})_\mu F_K (p_{K'} - p_K)^2,$$

where F_K is the electromagnetic form factor of K^0, $F_K(q^2) = q^2 g(q^2)$. Note $g(q^2)$ is not singular at $q^2 = 0$. Thus

$$t_{fi} = ie^2 \bar{u}^+(p_{e'}) \gamma^\mu u(p_e) g[(p_{K'} - p_K)^2](p_{K'} + p_K)_\mu$$

$$= ie^2 \bar{u}^+(\mathbf{p}_{e'} = 0) u(\mathbf{p}_e = 0) \cdot 2E_K g(0)$$

$$= \begin{cases} ie^2 2E_K g(0) & \text{if the initial and final electrons have the same spin,} \\ 0 & \text{if the initial and final electrons have different spins.} \end{cases}$$

Thus the forward scattering differential cross section has the energy dependence

$$\left. \frac{d\sigma}{d\Omega} \right|_0 \propto E_K^2.$$

3063

Inelastic neutrino scattering in the quark model. Consider the scattering of neutrinos on free, massless quarks. We will simplify things and discuss only strangeness-conserving reactions, i.e. transitions only between the u and d quarks.

(a) Write down all the possible charged-current elastic reactions for both ν and $\bar{\nu}$ incident on the u and d quarks as well as the $\bar{u}$ and $\bar{d}$ antiquarks. (There are four such reactions.)

(b) Calculate the cross section for one such process, e.g. $\frac{d\sigma}{d\Omega}(\nu d \to \mu^- u)$.

(c) Give helicity arguments to predict the angular distribution for each of the reactions.

(d) Assume that inelastic ν (or $\bar{\nu}$)-nucleon cross sections are given by the sum of the cross sections for the four processes that have been listed above. Derive the quark model prediction for the ratio of the total cross section for antineutrino-nucleon scattering compared with neutrino-nucleon scattering, $\sigma^{\bar{\nu}N}/\sigma^{\nu N}$.

(e) The experimental value is $\sigma^{\bar{\nu}N}/\sigma^{\nu N} = 0.37 \pm 0.02$. What does this value tell you about the quark/antiquark structure of the nucleon?

(*Princeton*)

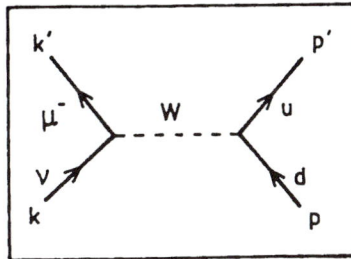

Fig. 3.19

Solution:

(a) The four charged-current interactions are (an example is shown in Fig. 3.19)

$$\nu_\mu d \to \mu^- u,$$

$$\bar{\nu}_\mu \bar{d} \to \mu^+ \bar{u},$$

$$\nu_\mu \bar{u} \to \mu^- \bar{d},$$

$$\bar{\nu}_\mu u \to \mu^+ d.$$

(b) For $\nu_\mu d \to \mu^- u$, ignoring m_μ, m_d, m_u and considering the reaction in the center-of-mass system, we have

$$\frac{d\sigma}{d\Omega} = \frac{1}{64\pi^2 S}|F|^2,$$

where the invariant mass squared is $S = -(k+p)^2 = -2kp$, and

$$|F|^2 = \frac{G_F^2}{2} \text{Tr} \left[k' \gamma^\mu (1-\gamma_5) k \gamma^\nu (1-\gamma_5) \right] \times \frac{1}{2} \text{Tr} \left[p' \gamma_\mu (1-\gamma_5) p \gamma_\nu (1-\gamma_5) \right] \cos^2 \theta_c ,$$

where θ_c is the Cabbibo mixing angle, and the factor $\frac{1}{2}$ arises from averaging over the spins of the initial muons. As

$$\text{Tr} \left[k' \gamma^\mu (1 - \gamma_5) k \gamma^\nu (1 - \gamma_5) \right] = \text{Tr} \left[k' \gamma^\mu k \gamma^\nu (1 - \gamma_5)^2 \right]$$

$$= 2\text{Tr} \left[k' \gamma^\mu k \gamma^\nu \right] - 2\text{Tr} \left[k' \gamma^\mu k \gamma^\nu \gamma_5 \right]$$

$$= 8 \left(k'^\mu k^\nu + k'^\nu k^\mu + \frac{q^2}{2} g^{\mu\nu} - i\varepsilon^{\mu\nu\gamma\delta} k'_\gamma k_\delta \right) ,$$

and similarly

$$\text{Tr} \left[p' (\gamma_\mu (1 - \gamma_5) p \gamma_\nu (1 - \gamma_5) \right] = 8 \left[p'_\mu p_\nu + p'_\nu p_\mu + \frac{q^2}{2} g_{\mu\nu} - i\varepsilon_{\mu\nu\alpha\beta} p'^\alpha p^\beta \right] ,$$

where $q^2 = -(k - k')^2 = -2kk'$ is the four-momentum transfer squared, we have

$$|F|^2 = 64 G_F^2 (k \cdot p)(k' \cdot p') \cos^2 \theta_c = 16 G_F^2 S^2 \cos^2 \theta_c ,$$

and so

$$\frac{d\sigma}{d\Omega} (\nu d \to \mu^- u)_{\text{cm}} = \frac{16 G_F^2 S^2 \cos^2 \theta_c}{64\pi^2 S} = \frac{G_F^2 S}{4\pi^2} \cos^2 \theta_c .$$

(c) In the weak interaction of hadrons, only the left-handed u, d quarks and e^-, μ^- and the right-handed quarks $\bar{u}$, $\bar{d}$ and e^+, μ^+ contribute. In the center-of-mass system, for the reactions $\nu d \to \mu^- u$ and $\bar{\nu} \bar{d} \to \mu^+ \bar{u}$, the orbital angular momentum is zero and the angular distribution is isotropic as shown.

In the reactions $\nu \bar{u} \to \bar{d} \mu^-$ and $\bar{\nu} u \to \mu^+ d$ (Fig. 3.20), the total spins of the incoming and outgoing particles are both 1 and the angular distributions are

$$\frac{d\sigma}{d\Omega} (\nu \bar{u} \to \mu^- \bar{d})_{\text{cm}} = \frac{G_F^2 S}{16\pi^2} \cos^2 \theta_c (1 - \cos\theta)^2 ,$$

$$\frac{d\sigma}{d\Omega} (\bar{\nu} u \to \mu^+ d)_{\text{cm}} = \frac{G_F^2 S}{16\pi^2} \cos^2 \theta_c (1 - \cos\theta)^2 .$$

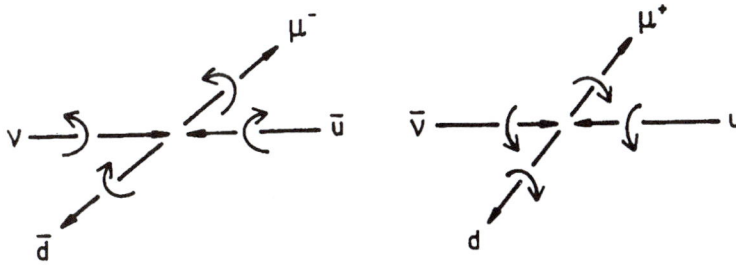

Fig. 3.20

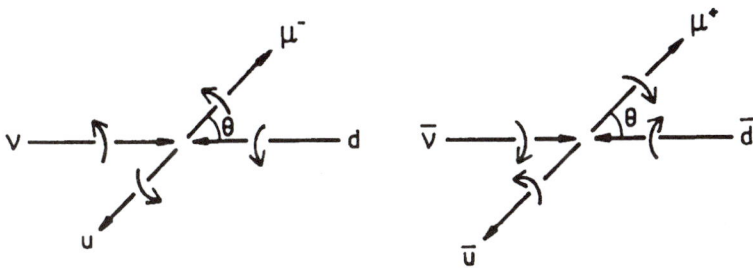

Fig. 3.21

(d) For the reactions $\nu d \to \mu^- u$ and $\bar{\nu}\bar{d} \to \mu^+ \bar{u}$ (Fig. 3.21) we have, similarly,

$$\frac{d\sigma}{d\Omega}(\nu d \to \mu^- u)_{cm} = \frac{G_F^2 S}{4\pi^2} \cos^2\theta_c ,$$

$$\frac{d\sigma}{d\Omega}(\bar{\nu}\bar{d} \to \mu^+ \bar{u})_{cm} = \frac{G_F^2 S}{4\pi^2} \cos^2\theta_c .$$

Integrating over the solid angle Ω we have

$$\sigma_1 = \sigma(\nu d \to \mu^- u)_{cm} = \frac{G_F^2 S}{\pi} \cos^2\theta_c ,$$

$$\sigma_2 = \sigma(\bar{\nu} u \to \mu^+ d)_{cm} = \frac{1}{3}\frac{G_F^2 S}{\pi} \cos^2\theta_c .$$

Neutron and proton contain quarks udd and uud respectively. Hence

$$\frac{\sigma(\nu n)}{\sigma(\bar{\nu}n)} = \frac{\sigma(\nu udd)}{\sigma(\bar{\nu}udd)} = \frac{2\sigma(\nu d)}{\sigma(\bar{\nu}u)} = \frac{2}{\left(\frac{1}{3}\right)} = 6\,,$$

$$\frac{\sigma(\nu p)}{\sigma(\bar{\nu}p)} = \frac{\sigma(\nu uud)}{\sigma(\bar{\nu}uud)} = \frac{\sigma(\nu d)}{2\sigma(\bar{\nu}u)} = \frac{1}{2 \times \frac{1}{3}} = \frac{3}{2}\,.$$

If the target contains the same number of protons and neutrons,

$$\frac{\sigma(\nu N)}{\sigma(\bar{\nu}N)} = \frac{\sigma(\nu p) + \sigma(\nu n)}{\sigma(\bar{\nu}p) + \sigma(\bar{\nu}n)} = \frac{\frac{3}{2}\sigma(\bar{\nu}p) + \sigma(\nu n)}{\sigma(\bar{\nu}p) + \frac{1}{6}\sigma(\nu n)} = \frac{\frac{3}{3} + 3}{1 + \frac{3}{6}} = 3\,,$$

where we have used $\sigma(\nu n) = 3\sigma(\bar{\nu}p)$.

(e) The experimental value $\sigma(\bar{\nu}N)/\sigma(\nu N) = 0.37 \pm 0.02$ is approximately the same as the theoretical value $1/3$. This means that nucleons consist mainly of quarks, any antiquarks present would be very small in proportion. Let the ratio of antiquark to quark in a nucleon be α, then

$$\frac{\sigma(\bar{\nu}N)}{\sigma(\nu N)} = \frac{3\sigma(\bar{\nu}u) + 3\alpha\sigma(\bar{\nu}\bar{d})}{3\sigma(\nu d) + 3\alpha\sigma(\nu\bar{u})} = \frac{3 \times \frac{1}{3} + 3\alpha \times 1}{3 \times 1 + 3\alpha \times \frac{1}{3}} = \frac{1 + 3\alpha}{3 + \alpha} = 0.37\,,$$

giving

$$\frac{1 + 3\alpha}{8} = \frac{0.37}{2.63}\,,$$

or

$$\alpha = 4 \times 10^{-2}\,.$$

3064

(a) According to the Weinberg-Salam model, the Higgs boson ϕ couples to every elementary fermion f (f may be a quark or lepton) in the form

$$\frac{em_f}{m_W}\phi\bar{f}f\,,$$

where m_f is the mass of the fermion f, e is the charge of the electron, and m_W is the mass of the W boson. Assuming that the Higgs boson decays primarily to the known quarks and leptons, calculate its lifetime in terms

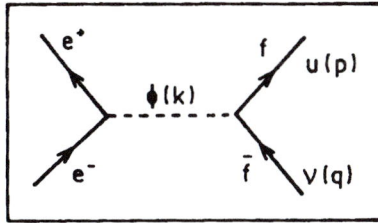

Fig. 3.22

of its mass m_H. You may assume that the Higgs boson is much heavier than the known quarks and leptons.

(b) Some theorists believe that the Higgs boson weighs approximately 10 GeV. If so do you believe it would be observed (in practice) as a resonance in e^+e^- annihilation (Fig. 3.22)? Roughly how large would the signal to background ratio be at resonance?

(Princeton)

Solution:

(a) Fermi's Golden Rule gives for decays into two fermions the transition probablity

$$\Gamma_f = \int \frac{d^3\mathbf{p}}{(2\pi)^3 2p_0} \frac{d^3\mathbf{q}}{(2\pi)^3 2q_0} \cdot \frac{(2\pi)^4}{2k_0} \delta^4(k - p - q)|M|^2 ,$$

where

$$|M|^2 = \mathrm{Tr} \sum_{s,t} \left[\left(\frac{em_f}{m_W}\right)^2 \bar{u}_s(p)v_t(q)\bar{\phi}\phi\bar{v}_t(q)u_s(p) \right]$$

$$= \left(\frac{em_f}{m_W}\right)^2 \mathrm{Tr}\, [\not{p}\not{q} - m_f^2]$$

$$= 4 \left(\frac{em_f}{m_W}\right)^2 (p \cdot q - m_f^2) .$$

As $p + q = k$, we have $p \cdot q = \frac{k^2 - p^2 - q^2}{2} = \frac{m_H^2 - 2m_f^2}{2}$,

$$|M|^2 = 4 \left(\frac{em_f}{m_W}\right)^2 \left(\frac{m_H^2 - 4m_f^2}{2}\right) = 2 \left(\frac{em_f}{m_W}\right)^2 m_H^2 \left(1 - \frac{4m_f^2}{m_H^2}\right)$$

in the rest system of the Higgs boson. Then

$$\Gamma_f = \int \frac{d^3\mathbf{p}\, d^3\mathbf{q}}{(2\pi)^6 4p_0 q_0} \frac{(2\pi)^4}{2m_H} \delta^4(k - p - q)|M|^2$$

$$= \frac{1}{(2\pi)^2} \int \frac{d^3\mathbf{p}}{4p_0 q_0} \cdot \frac{1}{2m_H} \delta^4(m_H - p_0 - q_0)|M|^2$$

$$= \frac{4\pi}{(2\pi)^2 \cdot 4q_0^2 \cdot 2m_H} \int q^2 dq \cdot \delta(m_H - 2q_0)|M|^2 \,.$$

With $qdq = q_0 dq_0$, we have

$$\Gamma_f = \frac{1}{8\pi m_H} \int \frac{q^2}{q_0^2} \frac{q_0}{q} dq_0 \delta(m_H - 2q_0)|M|^2$$

$$= \frac{1}{8\pi m_H} \frac{2}{m_H} \cdot \frac{1}{2} \left[\left(\frac{m_H}{2}\right)^2 - m_f^2 \right]^{1/2} \cdot 2 \frac{e^2 m_f^2 m_H^2}{m_W^2} \left(1 - \frac{4m_f^2}{m_H^2}\right)$$

$$= \frac{e^2 m_f^2 m_H}{4\pi m_W^2} \left(1 - \frac{4m_f^2}{m_H^2}\right)^{3/2} \cdot \frac{1}{2}$$

$$\approx \frac{e^2 m_f^2 m_H}{8\pi m_W^2} \quad \text{if} \quad m_H \gg m_f \,.$$

Then $\Gamma = \Sigma \Gamma_i = \frac{e^2 m_H}{8\pi m_W^2} \sum a_f m_f^2$, with $a_f = 1$ for lepton and $a_f = 3$ for quark. Assuming $m_H \approx 10$ GeV, $m_W \approx 80$ GeV, and with $m_u = m_d = 0.35$ GeV, $m_s = 0.5$ GeV, $m_c = 1.5$ GeV, $m_b = 4.6$ GeV, $m_e = 0.5 \times 10^{-3}$ GeV, $m_\mu = 0.11$ GeV, $m_\tau = 1.8$ GeV, we have

$$\sum a_f m_f^2 \left(1 - \frac{4m_f^2}{m_H^2}\right)^{3/2} \approx \sum_{f \neq b} a_f m_f^2 + 3m_b^2 \left(1 - \frac{4m_b^2}{m_H^2}\right)^{3/2}$$

$$= 0.005^2 + 0.11^2 + 1.8^2 + 3 \times (0.35^2 + 0.35^2 + 0.5^2 + 1.5^2)$$

$$+ 3 \times 4.6^2 \left[1 - 4 \times \left(\frac{4.6}{10}\right)^2\right]^{3/2}$$

$$= 15.3 \text{ GeV}^2 \,,$$

and hence

$$\Gamma = \frac{1}{8\pi} \left(\frac{e^2}{\hbar c}\right) \hbar c \cdot \frac{m_H}{m_W^2} \sum a_f m_f^2 \left(1 - \frac{4m_f^2}{m_H^2}\right)^{3/2}$$

$$\approx \frac{1}{8\pi \times 137} \times \frac{10}{80^2} \times 15.3 = 6.9 \times 10^{-6} \text{ GeV},$$

or

$$\tau = \Gamma^{-1} = 145 \text{ MeV}^{-1} = 6.58 \times 10^{-22} \times 145 \text{ } s = 9.5 \times 10^{-20} \text{ } s.$$

(b) The process $e^+ e^- \to \bar{f}f$ consists of the following interactions:

$$e^+ e^- \overset{\gamma, Z^0}{\to} \bar{f}f \quad \text{and} \quad e^+ e^- \overset{H}{\to} \bar{f}f.$$

When $\sqrt{S} = 10$ GeV, Z^0 exchange can be ignored. Consider $e^+ e^- \overset{\gamma}{\to} \bar{f}f$. The total cross section is given approximately by

$$\sigma_{\bar{f}f} \approx \frac{4\pi\alpha^2}{3S} Q_f^2,$$

where α is the fine structure constant and Q_f is the charge (in units of the electron charge) of the fermion. Thus

$$\sigma(^+ e^- \overset{\gamma}{\to} \bar{f}f) = \frac{4\pi\alpha^2}{3S} \sum Q_f^2 \cdot a_f,$$

where $a_f = 1$ for lepton, $a_f = 3$ for quark. As $\sum Q_f^2 a_f = (\frac{1}{9} + \frac{4}{9} + \frac{1}{9} + \frac{4}{9} + \frac{1}{9}) \times 3 + 1 + 1 + 1 = \frac{20}{3}$, $S = m_H^2$, we have

$$\sigma(e^+ e^- \overset{\gamma}{\to} \bar{f}f) = \frac{4\pi\alpha^2}{3S} \frac{20}{3} \approx 8\pi\alpha^2/S = \frac{8\pi\alpha^3}{m_H^2}.$$

For the $e^+ e^- \overset{H}{\to} \bar{f}f$ process we have at resonance ($J_H = 0$)

$$\sigma(e^+ e^- \overset{H}{\to} \bar{f}f) = \pi \lambda^2 \Gamma_{ee}/\Gamma \approx \pi p^{*-2} \Gamma_{ee}/\Gamma.$$

As a rough estimate, taking $\Gamma_{ee} \approx m^2$, i.e., $\Gamma_{ee}/\Gamma \approx (0.5 \times 10^{-3})^2/15.3 \approx 1.6 \times 10^{-8}$ and $p^{*2} = \frac{m_H^2}{4}$, we have

$$\sigma(e^+ e^- \overset{H}{\to} \bar{f}f) : \sigma(e^+ e^- \overset{\gamma}{\to} \bar{f}f) = \left(\frac{4\pi}{m_H^2} 1.6 \times 10^{-8}\right) \left(\frac{8\pi\alpha^2}{m_H^2}\right)^{-1}$$

$$\approx 0.8 \times 10^{-8}/\alpha^2 \approx 1.5 \times 10^{-4}.$$

In e^+e^- annihilation in the 10 GeV region, the background, which is mainly due to the photon process, is almost 10^4 times as strong as the H_0 resonance process. The detection of the latter is all but impossible.

3065

Parity Violation. Recently the existence of a parity-violating neutral current coupled to electrons was demonstrated at SLAC. The experiment involved scattering of polarized electrons off (unpolarized) protons.

(a) Why are polarized electrons required? What is the signature for the parity violation?

(b) Estimate the magnitude of the effect.

(c) How would this parity violation manifest itself in the passage of light through matter?

(Princeton)

Solution:

(a) To observe the parity violation, we must measure the contribution of the pseudoscalar terms to the interaction, such as the electron and hadron spinor terms. Hence we must study the interaction between electrons of fixed helicity and an unpolarized target (or conversely electrons and a polarized target, or electrons and target both polarized). The signature for parity violation is a measureable quantity relating to electron helicity, such as the dependence of scattering cross section on helicity, etc.

(b) Electron-proton scattering involves two parts representing electromagnetic and weak interactions, or specifically scattering of the exchanged photons and exchanged Z^0 bosons. Let their amplitudes be A and B. Then

$$\sigma \approx A^2 + |A \cdot B| + B^2 \,.$$

In the energy range of the experiment, $A^2 \gg B^2$. As parity is conserved in electromagnetic interaction, parity violation arises from the interference term (considering only first order effect):

$$\frac{|A \cdot B|}{A^2 + B^2} \approx \frac{|A \cdot B|}{A^2} \approx \frac{|B|}{|A|} \approx \frac{G_f}{e^2/q^2} \,,$$

where G_F is the Fermi constant, e is the electron charge, q^2 is the square of the four-momentum transfer. We have

$$\frac{G_F}{e^2/q^2} \approx \frac{10^{-5}m_p^{-2}}{4\pi/137}q^2 \approx 10^{-4}q^2/m_p^2 \approx 10^{-4}q^2 \text{ GeV}^{-2}.$$

as $m_p \approx 1$ GeV. In the experiment at SLAC, $E_e \approx 20$ GeV, $q^2 \approx 10 \sim 20$ GeV^2, and the parity violation should be of order of magnitude 10^{-3}. The experiment specifically measured the scattering cross sections of electrons of different helicities, namely the asymmetry

$$A = \frac{\sigma(\lambda = 1/2) - \sigma(\lambda = -1/2)}{\sigma(\lambda = 1/2) + \sigma(\lambda = -1/2)} \approx q^2[a_1 + a_2 f(y)],$$

where a_1 and a_2 involve A_e, V_Q and A_Q, V_e respectively, being related to the quark composition of proton and the structure of the weak neutral current, $\sigma(\lambda = 1/2)$ is the scattering cross section of the incoming electrons of helicity $1/2$, $y = (E-E')/E$, E and E' being the energies of the incoming and outgoing electrons respectively. From the experimental value of A, one can deduce the weak neutral current parameter.

(c) Parity violation in atomic range manifests itself as a slight discrepancy in the refractive indices of the left-handed and right-handed circularly polarized lights passing through a high-nuclear-charge material. For a linearly polarized light, the plane of polarization rotates as it passes through matter by an angle

$$\phi = \left(\frac{\omega L}{2c}\right) \text{Re}(n_+ - n_-),$$

where L is the thickness of the material, ω is the angular frequency of the light, n_+ and n_- are the refractive indices of left-handed and right-handed circularly polarized lights

3066

There are now several experiments searching for proton decay. Theoretically, proton decay occurs when two of the quarks inside the proton exchange a heavy boson and become an antiquark and an antilepton. Suppose this boson has spin 1. Suppose, further, that its interactions conserve charge, color and the SU(2)$\times$U(1) symmetry of the Weinberg–Salam model.

(a) It is expected that proton decay may be described by a fermion effective Lagrangian. Which of the following terms may appear in the effective

Lagrangian? For the ones which are not allowed, state what principle or facts forbid them, e.g., charge conservation.

$$(1)\ u_R u_L d_R e_L^- \qquad\qquad (2)\ u_R d_R d_L \nu_L$$

$$(3)\ u_R u_L d_L e_R^- \qquad\qquad (4)\ u_L d_L d_L \nu_L$$

$$(5)\ u_R u_R d_R e_R^- \qquad\qquad (6)\ u_L u_L d_R e_R^-$$

$$(7)\ u_L d_L d_R \nu_L \qquad\qquad (8)\ u_L u_R d_R \nu_L$$

All Fermions are incoming.

(b) Consider the decay $p \to e^+ H$, where H is any hadronic state with zero strangeness. Show that the average positron polarization defined by the ratio of the rates

$$P = \frac{\Gamma(p \to e_L^+ H) - \Gamma(p \to e_R^+ H)}{\Gamma(p \to e_L^+ H) + \Gamma(p \to e_R^+ H)}$$

is independent of the hadronic state H.

(c) If the spin-one boson has a mass of 5×10^{14} GeV and couples to fermions with electromagnetic strength (as predicted by grand unified theories), give a rough estimate of the proton lifetime (in years).

(Princeton)

Solution:

(a) (1), (2), (3), (4), (5) are allowed, (6), (7), (8) are forbidden. Note that (6) is forbidden because $u_L u_L$ is not an isospin singlet, (7) is forbidden because it does not contain ν_R (8) is forbidden because total charge is not zero.

(b) The decay process $p \to e^+ H$ can be described with the equivalent interaction Lagrangian

$$L_{\text{eff}} = [g_1(\bar{d}_{\alpha R}^c \mu_{\beta R})(\bar{\mu}_{\gamma L}^c e_L - \bar{d}_{\gamma L}^c \nu_L) + g_2(\bar{d}_{\alpha L}^c \mu_{\beta L})(\bar{\mu}_{\gamma R}^c e_R)]\varepsilon_{\alpha\beta\gamma},$$

where g_1, g_2 are equivalent coupling coefficients, c denotes charge conjugation, α, β, γ are colors signatures, $\varepsilon_{\alpha\beta\gamma}$ is the antisymmetric matrix. Thus the matrix element of $p \to e_L^+ H$ is proportional to g_1, that of $e_R^+ H$ is proportional to g_2, both having the same structure. Hence

$$P = \frac{|g_1|^2 - |g_2|^2}{|g_1|^2 + |g_2|^2}$$

and is independent of the choice of the H state.

(c) An estimate of the lifetime of proton may be made, mainly on the basis of dimensional analysis, as follows. A massive spin-1 intermediate particle contributes a propagator $\sim m^{-2}$, where m is its mass. This gives rise to a transition matrix element of $\mathcal{M} \sim m^{-2}$. The decay rate of proton is thus

$$\Gamma_p \propto |\mathcal{M}|^2 \sim m^{-4},$$

or

$$\Gamma_p \sim \frac{C\alpha^2}{m^4},$$

where $\alpha = e^2/\hbar c$ is the dimensionless coupling constant for electromagnetic interaction (**Problem 3001**), and C is a constant. The lifetime of proton τ_p has dimension

$$[\tau_p] = M^{-1},$$

since in our units $Et \sim \hbar = 1$ and so $[t] = [E]^{-1} = M^{-1}$. This means that

$$[C] = M^4 M^1 = M^5.$$

For a rough estimate we may take $C \sim m_p^5$, m_p being the proton mass. Hence, with $m \approx 5 \times 10^{14}$ GeV, $m_p \approx 1$ GeV,

$$\tau_p = \Gamma_p^{-1} \sim \frac{m^4}{\alpha^2 m_p^5} = 1.2 \times 10^{63} \text{ GeV}^{-1},$$

or, in usual units,

$$\tau_p \sim \frac{1.2 \times 10^{63} \hbar}{365 \times 24 \times 60 \times 60} = 3 \times 10^{31} \text{years}.$$

3067

It is generally recognized that there are at least three different kinds of neutrino. They can be distinguished by the reactions in which the neutrinos are created or absorbed. Let us call these three types of neutrino ν_e, ν_μ and ν_τ. It has been speculated that each of the neutrinos has a small but finite rest mass, possibly different for each type.

Let us suppose, for this question, that there is a small perturbing interaction between these neutrino types, in the absence of which all three types have the same nonzero rest mass M_0. Let the matrix element of this perturbation have the same real value $\hbar\omega_1$ between each pair of neutrino types. Let it have zero expectation value in each of the states ν_e, ν_μ and ν_τ.

(a) A neutrino of type ν_e is produced at rest at time zero. What is the probability, as a function of time, that the neutrino will be in each of the other states?

(b) (Can be answered independently of (a).) An experiment to detect these "neutrino oscillations" is being performed. The flight path of the neutrinos is 2000 meters. Their energy is 100 GeV. The sensitivity is such that the presence of 1% of neutrinos of one type different from that produced at the start of flight path can be measured with confidence. Take M_0 to be 20 electron volts. What is the smallest value of $\hbar\omega_1$ that can be detected? How does this depend on M_0?

(UC, Berkeley)

Solution:

(a) Let $|\psi\rangle = a_1(t)|\nu_e\rangle + a_2(t)|\nu_\mu\rangle + a_3(t)|\nu_\tau\rangle$. Initially the interaction Hamiltonian is zero. Use of the perturbation matrix

$$H' = \begin{pmatrix} 0 & \hbar\omega_1 & \hbar\omega_1 \\ \hbar\omega_1 & 0 & \hbar\omega_1 \\ \hbar\omega_1 & \hbar\omega_1 & 0 \end{pmatrix}$$

in the time-dependent Schrödinger equation

$$i\hbar\frac{\partial}{\partial t}\begin{pmatrix} a_1 \\ a_2 \\ a_3 \end{pmatrix} = \hbar\omega_1\begin{pmatrix} 0 & 1 & 1 \\ 1 & 0 & 1 \\ 1 & 1 & 0 \end{pmatrix}\begin{pmatrix} a_1 \\ a_2 \\ a_3 \end{pmatrix}$$

gives

$$\begin{cases} i\dot{a}_1 = \omega_1(a_2 + a_3), \\ i\dot{a}_2 = \omega_1(a_1 + a_3), \\ i\dot{a}_3 = \omega_1(a_1 + a_2). \end{cases}$$

Eliminating a_1 from the last two equations gives

$$i(\dot{a}_3 - \dot{a}_2) = -\omega_1(a_3 - a_2),$$

or

$$a_3(t) - a_2(t) = Ae^{i\omega_1 t}.$$

At time $t = 0$, $a_2(0) = a_3(0) = 0$, so $A = 0$, $a_2 = a_3$, with which the system of equations becomes

$$\begin{cases} i\dot{a}_1 = 2\omega_1 a_2, \\ i\dot{a}_2 = \omega_1(a_1 + a_2). \end{cases}$$

Eliminating a_1 again, we have

$$\ddot{a}_2 + i\omega_1\dot{a}_2 + 2\omega_1^2 a_2 = 0,$$

whose solution is $a_2(t) = A_1 e^{i\omega_1 t} + A_2 e^{-i2\omega_1 t}$. At time $t = 0$, $a_2(0) = 0$, giving

$$A_1 + A_2 = 0, \quad \text{or} \quad a_2 = A_1(e^{i\omega_1 t} - e^{-i2\omega_1 t}).$$

Hence

$$\dot{a}_1 = -i2\omega_1 A_1(e^{i\omega_1 t} - e^{-i2\omega_1 t}),$$

or

$$a_1 = -2A_1 e^{i\omega_1 t} - A_1 e^{-i2\omega_1 t}.$$

Initially only $|\nu_e\rangle$ is present, so

$$a_1(0) = 1.$$

Thus $A_1 = -1/3$, and

$$a_2 = a_3 = \frac{1}{3}(e^{-i2\omega_1 t} - e^{i\omega_1 t}).$$

The probability that the neutrino is in $|\nu_\mu\rangle$ or $|\nu_\tau\rangle$ at time t is

$$P(|\nu_\mu\rangle) = P(|\nu_\tau\rangle) = |a_2|^2 = \frac{1}{9}(e^{-i2\omega_1 t} - e^{i\omega_1 t})(e^{i2\omega_1 t} - e^{-i\omega_1 t})$$

$$= \frac{2}{9}[1 - \cos(3\omega_1 t)].$$

(b) For simplicity consider the oscillation between two types of neutrino only, and use a maximum mixing angle of $\theta = 45°$. From **Problem 3068** we have

$$P(\nu_1 \to \nu_2, t) = \sin^2 2\theta \sin^2\left(\frac{E_1 - E_2}{2}t\right) = \sin^2\left[1.27\left(\frac{l}{E}\Delta m^2\right)\right],$$

where l is in m, E in MeV, and Δm^2 in eV^2. For detection of ν_2 we require $P \geq 0.01$, or $\sin\left[1.27\left(\frac{l}{E}\Delta m^2\right)\right] \geq 0.1$, giving

$$\Delta m^2 \geq \frac{100 \times 10^3}{1.27 \times 2000} \times \arcsin 0.1 = 3.944 \text{ eV}^2 .$$

As $\Delta m^2 = (M_0 + \hbar\omega_1)^2 - M_0^2 \approx 2M_0\hbar\omega_1$, we require

$$\hbar\omega_1 \geq \frac{3.944}{2 \times 20} = 9.86 \times 10^{-2} \text{ eV} \approx 0.1 \text{ eV}$$

Note that the minimum value of $\hbar\omega_1$ varies as M_0^{-1} if $M_0 \gg \hbar\omega_1$.

3068

Suppose that ν_e and ν_μ, the Dirac neutrinos coupled to the electron and the muon, are a mixture of two neutrinos ν_1 and ν_2 with masses m_1 and m_2:

$$\nu_e = \nu_1 \cos\theta + \nu_2 \sin\theta ,$$

$$\nu_\mu = -\nu_1 \sin\theta + \nu_2 \cos\theta ,$$

θ being the mixing angle.

The Hamiltonian has a mass term $H = m_1\bar{\nu}_1\nu_1 + m_2\bar{\nu}_2\nu_2$.

(a) Express the stationary-state masses m_1 and m_2, and the mixing angle θ in terms of the mass matrix elements of the Hamiltonian in the ν_e, ν_μ representation:

$$H = \bar{\nu}_l M_{ll'} \nu_{l'} \quad \text{with} \quad l, l' = e, \mu .$$

(b) Specify under what conditions there is maximal mixing or no mixing.

(c) Suppose that at $t = 0$ one has pure ν_e. What is the probability for finding a ν_μ at time t?

(d) Assuming that p (the neutrino momentum) is $\gg m_1$ and m_2, find the oscillation length.

(e) If neutrino oscillations were seen in a detector located at a reactor, what would be the order of magnitude of the oscillation parameter

$\Delta = |m_1^2 - m_2^2|$? (Estimate the particle energies and the distance between the source and the detector.)

(f) Answer (e) for the case of neutrino oscillations observed at a 100 GeV proton accelerator laboratory.

(Princeton)

Solution:

(a) In the ν_e, ν_μ representation the Hamiltonian is

$$H = \begin{pmatrix} M_{ee} & M_{e\mu} \\ M_{\mu e} & M_{\mu\mu} \end{pmatrix} .$$

For simplicity assume $M_{\mu e} = M_{e\mu}$. Then the eigenvalues are the solutions of

$$\begin{vmatrix} M_{ee} - m & M_{\mu e} \\ M_{\mu e} & M_{\mu\mu} - m \end{vmatrix} = 0 ,$$

i.e.

$$m^2 - (M_{ee} + M_{\mu\mu})m + (M_e M_\mu - M_{\mu e}^2) = 0 .$$

Solving the equation we have the eigenvalues

$$m_1 = \frac{1}{2}\left[(M_{ee} + M_{\mu\mu}) - \sqrt{(M_{ee} - M_{\mu\mu})^2 + 4M_{\mu e}^2} \right] ,$$

$$m_2 = \frac{1}{2}\left[(M_{ee} + M_{\mu\mu}) + \sqrt{(M_{ee} - M_{\mu\mu})^2 + 4M_{\mu e}^2} \right] .$$

In the ν_e, ν_μ representation let $\nu_2 = \begin{pmatrix} a_1 \\ a_2 \end{pmatrix}$. The operator equation

$$H\nu_2 = m_2 \nu_2 ,$$

i.e.,

$$\begin{pmatrix} M_{ee} & M_{\mu e} \\ M_{\mu e} & M_{\mu\mu} \end{pmatrix} \begin{pmatrix} a_1 \\ a_2 \end{pmatrix} = m_2 \begin{pmatrix} a_1 \\ a_2 \end{pmatrix} ,$$

gives, with the normalization condition $a_1^2 + a_2^2 = 1$,

$$a_1 = \frac{M_{\mu e}}{\sqrt{M_{\mu e}^2 + (m_2 - M_{ee})^2}} ,$$

$$a_2 = \frac{m_2 - M_{ee}}{\sqrt{M_{\mu e}^2 + (m_2 - M_{ee})^2}} .$$

The mixing equations

$$\nu_e = \nu_1 \cos\theta + \nu_2 \sin\theta\,,$$

$$\nu_\mu = -\nu_1 \sin\theta + \nu_2 \cos\theta$$

can be written as

$$\nu_1 = \nu_e \cos\theta - \nu_\mu \sin\theta\,,$$

$$\nu_2 = \nu_e \sin\theta + \nu_\mu \cos\theta\,.$$

However, as $\nu_2 = a_1 \nu_e + a_2 \nu_\mu$,

$$\tan\theta = \frac{a_1}{a_2} = \frac{M_{\mu e}}{m_2 - M_{ee}}$$

$$= \frac{2M_{\mu e}}{M_{\mu\mu} - M_{ee} + \sqrt{(M_{ee} - M_{\mu\mu})^2 + 4M_{\mu e}^2}}\,,$$

or

$$\theta = a = \arctan\left(\frac{2M_{\mu e}}{M_{\mu\mu} - M_{ee} + \sqrt{(M_{ee} - M_{\mu\mu})^2 + 4M_{\mu e}^2}}\right).$$

(b) When $M_{\mu\mu} = M_{ee}$, mixing is maximum and the mixing angles is $\theta = 45°$. In this case ν_1 and ν_2 are mixed in the ratio $1 : 1$. When $M_{\mu e} = 0$, $\theta = 0$ and there is no mixing.

(c) At $t = 0$, the neutrinos are in a pure electron-neutrino state ν_e which is a mixture of states ν_1 and ν_2:

$$\nu_e = \nu_1 \cos\theta + \nu_2 \sin\theta\,.$$

The state ν_e changes with time. Denote it by $\psi_e(t)$. Then

$$\psi_e(t) = \nu_1 e^{-iE_1 t}\cos\theta + \nu_2 e^{-iE_2 t}\sin\theta$$

$$= (\nu_e \cos\theta - \nu_\mu \sin\theta)e^{-iE_1 t}\cos\theta + (\nu_e \sin\theta + \nu_\mu \cos\theta)e^{-iE_2 t}\sin\theta$$

$$= (\cos^2\theta e^{-iE_1 t} + \sin^2\theta e^{-iE_2 t})\nu_e + \sin\theta\cos\theta(e^{-iE_1 t} + e^{-iE_2 t})\nu_\mu\,.$$

So the probability of finding a ν_μ at time t is

$$P = |\langle \nu_\mu | \psi_e(t) \rangle|^2$$

$$= \sin^2 \theta \cos^2 \theta | - e^{-iE_1 t} + e^{-iE_2 t}|^2$$

$$= \frac{1}{2} \sin^2(2\theta)\{1 - \cos[(E_1 - E_2)t]\}$$

$$= \sin^2(2\theta) \sin^2\left(\frac{E_1 - E_2}{2}t\right),$$

where E_1, E_2 are the eigenvalues of the states $|\nu_1\rangle$, $|\nu_2\rangle$ respectively

(d) As $E_1 - E_2 = \frac{E_1^2 - E_2^2}{E_1 + E_2} = \frac{1}{2E}[p_1^2 + m_1^2 - p_2^2 - m_2^2] \approx \frac{\Delta m^2}{2E}$ with $\Delta m^2 = m_1^2 - m_2^2$, $E = \frac{1}{2}(E_1 + E_2)$,

$$P = \sin^2(2\theta) \sin^2\left(\frac{\Delta m^2}{4E}t\right) = \sin^2(2\theta) \sin^2\left(\frac{\Delta m^2}{4E}l\right),$$

since $l = t\beta \approx t$, the neutrino velocity being $\beta \approx 1$ as $p \gg m$. In ordinary units the second argument should be

$$\frac{\Delta m^2}{4E\hbar}\frac{l}{c} = \frac{10^{-12}}{4 \times 197 \times 10^{-13}}\frac{\Delta m^2}{E}l = \frac{1.27l\Delta m^2}{E}$$

with l in m, Δm^2 in eV^2, E in MeV.

Thus

$$P = \sin^2(2\theta) \sin^2(1.27l\Delta m^2/E),$$

and the oscillation period is $1.27l\Delta m^2/E \approx 2\pi$. Hence $\Delta m^2 l/E \ll 1$ gives the non-oscillation region, $\Delta m^2 l/E \approx 1$ gives the region of appreciable oscillation, and $\Delta m^2 l/E \gg 1$ gives the region of average effect.

(e) Neutrinos from a reactor have energy $E \approx 1$ MeV, and the distance between source and detector is several meters. As oscillations are observed,

$$\Delta m^2 = E/l \approx 0.1 \sim 1 \text{ eV}^2.$$

(f) With protons of 100 GeV, the pions created have energy E_π of tens of GeV. Then the neutrino energy $E_\nu \geq 10$ GeV. With a distance of observation 100 m,

$$\Delta m^2 \approx E/l \approx 10^2 \sim 10^3 \text{ eV}^2.$$

For example, for an experiment with $E_\nu \approx 10$ GeV, $l = 100$ m,

$$\Delta m^2 = \frac{2\pi E}{1.27l} \approx 5 \times 10^2 \text{ eV}^2 .$$

3069

(a) Neutron n and antineutron $\bar{n}$ are also neutral particle and antiparticle just as K^0 and $\bar{K}^0$. Why is it not meaningful to introduce linear combinations of n_1 and n_2, similar to the K_1^0 and K_2^0? Explain this.

(b) How are the pions, muons and electrons distringuished in photographic emulsions and in bubble chambers? Discuss this briefly.

(SUNY Buffalo)

Solution:

(a) n and $\bar{n}$ are antiparticles with respect to each other with baryon numbers 1 and -1 respectively. As the baryon number B is conserved in any process, n and $\bar{n}$ are eigenstates of strong, electromagnetic and weak interactions. If they are considered linear combinations of n_1 and n_2 which are not eigenstates of strong, electromagnetic and weak interactions, as n and $\bar{n}$ have different B the linear combination is of no meaning. If some interaction should exist which does not conserve B, then the use of n_1 and n_2 could be meaningful. This is the reason for the absence of oscillations between neutron and antineutron.

(b) It is difficult to distinguish the charged particles e, μ, π over a general energy range merely by means of photographic emulsions or bubble chambers. At low energies ($E < 200 \sim 300$ MeV), they can be distinguished by the rate of ionization loss. The electron travels with the speed of light and causes minimum ionization. Muon and pion have different velocities for the same energy. As $-dE/dx \sim v^{-2}$, we can distinguish them in principle from the different ionization densities of the tracks in the photographic emulsion. However, it is difficult in practice because their masses are very similar.

At high energies, ($E > 1$ GeV), it is even more difficult to distinguish them as they all have velocity $v \approx c$. Pions may be distinguished by their interaction with the nuclei of the detecting medium. However the Z values of the materials in photographic emulsions and bubble chambers are rather low and the probability of nuclear reaction is not large. Muons and electrons do not cause nuclear reactions and cannot be distinguished this way. With bubble chambers, a transverse magnetic field is usually applied and the curvatures of the tracks can be used to distinguish the particles, provided the energy is not too high. For very low energies, muons and pions can be distinguished by their characteristic decays.

3070

Neutron-Antineutron Oscillations. If the baryon number is conserved, the transition $n \leftrightarrow \bar{n}$, know as "neutron oscillation" is forbidden. The experimental limit on the time scale of such oscillations in free space and zero magnetic field is $\tau_{n-\bar{n}} \geq 3 \times 10^6$ s. Since neutrons occur abundantly in stable nuclei, one would naively think it possible to obtain a much better limit on $\tau_{n-\bar{n}}$. The object of this problem is to understand why the limit is so poor.

Let H_0 be the Hamiltonian of the world in the absence of any interaction which mixes n and $\bar{n}$. Then

$$H_0|n\rangle = m_n c^2 |n\rangle, \quad H_0|\bar{n}\rangle = m_n c^2 |\bar{n}\rangle$$

for states at rest. Let H' be the interaction which turns n into $\bar{n}$ and vice versa:

$$H'|n\rangle = \varepsilon|\bar{n}\rangle, \quad H'|\bar{n}\rangle = \varepsilon|n\rangle,$$

where ε is real and H' does not flip spin.

(a) Start with a neutron at $t = 0$ and calculate the probability that it will be observed to be an antineutron at time t. When the probability is first equal to 50%, call that time $\tau_{n-\bar{n}}$. In this way convert the experimental limit on $\tau_{n-\bar{n}}$ into a limit on ε. Note $m_n c^2 = 940$ MeV.

(b) Now reconsider the problem in the presence of the earth's magnetic field $B_0 = 0.5$ Gs. The magnetic moment of the neutron is $\mu_n \approx -6 \times 10^{-18}$ MeV/Gs. The magnetic moment of the antineutron is opposite. Begin with a neutron at $t = 0$ and calculate the probability it will be observed to be an antineutron at time t. Ignore possible radioactive transitions. [Hint: work to lowest order in small quantities.]

(c) Nuclei with spin have non-vanishing magnetic fields. Explain briefly and qualitatively, in light of part (b), how neutrons in such nuclei can be so stable while $\tau_{n-\bar{n}}$ is only bounded by $\tau_{n-\bar{n}} \geq 3 \times 10^6$ sec.

(d) Nuclei with zero spin have vanishing average magnetic field. Explain briefly why neutron oscillation in such nuclei is also suppressed.

(*MIT*)

Solution:

(a) Consider the Hamiltonian $H = H_0 + H'$. As (using units where $c = 1$, $\hbar = 1$)

$$H(|n\rangle + |\bar{n}\rangle) = m_n(|n\rangle + |\bar{n}\rangle) + \varepsilon(|n\rangle + |\bar{n}\rangle) = (m_n + \varepsilon)(|n\rangle + |\bar{n}\rangle),$$

$$H(|n\rangle - |\bar{n}\rangle) = m_n(|n\rangle - |\bar{n}\rangle) - \varepsilon(|n\rangle - |\bar{n}\rangle) = (m_n - \varepsilon)(|n\rangle - |\bar{n}\rangle),$$

$|n\rangle \pm |\bar{n}\rangle$ are eigenstates of H. Denote these by $|n_\pm\rangle$.

Let Φ_0 be the wave function at $t = 0$. Then

$$\Phi_0|n\rangle = \frac{1}{2}(|n_+\rangle + |n_-\rangle),$$

and the wave function at the time t is

$$\begin{aligned}
\Phi &= \frac{1}{2}(|n_+\rangle e^{-i(m_n+\varepsilon)t} + |n_-\rangle e^{-i(m_n-\varepsilon)t}) \\
&= \frac{1}{2}e^{-im_nt}[(e^{-i\varepsilon t} + e^{i\varepsilon t})|n\rangle + (e^{-i\varepsilon t} - e^{i\varepsilon t})|\bar{n}\rangle \\
&= e^{-im_nt}(\cos\varepsilon t|n\rangle - i\sin\varepsilon t|\bar{n}\rangle).
\end{aligned}$$

The probability of observing an antineutron at time t is $P = \sin^2 \varepsilon t$. As at $t = \tau_{n-\bar{n}}$, $\sin^2 \varepsilon t|_{n-\bar{n}} = \sin^2 \varepsilon\tau_{n-\bar{n}} = 1/2$,

$$\varepsilon\tau_{n-\bar{n}} = \pi/4.$$

Hence

$$\varepsilon \le \frac{\pi}{4} \cdot \frac{1}{3 \times 10^6} = 2.62 \times 10^{-7}\ s^{-1} = 2.62 \times 10^{-7}\hbar = 1.73 \times 10^{-28}\ \text{MeV}.$$

(b) The Hamiltonian is now $H = H_0 + H' - \boldsymbol{\mu} \cdot \mathbf{B}$. Then

$$H|n\rangle = m_n|n\rangle + \varepsilon|\bar{n}\rangle - \mu_n B|n\rangle = (m_n - \mu_n B)|n\rangle + \varepsilon|\bar{n}\rangle,$$

$$H|\bar{n}\rangle = m_n|\bar{n}\rangle + \varepsilon|n\rangle + \mu_n B|\bar{n}\rangle = (m_n + \mu_n B)|\bar{n}\rangle + \varepsilon|n\rangle.$$

Here we assume that $n, \bar{n}$ are polarized along z direction which is the direction of $\mathbf{B}$, i.e., $s_z(n) = 1/2$, $s_z(\bar{n}) = 1/2$. Note this assumption does not affect the generality of the result.

Let the eigenstate of H be $a|n\rangle + b|\bar{n}\rangle$. As

$$H(a|n\rangle + b|\bar{n}\rangle) = aH|n\rangle + bH|\bar{n}\rangle$$

$$= [a(m_n - \mu_n B) + b\varepsilon)]|n\rangle + [b(m_n + \mu_n B) + a\varepsilon]|\bar{n}\rangle\,,$$

we have

$$\frac{a(m_n - \mu_n B) + b\varepsilon}{a} = \frac{b(m_n + \mu_n B) + a\varepsilon}{b}\,,$$

or

$$b^2 - a^2 = \frac{2\mu_n B}{\varepsilon}ab = Aab\,,$$

where $A = \frac{2\mu_n B}{\varepsilon} \approx \frac{6\times 10^{-18}}{1.73\times 10^{-28}} = 3.47 \times 10^{10}$, and $b^2 + a^2 = 1$. Solving for a and b we have either

$$\begin{cases} a \approx 1, \\ b \approx -1/A, \end{cases} \quad \text{or} \quad \begin{cases} a \approx 1/A, \\ b \approx 1. \end{cases}$$

Hence the two eigenstates of H are

$$|n_+\rangle = \frac{1}{A}|n\rangle + |\bar{n}\rangle, \quad |n_-\rangle = |n\rangle - \frac{1}{A}|\bar{n}\rangle\,.$$

At $t = 0$, $\Phi_0 = |n\rangle = \frac{|n_+\rangle + A|n_-\rangle}{A + \frac{1}{A}} = \frac{A}{1+A^2}|n_+\rangle + \frac{A^2}{1+A^2}|n_-\rangle$.

At time t the wave function is

$$\Phi = \frac{A}{1+A^2}|n_+\rangle e^{-iE_+ t} + \frac{A^2}{1+A^2}|n_-\rangle e^{-iE_- t}\,,$$

where $E_+ = m_n - \mu_n B + A\varepsilon$, $E_- = m_n - \mu_n B - \varepsilon/A$. So

$$\Phi = e^{-i(m_n - \mu_n B)t}\left(\frac{A}{1+A^2}|n_+\rangle e^{-iA\varepsilon t} + \frac{A^2}{1+A^2}|n_-\rangle e^{-i\frac{\varepsilon}{A}t}\right)$$

$$= \frac{1}{1+A^2}e^{-i(m_n - \mu_n B)t}[(e^{-iA\varepsilon t} + A^2 e^{i\frac{\varepsilon}{A}t})|n\rangle + (Ae^{-iA\varepsilon t} - Ae^{i\frac{\varepsilon}{A}t})|\bar{n}\rangle]\,.$$

The probability of observing an $\bar{n}$ at time t is

$$P = \frac{A^2}{(1+A^2)^2} |e^{-iA\varepsilon t} - e^{i\frac{\varepsilon}{A}t}|^2$$

$$= \frac{A^2}{(1+A^2)^2} \left[2 - 2\cos\left(A\varepsilon - \frac{\varepsilon}{A}\right)t \right]$$

$$= \frac{4A^2}{(1+A^2)^2} \sin^2\left(\frac{A^2-1}{2A}\varepsilon t\right)$$

$$\approx \frac{4}{A^2} \sin^2\left(\frac{A}{2}\varepsilon t\right) .$$

(c) Nuclei with spin have non-vanishing magnetic fields and so the results of (b) are applicable. For $\tau_{n-\bar{n}} \geq 3 \times 10^6$ s, or $\varepsilon \leq 1.73 \times 10^{-28}$ MeV, $A = \frac{2\mu_n B}{\varepsilon}$ is quite a large number, so the probability of observing an $\bar{n}$ is almost zero ($\approx 1/A^2$). Thus there is hardly any oscillation between n and $\bar{n}$; the nuclei are very stable.

(d) While nuclei with zero spin have zero mean magnetic field $\langle B \rangle$, the mean square of B, $\langle B^2 \rangle$, is not zero because the magnetic field is not zero everywhere in a nucleus. The probability of observing an $\bar{n}$, $P \approx 1/\langle A^2 \rangle = \frac{\varepsilon^2}{4\mu_n^2 \langle B^2 \rangle}$, is still small and almost zero. Hence neutron oscillation in such nuclei is also suppressed.

3071

It has been conjectured that stable magnetic monopoles with magnetic charge $g = c\hbar/e$ and mass $\approx 10^4$ GeV might exist.

(a) Suppose you are supplied a beam of such particles. How would you establish that the beam was in fact made of monopoles? Be as realistic as you can.

(b) Monopoles might be pair-produced in cosmic ray collisions. What is the threshold for this reaction ($p + p \to M + \bar{M} + p + p$)?

(c) What is a practical method for recognizing a monopole in a cosmic ray event?

(Princeton)

Solution:

(a) The detection of magnetic monopoles makes use of its predicted characteristics as follows:

(1) Magnetic monopole has great ionizing power. Its specific ionization $-\frac{dE}{dx}$ is many times larger than that of a singly charged particle when it passes through matter, say a nuclear track detector like nuclear emulsion or cloud chamber.

(2) A charge does not suffer a force when moving parallel to a magnetic field, whereas a magnetic monopole is accelerated or decelerated (depending on the sign of its magnetic charge) when moving parallel to a magnetic field. A magnetic monopole can acquire an energy of 400 MeV when passing through a magnetic field of 10 kGs, whereas the energy of a charge does not change in the same process.

(3) When a magnetic monopole passes though a closed circuit, it would be equivalent to a large magnetic flux passing through the coil and a large current pulse would be induced in the circuit.

(4) When a charge and a magnetic monopole pass through a transverse magnetic field, they would suffer different deflections. The former is deflected transversely in the direction of $\mathbf{F} = \frac{1}{c}\mathbf{v} \times \mathbf{B}$, while the latter is deflected parallel or antiparallel to the magnetic field direction.

(b) Consider the process $p + p \to M + \bar{M} + p + p$, where one of the initial protons is assumed at rest, as is generally the case. As $E^2 - P^2$ is invariant and the particles are produced at rest in the center-of-mass system at threshold, we have

$$(E + m_p)^2 - P^2 = (2m_M + 2m_p)^2,$$

where $E^2 - P^2 = m_p^2$, or

$$E = \frac{(2m_M + 2m_p)^2 - 2m_p^2}{2m_p}.$$

Taking $m_M = 10^4$ GeV, $m_p = 1$ GeV, we have $E \approx 2 \times 10^8$ GeV as the laboratory threshold energy.

If in the reaction the two initial protons have the same energy and collide head-on as in colliding beams, the minimum energy of each proton is given by

$$2E = 2m_M + 2m_p,$$

Hence $E \approx m_M = 10^4$ GeV.

(c) To detect magnetic monopoles in cosmic ray events, in principle, any one of the methods in (a) will do. A practical one is to employ a solid track detector telescope. When a particle makes a thick track in the system of detectors, several of the detectors together can distinguish the track due to a multiply charged particle from that due to a magnetic monopole, as in the former case the track thickness is a function of the particle velocity, but not in the latter case. Particle idetification is more reliable if a magnetic field is also used.

If magnetic monopoles are constantly created in cosmic-ray collisions above the earth they may be detected as follows. As a monopole loses energy rapidly by interacting with matter it eventually drops to the earth's surface. Based on their tendency of moving to the magnetic poles under the action of a magnetic field, we can collect them near the poles. To detect monopoles in a sample, we can place a coil and the sample between the poles of a strong magnet (Fig. 3.23). As a magnetic monopole moves from the sample to a pole a current pulse will be produced in the coil.

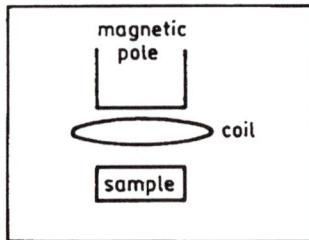

Fig. 3.23

3. STRUCTURE OF HADRONS AND THE QUARK MODEL (3072–3090)

3072

Describe the evidence (one example each) for the following conclusions:

(a) Existence of quarks (substructure or composite nature of mesons and baryons).

(b) Existence of the "color" quantum number.

(c) Existence of the "gluon".

(Wisconsin)

Solution:

(a) The main evidence supporting the quark theory is the non-uniform distribution of charge in proton and neutron as seen in the scattering of high energy electrons on nucleons, which shows that a nucleon has internal structure. Gell-Mann *et al.* discovered in 1961 the SU(3) symmetry of hadrons, which indicates the inner regularity of hadronic structure. Basing on these discoveries, Gell-Mann and Zweig separately proposed the quark theory. In it they assumed the existence of three types of quark, u, d, s and their antiparticles, which have fractional charges and certain quantum numbers, as constituents of hadrons: a baryon consists of three quarks; a meson, a quark and an antiquark. The quark theory was able to explain the structure, spin and parity of hadrons. It also predicted the existence of the Ω particle, whose discovery gave strong support to the quark theory. Later, three types of heavy quarks c, b, t were added to the list of quarks.

(b) The main purpose of postulating the color quantum number was to overcome the statistical difficulty that according to the quark theory Δ^{++}, a particle of spin 3/2, should consist of three u quarks with parallel spins, while the Pauli exclusion principle forbids three ferminions of parallel spins in the same ground state. To get over this Greenberg proposed in 1964 the color dimension for quarks. He suggested that each quark could have one of three colors. Although the three quarks of Δ^{++} have parallel spins, they have different colors, thus avoiding violation of the Pauli exclusion principle. The proposal of the 'color' freedom also explained the relative cross section R for producing hadrons in e^+e^- collisions. Quantum electrodynamics gives, for $E_{cm} < 3$ GeV, $R = \sum_i Q_i^2$, where Q_i is the charge of the ith quark, summing over all the quarks that can be produced at that energy. Without the color freedom, $R = 2/3$. Including the contribution of the color freedom, $R = 2$, in agreement with experiment.

(c) According to quantum chromodynamics, strong interaction takes place through exchange of gluons. The theory predicts the emission of hard gluons by quarks. In the electron-positron collider machine PETRA in DESY the "three-jet" phenomenon found in the hadronic final state provides strong evidence for the existence of gluons. The phenomenon is interpreted as an electron and a position colliding to produce a quark-antiquark pair, one of which then emits a gluon. The gluon and the two original quarks separately fragment into hadron jets, producing three jets in the final state. From the observed rate of three-jets events the coupling constant α_s for strong interaction can be deduced.

3073

Explain why each of the following particles cannot exist according to the quark model.

(a) A baryon of spin 1.
(b) An antibaryon of electric charge +2.
(c) A meson with charge +1 and strangeness −1.
(d) A meson with opposite signs of charm and strangeness.

(Wisconsin)

Solution:

(a) According to the quark model, a baryon consists of three quarks. Since the quark spin is 1/2, they cannot combine to form a baryon of spin 1.

(b) An antibaryon consists of three antiquarks. To combine three anti-quarks to form an antribaryon of electric charge +2, we require antiquarks of electric charge +2/3. However, there is no such antiquark in the quark model.

(c) A meson consists of a quark and an antiquark. As only the s quark $(S = -1, Z = -\frac{1}{3})$ has nonzero strangeness, to form a meson of strangeness −1 and electric charge 1, we need an s quark and an antiquark of electric charge 4/3. There is, however, no such an antiquark.

(d) A meson with opposite signs of strangeness and charm must consist of a strange quark (antistrange quark) and anticharmed quark (charmed quark). Since the strangeness of strange quark and the charm of charm quark are opposite in sign, a meson will always have strangeness and charm of the same sign. Therefore there can be no meson with opposite signs of strangeness and charm.

3074

The Gell-Mann–Nishijima relationship which gives the charge of mesons and baryons in terms of certain quantum numbers is

$$q = e(I_3 + B/2 + S/2).$$

(a) Identify the terms I_3, B and S, and briefly explain their usefulness in discussing particle reactions.

(b) Make a table of the values of these quantum numbers for the family: proton, antiproton, neutron, antineutron.

(Wisconsin)

Solution:

(a) I_3 is the third component of isospin and denotes the electric charge state of the isospin I. In strong and electromagnetic interactions I_3 is conserved, while in weak interaction it is not.

B is the baryon number. $B = 0$ for a meson and $B = 1$ for a baryon. $\Delta B = 0$ for any interaction. The conservation of baryon number means that proton is stable.

Table 3.7

Quantum number	p	$\bar{p}$	n	$\bar{n}$
I_3	1/2	−1/2	−1/2	1/2
B	1	−1	1	−1
S	0	0	0	0

S is the strangeness, introduced to account for the associated production of strange particles. S is conserved in strong and electromagnetic interactions, which implies that strange particles must be produced in pairs. S is not conserved in weak interaction, so a strange particle can decay through weak interaction to ordinary particles.

(b) The I_3, B, and S values of for nucleons are listed in Table 3.7.

3075

Give the quantum numbers and quark content of any 5 different hadrons.

(Wisconsin)

Solution:

The quantum numbers and quark content of five most common hadrons are listed in Table 3.8

Table 3.8

Hadron	Electric charge (Q)	Baryon number(B)	Spin(J)	Isospin(I)	I_3	quark content
n	0	1	1/2	1/2	$-1/2$	udd
p	1	1	1/2	1/2	1/2	uud
π^-	-1	0	0	1	-1	$d\bar{u}$
π^0	0	0	0	1	0	$\frac{1}{\sqrt{2}}(u\bar{u} - d\bar{d})$
π^+	1	0	0	1	1	$u\bar{d}$

3076

Give a specific example of an SU(3) octet by naming all 8 particles. What is the value of the quantum numbers that are common to all the particles of the octet you have selected?

(Wisconsin)

Solution:

Eight nucleons and hyperons form an SU(3) octet, shown in Fig. 3.24. Their common quantum numbers are $J^) = \frac{1^+}{2}$, $B = 1$.

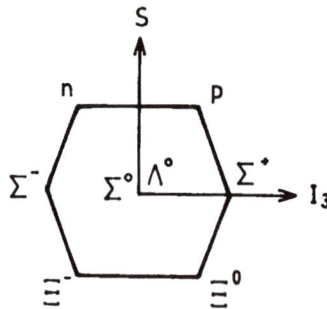

Fig. 3.24

3077

Calculate the ratio $R = \frac{\sigma(e^+e^- \to \text{hadrons})}{\sigma(e^+e^- \to \mu^+\mu^-)}$

(a) just below the threshold for "charm" production,

(b) above that threshold but below the b quark production threshold.

(*Wisconsin*)

Solution:

Quantum electrodynamics (QED) gives

$$\sigma(e^+e^- \to q_i\bar{q}_i \to \text{hadrons}) = \frac{4\pi\alpha^2}{S}Q_i^2,$$

where S is the square of the energy in the center-of-mass frame of e^+, e^-, α is the coupling constant, and Q_i is the electric charge (unit e) of the ith quark, and

$$\sigma(e^+e^- \to \mu^+\mu^-) = \frac{4\pi\alpha^2}{3S}.$$

Hence

$$R = \frac{\sigma(e^+e^- \to \text{hadrons})}{\sigma(e^+e^- \to \mu^+\mu^-)} = \sum_i \frac{\sigma(e^+e^- \to q_i\bar{q}_i \to \text{hadrons})}{\sigma(e^+e^- \to \mu^+\mu^-)} = 3\sum_i Q_i^2,$$

where $\sum_i$ sums over all the quarks which can be produced with the given energy.

(a) With such an energy the quarks which can be produced are u, d and s. Thus

$$R = 3\sum_i Q_i^2 = 3 \times \left(\frac{4}{9} + \frac{1}{9} + \frac{1}{9}\right) = 2.$$

(b) The quarks that can be produced are now u, d, s and c. As the charge of c quark is $2/3$,

$$R = 3\sum_i Q_i^2 = 3 \times \left(\frac{4}{9} + \frac{1}{9} + \frac{1}{9} + \frac{4}{9}\right) = \frac{10}{3}.$$

3078

(a) It is usually accepted that hadrons are bound states of elementary, strongly-interacting, spin-1/2 fermions called quarks. Briefly describe some evidence for this belief.

The lowest-lying mesons and baryons are taken to be bound states of the u, d, and s (or p, n, and λ in an alternative notation) quarks, which form an SU(3) triplet.

(b) Define what is meant by the approximate Gell-Mann–Neeman global SU(3) symmetry of strong interactions. How badly is this symmetry broken?

(c) Construct the lowest-lying meson and baryon SU(3) multiplets, giving the quark composition of each state and the corresponding quantum numbers J, P, I, Y, S, B and, where appropriate, G.

(d) What is the evidence for another quantum number "color", under which the strong interactions are exactly symmetric? How many colors are there believed to be? What data are used to determine this number?

(e) It is by now well established that there is a global SU(3) singlet quark c with charge 2/3 and a new quantum number C preserved by the strong interactions. Construct the lowest-lying $C = 1$ meson and baryon states, again giving J, P, I, Y, S and B.

(f) What are the main semileptonic decay modes (i.e., those decays that contain leptons and hadrons in the final state) of the $C = 1$ meson?

(g) Denoting the strange $J = 1$ and $J = 0$ charmed mesons by F^* and F respectively and assuming that $m_{F^*} > m_F + m_\pi$ (something not yet established experimentally), what rate do you expect for $F^* \to F\pi$. What might be the main decay mode of the F^*?

(Princeton)

Solution:

(a) The evidence supporting the quark model includes the following: (1) The deep inelastic scattering data of electrons on nucleons indicate that nucleon has substructure. (2) The SU(3) symmetry of hadrons can be explained by the quark model. (3) The quark model gives the correct cross-section relationship of hadronic reactions. (4) The quark model can explain the abnormal magnetic moments of nucleons.

(b) The approximate SU(3) symmetry of strong interactions means that isospin multiplets with the same spin and parity, i.e., same J^P, but different strangeness numbers can be transformed into each other. They are considered as the supermultiplet states of the same original particle U with different electric charges (I_3) and hypercharges (Y).

If SU(3) symmetry were perfect, particles of the same supermultiplet should have the same mass. In reality the difference of their masses can be quite large, which shows that such a supersymmetry is only approximate. The extent of the breaking of the symmetry can be seen from the difference between their masses, e.g., for the supermultiplet of 0^- mesons, $m_{\pi^0} = 135$ MeV, $m_{K^0} = 498$ MeV.

(c) The lowest-lying SU(3) multiplets of mesons and baryons formed by u, d and s quarks are as follows.

For mesons, the quarks can form octet and singlet of J^P equal to 0^- and 1^-. They are all ground states with $l = 0$, with quark contents and quantum numbers as listed in Table 3.9.

For baryons, which consist of three quarks each, the lowest-lying states are an octet of $J^P = \frac{1}{2}^+$ and a decuplet of $J^P = \frac{3}{2}^+$. They are ground states with $l = 0$ and other characteristics as given in Table 3.10.

(d) The purpose of introducing the color freedom is to overcome statistical difficulties. In the quark model, a quark has spin 1/2 and so must obey the Fermi statistics, which requires the wave function of a baryon to be antisymmetric for exchanging any two quarks. In reality, however, there are some baryons having quark contents sss or uuu, for which the wave functions are symmetric for quark exchange. To get over this contradiction, it is

Table 3.9 Quantum numbers and quark contents of meson supermultiplets of $J^P = 0^-, 1^-$

	0^-	1^-	quark content	I	I_3	Y	B	S	G
	π^+	ρ^+	$\bar{d}u$	1	$+1$	0	0	0	-1
	π^0	ρ^0	$(u\bar{u} - d\bar{d})/\sqrt{2}$	1	0	0	0	0	-1
	π^-	ρ^-	$\bar{u}d$	1	-1	0	0	0	-1
	K^+	K^{*+}	$\bar{s}u$	1/2	1/2	1	0	0	
octet	K^-	K^{*-}	$s\bar{u}$	1/2	$-1/2$	-1	0	0	
	K^0	K^{*0}	$\bar{s}d$	1/2	$-1/2$	1	0	0	
	$\bar{K}^0$	$\bar{K}^{*0}$	$s\bar{d}$	1/2	1/2	-1	0	0	
	$\eta(549)$		$\frac{(u\bar{u}+d\bar{d}-2\bar{s}s)}{\sqrt{6}}$	0	0	0	0	0	$+1$
		$\omega(783)$	$\frac{u\bar{u}+d\bar{d}}{\sqrt{2}}$	0	0	0	0	0	-1
singet	$\eta(958)$		$\frac{(u\bar{u}+d\bar{d}+\bar{s}s)}{\sqrt{3}}$	0	0	0	0	0	$+1$
		$\psi(1020)$	$\bar{s}s$	0	0	0	0	0	-1

Table 3.10 Characteristics of baryon octet ($\frac{1}{2}^+$) and decuplet ($\frac{3}{2}^+$)

J^P	particles	the quark content	I	I_3	Y	B	S
	p	uud	$1/2$	$1/2$	1	1	0
	n	udd	$1/2$	$-1/2$	1	1	0
	Σ^+	uus	1	1	0	1	-1
$\frac{1}{2}^+$	Σ^0	$s(ud+du)/\sqrt{2}$	1	0	0	1	-1
	Σ^-	dds	1	-1	0	1	-1
	Ξ^0	uss	$1/2$	$1/2$	-1	1	-2
	Ξ^-	dss	$1/2$	$-1/2$	-1	1	-2
	Λ^0	$s(du-ud)/\sqrt{2}$	0	0	0	1	-1
	Δ^-	ddd	$3/2$	$-3/2$	1	1	0
	Δ^0	ddu	$3/2$	$-1/2$	1	1	0
	Δ^+	duu	$3/2$	$1/2$	1	1	0
	Δ^{++}	uuu	$3/2$	$3/2$	1	1	0
$\frac{3}{2}^+$	Σ^{*-}	sdd	1	-1	0	1	-1
	Σ^{*0}	sdu	1	0	0	1	-1
	Σ^{*+}	suu	1	1	0	1	-1
	Ξ^{*-}	ssd	$1/2$	$-1/2$	-1	1	-2
	Ξ^{*0}	ssu	$1/2$	$1/2$	-1	1	-2
	Ω	sss	0	0	-2	1	-3

assumed that there is an additional quantum number called "color" which has three values. The hypothesis of color can be tested by the measurement of R in high-energy e^+e^- collisions, which is the ratio of the cross sections for producing hadrons and for producing a muonic pair

$$R = \frac{\sigma(e^+e^- \to \text{hadrons})}{\sigma(e^+e^- \to \mu^+\mu^-)} .$$

Suppose the energy of e^+e^- system is sufficient to produce all the three flavors of quarks. If the quarks are colorless,

$$R = \sum_i Q_i^2 = \left(\frac{4}{9} + \frac{1}{9} + \frac{1}{9}\right) = \frac{2}{3} ;$$

if each quark can have three colors,

$$R = 3\sum_i Q_i^2 = 3 \times \left(\frac{4}{9} + \frac{1}{9} + \frac{1}{9}\right) = 2 ,$$

The latter is in agreement with experiments.

(e) A $c(\bar{c})$ quark and an ordinary antiquark (quark) can combine into a charmed meson which can have J^P equal to 0^- or 1^-. The characteristics of charmed mesons are listed in Table 3.11. They can be regarded as the result of exchanging an $u(\bar{u})$ quark for a c ($\bar{c}$) quark in an ordinary meson. There are six meson states with $C = 1$, namely $D^+, D^0, F^+, D^{*+}, D^{*0}$ and F^{*+}. Also, a c quark and two ordinary quarks can combine into a charmed baryon of $J^P = \frac{1}{2}^+$ or $\frac{3}{2}^+$. Theoretically there should be 9 charmed baryons of $J^P = \frac{1}{2}^+$, whose characteristics are included in Table 3.12. Experimentally, the first evidence for charmed baryons Λ_c^+, Σ_c^{++} appeared in 1975, that for charmed mesons D^+, D^0, F^+ appeared in 1976–77.

Correspondingly, baryons with $C = 1$ and $J^P = \frac{3}{2}^+$ should exist. Theoretically there are six such baryons, with quark contents (ddc), (duc), (uuc), (cds), (css), (cus). Their expected quantum numbers, except for $J = 3/2$, have not been confirmed experimentally, but they should be the same as those of Σ_c^0, Σ_c^+, Σ_c^{++}, S^0, T^0 and S^+, respectively.

(f) The semileptonic decay of a meson with $C = 1$ actually arises from the semileptonic decay of its c quark:

$$c \to s\, l^+\, \nu_e, \quad \text{with amplitude} \sim \cos\theta_c,$$

$$c \to d\, l^+\, \nu_e, \quad \text{with amplitude} \sim \sin\theta_c,$$

Table 3.11 Characteristics of mesons with charmed quarks

J^P	particle	quark content	I	I_3	Y	S	C	B
	D^0	$\bar{u}c$	1/2	$-1/2$	1	0	1	0
	D^+	$\bar{d}c$	1/2	1/2	1	0	1	0
	$\bar{D}^0$	$\bar{c}u$	1/2	1/2	-1	0	-1	0
0^-	D^-	$\bar{c}d$	1/2	$-1/2$	-1	0	-1	0
	F^+	$\bar{s}c$	0	0	2	1	1	0
	F^-	$\bar{c}s$	0	0	-2	-1	-1	0
	η_0	$\bar{c}c$	0	0	0	0	0	0
	D^{*0}	$\bar{u}c$	1/2	$-1/2$	1	0	1	0
	D^{*+}	$\bar{d}c$	1/2	1/2	1	0	1	0
	$\bar{D}^{*0}$	$\bar{c}u$	1/2	1/2	-1	0	-1	0
1^-	D^{*-}	$\bar{c}d$	1/2	$-1/2$	-1	0	-1	0
	F^{*+}	$\bar{s}c$	0	0	2	1	1	0
	F^{*-}	$\bar{c}s$	0	0	-2	-1	-1	0
	J/ψ	$\bar{c}c$	0	0	0	0	0	0

Table 3.12 Characteristics of charmed baryons ($C = 1$) of $J^P = \frac{1}{2}^+$

Particle	Quark content	I	I_3	Y	S	C	B
Σ_c^{++}	cuu	1	1	2	0	1	1
Σ_c^{+}	$c(ud + du)/\sqrt{2}$	1	0	2	0	1	1
Σ_c^{0}	cdd	1	−1	2	0	1	1
S^{+}	$c(us + su)/\sqrt{2}$	1/2	1/2	1	−1	1	1
S^{0}	$c(ds + sd)/\sqrt{2}$	1/2	−1/2	1	−1	1	1
T^{0}	css	0	0	0	−2	1	1
Λ_c^{+}	$c(ud - du)/\sqrt{2}$	0	0	2	0	1	1
A^{+}	$c(us - su)/\sqrt{2}$	1/2	1/2	1	−1	1	1
A^{0}	$c(ds - sd)/\sqrt{2}$	1/2	−1/2	1	−1	1	1

where θ_c is the Cabibbo angle. For example, the reaction $D^0 \to K^- e^+ \nu_e$, is a Cabibbo-allowed decay, and $D^0 \to \pi^- e^+ \nu_e$, is a Cabibbo-forbidden decay.

(g) If F^* exists and $m_{F^*} > m_F + m_\pi$, then $F^* \to \pi^0 F$ is a strong decay and hence the main decay channel, as it obeys all the conservation laws. For example, F^* has $J^{PC} = 1^{--}$, F has $J^{PC} = 0^{-+}$, pion has $J^{PC} = 0^{-+}$. In the decay $F^* \to \pi^0 F$, the orbital angular momentum of the πF system is $l = 1$, the parity of the final state is $P(\pi^0)P(F)(-1)^l = -1$. Also, $C(\pi^0)C(F) = 1$. Thus the final state has $J^P = 1^-$, same as $J^P(F^*)$.

Another competiting decay channel is $F^* \to \gamma + F$, which is an electromagnetic decay with the relative amplitude determined by the interaction constant and the phase-space factor.

3079

Imagine that you have performed an experiment to measure the cross sections for the "inclusive" process

$$a + N \to \mu^+ + \mu^- + \text{anything}$$

where $a = p$, π^+ or π^-, and N is a target whose nuclei have equal numbers of protons and neutrons.

You have measured these three cross sections as a function of m, the invariant mass of the muon pair, and of s, the square of the energy in the center of mass.

The following questions are designed to test your understanding of the most common model used to describe these processes, the quark-antiquark annihilation model of Drell and Yan.

(a) In the simplest quark picture (baryons being composed of three quarks and mesons of a quark-antiquark pair), what is the predicted ratio

$$\frac{d\sigma_{pN}(s,m)}{dm} : \frac{d\sigma_{\pi^+N}(s,m)}{dm} : \frac{d\sigma_{\pi^-N}(s,m)}{dm} ?$$

(b) An accurate measurement shows each element of the ratio to be nonzero. How do you modify you answer to (a) to account for this? (A one or two sentence answer is sufficient.)

(c) Given this modification, how do you expect the ratio to behave with m (for fixed s)? (Again, a one or two sentence qualitative answer is sufficient.)

(d) How would the predicted values of the three cross sections change if the concept of color were introduced into the naive model?

(e) An important prediction of Drell and Yan is the concept of scaling. Illustrate this with a formula or with a sketch (labeling the ordinate and the abscissa).

(f) How would you determine the quark structure of the π^+ from your data?

(g) How would you estimate the antiquark content of the proton?

(*Princeton*)

Solution:

(a) According to the model of Drell and Yan, these reactions are processes of annihilation of a quark and an antiquark with emission of a leptonic pair. QED calculations show that if the square of the energy in the center of mass of the muons $s_{\mu u} \gg m_\mu^2, m_q^2$, the effect of m_μ, m_q can be neglected, yielding

$$\sigma(\mu^+\mu^- \to \gamma \to q_i\bar{q}_i) = \frac{4\pi}{3s_{\mu\mu}}\alpha^2 Q_i^2 ,$$

where Q_i is the charge number of the i quark, α is the fine structure constant. Making use of the principle of detailed balance, we find

$$\sigma(q_i\bar{q}_i \to \gamma \to \mu^+\mu^-) = \frac{4\pi}{3s}\alpha^2 Q_i^2 = Q_i^2\sigma_0 ,$$

where s is the square of total energy in the center-of-mass system of the two quarks, i.e., $s = s_{\mu\mu} = m^2$, m being the total energy in the center-of-mass system of the $\mu^+\mu^-$ (i.e. in the c.m.s. of $q_i \bar{q}_i$). Thus in the simplest quark picture,

$$\sigma(d\bar{d} \to \mu^+\mu^-) \approx \frac{1}{9}\sigma_0 ,$$

$$\sigma(u\bar{u} \to \mu^+\mu^-) \approx \frac{4}{9}\sigma_0 ,$$

For $pN \to \mu^+\mu^- + X$, as there is no antiquark in the proton and in the neutron,

$$\frac{d\sigma(s,m)}{dm} = 0 .$$

For the same s and m, recalling the quark contents of p, n, π^+ and π^- are uud, udd, $u\bar{d}$, $\bar{u}d$ respectively, we find

$$\sigma(\pi^+ N) = \sigma\left[(u\bar{d}) + \frac{1}{2}(uud + udd)\right] = \frac{1}{2}\sigma(d\bar{d})(1+2) \approx \frac{1}{6}\sigma_0 ,$$

$$\sigma(\pi^- N) = \sigma\left[(\bar{u}d) + \frac{1}{2}(uud + udd)\right] = \frac{1}{2}\sigma(u\bar{u})(2+1) \approx \frac{2}{3}\sigma_0 ,$$

and hence

$$\frac{d\sigma_{pN}(s,m)}{dm} : \frac{d\sigma_{\pi^+ N}(s,m)}{dm} : \frac{d\sigma_{\pi^- N}(s,m)}{dm} = 0 : 1 : 4 .$$

(b) The result that $\frac{d\sigma_{pN}(s,m)}{dm}$ is not zero indicates that there are anti-quarks in proton and neutron. Let the fraction of antiquarks in a proton or a neutron be α, where $\alpha \ll 1$. Then the fraction of quark is $(1-\alpha)$ and so

$$\sigma_{pN} = \sigma\left\{ 2\alpha\bar{u} + \frac{1}{2}[2(1-\alpha)u + (1-\alpha)u] + 2(1-\alpha)u \right.$$

$$+ \frac{1}{2}(2\alpha\bar{u} + \alpha\bar{u}) + \alpha\bar{d} + \frac{1}{2}[(1-\alpha)d + 2(1-\alpha)d] + (1-\alpha)d$$

$$\left. + \frac{1}{2}(\alpha\bar{d} + 2\alpha\bar{d})] \right\}$$

$$= \sigma(u\bar{u})[3\alpha(1-\alpha) + 3(1-\alpha)\alpha] + \sigma(d\bar{d})\left[\frac{3}{2}\alpha(1-\alpha) + \frac{3}{2}(1-\alpha)\alpha\right]$$

$$= 6\alpha(1-\alpha)\sigma(u\bar{u}) + 3\alpha(1-\alpha)\sigma(d\bar{d})$$

$$= 3\alpha(1-\alpha)[2\sigma(u\bar{u}) + \sigma(d\bar{d})] \approx 3\alpha(1-\alpha)\sigma_0 ,$$

$$\sigma_{\pi^+ N} = \sigma\left\{\bar{d} + \frac{1}{2}[(1-\alpha)d + 2(1-\alpha)d] + u + \frac{1}{2}(2\alpha\bar{u} + \alpha\bar{u})\right\}$$

$$= \frac{3}{2}(1-\alpha)\sigma(d\bar{d}) + \frac{3}{2}\alpha\sigma(u\bar{u})$$

$$= \frac{3}{2}\sigma(d\bar{d}) + \frac{3}{2}\alpha[\sigma(u\bar{u}) - \sigma(d\bar{d})] \approx \frac{1}{6}(1+3\alpha)\sigma_0 ,$$

$$\sigma_{\pi^- N} = \sigma\left\{\bar{u} + \frac{1}{2}[2(1-\alpha)u + (1-\alpha)u] + d + \frac{1}{2}(\alpha\bar{d} + 2\alpha\bar{d})\right\}$$

$$= \frac{3}{2}(1-\alpha)\sigma(u\bar{u}) + \frac{3}{2}\alpha\sigma(d\bar{d})$$

$$= \frac{3}{2}\sigma(u\bar{u}) + \frac{3}{2}\alpha[\sigma(d\bar{d}) - \sigma(u\bar{u})] \approx \frac{1}{6}(4-3\alpha)\sigma_0 .$$

Hence

$$\frac{d\sigma_{pN}(s,m)}{dm} : \frac{d\sigma_{\pi^+ N}(s,m)}{dm} : \frac{d\sigma_{\pi^- N}(s,m)}{dm} = 18\alpha(1-\alpha) : (1+3\alpha) : (4-3\alpha).$$

For example if $\alpha = 0.01$, the cross sections are in the ratio $0.17 : 1 : 3.85$. Thus the cross sections, especially $\frac{d\sigma_{pN}(s,m)}{dm}$, is extremely sensitive to the fraction of antiquarks in the nucleon.

(c) An accurate derivation of the ratio is very complicated, as it would involve the structure functions of the particles (i.e., the distribution of quarks and their momenta in the nucleon and meson). If we assume that the momenta of the quarks in a nucleon are the same, then the cross section in the quark-antiquark center-of-mass system for a head-on collision is

$$\sigma(q_i \bar{q}_i \to \mu^+ \mu^-) = \frac{4\pi}{3m^2}\alpha^2 Q_i^2 ,$$

or

$$\frac{d\sigma}{dm} \sim m^{-3}\alpha^2 Q_i^2 .$$

Hence σ is proportional to m^{-2}, in agreement with experiments.

(d) The ratio would not be affected by the introduction of color.

(e) Scaling means that in a certain energy scale the effect on Drell-Yan process of smaller energies can be neglected. For instance, for second order electromagnetic processes, we have the general formula $d\sigma_{em} = \alpha^2 f(s, q^2, m_l)$, where s is the square of energy in the center-of-mass system, q^2 is the square of the transferred 4-momentum, and m_l is the mass of the charged particle. If s and $|q^2| \gg m_l^2$, it is a good approximation to set $m_l = 0$, yielding

$$d\sigma = \alpha^2 f(s, q^2).$$

Thus, for example, in the process $q_i \bar{q}_i \to \mu^+ \mu^-$, if $m \gg m_\mu, m_q$ we can let $m_\mu \approx m_q \approx 0$ and obtain

$$\sigma(q_i \bar{q}_i \to \mu^+ \mu^-) \propto Q_i^2 / m^2.$$

(f) The good agreement between the calculated result

$$\frac{d\sigma_{\pi^+ N}(s, m)}{dm} : \frac{d\sigma_{\pi^- N}(s, m)}{dm} = 1 : 4$$

and experiment supports the assumption of quark contents of $\pi^+(u\bar{d})$ and $\pi^-(\bar{u}d)$.

(g) By comparing the calculation in (b) with experiment we can determine the fraction α of antiquark in the quark content of proton.

3080

The bag model of hadron structure has colored quarks moving as independent spin-$\frac{1}{2}$ Dirac particles in a cavity of radius R. The confinement of the quarks to this cavity is achieved by having the quarks satisfy the free Dirac equation with a mass that depends on position: $m = 0$ for $r < R$ and $m = \infty$ for $r > R$. The energy operator for the quarks contains a term $\int d^3 r m(r) \bar{\psi} \psi$. In order for this term to give a finite contribution to the energy, the allowed Dirac wave functions must satisfy $\bar{\psi}\psi = 0$ where $m = \infty$ (i.e. for $r > R$), This is achieved by choosing a boundary condition at R on the solution of the Dirac equation.

(a) Show that the boundary conditions

(1) $\psi(|\mathbf{x}| = R) = 0$, (2) $i\hat{\mathbf{x}} \cdot \boldsymbol{\gamma}\psi(|\mathbf{x}| = R) = \psi(|\mathbf{x}| = R)$, where $\hat{\mathbf{x}}$ is the unit radial vector from the center of the cavity, both achieve the effect of setting $\bar{\psi}\psi = 0$ at $|\mathbf{x}| = R$. Which condition is physically acceptable?

(b) The general s-wave solution to the free massless Dirac equation can be written (using Bjorken–Drell conventions) as

$$\psi = N \begin{pmatrix} j_0(kR)x \\ i\boldsymbol{\sigma} \cdot \hat{\mathbf{x}} j_1(kR)x \end{pmatrix},$$

where $x = $ 2-component spinor, $j_l = $ spherical Bessel function, $N = $ normalization constant. (Our convention is that $\gamma_0 = \begin{pmatrix} I & 0 \\ 0 & -I \end{pmatrix}$, $\boldsymbol{\gamma} = \begin{pmatrix} 0 & \boldsymbol{\sigma} \\ -\boldsymbol{\sigma} & 0 \end{pmatrix}$, $\boldsymbol{\sigma} = $ Pauli matrices). Use the boundary condition at $|\mathbf{x}| = R$ to obtain a condition that determines k (do not try to solve the equation).

(*Princeton*)

Solution:

(a) Clearly, the condition (1), $\psi(X = R) = 0$, satisfies the condition $\bar{\psi}\psi|_{X=R} = 0$. For condition (2), we have (at $X = R$)

$$\bar{\psi}\psi = (i\hat{\mathbf{x}} \cdot \boldsymbol{\gamma}\psi)^+ \beta(i\hat{\mathbf{x}} \cdot \boldsymbol{\gamma}\psi)$$

$$= (-i\psi^+\hat{\mathbf{x}} \cdot \beta\boldsymbol{\gamma}\beta)\beta(i\hat{\mathbf{x}} \cdot \boldsymbol{\gamma}\psi)$$

$$= \psi^+ \beta(\hat{\mathbf{x}} \cdot \boldsymbol{\gamma})(\hat{\mathbf{x}} \cdot \boldsymbol{\gamma})\psi.$$

As

$$(\hat{\mathbf{x}} \cdot \boldsymbol{\gamma})(\hat{\mathbf{x}} \cdot \boldsymbol{\gamma}) = \begin{pmatrix} 0 & \boldsymbol{\sigma} \cdot \hat{\mathbf{x}} \\ -\boldsymbol{\sigma} \cdot \hat{\mathbf{x}} & 0 \end{pmatrix}\begin{pmatrix} 0 & \boldsymbol{\sigma} \cdot \hat{\mathbf{x}} \\ -\boldsymbol{\sigma} \cdot \hat{\mathbf{x}} & 0 \end{pmatrix} = -1,$$

we have

$$\bar{\psi}\psi = -\psi^+ \beta\psi = -\bar{\psi}\psi,$$

and hence

$$\bar{\psi}\psi|_{X=R} = 0.$$

The second condition is physically acceptable. The Dirac equation consists of four partial differential equations, each of which contains first partial differentials of the coordinates. Hence four boundary conditions are needed.

The requirement that wave functions should tend to zero at infinity places restriction on half of the solutions. This is equivalent to two boundary conditions, and we still need two more boundary conditions. $\Psi(X = R) = 0$ is equivalent to four boundary conditions, while the condition

$$i\hat{\mathbf{x}} \cdot \gamma\psi(X = R) = \psi(X = R),$$

i.e.,

$$i\begin{pmatrix} 0 & \boldsymbol{\sigma} \cdot \hat{\mathbf{x}} \\ -\boldsymbol{\sigma} \cdot \hat{\mathbf{x}} & 0 \end{pmatrix}\begin{pmatrix} \alpha \\ \beta \end{pmatrix} = \begin{pmatrix} \alpha \\ \beta \end{pmatrix},$$

or

$$i(\boldsymbol{\sigma} \cdot \hat{\mathbf{x}})\beta = \alpha,$$

only has two equations which give the relationship between the major and minor components. Therefore, only the condition (2) is physically acceptable. We can see from the explicit expression of the solution in (b) that the major and minor components of the Dirac spinor contain Bessel functions of different orders and so cannot both be zero at $X = R$. Condition (1) is thus not appropriate.

(b) The condition $\alpha = i(\boldsymbol{\sigma} \cdot \hat{\mathbf{x}})\beta$ gives

$$j_0(kR)x = i \cdot i(\boldsymbol{\sigma} \cdot \hat{\mathbf{x}})(\boldsymbol{\sigma} \cdot \hat{\mathbf{x}})j_1(kR)x,$$

or

$$j_0(kR) = -j_1(kR),$$

which determines k.

3081

The bag model of hadron structure has colored quarks moving as independent spin-half Dirac particles within a spherical cavity of radius R. To obtain wave functions for particular hadron states, the individual quark "orbitals" must be combined to produce states of zero total color and the appropriate values of the spin and flavor (isospin, charge, strangeness) quantum numbers.

In the very good approximation that the "up" and "down" quarks are massless one can easily obtain the lowest energy (s-wave) bag orbitals. These are given by the Dirac spinor

$$\psi = N \begin{pmatrix} j_0(kr)x \\ i\boldsymbol{\sigma} \cdot \hat{\mathbf{r}} j_1(kr)x \end{pmatrix},$$

where x is a 2-component spinor, $k = 2.04/R$, $j_l =$ spherical Bessel function.

(a) The lowest-lying baryons (proton and neutron) are obtained by putting three quarks in this orbital. How would you construct the wave function for the proton and for the neutron, i.e., which quarks would be combined and what is the structure of the spin wave function consistent with the quantum numbers of proton and neutron and Pauli's principle?

(b) The magnetic moment operator is defined as $\boldsymbol{\mu} = \int_{|\mathbf{x}|<R} d^3\mathbf{x} \frac{1}{2}\mathbf{r} \times \mathbf{J}_{EM}$, where $\mathbf{J}_{EM}$ is the usual Dirac electric current operator. Find an expression for this operator in terms of the spin operators of the constituent quarks. (You may leave integrals over Bessel functions undone.)

(c) Show that $\mu_n/\mu_p = -2/3$.

You may need the following Clebsch–Gordon coefficients:

$$\langle 1/2, 1/2 | 1, 1; 1/2, 1/2 \rangle = (2/3)^{1/2},$$

$$\langle 1/2, 1/2 | 1, 0; 1/2, 1/2 \rangle = -(1/3)^{1/2}.$$

(Princeton)

Solution:

(a) If we neglect "color" freedom, the lowest states of a baryon (p and n) are symmetric for quark exchange. Since the third component of the isospin of p is $I_3 = 1/2$, while u has $I_3 = \frac{1}{2}$, d has $I_3 = -\frac{1}{2}$, its quark content must be uud. As the system has isospin $\frac{1}{2}$ it cannot be completely symmetric for ud exchange, (i.e., the wave function cannot be in the form $uud + udu + duu$ as this would result in a decuplet with $I = 3/2$). Thus the wave function must have components of the form $uud - udu$. But, as mentioned above, the lowest-state baryon is perfectly symmetric for quark exchange. We have to multiply such forms with a spin wave function antisymmetric for exchanging the second and third quarks ($\uparrow\uparrow\downarrow - \uparrow\downarrow\uparrow$) to yield a wave function symmetric with respect to such an exchange:

$$u \uparrow (1)u \uparrow (2)d \downarrow (3) - u \uparrow (1)d \uparrow (2)u \downarrow (3)$$

$$- u \uparrow (1)u \downarrow (2)d \uparrow (3) + u \uparrow (1)d \downarrow (2)u \uparrow (3).$$

Note this also satisfies the isospin conditions. Then use the following procedure to make the wave function symmetric for exchanging the first and second quarks, and the first and third quarks. Exchanging the first and second quarks gives

$$u \uparrow u \uparrow d \downarrow -d \uparrow u \uparrow u \downarrow -u \downarrow u \uparrow d \uparrow +d \downarrow u \uparrow u \uparrow,$$

and exchanging the first and third quarks gives

$$d \downarrow u \uparrow u \uparrow -u \downarrow d \uparrow u \uparrow -d \uparrow u \downarrow u \uparrow +u \uparrow d \downarrow u \uparrow.$$

Combining the above three wave functions and normalizing, we have

$$\frac{1}{\sqrt{18}}(2u \uparrow u \uparrow d \downarrow +2u \uparrow d \downarrow u \uparrow +2d \downarrow u \uparrow u \uparrow -u \uparrow u \downarrow d \uparrow -u \uparrow d \uparrow u \downarrow$$

$$- u \downarrow u \uparrow d \uparrow -u \downarrow d \uparrow u \uparrow -d \uparrow u \uparrow u \downarrow -d \uparrow u \downarrow u \uparrow).$$

The color wave function antisymmetric for exchanging any two quarks takes the form

$$\frac{1}{\sqrt{6}}(RGB - RBG + GBR - GRB + BRG - BGR).$$

Let

$$\psi_\uparrow = \begin{pmatrix} j_0(kr)x(\uparrow) \\ ij_1(kr)\boldsymbol{\sigma} \cdot \hat{\mathbf{r}}x(\uparrow) \end{pmatrix}, \qquad x(\uparrow) = \begin{pmatrix} 1 \\ 0 \end{pmatrix},$$

$$\psi_\downarrow = \begin{pmatrix} j_0(kr)x(\downarrow) \\ ij_1(kr)\boldsymbol{\sigma} \cdot \hat{\mathbf{r}}x(\downarrow) \end{pmatrix}, \qquad x(\downarrow) = \begin{pmatrix} 0 \\ 1 \end{pmatrix}.$$

To include in the orbital wave functions, we need only to change $\uparrow$ to $\psi \uparrow$, and $\downarrow$ to $\psi \downarrow$. Then the final result is

$$\frac{1}{6\sqrt{3}}(RGB - RBG + GBR - GRB + BRG - BGR)$$

$$\times (2u\psi \uparrow u\psi \uparrow d\psi \downarrow +2u\psi \uparrow d\psi \downarrow u\psi \uparrow +2d\psi \downarrow u\psi \uparrow u\psi \uparrow$$

$$- u\psi \uparrow u\psi \downarrow d\psi \uparrow -u\psi \uparrow d\psi \uparrow u\psi \downarrow -u\psi \downarrow u\psi \uparrow d\psi \uparrow$$

$$- u\psi \downarrow d\psi \uparrow u\psi \uparrow -d\psi \uparrow u\psi \uparrow u\psi \downarrow -d\psi \uparrow u\psi \downarrow u\psi \uparrow).$$

The neutron wave function can be obtained by applying the isospin-flip operator on the proton wave function $(u \leftrightarrow d)$, resulting in

$$\frac{1}{6\sqrt{3}}(RGB - RBG + GBR - GRB + BRG - BGR)$$

$$\times \ (2d\psi \uparrow d\psi \uparrow u\psi \downarrow + 2d\psi \uparrow u\psi \downarrow d\psi \uparrow + 2u\psi \downarrow d\psi \uparrow d\psi \uparrow$$

$$- \ d\psi \uparrow d\psi \downarrow u\psi \uparrow - d\psi \uparrow u\psi \uparrow d\psi \downarrow - d\psi \downarrow d\psi \uparrow u\psi \uparrow$$

$$- \ d\psi \downarrow u\psi \uparrow d\psi \uparrow - u\psi \uparrow d\psi \uparrow d\psi \downarrow - u\psi \uparrow d\psi \downarrow d\psi \uparrow).$$

The above wave functions are valid only for spin-up proton and neutron. For spin-down nucleons, the wave functions can be obtained by changing $\uparrow$ into $\downarrow$, $\downarrow$ into $\uparrow$ in the spin-up wave function.

(b) The Dirac current operator is defined as

$$\mathbf{J} = Q\bar{\psi}^*\gamma\psi = Q\bar{\psi}^*\beta\boldsymbol{\alpha}\psi = Q\psi_{\downarrow}^{\dagger}\boldsymbol{\alpha}\psi = Q\psi_{\downarrow}^{\dagger}\begin{pmatrix} 0 & \boldsymbol{\sigma} \\ \boldsymbol{\sigma} & 0 \end{pmatrix}\psi .$$

where $\boldsymbol{\sigma}$ is the Pauli matrix. Inserting the expression of ψ into the above, we have

$$\mathbf{J}_{EM} = QN^+N(j_0(kr)x_{\downarrow}^+, -ij_1(kr)x_{\downarrow}^+\boldsymbol{\sigma} \cdot \hat{\mathbf{r}})$$

$$\times \begin{pmatrix} 0 & \boldsymbol{\sigma} \\ \boldsymbol{\sigma} & 0 \end{pmatrix}\begin{pmatrix} j_0(kr)x \\ ij_1(kr)\boldsymbol{\sigma} \cdot \hat{\mathbf{r}}x \end{pmatrix}$$

$$= iQ|N|^2 j_0(kr)j_1(kr)x_{\downarrow}^+[\boldsymbol{\sigma}, \boldsymbol{\sigma} \cdot \hat{\mathbf{r}}]x$$

$$= iQ|N|^2 j_0(kr)j_1(kr)x_{\downarrow}^+(-2i\boldsymbol{\sigma} \times \hat{\mathbf{r}})x$$

$$= 2Q|N|^2 j_0(kr)j_1(kr)x_{\downarrow}^+(\boldsymbol{\sigma} \times \hat{\mathbf{r}})x ,$$

and hence

$$\mu = \int_{|X|<R}\frac{1}{2}\mathbf{r} \times J_{EM}d^3X$$

$$= \int_{|X|<R}Q|N|^2 j_0(kr)j_1(kr)x_{\downarrow}^+[r\boldsymbol{\sigma} - (\boldsymbol{\sigma} \cdot \mathbf{r})\mathbf{r}]xd^3X .$$

When we integrate this over the angles the second term in the brackets gives zero. Thus

$$\mu = 4\pi Q|N|^2 \left[\int_{r<R} r^3 j_0(kr) j_1(kr) \right] x_\downarrow^+ \sigma x dr .$$

(c) The expected value of the magnetic moment of a spin-up proton is

$$\langle p \uparrow |\mu|p \uparrow\rangle = \frac{Q}{18}\left[4\left(\frac{2}{3}+\frac{2}{3}+\frac{1}{3}\right) + 4\left(\frac{2}{3}+\frac{2}{3}+\frac{1}{3}\right) \right.$$

$$+ 4\left(\frac{2}{3}+\frac{2}{3}+\frac{1}{3}\right) + \left(\frac{2}{3}-\frac{2}{3}-\frac{1}{3}\right)$$

$$+ \left(\frac{2}{3}-\frac{2}{3}-\frac{1}{3}\right) + \left(\frac{2}{3}-\frac{2}{3}-\frac{1}{3}\right)$$

$$\left. + \left(\frac{2}{3}-\frac{2}{3}-\frac{1}{3}\right) + \left(\frac{2}{3}-\frac{2}{3}-\frac{1}{3}\right) + \left(\frac{2}{3}-\frac{2}{3}-\frac{1}{3}\right) \right]$$

$$= Q .$$

Similarly,

$$\langle n \uparrow |\mu|n \uparrow\rangle = \frac{Q}{18}\left[3\times 4\left(-\frac{1}{3}-\frac{1}{3}-\frac{2}{3}\right) + 6\left(-\frac{1}{3}+\frac{1}{3}-\frac{2}{3}\right) \right]$$

$$= -\frac{2}{3}Q .$$

Therefore

$$\frac{\mu_n}{\mu_p} = \frac{\langle n \uparrow |\mu|n \uparrow\rangle}{\langle p \uparrow |\mu|p \uparrow\rangle} = -\frac{2}{3} .$$

3082

Recent newspaper articles have touted the discovery of evidence for gluons, coming from colliding beam e^+e^- experiments. These articles are inevitably somewhat garbled and you are asked to do better.

(a) According to current theoretical ideas of quantum chromodynamics (based on gauge group SU(3)): What are gluons? How many different kinds are there? What are their electrical charge? What is spin of a gluon?

(b) One speaks of various quark types or 'flavors', e.g., 'up' quarks, 'down' quarks, etc. According to QCD how many types of quark are there for each flavor? What are their charges? Does QCD say anything about the number of different flavors? According to currently available evidence how many different flavors are in fact recently well established? Discuss the evidence. Discuss also what weak interaction ideas say about whether, given the present flavors, there is reason to expect more, and characterize the "morez". How do results on the inclusive cross section for $e^+ + e^- +$ hadrons, at various energies, bear on the number of flavors?

(c) At moderately high energies one finds that the hadrons coming from e^+e^- collisions form two 'jets (Fig. 3.25). This has made people happy. How does one account for this two-jet phenomenon on the quark-gluon picture? At still higher energies one occasionally sees three jets. This has also made people happy. Account for this three-jet phenomenon.

(Princeton)

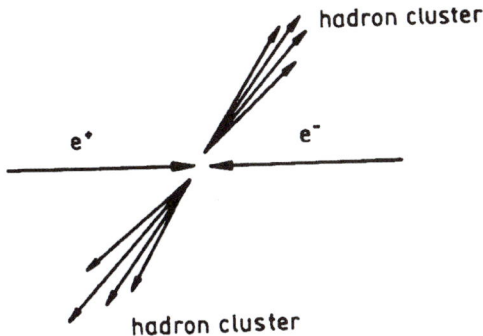

Fig. 3.25

Solution:

(a) According to QCD, hadrons consist of quarks and interactions between quarks are mediated by gluon field. Similar to the role of photons in electromagnetic interaction, gluons are propagators of strong interaction. There are eight kinds of gluons, all vector particles of electric charge zero, spin 1.

(b) In QCD theory, each kind of quark can have three colors, and quarks of the same flavor and different colors carry the same electric charge. An important characteristic of quarks is that they have fractional charges. QCD gives a weak limitation to the number of quarks, namely, if the number of quark flavors is larger than 16, asymototic freedom will be violated. The weak interaction does not restrict the number of quark flavors. However, cosmology requires the types of neutrino to be about 3 or 4 and the symmetry between leptons and quarks then restricts the number of flavors of quarks to be not more than 6 to 8. At various energies the relative total cross section for hadron production

$$R = \frac{\sigma(e^+e^- \to \text{hadrons})}{\sigma(e^+e^- \to \mu^+\mu^-)}$$

has been found to agree with

$$R(E) = 3\sum_i Q_i^2,$$

where the summation is over all quarks that can be produced at energy E, Q_i is the electric charge of the ith quark, and the factor 3 accounts for the three colors (**Problem 3078 (d)**).

(c) The two-jet phenomenon in e^+e^- collisions can be explained by the quark model. The colliding high energy e^+, e^- first produce a quark-antiquark pair of momenta $\mathbf{p}$ and $-\mathbf{p}$. When each quark fragments into hadrons, the sum of the hadron momenta in the direction of $\mathbf{p}$ is $\sum p_{||} = |\mathbf{p}|$, and in a transverse direction of $\mathbf{p}$ is $\sum p_\perp = 0$. In other words, the hadrons produced in the fragmentation of the quark and antiquark appear as two jets with axes in the directions of $\mathbf{p}$ and $-\mathbf{p}$. Measurements of the angular distribution of the jets about the electron beam direction have shown that quarks are ferminions of spin $1/2$.

The three-jet phenomenon can be interperated as showing hard gluon emission in the QCD theory. At high energies, like electrons emitting photons, quarks can emit gluons. In e^+e^- collisions a gluon emitted with the quark pair can separately fragment into a hadron jet. From the rate of three-jet events it is possible to calculate α_s, the coupling constant of strong interaction.

3083

The observation of narrow long-lived states $(J/\psi, \psi')$ suggested the existence of a new quantum number (charm). Recently a new series of massive states has been observed through their decay into lepton pairs ($\Upsilon, \Upsilon', \ldots$ with masses ~ 10 GeV/c^2). Suppose the observation is taken to imply yet another quantum number (beauty).

(a) Estimate roughly the mass of the beauty quark.

(b) If this quark has an electric charge of $-1/3$ indicate how the Gell-Mann–Nishijima formula should be modified to incorporate the new quantum number.

(c) In the context of the conventional (colored) quark model, estimate the value of the ratio

$$R = \frac{\sigma(e^+ e^- \rightarrow \text{hadrons})}{\sigma(e^+ e^- \rightarrow \mu^+ \mu^-)}$$

in the region well above the threshold for the production of the beauty.

(d) How would you expect the cross section for production of an Υ ($b\bar{b}$ bound state) in colliding $e^+ e^-$ beams to change if the charge of b quark is $+2/3$ instead of $-1/3$? How would the branching ratio to lepton pairs change? What might be the change in its production cross section in hadronic collisions? Discuss this last answer briefly.

(Princeton)

Solution:

(a) The heavy meson Υ is composed of $b\bar{b}$. Neglecting the binding energy of b quarks, we have roughly

$$m_b \approx \frac{1}{2} M_\Upsilon \approx 5 \text{ GeV}/e^2 .$$

(b) For u, d, and s quarks, the Gell-Mann–Nishijima formula can be written as

$$Q = I_3 + \frac{1}{2}(B + S) .$$

Let the charm c of c quark be 1, the beauty b of b quark be -1. Then the Gell-Mann–Nishijima formula can be generalized as

$$Q = I_3 + \frac{1}{2}(B + S + c + b) ,$$

which gives for c quark, $Q(c) = 0 + \frac{1}{2}(\frac{1}{3} + 0 + 1 + 0) = \frac{2}{3}$; for b quark $Q(b) = 0 + \frac{1}{2}(\frac{1}{3} + 0 + 0 - 1) = -\frac{1}{3}$

(c) If a certain quark q_i can be produced, its contribution to R is

$$R = \frac{\sigma(e^+e^- \to q_i \bar{q}_i)}{\sigma(e^+e^- \to \mu^+\mu^-)} = 3Q_i^2 \,,$$

where Q_i is its charge, and the factor 3 accounts for the three colors. If the c.m.s energy is above the threshold for producing beauty, the five flavors of quarks u, d, s, c, and b can be produced. Hence

$$R(E) = 3\sum_i Q_i^2 = 3\left[3 \times \left(\frac{1}{3}\right)^2 + 2 \times \left(\frac{2}{3}\right)^2\right] = \frac{11}{3} \,.$$

(d) The cross section for the resonance state Υ is given by

$$\sigma = \frac{\pi(2J+1)}{m^2} \frac{\Gamma_{ee}\Gamma}{(E-m)^2 + \frac{\Gamma^2}{4}} \,,$$

where J and m are the spin and mass of Υ respectively, Γ is the total width of the resonance state, Γ_{ee} is the partial width of the e^+e^- channel. The partial width of $\Upsilon \to e^+e^-$ is

$$\Gamma_{ee}(\Upsilon \to e^+e^-) = 16\pi \frac{\alpha^2 Q_b^2}{m_b^2} |\psi(0)|^2 \,,$$

where $\Psi(0)$ is the ground state wave function, Q_b and m_b are the charge and mass of b respectively, and α is the fine structure constant. At $E \approx m$,

$$\sigma = \frac{12\pi\Gamma_{ee}}{m^2\Gamma} \propto Q_b^2 \,,$$

as Υ has spin $J = 1$.

When the charge of b quark changes from $-\frac{1}{3}$ to $\frac{2}{3}$, Q_p^2 changes from $\frac{1}{9}$ to $\frac{4}{9}$ and σ increases by 3 times. This means that both the total cross section and the partial width for the leptonic channel increase by 3 times.

There is no resonance in the production cross section in hadron collisions, because the hadron collision is a reaction process $h+\bar{h} \to \Upsilon+X$, but not a production process as $e^+ + e^- \to \Upsilon$. However, in the invariant mass spectrum of μ pairs (or e pairs) in hadron collisions we can see a small peak at the invariant mass $m(\mu\mu) = m_\Upsilon$. The height of this peak will increase by 3 times also.

3084

The recently discovered $\psi(M = 3.1 \text{ GeV}/c^2)$ and $\psi^*(M = 3.7 \text{ GeV}/c^2)$ particles are both believed to have the following quantum numbers:

$$J^P = 1^-,$$

$$C = -1 (\text{charge conjugation}),$$

$$I = 0 (I\text{-spin}),$$

$$Q = 0.$$

Indicate which of the following decay modes are allowed by strong interaction, which by electromagnetic and which by weak interaction, and which are strictly forbidden. If strong decay is forbidden or if the decay is strictly forbidden, state the selection rule.

$$\psi \to \mu^+ \mu^-$$

$$\psi \to \pi^0 \pi^0$$

$$\psi^* \to \psi \pi^+ \pi^-$$

$$\psi^* \to \psi + \eta'(0.96 \text{ GeV}/c^2)$$

(Wisconsin)

Solution:

The process $\psi \to \mu^+ \mu^-$ is the result of electromagnetic interaction, and the decay $\psi^* \to \psi \pi^+ \pi^-$ is a strong interaction process. The decay mode $\psi \to \pi^0 \pi^0$ by strong interaction is forbidden since the C-parity of ψ is -1 and that of the two π^0 in the final state is $+1$, violating the conservation of C-parity in strong interaction. The decay mode $\psi^* \to \psi + \eta'(0.96 \text{ GeV}/c^2)$ is strictly forbidden as it violates the conservation of energy.

3085

At SPEAR (e^+e^- colliding-beam storage ring) several states called ψ, χ have been observed. The ψ's have quantum numbers of the photon ($J^P = 1^-$, $I^G = 0^-$) and have massses at 3.1 and 3.7 GeV/c². Suppose the following reaction was observed:

$$e^+e^- \to \psi(3.7) \to \gamma + \chi$$
$$\, \rightharpoondown \pi^+\pi^-$$

where $E_\gamma^* = 0.29$ GeV. What are the mass, spin, parity, isotopic spin, G-parity and charge conjugation possibilities for the χ? Assume an electric dipole $E1$ transition for the γ-ray emission and strong decay of the χ to 2π.

(Wisconsin)

Solution:

First we find the mass of χ. In the ψ rest frame

$$E_\chi + E_\gamma = m_\psi \,,$$

or

$$E_\chi = 3.7 \text{ GeV} - 0.29 \text{ GeV} = 3.41 \text{ GeV} \,.$$

Momentum conservation gives

$$p_\chi = p_\gamma = 0.29 \text{ GeV}/c \,.$$

As

$$E_\chi^2 = p_\chi^2 + m_\chi^2 \,,$$

we obtain

$$m_\chi = \sqrt{E_\chi^2 - p_\chi^2} = \sqrt{3.4^2 - 0.29^2} = 3.40 \text{ GeV}/c^2 \,.$$

Now to find the other quantum numbers of χ. As $\psi(3.7) \to \gamma\chi$ is an $E1$ transition, we see from its selection rules that the parities of ψ and χ are opposite and the change of spin is 0 or ±1. Then the possible spin values of χ are $J = 0, 1, 2$ and its parity is positive, as ψ has $J^P = 1^-$.

Consider the strong decay $\chi \to \pi^+ \pi^-$. As the parity of χ is $+1$, parity conservation requires $P(\pi^+)P(\pi^-)(-1)^l = (-1)^{2+l} = (-1)^l = +1$, giving $l = 0$ or 2. Thus the spin of χ can only be $J = 0$ or 2. Furthermore,

$$C(\chi) = (-1)^{l+s} = (-1)^l = +1.$$

As π has positive G-parity, conservation of G-parity requires

$$G(\chi) = G(\pi^+)G(\pi^-) = +1.$$

Now for mesons with C-parity, G-parity and C-parity are related through isospin I:

$$G(\chi) = (-1)^I C(\chi).$$

As

$$G(\chi) = C(\chi) = 1,$$

$(-1)^I = +1$, giving $I = 0$ or 2 for χ.

Up to now no meson with $I = 2$ has been discovered, so we can set $I = 0$. Hence the quantum numbers of χ can be set as

$$m_\chi = 3.40 \text{ GeV}/c^2, \quad I^G(J^P)C = 0^+(0^+) + \quad \text{or} \quad 0^+(2^+) + .$$

The angular distribution of γ emitted in ψ decay indicates that the spin of χ (3.40) is probably $J = 0$.

3086

It is well established that there are three $c\bar{c}$ states intermediate in mass between the $\psi(3095)$ and $\psi'(3684)$, namely,

$$\chi_0(3410): \quad J^{PC} = 0^{++},$$

$$\chi_1(3510): \quad J^{PC} = 1^{++},$$

$$\chi_2(3555): \quad J^{PC} = 2^{++}.$$

The number in parentheses is the mass in MeV/c^2.

(a) What electric and magnetic multipoles are allowed for each of the three radioactive transitions:

$$\psi' \to \gamma + \chi_{0,1,2}?$$

(b) Suppose that the ψ' is produced in e^+e^- collisions at an electron-positron storage ring. What is the angular distribution of the photons relative to the beam direction for the decay $\psi' \to \gamma + \chi_0$?

(c) In the condition of part (b), could one use the angular distribution of the photons to decide the parity of the χ_0?

(d) For χ_0 and χ_1 states separately, which of the following decay modes are expected to be large, small, or forbidden?

$$\pi^0\pi^0, \quad \gamma\gamma, \quad p\bar{p}, \quad \pi^+\pi^-\pi^0, \quad 4\pi^0, \quad D^0\bar{K}^0, \quad e^+e^-, \quad \psi\eta^0 .$$

$$(DATA: M_p = 938 \text{ MeV}/c^2; \quad M_{\pi^0} = 135 \text{ MeV}/c^2, \quad M_\eta = 549 \text{ MeV}/c^2)$$

(e) The strong decays of the χ states are pictured as proceeding through an intermediate state consisting of a small number of gluons which then interact to produce light quarks, which further interact and materialize as hadrons. If gluons are massless and have $J^P = 1^-$, what is the minimum number of gluons allowed in the pure gluon intermediate state of each of the $\chi_{0,1,2}$? What does this suggest about the relative hadronic decay widths for these three states?

(Princeton)

Solution:

(a) As γ and ψ both have $J^P = 1^-$, in the decay $|\Delta J| = 0, 1$ and parity changes. Hence it is an electric dipole transition.

(b) The partial width of the electric dipole transition is given by

$$\Gamma(2^3S_1 \to \gamma 2^3 P_J) = \left(\frac{16}{243}\right) \alpha(2J + 1)k^3 |\langle 2P|\gamma|2S\rangle|^2 ,$$

where α is the fine structure constant. Thus

$$\Gamma(2^3S_1 \to \gamma_0 2^3 P_0) : \Gamma(2^3S_1 \to \gamma_1 2^3 P_1) : \Gamma(2^3S_1 \to \gamma_2 2^3 P_2)$$

$$= k_0^3 : 3k_1^3 : 5k_2^3 ,$$

where k is the momentum of the emitted photon (setting $\hbar = 1$). The angular distributions of the photons are calculated to be

$$1 + \cos^2 \theta \quad \text{for process } \psi' \to \gamma_0 + \chi_0 ,$$

$$1 - (1/3) \cos^2 \theta \quad \text{for process } \psi' \to \gamma_1 + \chi_1 ,$$

$$1 + (1/13) \cos^2 \theta \quad \text{for process } \psi' \to \gamma_2 + \chi_2 .$$

(c) As the angular distributions of γ_1, γ_2, and γ_3 are different, they can be measured experimentally and used to determine the spin of χ_i. The other quantum numbers of χ_i may also be decided by the modes of their decay. For example, the χ_0 state decays to $\pi^+\pi^-$ or K^+K^-, and so $J^P = 0^+, 1^-, 2^+ \cdots$. Then from the angular distribution, we can set its $J^P = 0^+$. As $C(\pi^+\pi^-) = (-1)^l$, $J^{PC} = 0^{++}$. For the χ_1 state, $\pi^+\pi^-$ and K^+K^- are not among the final states so $J^P = 0^-, 1^+, 2^-$. The angular distribution then gives $J = 1$ and so $J^P = 1^+$. It is not possible to determine their J^P by angular distributions alone.

(d) $\chi_1 \to \pi^0\pi^0$ is forbidden. As π^0 has $J^P = 0^-$, $\pi^0\pi^0$ can only combine into states $0^+, 1^-, 2^+$. As χ_1 has $J^P = 1^+$, angular momentum and parity cannot both be conserved.

$\chi_0 \to \pi^0\pi^0$ satisfies all the conservation laws. However, it is difficult to detect. The whole process is $\psi' \to \gamma\chi_0 \to \gamma\pi^0\pi^0 \to \gamma\gamma\gamma\gamma\gamma$ and one would have to measure the five photons and try many combinations of invariant masses simultaneously to check if the above mode is satisfied. This mode has yet to be detected. Similarly we have the following:

$\chi_1 \to \gamma\gamma$ is forbidden. $\chi_0 \to \gamma\gamma$ is an allowed electromagnetic transition. However as χ_0 has another strong decay channel, the branching ratio of this decay mode is very small.

$\chi_0, \chi_1 \to p\bar{p}$ are allowed decays. However, their phase spaces are much smaller than that of $\chi_0 \to \pi^0\pi^0$, and so are their relative decay widths.

$\chi_0, \chi_1 \to \pi^+\pi^-\pi^0$ are forbidden as G-parity is not conserved;

$\chi_0, \chi_1 \to \pi^0\psi$ are forbidden as C-parity is not conserved;

$\chi_0, \chi_1 \to D^0\bar{K}^0$ are weak decays with very small branching ratios.

$\chi_0 \to e^+e^-$ is a high order electromagnetic decay with a very small branching ratio.

$\chi_1 \to e^+e^-$ is an electromagnetic decay. It is forbidden, however, by conservation of C-parity.

$\chi_0, \chi_1 \to \eta\psi$ are forbidden for violating energy conservation.

(e) As gluon has $J^P = 1^-$, it is a vector particle and the total wave function of a system of gluons must be symmetric. As a two-gluon system can only have states with 0^{++} or 2^{++}, a three-gluon system can only have states with 1^{++}, χ_0 and χ_2 have strong decays via a two-gluon intermediate state and χ_1 has strong decay via a three-gluons intermediate state. Then as decay probability is proportional to α_s^n, where α_s is the strong interaction constant ($\alpha_s \approx 0.2$ in the energy region of J/ψ) and n is the number of gluons in the intermediate state, the strong decay width of χ_1 is α_s times smaller than those of χ_0, χ_2. The result given by QCD is $\Gamma(\chi_0 \to hadrons)$: $\Gamma(\chi_2 \to hadrons) : \Gamma(\chi_1 \to hadrons) = 15 : 4 : 0.5$.

3087

Particles carrying a new quantum number called charm have recently been discovered. One such particle, D^+, was seen produced in e^+e^- anni-hilation, at center-of-mass energy $E = 4.03$ GeV, as a peak in the $K^-\pi^+\pi^+$ mass spectrum at $M_{K\pi\pi} = 1.87$ GeV. The Dalitz plot for the three-body decay shows nearly uniform population.

(a) Using the simplest quark model in which mesons are bound states of a quark and an antiquark, show that D^+ cannot be an ordinary strange particle resonance (e.g. K^{*+}).

(b) What are the spin and parity (J^P) of the $K\pi\pi$ final state?

(c) Another particle, D^0, was seen at nearly the same mass in the $K^-\pi^+$ mass spectrum from the same experiment. What are the allowed J^P assignments for the $K\pi$ state?

(d) Assume that these two particles are the same isospin multiplet, what can you infer about the type of interaction by which they decay?

(e) Suppose the $K_s \to 2\pi$ decay to be typical of strangeness-changing charm-conserving weak decays. Estimate the lifetime of D^0, assuming that the branching ratio $(D^0 \to K^-\pi^+)/(D^0 \to all) \approx 5\%$. The lifetime of K_s is $\sim 10^{-10}$ sec.

(Princeton)

Solution:

(a) According to the simplest quark model, K meson consists of an $\bar{s}$ quark and a u quark. All strange mesons are composed of an $\bar{s}$ and an ordinary quark, and only weak decays can change the quark flavor. If the s quark in a strange meson changes into a u or d quark, the strange meson will become an ordinary meson. On the other hand, strong and electromagnetic decays cannot change quark flavor. $D^+ \to K\pi\pi$ is a weak decay. So if there is an $\bar{s}$ quark in D^+, its decay product cannot include K meson, which also has an $\bar{s}$ quark. Hence there is no $\bar{s}$ but a quark of a new flavor in D^+, which changes into $\bar{s}$ in weak decay, resulting in a K meson in the final state.

(b) The Dalitz plot indicated $J = 0$ for a $K\pi\pi$ system. As the total angular momentum of the three particles is zero, the spin of D^+ is zero. Let the relative orbital angular momentum of the two π system be l, the orbital angular momentum of the K relative to the two π be l'. Since the spins of K, π are both zero, $\mathbf{J} = \mathbf{l} + \mathbf{l'} = 0$, i.e., $\mathbf{l} = -\mathbf{l'}$, or $|\mathbf{l}| = |\mathbf{l'}|$. Hence

$$P(K\pi\pi) = (-1)^{l+l'} P^2(\pi)P(K) = (-1)^2(-1) = (-1)^3 = -1.$$

Thus the $K\pi\pi$ final state has $J^P = 0^-$.

(c) For the $K\pi$ state,

$$P(K\pi) = (-1)^l P(\pi)P(K) = (-1)^l, \qquad J = 0 + 0 + l.$$

Hence

$$J^P = 0^+, 1^-, 2^+ \cdots.$$

If $J(D) = 0$, then $l = 0$ and $J^P = 0^+$.

(d) If D^+, D^0 belong to an isospin multiplet, they must have the same J^P. As the above-mentioned $K\pi\pi$ and $K\pi$ systems have odd and even parities respectively, the decays must proceed through weak interaction in which parity is not conserved.

(e) Quark flavor changes in both the decays $D^0 \to K\pi$ and $K_s^0 \to \pi^+\pi^-$, which are both Cabibbo-allowed decays. If we can assume their matrix elements are roughly same, then the difference in lifetime is due to the difference in the phase-space factor. For the two-body weak decays, neglecting the difference in mass of the final states, we have

$$\Gamma(D_1^0 \to K^-\pi^+) = f_D^2 \cdot m_D \cdot m_K^2 \left(1 - \frac{m_K^2}{m_D^2}\right)^2 = f_D^2 \cdot \frac{m_K^2}{m_D^3}(m_D^2 - m_K^2)^2,$$

$$\Gamma(K_S \to 2\pi) = f_K^2 \cdot m_K \cdot m_\pi^2 \left(1 - \frac{m_\pi^2}{m_K^2}\right)^2 = f_K^2 \cdot \frac{m_\pi^2}{m_K^3}(m_K^2 - m_\pi^2)^2,$$

where f_D and f_K are coupling constants associated with the decays. Take $f_D = f_K$ and assume the branching ratio of $K_S^0 \to 2\pi$ is nearly 100%, we have

$$\frac{\tau_{D^0}}{\tau_K} = \frac{\Gamma(K \to 2\pi)}{\Gamma(D \to all)} = \frac{\Gamma(K \to 2\pi)}{20\Gamma(D \to K\pi)} = \frac{m_\pi^2 m_D^3 (m_K^2 - m_\pi^2)^2}{20 m_K^5 (m_D^2 - m_K^2)^2},$$

and hence

$$\tau_{D^0} = \frac{140^2 \times 1870^3}{20 \times 494^5} \left(\frac{494^2 - 140^2}{1870^2 - 494^2} \right)^2 \times 10^{-10} = 1.0 \times 10^{-13} \ s,$$

which may be compared with the experimental value

$$\tau_{D^0} = \left(4.4 \, {}^{+0.8}_{-0.6} \right) \times 10^{-13} \ s.$$

3088

A recent development in elementary particle physics is the discovery of charmed nonstrange mesons (called D^+, D^0, and their charge conjugates) with masses around 1870 MeV/c^2.

(a) Knowing the charge of charmed quark to be 2/3, give the quark contents of the D^+ and D^0 mesons.

(b) The D mesons decay weakly into ordinary mesons $(\pi, K, \cdots)$. Give estimates (with your reasoning) for the branching ratios of the following two-body decays:

$$\frac{BR(D^0 \to K^+ K^-)}{BR(D^0 \to K^- \pi^+)}, \qquad \frac{BR(D^0 \to \pi^+ \pi^-)}{BR(D^0 \to K^- \pi^+)}, \qquad \frac{BR(D^0 \to K^+ \pi^-)}{BR(D^0 \to K^- \pi^+)}.$$

(c) How would you show that the decay of D mesons is by means of weak interaction?

(d) In a colliding beam at c.m. energy 4.03 GeV, a D^+ meson (mass = 1868.3 MeV/c^2) and a D^{*-} meson (mass = 2008.6 MeV/c^2) are produced. The D^{*-} decays into a $\bar{D}^0$ (mass = 1863.3 MeV/c^2) and a π^-. What is the maximum momentum in the laboratory of the D^{*-}? of the π?

(*Princeton*)

Solution:

(a) A D meson consists of a charmed quark c (charge $\frac{2}{3}$) and the antiparticle of a light quark u (charge $\frac{2}{3}$) or d (charge $-\frac{1}{3}$). To satisfy the charge requirements, the quark contents of D^+ and D^0 are $c\bar{d}$ and $c\bar{u}$ respectively.

(b) The essence of D meson decay is that one of its quarks changes flavor via weak interaction, the main decay modes arising from decay of the c quark as shown in Fig. 3.26.

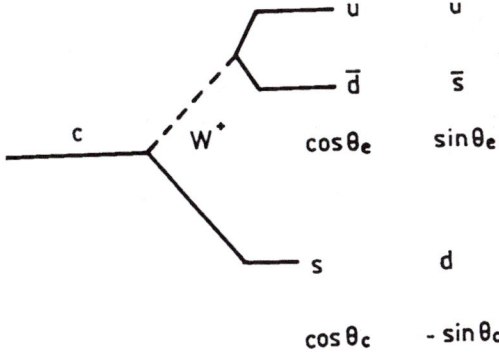

Fig. 3.26

Let θ_c be the Cabibbo mixing angle. We have

$c \to su\bar{d}$, amplitude $\sim \cos^2 \theta_c$,
$c \to su\bar{s}$, amplitude $\sim \sin \theta_c \cos \theta_c$,
$c \to du\bar{d}$, amplitude $\sim -\sin \theta_c \cos \theta_c$,
$c \to du\bar{s}$, amplitude $\sim \sin^2 \theta_c$,

and correspondingly

$D^0 \to K^- + \pi^-$,   Cabibbo allowed,

$D^0 \to K^- + K^+$,   first order Cabibbo forbidden,

$D^0 \to \pi^+ + \pi^-$,   first order Cabibbo forbidden,

$D^0 \to K^+ + \pi^-$,   second order Cabibbo forbidden.

The value of θ_c has been obtained by experiment to be $\theta_c = 13.1^0$. Hence

$$\frac{BR(D^0 \to K^+ K^-)}{BR(D^0 \to K^- \pi^+)} = \tan^2 \theta_c \approx 0.05,$$

$$\frac{BR(D^0 \to \pi^+ \pi^-)}{BR(D^0 \to K^- \pi^+)} = \tan^2 \theta_c \approx 0.05,$$

$$\frac{BR(D^0 \to K^+ \pi^-)}{BR(D^0 \to K^- \pi^+)} = \tan^4 \theta_c \approx 2.5 \times 10^{-3}.$$

(c) In D^0 decay, the charm quantum number C changes. As only weak decays can change the flavor of a quark, the decays must all be weak decays.

(d) In the head-on collision of colliding beams the laboratory frame is the same as the center-of-mass frame. Let the masses, energies and momenta of D^{*-} and D^+ be m^*, m, E^*, E, p^*, p respectively and denote the total energy as E_0. Momentum and energy conservation gives

$$p^* = p, \qquad E^* + \sqrt{p^2 + m^2} = E_0 .$$

Thus

$$E^{*2} + E_0^2 - 2E^* E_0 = p^{*2} + m^2 .$$

With $E^{*2} = p^{*2} + m^{*2}$, we have

$$E^* = \frac{m^{*2} - m^2 + E_0^2}{2E_0} = \frac{2.0086^2 - 1.8683^2 + 4.03^2}{2 \times 4.03} = 2.08 \text{ GeV} ,$$

$$p^* = \sqrt{2.08^2 - 2.008^2} = 0.54 \text{ GeV}/c ,$$

giving

$$\beta = p^*/E^* = 0.26, \qquad \gamma = E^*/m^* = 1.04 .$$

In the D^{*-} rest frame, the decay takes place at rest and the total energy is equal to m^*. Using the above derivation we have

$$\bar{E}_\pi = \frac{m_\pi^2 - m_D^2 + m^{*2}}{2m^*} = 0.145 \text{ GeV} ,$$

$$\bar{p}_\pi = \sqrt{\bar{E}_\pi^2 - m_\pi^2} = 38 \text{ MeV}/c .$$

In the laboratory, the π meson will have the maximum momentum if it moves in the direction of D^{*-}. Let it be $P_{\max}$. Then

$$p_{\max} = \gamma(\bar{p}_\pi + \beta \bar{E}_\pi)$$

$$= 1.04(38 + 0.26 \times 145) = 79 \text{ MeV}/c .$$

Hence the maximum momenta of D^{*-} and π^- are 540 MeV/c and 79 MeV/c respectively.

3089

In $e^+ e^-$ annihilation experiments, a narrow resonance (of width less than the intrinsic energy spread of the two beams) has been observed at

$E_{CM} = 9.5$ GeV for both

$$e^+e^- \to \mu^+\mu^-$$

and

$$e^+e^- \to \text{hadrons}.$$

The integrated cross sections for these reactions are measured to be

$$\int \sigma_{\mu\mu}(E)dE = 8.5 \times 10^{-33} \text{ cm}^2 \cdot \text{MeV},$$

$$\int \sigma_h(E)dE = 3.3 \times 10^{-31} \text{ cm}^2 \cdot \text{MeV}.$$

Use the Breit–Wigner resonance formula to determine the partial widths $\Gamma_{\mu\mu}$ and Γ_h for the $\mu\mu$ and hadronic decays of the resonance.

Solution:

The Breit–Wigner formula can be written for the two cases as

$$\sigma_\mu(E) = \frac{\pi(2J+1)}{M^2} \frac{\Gamma_{ee}\Gamma_{\mu\mu}}{(E-M)^2 + \frac{\Gamma^2}{4}},$$

$$\sigma_h(E) = \frac{\pi(2J+1)}{M^2} \frac{\Gamma_{ee}\Gamma_h}{(E-M)^2 + \frac{\Gamma^2}{4}},$$

where M and J are the mass and spin of the resonance state, Γ, Γ_{ee}, Γ_h and $\Gamma_{\mu\mu}$ are the total width, and the partial widths for decaying into electrons, hadrons, and muons respectively. We have

$$\Gamma = \Gamma_{ee} + \Gamma_{\tau\tau} + \Gamma_{\mu\mu} + \Gamma_h,$$

where $\Gamma_{\tau\tau}$ is the partial width for decaying into τ particles. Because of the universality of lepton interactions, if we neglect the difference phase space factors, we have $\Gamma_{ee} = \Gamma_{\tau\tau} = \Gamma_{\mu\mu}$, and so

$$\Gamma = 3\Gamma_{\mu\mu} + \Gamma_h.$$

For the resonance at $M = 9.5$ GeV, $J = 1$. Therefore

$$\int \sigma_{\mu\mu}(E)dE = \frac{3\pi\Gamma_{\mu\mu}^2}{M^2} \int \frac{dE}{(E-M)^2 + \frac{\Gamma^2}{4}}$$

$$= \frac{6\pi^2\Gamma_{\mu\mu}^2}{M^2\Gamma} = 8.5 \times 10^{-33} \text{ cm}^2 \cdot \text{MeV},$$

$$\int \sigma_h(E)dE = \frac{3\pi\Gamma_{\mu\mu}\Gamma_h}{M^2} \int \frac{dE}{(E-M)^2 + \frac{\Gamma^2}{4}}$$

$$= \frac{6\pi^2\Gamma_{\mu\mu}\Gamma_h}{M^2\Gamma} = 3.3 \times 10^{-31} \text{ cm}^2 \cdot \text{MeV},$$

whose ratio gives

$$\Gamma_h = 38.8\Gamma_{\mu\mu}.$$

Hence

$$\Gamma = \Gamma_h + 3\Gamma_{\mu\mu} = 41.8\Gamma_{\mu\mu},$$

and

$$\Gamma_{\mu\mu} = \frac{M^2}{6\pi^2} \frac{\Gamma}{\Gamma_h} \times 3.3 \times 10^{-31} = 5.42 \times 10^{-26} \text{ MeV}^3\text{cm}^2.$$

To convert it to usual units, we note that

$$1 = \hbar c = 197 \times 10^{-13} \text{ MeV} \cdot \text{cm},$$

or

$$1 \text{ cm} = \frac{1}{197 \times 10^{-13}} \text{ MeV}^{-1}.$$

Thus

$$\Gamma_{\mu\mu} = 1.40 \times 10^{-3} \text{ MeV},$$

and

$$\Gamma_h = 38.8\Gamma_{\mu\mu} = 5.42 \times 10^{-2} \text{ MeV},$$

$$\Gamma = 41.8\Gamma_{\mu\mu} = 5.84 \times 10^{-2} \text{ MeV}.$$

3090

Suppose nature supplies us with massive charged spin-1 'quark' Q^+ and antiquark $\bar{Q}^-$. Using a model like the nonrelativistic charmonium model which successfully describes the J/ψ family, predict the spectrum of the neutral $Q\bar{Q}$ resonance. Make a diagram of the lowest few expected states, indicating the spin, charge conjugation parities, and allowed electromagnetic transitions, as well as the expected ordering of levels.

(Princeton)

Solution:

The current nonrelativistic model for dealing with heavy quarks employs a strong-interaction potential, approximated by a central potential. Then the angular part of the wave functions takes the form of spherical harmonic functions. To take account of quark confinement, a better potential is given by the Cornell model as $V(r) = -k/r + r/a^2$, which is a Coloumb potential superposed on a linear potential, with the former implying asymptotic freedom, and the latter quark confinement. By considering spin correlation the order of levels can be calculated numerically. For the quark-antiquark system,

spin: $\mathbf{J} = \mathbf{S} + \mathbf{l}$, where $\mathbf{S} = \mathbf{s}_1 + \mathbf{s}_2$, $s_1 = s_2 = 1$,

P-parity: $P(Q^+Q^-) = P(Q^+)P(Q^-)(-1)^l = (-1)^l$, as for a boson of spin 1, $P(\bar{Q}) = P(Q)$,

C-parity: $C(Q^+Q^-) = (-1)^{l+S}$.

Thus the system can have J^{PC} as follows:

$$l = 0, \quad S = |\mathbf{s}_1 + \mathbf{s}_2| = 0, \quad n^1 S_0 \quad J^{PC} = 0^{++}$$

$$S = |\mathbf{s}_1 + \mathbf{s}_2| = 1 \quad n^3 S_1 \quad 1^{+-}$$

$$S = |\mathbf{s}_1 + \mathbf{s}_2| = 2 \quad n^5 S_2 \quad 2^{++}$$

$$l = 1, \quad S = |\mathbf{s}_1 + \mathbf{s}_2| = 0 \quad n^1 P_1 \quad 1^{--}$$

$$S = |\mathbf{s}_1 + \mathbf{s}_2| = 1 \quad n^3 P_0 \quad 0^{-+}$$

$$n^3 P_1 \quad 1^{-+}$$

$$n^3 P_2 \quad 2^{-+}$$

$$S = |\mathbf{s}_1 + \mathbf{s}_2| = 2 \quad n^5 P_1 \quad 1^{--}$$

$$n^5 P_2 \quad 2^{--}$$

$$n^5 P_3 \quad 3^{--}$$

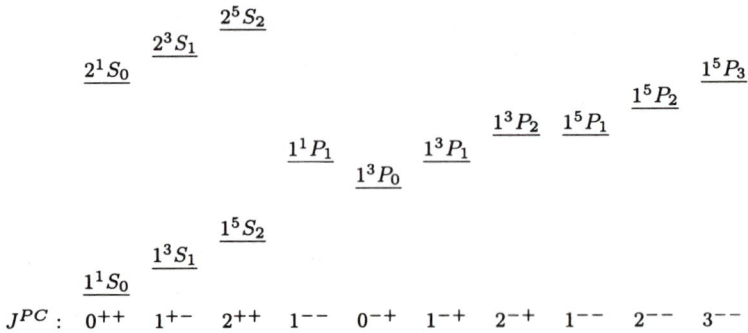

$$
\begin{array}{c}
2^5 S_2 \\
2^3 S_1 \\
2^1 S_0
\end{array}
\qquad\qquad
1^5 P_3
$$

$$
1^5 P_2
$$

$$
1^3 P_2 \quad 1^5 P_1
$$

$$
1^1 P_1 \qquad 1^3 P_1
$$

$$
1^3 P_0
$$

$$
\begin{array}{c}
1^5 S_2 \\
1^3 S_1 \\
1^1 S_0
\end{array}
$$

$$
J^{PC}: \quad 0^{++} \quad 1^{+-} \quad 2^{++} \quad 1^{--} \quad 0^{-+} \quad 1^{-+} \quad 2^{-+} \quad 1^{--} \quad 2^{--} \quad 3^{--}
$$

Fig. 3.27

In the above we have used spectroscopic symbols $n^{2S+1}S_J$, $n^{2S+1}P_J$ etc. to label states, with n denoting the principal quantum number, $2S + 1$ the multiplicily, singlet, triplet or quintuplet, and J the total angular momentum. The order of the levels is shown in Fig. 3.27 (only S and P states are shown).

As the order of P states is related to the spin-correlation term, the order given here is only a possible one. The true order must be calculated using the assumed potential. Even the levels given here are seen more complicated than those for a spin-1/2 charm-anticharm system, with the addition of the $^5S_2$ and ^5P_J spectra. In accordance with the selection

Table 3.13. Possible γ transitions.

Transition	ΔJ	ΔP	ΔC	Type of transition
$2^3 S_1 \to 1^3 P_J$	$0, 1$	-1	-1	$E1$
$1^3 P_J \to 1^3 S_1$	$0, 1$	-1	-1	$E1$
$2^5 S_2 \to 2^3 S_1 \to 2^1 S_0$	1	1	-1	$M1(E2)$
$1^5 S_2 \to 1^3 S_1 \to 1^1 S_0$	1	1	-1	$M1(E2)$
$2^5 S_2 \to 1^3 S_1$	1	1	-1	$M1(E2)$
$2^3 S_1 \to 1^1 S_1$	1	1	-1	$M1(E2)$
$2^5 S_2 \to 1^5 P_J$	$0, 1$	-1	-1	$E1$
$1^5 P_J \to 1^5 S_2$	$0, 1$	-1	-1	$E1$
$2^1 S_0 \to 1^1 P_1, 1^5 P_1$	1	-1	-1	$E1$
$1^1 P_1, 1^5 P_1 \to 1^1 S_0$	1	-1	-1	$E1$
$2^5 S_2 \to 1^1 P_1$	1	-1	-1	$E1$
$1^1 P_1 \to 1^5 S_2$	1	-1	-1	$E1$

rules of electromagnetic transitions, the possible transitions are listed in Table 3.13.

Note that electromagnetic transitions between the P states are not included in the table because the level order cannot be ascertained. Higher order transitions ($M2, E3$, etc.) between $2^1S_0 \rightarrow 1^5P_{2,3}$ are also excluded. The transitions $^5S_2 \rightarrow {}^1S_0$, $^3S_1 \rightarrow {}^5P_2, {}^1P_1$, etc. are C-parity forbidden and so excluded.

PART IV

EXPERIMENTAL METHODS AND
MISCELLANEOUS TOPICS

1. KINEMATICS OF HIGH-ENERGY PARTICLES (4001–4061)

4001

An accelerator under study at SLAC has as output bunches of electrons and positrons which are made to collide head-on. The particles have 50 GeV in the laboratory. Each bunch contains 10^{10} particles, and may be taken to be a cylinder of uniform charge density with a radius of 1 micron and a length of 2 mm as measured in the laboratory.

(a) To an observer traveling with a bunch, what are the radius and length of its bunch and also the one of opposite sign?

(b) How long will it take the two bunches to pass completely through each other as seen by an observer traveling with a bunch?

(c) Draw a sketch of the radial dependence of the magnetic field as measured in the laboratory when the two bunches overlap. What is the value of B in gauss at a radius of 1 micron?

(d) Estimate in the impulse approximation the angle in the laboratory by which an electron at the surface of the bunch will be deflected in passing through the other bunch. (Ignore particle-particle interaction.)

(UC, Berkeley)

Solution:

(a) Consider a particle P in the bunch traveling with the observer. Let Σ, Σ_0 be the reference frames attached to the laboratory and the observer respectively, taking the direction of motion of P as the x direction. The Lorentz factor of P, and hence of Σ_0, in Σ is

$$\gamma = \frac{E}{mc^2} = \frac{50 \times 10^9}{0.5 \times 10^6} = 1 \times 10^5 \,.$$

To an observer in Σ, the bunch is contracted in length:

$$L = \frac{1}{\gamma} L_0 \,,$$

where L_0 is its length in Σ_0. Thus

$$L_0 = \gamma L = 1 \times 10^5 \times 2 \times 10^{-3} = 200 \text{ m} \,.$$

The radius of the bunch is

$$r_0 = r = 1 \ \mu m,$$

as there is no contraction in a transverse direction.

The bunch of opposite charge travels with velocity $-\beta c$ in Σ, where β is given by

$$\gamma^2 = \frac{1}{1 - \beta^2}.$$

Its velocity in Σ_0 is obtained by the Lorentz transformation for velocity:

$$\beta' = \frac{-\beta - \beta}{1 - \beta(-\beta)} = -\frac{2\beta}{1 + \beta^2}.$$

Its length in Σ_0 is therefore

$$L' = \frac{1}{\gamma'} L_0 = L_0 \sqrt{1 - \beta'^2} = L_0 \sqrt{1 - \left(\frac{2\beta}{1 + \beta^2}\right)^2}$$

$$= L_0 \left(\frac{1 - \beta^2}{1 + \beta^2}\right) = \frac{L_0}{2\gamma^2 - 1}$$

$$= \frac{200}{2 \times 10^{10} - 1} \approx 10^{-8} = 0.01 \ \mu m.$$

(b) To an observer in Σ_0 the time taken for the two bunches to pass through each other completely is

$$t' = \frac{L_0 + L'}{\beta' c}.$$

As

$$\beta = \sqrt{1 - \frac{1}{\gamma^2}} \approx 1$$

and so

$$\beta' = \frac{2\beta}{1 + \beta^2} \approx 1,$$

$$t' \approx \frac{200 + 10^{-8}}{c}$$

$$= \frac{200}{3 \times 10^8} = 6.67 \times 10^{-7} \ \text{s}.$$

(c) Consider the bunch of positrons and let its length, radius, number of particles, and charge density be l, r_0, N and ρ respectively. Then

$$\rho = \frac{eN}{\pi r_0^2 l} .$$

The two bunches of positrons and electrons carry opposite charges and move in opposite directions, and so the total charge density is

$$J = 2\rho\beta c ,$$

where βc is the speed of the particles given by

$$\gamma = \frac{E}{mc^2} = (1 - \beta^2)^{-\frac{1}{2}} .$$

Applying Ampére's circuital law

$$\oint_c \mathbf{B} \cdot d\mathbf{l} = \mu_0 I ,$$

we find for $r > r_0$,

$$2\pi r B = \mu_0 \cdot \frac{2eN}{\pi r_0^2 l}\beta c \cdot \pi r_0^2 ,$$

or

$$B = \frac{\mu_0 eN}{\pi l} \frac{\beta c}{r} ;$$

for $r < r_0$,

$$2\pi r B = \mu_0 \cdot \frac{2eN}{\pi r_0^2 l}\beta c \pi r^2$$

or

$$B = \frac{\mu_0 eN}{\pi l} \frac{\beta c r}{r_0^2} .$$

Figure 4.1 shows the variation of B with r. At $r = r_0 = 1$ μm,

$$B = \frac{4\pi \times 10^{-7} \times 1.6 \times 10^{-19} \times 10^{10}}{\pi \times 2 \times 10^{-3} \times 10^{-6}} \times 1 \times 3 \times 10^8 = 96 \text{ T}$$

$$= 9.6 \times 10^5 \text{ Gs} .$$

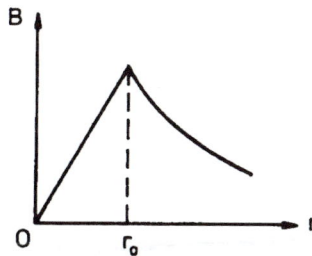

Fig. 4.1

(d) The magnetic field exerts a force vB perpendicular to the motion of an electron. If Δt is the duration of encounter with the opposite bunch, it will acquire a transverse momentum of

$$p_\perp = evB\Delta t.$$

Hence

$$\theta \approx \frac{p_\perp}{p} = \frac{evB\Delta t}{m\gamma v} = \frac{eBl}{m\gamma v} = \frac{eBcl}{pc}$$

$$= \frac{1.6 \times 10^{-19} \times 96 \times 3 \times 10^8 \times 2 \times 10^{-3}}{50 \times 10^9 \times 1.6 \times 10^{-19}} = 1.15 \times 10^{-3} \text{ rad} = 39.6'.$$

4002

A certain elementary process is observed to produce a relativistic meson whose trajectory in a magnetic field B is found to have a curvature given by $(\rho B)_1 = 2.7$ Tesla-meters.

After considerable energy loss by passage through a medium, the same meson is found to have $(\rho B)_2 = 0.34$ Telsa-meters while a time-of-flight measurement yields a speed of $v_2 = 1.8 \times 10^8$ m/sec for this 'slow' meson.

(a) Find the rest mass and the kinetic energies of the meson (in MeV) before and after slowing down (2-figure accuracy).

(b) If this 'slow' meson is seen to have a 50% probability of decaying in a distance of 4 meters, compute the intrinsic half life of this particle in its own rest frame, as well as the distance that 50% of the initial full-energy mesons would travel in the laboratory frame.

(UC, Berkeley)

Solution:

(a) As $evB = \frac{\gamma m v^2}{\rho}$, or $\rho B = \frac{\gamma \beta m c}{e}$, we have for the meson

$$\frac{(\rho B)_1}{(\rho B)_2} = \frac{\gamma_1 \beta_1}{\gamma_2 \beta_2}.$$

At $\beta_2 = \frac{v_2}{c} = \frac{1.8 \times 10^8}{3 \times 10^8} = 0.6$, or $\gamma_2 \beta_2 = \frac{\beta_2}{\sqrt{1-\beta_2^2}} = 0.75$, we have

$$p_2 c = \gamma_2 \beta_2 m c^2 = ec(\rho\beta)_2$$

$$= 1.6 \times 10^{-19} \times 0.34c \text{ Joule}$$

$$= 0.34 \times 3 \times 10^8 \text{ eV}$$

$$= 0.102 \text{ GeV}.$$

The rest mass of the meson is therefore

$$m = \frac{p_2 c}{\gamma_2 \beta_2 c^2} = \frac{0.102}{0.75} \text{ GeV}/c^2 = 0.14 \text{ GeV}/c^2.$$

Before slowing down, the meson has momentum

$$p_1 c = ec(\rho B)_1 = 2.7 \times 0.3 = 0.81 \text{ GeV},$$

and hence kinetic energy

$$T_1 = \sqrt{p_1^2 c^2 + m^2 c^4} - mc^2 = \sqrt{0.81^2 + 0.14^2} - 0.14 = 0.68 \text{ GeV}.$$

After slowing down, the meson has kinetic energy

$$T_2 = \sqrt{p_2^2 c^2 + m^2 c^4} - mc^2 = \sqrt{0.102^2 + 0.14^2} - 0.14 = 0.033 \text{ GeV}.$$

(b) The half life τ is defined by

$$\exp\left(-\frac{t}{\tau}\right) = \exp\left(-\frac{l}{\beta c \tau}\right) = \frac{1}{2},$$

or

$$\tau = \frac{l}{\beta c \ln 2}.$$

In the rest frame of the meson, on account of time dilation, the half-life is

$$\tau_0 = \frac{\tau}{\gamma_2} = \frac{l_2}{\gamma_2 \beta_2 c \ln 2} = \frac{4}{0.75 \times 3 \times 10^8 \ \ln 2} = 2.6 \times 10^{-8} \text{ s}.$$

In the laboratory frame, the distance full-energy mesons travel before their number is reduced by 50% is given by

$$l_1 = \tau_1 \beta_1 c \ln 2 = \tau_0 \gamma_1 \beta_1 c \ln 2.$$

As

$$\gamma_1 \beta_1 = \frac{p_1 c}{mc^2} = \frac{0.81}{0.14} = 5.8 \,,$$

$$l_1 = 2.6 \times 10^{-8} \times 5.8 \times 3 \times 10^8 \times \ln 2 = 31 \text{ m} \,.$$

4003

The Princeton synchrotron (PPA) has recently been used to accelerate highly charged nitrogen ions. If the PPA can produce protons of nominal total energy 3 GeV, what is the maximum kinetic energy of charge 6^+ ^{14}N ions?

(Wisconsin)

Solution:

After the ions enter the synchrotron, they are confined by magnetic field and accelerated by radio frequency accelerator. The maximum energy attainable is limited by the maximum value B_m of the magnetic field. The maximum momentum p_m is given by

$$p_m = |q| \rho B_m$$

where $|q|$ is the absolute charge of the ion and ρ the radius of its orbit. Considering protons and nitrogen ions we have

$$\frac{p_p}{p_N} = \frac{|q|_p}{|q|_N} \,, \qquad p_N = 6 p_p \,.$$

As

$$\sqrt{p_p^2 + m_p^2} = \sqrt{p_p^2 + 0.938^2} = 3 \,,$$

we have

$$p_p = 2.85 \text{ GeV}/c \,,$$

and

$$p_N = 17.1 \text{ GeV}/c \,.$$

Hence the maximum kinetic energy of the accelerated nitrogen ions is

$$T = \sqrt{17.1^2 + (0.938 \times 14)^2} - 0.938 \times 14 = 8.43 \text{ GeV} \,.$$

4004

(a) A muon at rest lives 10^{-6} sec and its mass is 100 MeV/c^2. How energetic must a muon be to reach the earth's surface if it is produced high in the atmosphere (say $\sim 10^4$ m up)?

(b) Suppose to a zeroth approximation that the earth has a 1-gauss magnetic field pointing in the direction of its axis, extending out to 10^4 m. How much, and in what direction, is a muon of energy E normally incident at the equator deflected by the field?

(c) Very high-energy protons in cosmic rays can lose energy through collision with 3-K radiation (cosmological background) in the process $p + \gamma \to p + \pi$. How energetic need a proton be to be above threshold for this reaction?

(Princeton)

Solution:

(a) Let the energy of the muons be $E \equiv \gamma m$, where m is their rest mass. In the laboratory frame the lifetime is $\tau = \tau_0 \gamma$, τ_0 being the lifetime in the muon rest frame. Then

$$l = \tau \beta c = \tau_0 \gamma \beta c \,,$$

giving

$$E = \frac{lm}{\beta \tau_0 c} \approx \frac{lm}{\tau_0 c} = \frac{10^4 \times 0.1}{10^{-6} \times 3 \times 10^8} = 3.3 \text{ GeV} \,.$$

(b) Consider a high energy μ^+ in the earth's magnetic field. The force exerted by the latter is balanced by the centripetal force:

$$evB = \frac{m \gamma v^2}{R} \,,$$

giving

$$R = \frac{pc}{ecB} \approx \frac{E}{ecB} \,,$$

where p and E are the momentum and total energy of the muon. With E in GeV and R in m,

$$R \approx \frac{1.6 \times 10^{-10} \, E}{1.6 \times 10^{-19} \times 3 \times 10^8 \times 10^{-4}}$$

$$= \frac{10^5}{3} \times E \,.$$

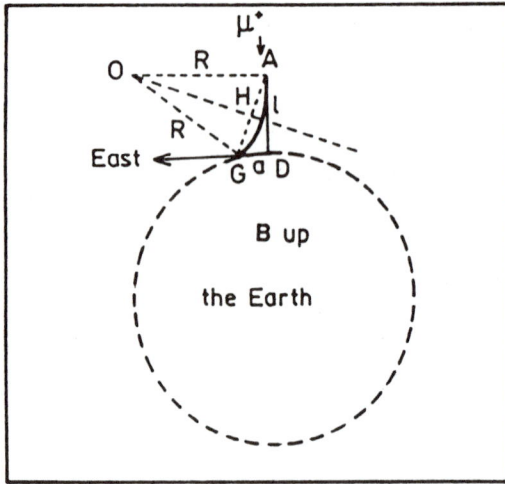

Fig. 4.2

A μ^+ incident vertically is deflected to the east and enters the earth's surface at a from the original path AD (Fig. 4.2). Let O be the center of curvature of the muon orbit and note that AD is tangential to the orbit. As $\angle OAD = \frac{\pi}{2}$, we have $\angle GAD = \angle AOH$. Hence $\triangle GAD$ and $\triangle AOH$ are similar and so

$$\frac{a}{\sqrt{l^2 + a^2}} = \frac{\sqrt{l^2 + a^2}}{2R},$$

or

$$a^2 - 2aR + l^2 = 0,$$

giving

$$a = \frac{2R \pm \sqrt{4R^2 - 4l^2}}{2} \approx \frac{l^2}{2R}$$

as $a \ll l \ll R$. Thus

$$a \approx \frac{3 \times 10^8}{2 \times 10^5 \times E} = \frac{1.5 \times 10^3}{E}.$$

For example, $a \approx 455$ m if $E = 3.3$ GeV; $a \approx 75$ m if $E = 20$ GeV.

As the earth's magnetic field points to the north, the magnetic force on a μ^+ going vertically down points to the east. It will be deflected to the east, while a μ^- will be deflected to the west.

(c) Radiation at $T = 3$ K consists of photons of energy $E = 3kT/2$, where $k = 8.6 \times 10^{-5}$ eV/K is the Boltzmann constant. Thus

$$E_\gamma = 8.6 \times 10^{-5} \times 3/2 \times 3 = 3.87 \times 10^{-4} \text{ eV}.$$

Consider the reaction $\gamma + p = p + \pi$. For head-on collision at threshold, taking $c = 1$ we have

$$(E_p + E_\gamma)^2 - (p_p - E_\gamma)^2 = (m_p + m_\pi)^2.$$

With $E_p^2 - p_p^2 = m_p^2$, and $p_p \approx E_p$ for very high energy protons, this becomes

$$E_p \approx \frac{m_\pi^2 + 2m_p m_\pi}{4E_\gamma}.$$

As $m_p = 0.938$ GeV, $m_\pi = 0.140$ GeV, $E_\gamma = 3.87 \times 10^{-13}$ GeV, the threshold energy is

$$E_p \approx \frac{0.14^2 + 2 \times 0.938 \times 0.14}{4 \times 3.87 \times 10^{-13}} = 1.82 \times 10^{11} \text{ GeV}.$$

4005

The mass of a muon is approximately $100 \text{ MeV}/c^2$ and its lifetime at rest is approximately two microseconds. How much energy would a muon need to circumnavigate the earth with a fair chance of completing the journey, assuming that the earth's magnetic field is strong enough to keep it in orbit? Is the earth's field actually strong enough?

(Columbia)

Solution:

To circumnavigate the earth, the life of a moving muon should be equal to or larger than the time required for the journey. Let the proper life of muon be τ_0. Then

$$\tau_0 \gamma \geq \frac{2\pi R}{\beta c},$$

where R is the earth's radius, βc is the muon's velocity and $\gamma = (1 - \beta^2)^{-\frac{1}{2}}$. The minimum momentum required by the muon is therefore

$$pc = m\gamma\beta c = \frac{2\pi R m c}{\tau_0},$$

and the minimum energy required is

$$E = \sqrt{m^2 c^4 + p^2 c^2} = mc^2 \sqrt{1 + \left(\frac{2\pi R}{\tau_0 c}\right)^2}$$

$$= 100 \times \sqrt{1 + \left(\frac{2\pi \times 6400 \times 10^3}{2 \times 10^{-6} \times 3 \times 10^8}\right)^2} = 6.7 \times 10^6 \text{ MeV}.$$

To keep the meson in orbit, we require

$$evB \geq \frac{m\gamma v^2}{R},$$

or

$$B \geq \frac{pc}{eRc} = \frac{6.7 \times 10^6 \times 1.6 \times 10^{-13}}{1.6 \times 10^{-19} \times 6400 \times 10^3 \times 3 \times 10^8}$$

$$= 3.49 \times 10^{-3} \ T \approx 35 \text{ Gs}.$$

As the average magnetic field on the earth's surface is about several tenths of one gauss, it is not possible to keep the muon in this orbit.

4006

(a) A neutron 5000 light-years from earth has rest mass 940 MeV and a half life of 13 minutes. How much energy must it have to reach the earth at the end of one half life?

(b) In the spontaneous decay of π^+ mesons at rest,

$$\pi^+ \rightarrow \mu^+ + \nu_\mu,$$

the μ^+ mesons are observed to have a kinetic energy of 4.0 MeV. The rest mass of the μ^+ is 106 MeV. The rest mass of neutrino is zero. What is the rest mass of π^+?

(Wisconsin)

Solution:

(a) Let the energy of the neutron be E, its velocity be βc, the half life in its rest frame be $\tau_{1/2}$. Then its half life in the earth's frame is $\tau_{1/2}\gamma$, where $\gamma = (1 - \beta^2)^{-\frac{1}{2}}$. For the neutron to reach the earth, we require

$$\gamma \beta c \tau_{\frac{1}{2}} = 5000 \times 365 \times 24 \times 60c,$$

or

$$\gamma \beta = 2.02 \times 10^8.$$

The energy of neutron is

$$E = \sqrt{m_0^2 + p^2} = m_0 \sqrt{1 + \gamma^2 \beta^2} = 1.9 \times 10^{11} \text{ MeV} .$$

(b) Consider the decay $\pi^+ \to \mu^+ + \nu_\mu$ at rest. Conservation of momentum requires the momenta of μ and ν_μ be $\mathbf{p}$ and $-\mathbf{p}$ respectively. Then their energies are $E_\mu = \sqrt{m_\mu^2 + p^2}$, $E_\nu = p$ respectively. As $m_\mu = 106$ MeV, $E_\mu = 4 + 106 = 110$ MeV, we have

$$p = \sqrt{E_\mu^2 - m_\mu^2} = 29.4 \text{ MeV} .$$

Hence

$$m_\pi = E_\mu + E_\nu = 110 + 29.4 = 139.4 \text{ MeV} .$$

4007

A certain electron-positron pair produced cloud chamber tracks of radius of curvature 3 cm lying in a plane perpendicular to the applied magnetic field of magnitude 0.11 Tesla (Fig. 4.3). What was the energy of the γ-ray which produced the pair?

(Wisconsin)

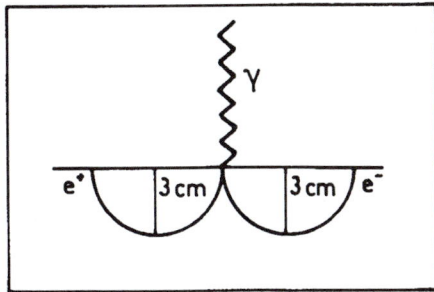

Fig. 4.3

Solution:

As

$$evB = \frac{m\gamma v^2}{\rho} = \frac{pv}{\rho} ,$$

we have

$$pc = ecB\rho$$

$$= \frac{1.6 \times 10^{-19} \times 3 \times 10^8}{1.6 \times 10^{-13}} B\rho$$

$$= 300 B\rho$$

with B in Tesla, ρ in meter and p in MeV/c. Hence, on putting $c = 1$, the momentum of the e^+ or e^- is

$$p = 300 B\rho = 300 \times 0.11 \times 0.03 = 0.99 \text{ MeV}/c,$$

and its energy is

$$E = \sqrt{p^2 + m_e^2} = \sqrt{0.99^2 + 0.51^2} = 1.1 \text{ MeV}.$$

Therefore the energy of the γ-ray that produced the e^+e^- pair is approximately

$$E_\gamma = 2E = 2.2 \text{ MeV}.$$

4008

Newly discovered D^0 mesons (mass = 1.86 GeV) decay by $D^0 \to K^+\pi^-$ in $\tau = 5 \times 10^{-13}$ sec. They are created with 18.6 GeV energy in a bubble chamber. What resolution is needed to observe more than 50% of the decays?

(a) 0.0011 mm.
(b) 0.44 mm.
(c) 2.2 mm.

                                                                          (CCT)

Solution:

As

$$I = I_0 e^{-t/\tau} \geq 0.5 I_0,$$

$$t \leq \tau \ln 2.$$

The mesons have $\gamma = \frac{18.6}{1.86} = 10$ and $\beta \approx 1$. Their proper lifetime is $\tau_0 = 5 \times 10^{-13}$ s, giving

$$\tau = \gamma\tau_0 = 5 \times 10^{-12} \text{ s}.$$

Thus the distance traveled by the mesons is

$$tc \leq \tau c \ln 2 = 5 \times 10^{-12} \times 3 \times 10^{11} \times \ln 2$$

$$= 1 \text{ mm}$$

Hence the resolution should be better than 1 mm, and the answer is (b).

4009

A collimated kaon beam emerges from an analyzing spectrometer with $E = 2$ GeV. At what distance is the flux reduced to 10% if the lifetime is 1.2×10^{-8} sec?

(a) 0.66 km.
(b) 33 m.
(c) 8.3 m.

(CCT)

Solution:

As $m_k = 0.494$ GeV, $\tau_0 = 1.2 \times 10^{-8}$ s, $E_k = 2$ GeV, we have

$$\gamma = \frac{2}{0.494} = 4.05, \qquad \beta = \sqrt{1 - \gamma^{-2}} = 0.97,$$

and the laboratory lifetime is

$$\tau = \gamma\tau_0 = 4.8 \times 10^{-8} \text{ s}.$$

The time t required to reduce the kaon flux from I_0 to $I_0/10$ is given by

$$I_0 e^{-t/\tau} = \frac{I_0}{10},$$

or

$$t = \tau \ln 10 = 11.05 \times 10^{-8} \text{ s}.$$

The distance traveled by the beam during t is

$$t\beta c = 11.05 \times 10^{-8} \times 0.97 \times 3 \times 10^8 = 32 \text{ m}.$$

Hence the answer is (b).

4010

The Compton wavelength of a proton is approximately

(a) 10^{-6} cm.
(b) 10^{-13} cm.
(c) 10^{-24} cm.

(CCT)

Solution:

The Compton wavelength of proton is

$$\lambda = \frac{2\pi\hbar}{m_p c} = \frac{2\pi\hbar c}{m_p c^2} = \frac{2\pi \times 197 \times 10^{-13}}{938} = 1.32 \times 10^{-13} \text{ cm}.$$

Hence the answer is (b).

4011

In a two-body elastic collision:

(a) All the particle trajectories must lie in the same plane in the center of mass frame.
(b) The helicity of a participant cannot change.
(c) The angular distribution is always spherically symmetric.

(CCT)

Solution:

Conservation of momentum requires all the four particles involved to lie in the same plane. Hence the answer is (a).

4012

In a collision between a proton at rest and a moving proton, a particle of rest mass M is produced, in addition to the two protons. Find the minimum energy the moving proton must have in order for this process to take place. What would be the corresponding energy if the original proton were moving towards one another with equal velocity?

(Columbia)

Solution:

At the threshold of the reaction

$$p + p \to M + p + p,$$

the particles on the right-hand side are all produced at rest. Let the energy and momentum of the moving proton be E_p and p_p respectively. The invariant mass squared of the system at threshold is

$$S = (E_p + m_p)^2 - p_p^2 = (2m_p + M)^2.$$

As

$$E_p^2 = m_p^2 + p_p^2,$$

the above gives

$$E_p = \frac{(2m_p + M)^2 - 2m_p^2}{2m_p}$$

$$= m_p + 2M + \frac{M^2}{2m_p}.$$

If the two protons move towards each other with equal velocity, the invariant mass squared at threshold is

$$S = (E_p + E_p)^2 - (p_p - p_p)^2 = (2m_p + M)^2,$$

giving

$$E_p = m_p + M/2.$$

4013

A relativistic particle of rest mass m_0 and kinetic energy $2m_0 c^2$ strikes and sticks to a stationary particle of rest mass $2m_0$.

(a) Find the rest mass of the composite.
(b) Find its velocity.

(SUNY, Buffalo)

Solution:

(a) The moving particle has total energy $3m_0$ and momentum

$$p = \sqrt{(3m_0)^2 - m_0^2} = \sqrt{8}m_0 \,.$$

The invariant mass squared is then

$$S = (3m_0 + 2m_0)^2 - p^2 = 17m_0^2 \,.$$

Let the rest mass of the composite particle be M. Its momentum is also p on account of momentum conservation. Thus

$$S = (\sqrt{M^2 + p^2})^2 - p^2 = M^2 \,,$$

giving

$$M = \sqrt{S} = \sqrt{17}m_0 \,.$$

(b) For the composite,

$$\gamma\beta = \frac{p}{M} = \sqrt{\frac{8}{17}} \,,$$

$$\gamma = \sqrt{\gamma^2\beta^2 + 1} = \sqrt{\frac{8}{17} + 1} = \frac{5}{\sqrt{17}} \,.$$

Hence

$$\beta = \frac{\gamma\beta}{\gamma} = \frac{\sqrt{8}}{5}$$

and the velocity is

$$v = \beta c = 1.7 \times 10^{10} \text{ cm/s} \,.$$

4014

Find the threshold energy (kinetic energy) for a proton beam to produce the reaction

$$p + p \to \pi^0 + p + p$$

with a stationary proton target.

(Wisconsin)

Solution:

Problem 4012 gives

$$E_p = m_p + 2m_\pi + \frac{m_\pi^2}{2m_p} = 938 + 2 \times 135 + \frac{135^2}{2 \times 938} = 1218 \text{ MeV}.$$

Hence the threshold kinetic energy of the proton is $T_p = 1218 - 938 = 280$ MeV.

4015

In high energy proton-proton collisions, one or both protons may "diffractively dissociate" into a system of a proton and several charged pions. The reactions are

(1) $p + p \rightarrow p + (p + n\pi)$,
(2) $p + p \rightarrow (p + n\pi) + (p + m\pi)$,

where n and m count the number of produced pions.

In the laboratory frame, an incident proton (the projectile) of total energy E strikes a proton (the target) at rest. Find the incident proton energy E that is

(a) the minimum energy for reaction 1 to take place when the target dissociates into a proton and 4 pions,

(b) the minimum energy for reaction 1 to take place when the projectile dissociates into a proton and 4 pions,

(c) the minimum energy for reaction 2 to take place when both protons dissociate into a proton and 4 pions. ($m_\pi = 0.140$ GeV, $m_p = 0.938$ GeV)

(Chicago)

Solution:

Let p_p be the momentum of the incident proton, n_p and n_π be the numbers of protons and pions, respectively, in the final state. Then the invariant mass squared of the system is

$$S = (E + m_p)^2 - p_p^2 = (n_p m_p + n_\pi m_\pi)^2,$$

giving

$$E = \frac{(n_p m_p + n_\pi m_\pi)^2 - 2m_p^2}{2m_p},$$

as

$$E^2 - p_p^2 = m_p^2 .$$

(a) For $p + p \rightarrow 2p + 4\pi$,

$$E = \frac{(2m_p + 4m_\pi)^2 - 2m_p^2}{2m_p} = 2.225 \ GeV .$$

(b) As the two protons are not distinguishable, the situation is identical with that of (a). Hence $E = 2.225$ GeV.

(c) For $p + p \rightarrow 2p + 8\pi$,

$$E = \frac{(2m_p + 8m_\pi)^2 - 2m_p^2}{2m_p} = 3.847 \ GeV .$$

4016

Protons from an accelerator collide with hydrogen. What is the minimum energy to create antiprotons?

(a) 6.6 GeV.
(b) 3.3 GeV.
(c) 2 GeV.

(CCT)

Solution:

The reaction to produce antiprotons is

$$\mathbf{p} + \mathbf{p} \rightarrow \bar{\mathbf{p}} + \mathbf{p} + \mathbf{p} + \mathbf{p} .$$

The hydrogen can be considered to be at rest. Thus at threshold the invariant mass squared is

$$(E + m_p)^2 - (E^2 - m_p^2) = (4m_p)^2 ,$$

or

$$E = 7m_p .$$

Hence the threshold energy is

$$E = 7m_p = 6.6 \ \text{GeV} ,$$

and the answer is (a).

4017

Determine the threshold energy for a gamma ray to create an electron-positron pair in an interaction with an electron at rest.

(Wisconsin)

Solution:

From the conservation of lepton number, the reaction is

$$\gamma + e^- \rightarrow e^+ + e^- + e^-.$$

At threshold the invariant mass squared is

$$S = (E_\gamma + m_e)^2 - p_\gamma^2 = (3m_e)^2.$$

With $E_\gamma = p_\gamma$, the above becomes

$$E_\gamma = 4m_e = 2.044 \text{ MeV}.$$

4018

Consider a beam of pions impinging on a proton target. What is the threshold for K^- production?

(Wisconsin)

Solution:

Conservation of strangeness requires a K^+ be also produced. Then the conservation of I_z requires that the p be converted to n as π^- has $I_z = -1$. Hence the reaction is

$$\pi^- + p \rightarrow K^- + K^+ + n.$$

Let the threshold energy and momentum of π^- be E_π and p_π respectively. (Conservation of the invariant mass squared $S = (\Sigma E)^2 - (\Sigma \mathbf{P})^2$ requires

$$S = (E_\pi + m_p)^2 - p_\pi^2 = (2m_K + m_n)^2.$$

With $E_\pi^2 - p_\pi^2 = m_\pi^2$, this gives

$$E_\pi = \frac{(2m_K + m_n)^2 - m_p^2 - m_\pi^2}{2m_p} = \frac{(2 \times 0.494 + 0.94)^2 - 0.938^2 - 0.14^2}{2 \times 0.938}$$

$$= 1.502 \text{ GeV}$$

4019

A particle of rest mass m whose kinetic energy is twice its rest energy collides with a particle of equal mass at rest. The two combine into a single new particle. Using only this information, calculate the rest mass such a new particle would have.

(*Wisconsin*)

Solution:

Let the mass of the new particle be M and that of the incident particle be m. The incident particle has total energy $E = m + T = 3m$. At threshold, M is produced at rest and the invariant mass squared is

$$S = (E + m)^2 - p^2 = M^2 .$$

With $E^2 - p^2 = m^2$, this gives

$$M^2 = 2Em + 2m^2 = 8m^2 ,$$

i.e.,

$$M = 2\sqrt{2}m .$$

4020

If a 1000 GeV proton hits a resting proton, what is the free energy to produce mass?

(a) 41.3 GeV.
(b) 1000 GeV.
(c) 500 GeV.

(*CCT*)

Solution:

Label the incident and target protons by 1 and 2 respectively. As the invariant mass squared

$$S = (\Sigma E_i)^2 - (\Sigma \mathbf{p})^2$$

is Lorentz-invariant,

$$(E_1 + m_p)^2 - p_1^2 = E^{*2},$$

where E^* is the total energy of the system in the center-of-mass frame. If the final state retains the two protons, the free energy for production of mass is

$$E^* - 2m_p = \sqrt{2m_p E_1 + 2m_p^2} - 2m_p$$

$$= \sqrt{2 \times 0.938 \times 1000 + 2 \times 0.938^2} - 2 \times 0.938$$

$$= 41.5 \text{ GeV}.$$

As $E_1 \gg m_p$, a rough estimate is

$$\sqrt{2m_p E_1} \approx \sqrt{2000} = 45 \text{ GeV}.$$

Thus the answer is (a).

4021

In the CERN colliding-beam storage ring, protons of total energy 30 GeV collide head-on. What energy must a single proton have to give the same center-of-mass energy when colliding with a stationary proton?

(Wisconsin)

Solution:

Consider a proton of energy E and momentum P incident on a stationary proton in the laboratory. This is seen in the center-of-mass frame as two protons each of energy $\bar{E}$ colliding head-on. The invariant mass squared S is Lorentz-invariant. Hence

$$S = (2\bar{E})^2 = (E + m_p)^2 - P^2 = 2m_p E + 2m_p^2,$$

giving

$$E = \frac{4\bar{E}^2 - 2m_p^2}{2m_p} = \frac{4 \times 30^2 - 2 \times 0.938^2}{2 \times 0.938}$$

$$= 1.92 \times 10^3 \text{ GeV}.$$

4022

Calculate the fractional change in the kinetic energy of an α-particle when it is scattered through $180°$ by an O^{16} nucleus.

(Wisconsin)

Solution:

Let E be the kinetic energy of the incident α-particle, p be its momentum, m_α be its mass, and let E' and p' represent the kinetic energy and momentum of the scattered α-particle respectively. In the nonrelativistic approximation,

$$p = \sqrt{2m_\alpha E}, \qquad p' = \sqrt{2m_\alpha E'}.$$

Let the recoil momentum of ^{16}O be P_0, conservation of momentum and of energy require

$$P_0 = p + p' = \sqrt{2m_\alpha E} + \sqrt{2m_\alpha E'},$$

$$E = E' + \frac{(\sqrt{2m_\alpha E} + \sqrt{2m_\alpha E'})^2}{2M},$$

where M is the mass of ^{16}O nucleus. With $M \approx 4m_\alpha$ the last equation gives

$$E = E' + \frac{1}{4}(\sqrt{E} + \sqrt{E'})^2 = \frac{5}{4}E' + \frac{1}{2}\sqrt{EE'} + \frac{1}{4}E,$$

or

$$(5\sqrt{E'} - 3\sqrt{E})(\sqrt{E'} + \sqrt{E}) = 0.$$

Thus $5\sqrt{E'} - 3\sqrt{E} = 0$, yielding $E' = \frac{9}{25}E$.

Therefore the fractional change in the kinetic energy of α-particle is

$$\frac{E' - E}{E} = -\frac{16}{25}.$$

4023

A beam of π^+ mesons of kinetic energy T yields some μ^+ going backward. The μ^+'s are products of the reaction

$$\pi^+ \to \mu^+ + \nu .$$

With

$$m_\pi c^2 = 139.57 \text{ MeV} ,$$

$$m_\mu c^2 = 105.66 \text{ MeV} ,$$

$$m_\nu c^2 = 0.0 \text{ MeV} .$$

for what range of T is this possible?

(*Wisconsin*)

Solution:

μ^+ from π^+ decay can go backward in the laboratory frame if its velocity in the center-of-mass frame (c.m.s.), which is also the rest frame of π^+, is greater than the velocity of π^+ in the laboratory frame. Denoting quantities in c.m.s. by a bar, we have

$$m_\pi = \sqrt{\bar{\mathbf{p}}_\mu^2 + m_\mu^2} + \bar{p}_\nu$$

since neutrino has zero rest mass. As $\bar{\mathbf{p}}_\mu = -\bar{\mathbf{p}}_\nu$, $\bar{p}_\mu = \bar{p}_\nu$ and the above gives

$$\bar{p}_\mu = \frac{m_\pi^2 - m_\mu^2}{2m_\pi} .$$

Hence

$$\bar{E}_\mu = \sqrt{\bar{p}_\mu^2 + m_\mu^2} = \frac{m_\pi^2 + m_\mu^2}{2m_\pi} ,$$

and so

$$\bar{\beta}_\mu = \frac{\bar{p}_\mu}{\bar{E}_\mu} = \frac{m_\pi^2 - m_\mu^2}{m_\pi^2 + m_\mu^2} .$$

We require $\beta_\pi \leq \bar{\beta}_\mu$ for some μ^+ to go backward. Hence

$$E_\pi \leq \frac{m_\pi}{\sqrt{1 - \bar{\beta}_\mu^{\,2}}} = \frac{m_\pi^2 + m_\mu^2}{2m_\mu} ,$$

or

$$T_\pi \leq E_\pi - m_\pi = \frac{(m_\pi - m_\mu)^2}{2m_\mu} = 5.44 \text{ MeV}.$$

4024

State whether the following processes are possible or impossible and prove your statement:

(a) A single photon strikes a stationary electron and gives up all its energy to the electron.

(b) A single photon in empty space is transformed into an electron and a positron.

(c) A fast positron and a stationary electron annihilate, producing only one photon.

(*Wisconsin*)

Solution:

All the three reactions cannot take place because in each case energy and momentum cannot be both conserved.

(a) For the process

$$\gamma + e \rightarrow e',$$

conservation of the invariant mass squared,

$$S = (E_\gamma + m_e)^2 - p_\gamma^2 = 2m_e E_\gamma + m_e^2 = E_{e'}^2 - p_{e'}^2 = m_e^2,$$

leads to $m_e E_\gamma = 0$, which contradicts the fact that neither E_γ nor m_e is zero.

(b) In the process $\gamma \rightarrow e^+ + e^-$, let the energies and momenta of the produced e^+ and e^- be E_1, E_2, $\mathbf{p}_1$, $\mathbf{p}_2$ respectively. The invariant mass squared of the initial state is

$$S(\gamma) = E_\gamma^2 - p_\gamma^2 = 0,$$

while for the final state it is

$$S(e^+ e^-) = (E_1 + E_2)^2 - (\mathbf{p}_1 + \mathbf{p}_2)^2$$

$$= 2m_e^2 + 2(E_1 E_2 - p_1 p_2 \cos \theta) \geq 2m_e^2,$$

where θ is the angle between $\mathbf{p}_1$ and $\mathbf{p}_2$. As $S(\gamma) \neq S(e^+e^-)$, its invariance is violated and the reaction cannot take place.

(c) The reaction is the inverse of that in (b). It similarly cannot take place.

4025

(a) Prove that an electron-positron pair cannot be created by a single isolated photon, i.e., pair production takes place only in the vicinity of a particle.

(b) Assuming that the particle is the nucleus of a lead atom, show numerically that we are justified in neglecting the kinetic energy of the recoil nucleus in estimating the threshold energy for pair production.

(Columbia)

Solution:

(a) This is not possible because energy and momentum cannot both be conserved, as shown in **Problem 4024**(b). However, if there is a particle in the vicinity to take away some momentum, it is still possible.

(b) Neglecting the kinetic energy of the recoiling nucleus, the threshold energy of the photon for e^+e^- pair production is

$$E_\gamma = 2m_e = 1.022 \text{ MeV}.$$

At most, the lead nucleus can take away all its momentum p_γ, i.e.,

$$p_{\text{Pb}} = p_\gamma = E_\gamma,$$

and the recoil kinetic energy of the Pb nucleus is

$$T_{\text{Pb}} = p_{\text{Pb}}^2/(2m_{\text{Pb}}) = \left(\frac{E_\gamma}{2m_{\text{Pb}}}\right) E_\gamma.$$

As $m_{\text{Pb}} \approx 200m_p = 1.88 \times 10^5 \text{ MeV}$,

$$T_{\text{Pb}} \approx \frac{1.022}{2 \times 1.88 \times 10^5} \times E_\gamma = 2.7 \times 10^{-6} \times E_\gamma.$$

Hence it is reasonable to neglect the kinetic energy of the recoiling nucleus.

4026

(a) Write the reaction equation for the decay of a negative muon. Identify in words all the particles involved.

(b) A mu-minus decays at rest. Could a lepton from this decay convert a proton at rest into a neutron? If so, how; and in particular will there be enough energy?

(Wisconsin)

Solution:

(a) The decay reaction for μ^- is

$$\mu^- \rightarrow e^- + \bar{\nu}_e + \nu_\mu \,,$$

where e^- represents electron, $\bar{\nu}_e$ electron-antineutrino, ν_μ muon-neutrino.

(b) If the energy of the electron or the electron-antineutrino is equal to or larger than the respective threshold energy of the following reactions, a proton at rest can be converted into a neutron.

$$e^- + p \rightarrow n + \nu_e \,, \tag{1}$$

$$\bar{\nu}_e + p \rightarrow e^+ + n \,. \tag{2}$$

The threshold energy for reaction (1) is

$$E_1 \approx m_n - m_p - m_e \approx 0.8 \text{ MeV} \,.$$

The threshold energy for reaction (2) is

$$E_2 \approx m_n - m_p + m_e \approx 1.8 \text{ MeV} \,.$$

Mu-minus decay releases quite a large amount of energy, about 105 MeV. The maximum energy ν_μ can acquire is about $m_\mu/2 \approx 53$ MeV. Then the combined energy of $\bar{\nu}_e$ and e^- is at the least about 53 MeV. In the reactions, as the mass of proton is much larger than that of muon or neutrino, the threshold energy in the center-of-mass system is approximately equal to that in the laboratory system. Therefore, at least one of the two leptons,

$\bar{\nu}_e$ or e^-, from μ^- decay has energy larger than the threshold of the above reactions and so can convert a proton at rest into a neutron.

4027

Two accelerator facilities are under construction which will produce the neutral intermediate vector boson Z^0 via the process

$$e^+ + e^- \rightarrow Z^0 .$$

The mass of the Z^0 is $M_Z = 92$ GeV.

(a) Find the energy of the electron beam needed for the colliding beam facility under construction.

Assume that a fixed target facility is to be built, such that a beam of e^+ will strike a target of e^- at rest.

(b) What is the required e^+ beam energy for this case?

(c) What is the energy and velocity of the Z^0 (in the laboratory) after production?

(d) Find the maximum energy in the laboratory frame of muons from the subsequent decay $Z^0 \rightarrow \mu^+ + \mu^-$.

(Columbia)

Solution:

(a) For the colliding-beam machine, the center-of-mass and laboratory frames are identical, and so the threshold electron energy for Z^0 production is $E = M_Z/2 = 46$ GeV.

(b) For the fixed target facility, conservation of the invariant mass gives

$$(E_{e^+} + m_e)^2 - p_{e^+}^2 = M_Z^2 .$$

With $E_{e^+}^2 - p_{e^+}^2 = m_e^2$, we find the threshold energy

$$E_{e^+} = \frac{M_Z^2 - 2m_e^2}{2m_e} \approx \frac{M_Z^2}{2m_e} = 8.30 \times 10^6 \text{ GeV} .$$

(c) In the center-of-mass frame (c.m.s.), total momentum is zero, total energy is $2\bar{E}$, $\bar{E}$ being the energy of e^+ or e^-. Invariance of the invariant mass squared,

$$S = (E_{e^+} + m_e)^2 - p_{e^+}^2 = (2\bar{E})^2 ,$$

gives

$$\bar{E} = \frac{\sqrt{2m_e E_{e^+} + 2m_e^2}}{2} \approx \sqrt{\frac{m_e E_{e^+}}{2}} = \frac{M_Z}{2} .$$

The Lorentz factor of c.m.s. is therefore

$$\gamma_0 = \frac{\bar{E}}{m_e} = \sqrt{\frac{E_{e^+}}{2m_e} + \frac{1}{2}} \approx \frac{M_Z}{2m_e} .$$

This is also the Lorentz factor of Z^0 as it is created at rest in c.m.s. Thus Z^0 has total energy $\gamma_0 M_z \approx \frac{M_Z^2}{2m_e} \approx E_{e^+}$ and velocity

$$\beta c = \left(1 - \frac{1}{\gamma_0^2}\right)^{\frac{1}{2}} c \approx \left[1 - \left(\frac{2m_e}{M_Z}\right)^2\right]^{\frac{1}{2}} c$$

$$\approx \left(1 - \frac{2m_e^2}{M_Z^2}\right) c .$$

(d) In the rest frame of Z^0 the angular distribution of the decay muons is isotropic. Those muons that travel in the direction of the incident e^+ have the maximum energy in the laboratory.

In c.m.s. Z^0 decays at rest into two muons, so that

$$\bar{E}_\mu = \frac{M_Z}{2} , \qquad \bar{\gamma}_\mu = \frac{\bar{E}_\mu}{m_\mu} = \frac{M_Z}{2m_\mu} .$$

For a muon moving in the direction of motion of e^+, inverse Lorentz transformation gives

$$\gamma_\mu = \gamma_0(\bar{\gamma}_\mu + \beta_0\bar{\gamma}_\mu\bar{\beta}_\mu) \approx 2\gamma_0\bar{\gamma}_\mu ,$$

as $\beta_0 \approx \beta_\mu \approx 1$. Hence the maximum laboratory energy of the decay muons is

$$E_\mu = \gamma_\mu m_\mu \approx 2\gamma_0\bar{\gamma}_\mu m_\mu = \frac{M_Z^2}{2m_e} \approx E_{e^+} .$$

This is to be expected physically as the velocity of the Z^0 is nearly equal to c. Compared to its kinetic energy, the rest mass of the muons produced in the reaction is very small. Thus the rest mass of the forward muon can be treated as zero, so that, like a photon, it takes all the momentum and energy of the Z^0.

4028

The following elementary-particle reaction may be carried out on a proton target at rest in the laboratory:

$$K^- + p \rightarrow \pi^0 + \Lambda^0 .$$

Find the special value of the incident K^- energy such that the Λ^0 can be produced at rest in the laboratory. Your answer should be expressed in terms of the rest masses m_{π^0}, m_{K^-}, m_p and m_{Λ^0}.

(MIT)

Solution:

The invariant mass squared $S = (\Sigma E)^2 - (\Sigma \mathbf{p})^2$ is conserved in a reaction. Thus

$$(E_K + m_p)^2 - p_K^2 = (E_\pi + m_\Lambda)^2 - p_\pi^2 .$$

As the Λ^0 is produced at rest, $p_\Lambda = 0$ and the initial momentum p_K is carried off by the π^0. Hence $p_\pi = p_K$ and the above becomes

$$E_K + m_p = E_\pi + m_\Lambda ,$$

or

$$E_\pi^2 = p_\pi^2 + m_\pi^2 = p_K^2 + m_\pi^2 = E_K^2 + (m_\Lambda - m_p)^2 - 2E_K(m_\Lambda - m_p) ,$$

or

$$2E_K(m_\Lambda - m_p) = m_K^2 - m_\pi^2 + (m_\Lambda - m_p)^2 ,$$

giving

$$E_K = \frac{m_K^2 - m_\pi^2 + (m_\Lambda - m_p)^2}{2(m_\Lambda - m_p)} .$$

4029

K^+ mesons can be photoproduced in the reaction

$$\gamma + p \rightarrow K^+ + \Lambda^0 .$$

(a) Give the minimum γ-ray energy in the laboratory, where p is at rest, that can cause this reaction to take place.

(b) If the target proton is not free but is bound in a nucleus, then the motion of the proton in the nucleus (Fermi motion) allows the reaction of part (a) to proceed with a lower incident photon energy. Assume a reasonable value for the Fermi motion and compute the minimum photon energy.

(c) The Λ^0 decays in flight into a proton and a π^- meson. If the Λ^0 has a velocity of $0.8c$, what is (i) the maximum momentum that the π^- can have in the laboratory, and (ii) the maximum component of laboratory momentum perpendicular to the Λ^0 direction?

$$(m_{K^+} = 494 \text{ MeV}/c^2, m_{\Lambda^0} = 1116 \text{ MeV}/c^2, m_{\pi^-} = 140 \text{ Mev}/c^2)$$

$$(CUSPEA)$$

Solution:

(a) Let P denote 4-momentum. We have the invariant mass squared

$$S = -(P_\gamma + P_p)^2 = (m_p + E_\gamma)^2 - E_\gamma^2 = m_p^2 + 2E_\gamma m_p = (m_K + m_\Lambda)^2,$$

giving

$$E_\gamma = \frac{(m_K + m_\Lambda)^2 - m_p^2}{2m_p} = 913 \text{ MeV}.$$

as the minimum γ energy required for the reaction to take place.

(b) If we assume that the proton has Fermi momentum $p_p = 200$ MeV/c then

$$S = -(P_\gamma + P_p)^2 = (E_\gamma + E_p)^2 - (\mathbf{p}_\gamma + \mathbf{p}_p)^2 = (m_K + m_\Lambda)^2.$$

With $E_\gamma = p_\gamma$, $E_p^2 - p_p^2 = m_p^2$, this gives

$$E_\gamma = \frac{(m_K + m_\Lambda)^2 - m_p^2 + 2\mathbf{p}_\gamma \cdot \mathbf{p}_p}{2E_p}.$$

The threshold energy E_γ is minimum when the proton moves opposite to the photon, in which case

$$E_\gamma = \frac{(m_K + m_\Lambda)^2 - m_p^2}{2(E_p + p_p)}$$

$$= \frac{(m_K + m_\Lambda)^2 - m_p^2}{2(\sqrt{p_p^2 + m_p^2} + p_p)} = 739 \text{ MeV}.$$

(c) In the rest frame of Λ^0, conservation of energy and of momentum give

$$\bar{E}_\pi + \bar{E}_p = m_\Lambda, \qquad \bar{\mathbf{p}}_\pi + \bar{\mathbf{p}}_p = 0.$$

Then

$$(m_\Lambda - \bar{E}_\pi)^2 = \bar{p}_p^2 + m_p^2 = \bar{p}_\pi^2 + m_p^2,$$

or

$$\bar{E}_\pi = \frac{m_\Lambda^2 + m_\pi^2 - m_p^2}{2m_\Lambda} = 173 \text{ MeV},$$

and so

$$\bar{p}_\pi = \sqrt{\bar{E}_\pi^2 - m_\pi^2} = 101 \text{ MeV}/c.$$

p_π is maximum in the laboratory if $\bar{p}_\pi$ is in the direction of motion of the Λ^0, which has $\beta_0 = 0.8$, $\gamma_0 = (1 - \beta^2)^{-\frac{1}{2}} = \frac{5}{3}$ in the laboratory. Thus

$$p_\pi = \gamma_0(\bar{p}_\pi + \beta_0 \bar{E}_\pi) = 399 \text{ MeV}/c.$$

As $(p_\pi)_\perp = (\bar{p}_\pi)_\perp$, the maximum momentum in the transverse direction is given by the maximum $(\bar{p}_\pi)_\perp$, i.e., 101 MeV/c.

4030

The ρ^- meson is a meson resonance with mass 769 MeV and width 154 MeV. It can be produced experimentally by bombarding a hydrogen target with a π^--meson beam,

$$\pi^- + p \to \rho^0 + n.$$

(a) What is the lifetime and mean decay distance for a 5 GeV ρ^0?

(b) What is the π^- threshold energy for producing ρ^0 mesons?

(c) If the production cross section is 1 mb $\equiv 10^{-27}$ cm^2 and the liquid hydrogen target is 30 cm long, how many ρ^0 are produced on the average per incident π^-? (The density of liquid hydrogen is 0.07 g/c.c.)

(d) ρ^0 mesons decay almost instantaneously into $\pi^+ + \pi^-$. Given that the ρ^0 is produced in the forward direction in the laboratory frame with an energy of 5 GeV, what is the minimum opening angle between the outgoing π^+ and π^- in the laboratory frame?

(Columbia)

Solution:

(a) The ρ^0 has Lorentz factor

$$\gamma_0 = \frac{E_\rho}{m_\rho} = \frac{5}{0.769} = 6.50 .$$

Its proper lifetime is

$$\tau_0 = \hbar/\Gamma = \frac{6.58 \times 10^{-22}}{154} = 4.27 \times 10^{-24} \text{ s} .$$

In laboratory frame the lifetime is

$$\tau = \gamma_0 \tau_0 = 2.78 \times 10^{-23} \text{ s} .$$

The mean decay distance for a 5 GeV ρ^0 is thus

$$d = \tau\beta c = \tau_0 \gamma_0 \beta c = \tau_0 c \sqrt{\gamma_0^2 - 1}$$

$$= 4.27 \times 10^{-24} \times 3 \times 10^{10} \times \sqrt{6.50^2 - 1}$$

$$= 8.23 \times 10^{-13} \text{ cm} .$$

(b) At threshold the invariant mass squared is

$$S = (E_\pi + m_p)^2 - p_\pi^2 = (m_\rho + m_n)^2 .$$

With $E_\pi^2 = m_\pi^2 + p_\pi^2$ this gives the threshold pion energy

$$E_\pi = \frac{(m_\rho + m_n)^2 - m_\pi^2 - m_p^2}{2m_p}$$

$$= \frac{(769 + 940)^2 - 140^2 - 938^2}{2 \times 938} = 1077 \text{ MeV} .$$

(c) The average number of ρ^- events caused by an incident π is

$$N = \rho l \sigma N_0 / A = 0.07 \times 30 \times 10^{-27} \times 6.02 \times 10^{23}$$

$$= 1.3 \times 10^{-3} ,$$

where $N_0 = 6.023 \times 10^{23}$ is the Avagadro number, $A = 1$ is the mass number of hydrogen, and ρ is the density of liquid hydrogen.

(d) In the rest frame $\bar{\Sigma}$ of the ρ^0, the pair of pions produced move in opposite directions with momenta $\bar{\mathbf{p}}_{\pi^+} = -\bar{\mathbf{p}}_{\pi^-}$ and energies $\bar{E}_{\pi^+} = \bar{E}_{\pi^-} = \frac{m_\rho}{2}$, corresponding to

$$\bar{\gamma}_\pi = \frac{\bar{E}_\pi}{m_\pi} = \frac{m_\rho}{2m_\pi}, \bar{\beta}_\pi = \sqrt{1 - \frac{1}{\bar{\gamma}_\pi^2}} = \frac{1}{m_\rho}\sqrt{m_\rho^2 - 4m_\pi^2} = 0.93.$$

$\bar{\Sigma}$ has Lorentz factor $\gamma_0 = 6.50$ in the laboratory, corresponding to

$$\beta_0 = \sqrt{1 - \frac{1}{6.50^2}} = 0.99.$$

Consider a pair of pions emitted in $\bar{\Sigma}$ parallel to the line of flight of ρ^0 in the laboratory. The forward-moving pion will move forward in the laboratory. As $\beta_0 > \bar{\beta}_\pi$, the backword-moving pion will also move forward in the laboratory. Hence the minimum opening angle between the pair is zero.

4031

(a) The Ω^- was discovered in the reaction $K^- + p \to \Omega^- + K^+ + K^0$. In terms of the masses of the various particles, what is the threshold kinetic energy for the reaction to occur if the proton is at rest?

(b) Suppose the K^0 travels at a speed of 0.8c. It decays in flight into two neutral pions. Find the maximum angle (in the laboratory frame) that the pions can make with the K^0 line of flight. Express your answer in terms of the π and K masses.

(Columbia)

Solution:

(a) At threshold the invariant mass squared is

$$S = (E_K + m_p)^2 - p_K^2 = (m_\Omega + 2m_K)^2.$$

With $E_K^2 = p_K^2 + m_K^2$, this gives

$$E_K = \frac{(m_\Omega + 2m_K)^2 - m_p^2 - m_K^2}{2m_p}.$$

Hence the threshold kinetic energy is

$$T_K = E_K - m_K = \frac{(m_\Omega + 2m_K)^2 - (m_p + m_K)^2}{2m_p} \, .$$

(b) Denote the rest frame of K^0 by $\bar{\Sigma}$ and label the two π^0 produced by 1 and 2. In $\bar{\Sigma}$,

$$\bar{\mathbf{p}}_1 = -\bar{\mathbf{p}}_2, \qquad \bar{E}_1 + \bar{E}_2 = m_K \, ,$$

and so

$$\bar{E}_1 = \bar{E}_2 = \frac{m_K}{2} \, ,$$

$$\bar{p}_1 = \bar{p}_2 = \sqrt{\bar{E}^2 - m_\pi^2}$$

$$= \frac{1}{2}\sqrt{m_K^2 - 4m_\pi^2} \, .$$

Consider one of the pions, say pion 1. Lorentz transformation

$$p_1 \cos\theta_1 = \gamma_0(\bar{p}_1 \cos\bar{\theta}_1 + \beta_0 \bar{E}_1) \, ,$$

$$p_1 \sin\theta_1 = \bar{p}_1 \sin\bar{\theta}_1 \, ,$$

gives

$$\tan\theta_1 = \frac{\sin\bar{\theta}_1}{\gamma_0\left(\cos\bar{\theta}_1 + \dfrac{\beta_0}{\bar{\beta}}\right)} \, ,$$

where γ_0 and β_0 are the Lorentz factor and velocity of the K^0 in laboratory and $\bar{\beta} = \frac{\bar{p}_1}{\bar{E}_1}$ is the velocity of the pion in $\bar{\Sigma}$.

To find maximum θ_1, let $\frac{d\tan\theta_1}{d\bar{\theta}_1} = 0$, which gives

$$\cos\bar{\theta}_1 = -\frac{\bar{\beta}}{\beta_0} \, .$$

Note that under this condition $\frac{d^2\tan\theta_1}{d\bar{\theta}_1^2} < 0$. Also, we have $\beta_0 = 0.8$,

$$\bar{\beta} = \frac{\bar{p}_1}{\bar{E}_1} = \frac{1}{m_K}\sqrt{m_K^2 - 4m_\pi^2}$$

$$= \sqrt{494^2 - 135^2 \times 4}/494 = 0.84 \, .$$

As $|\cos\bar{\theta}_1| \leq 1$, the condition cannot be satisfied. However, we see that as $\bar{\theta}_1 \to \pi$, $\cos\bar{\theta}_1 \to -1$, $\sin\bar{\theta}_1 \to 0$ and $\tan\theta_1 \to 0$, or $\theta_1 \to \pi$. Thus the maximum angle a pion can make with the line of flight of K^0 is π.

4032

The reaction

$$p + p \to \pi^+ + D, \tag{1}$$

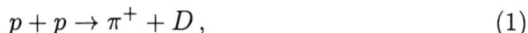

in which energetic protons from an accelerator strike resting protons to produce positive pi-meson-deuteron pairs, was an important reaction in the "early days" of high-energy physics.

(a) Calculate the threshold kinetic energy T in the laboratory for the incident proton. That is, T is the minimum laboratory kinetic energy allowing the reaction to proceed. Express T in terms of the proton mass m_p, the pion mass m_π, and the deuteron mass m_D. Evaluate T, taking $m_p = 938$ MeV/c^2, $m_D = 1874$ MeV/c^2, $m_\pi = 140$ MeV/c^2.

(b) Assume that the reaction is isotropic in the center-of-mass system. That is, the probability of producing a π^+ in the solid angle element $d\Omega^* = d\phi^* d(\cos\theta^*)$ is constant, independent of angle. Find an expression for the normalized probability of the π^+ per unit solid angle in the laboratory, in terms of $\cos\theta_{\text{lab}}$, the velocity $\bar{\beta}c$ of the center of mass, the π^+ velocity βc in the laboratory, and the momentum p^* in the center of mass.

(c) In 2-body endothermic reactions such as (1) it can happen that the probability per unit solid angle in the laboratory for a reaction product can be singular at an angle $\theta \neq 0$. How does this relate to the result derived in (b)? Comment briefly but do not work out all of the relevant kinematics.

(CUSPEA)

Solution:

(a) At threshold the invariant mass squared is

$$(E + m_p)^2 - p^2 = (m_\pi + m_D)^2,$$

where E and p are the energy and momentum of the incident proton in the laboratory. With

$$E^2 = p^2 + m_p^2$$

this gives

$$E = \frac{(m_\pi + m_D)^2 - 2m_p^2}{2m_p},$$

or the threshold kinetic energy

$$T = E - m_p = \frac{(m_\pi + m_D)^2 - 4m_p^2}{2m_p} = 286.2 \text{ MeV}.$$

(b) Let the normalized probability for producing a π^+ per unit solid angle in the center-of-mass and laboratory frames be $\frac{dP}{d\Omega^*}$ and $\frac{dP}{d\Omega}$ respectively. Then

$$\frac{dP}{d\Omega^*} = \frac{1}{4\pi},$$

$$\frac{dP}{d\Omega} = \frac{dP}{d\Omega^*}\frac{d\Omega^*}{d\Omega} = \frac{1}{4\pi}\frac{d\cos\theta^*}{d\cos\theta},$$

where the star denotes quantities in the center-of-mass frame.

The Lorentz transformation for the produced π^+

$$p^* \sin\theta^* = p\sin\theta, \tag{1}$$

$$p^* \cos\theta^* = \bar\gamma(p\cos\theta - \bar\beta E), \tag{2}$$

$$E^* = \bar\gamma(E - \bar\beta p\cos\theta), \tag{3}$$

where $\bar\gamma$ and $\bar\beta$ are the Lorentz factor and velocity of the center of mass in the laboratory. Differentiating Eq. (2) with respect to $\cos\theta$, as p^* and E^* are independent of θ^* and hence of θ, we have

$$p^* \frac{d\cos\theta^*}{d\cos\theta} = \bar\gamma\left(p + \cos\theta\frac{dp}{d\cos\theta} - \bar\beta\frac{dE}{dp}\frac{dp}{d\cos\theta}\right).$$

As $E = (m^2 + p^2)^{1/2}$, $dE/dp = p/E = \beta$ and the above becomes

$$p^* \frac{d\cos\theta^*}{d\cos\theta} = \bar\gamma\left(p + \cos\theta\frac{dp}{d\cos\theta} - \bar\beta\beta\frac{dp}{d\cos\theta}\right). \tag{4}$$

Differentiate Eq. (3) with respect to $\cos\theta$, we find

$$0 = \bar\gamma\left(\frac{dE}{d\cos\theta} - \bar\beta p - \bar\beta\cos\theta\frac{dp}{d\cos\theta}\right)$$

$$= \bar\gamma\left(\beta\frac{dp}{d\cos\theta} - \bar\beta p - \bar\beta\cos\theta\frac{dp}{d\cos\theta}\right),$$

or
$$\frac{dp}{d\cos\theta} = \frac{p\bar{\beta}}{\beta - \bar{\beta}\cos\theta}.$$

Substituting this in Eq. (4) gives

$$p^* \frac{d\cos\theta^*}{d\cos\theta} = \bar{\gamma}\left[p + \frac{(\cos\theta - \bar{\beta}\beta)\bar{\beta}p}{\beta - \bar{\beta}\cos\theta}\right]$$

$$= \frac{(1 - \bar{\beta}^2)\bar{\gamma}\beta p}{\beta - \bar{\beta}\cos\theta} = \frac{p}{\bar{\gamma}(1 - \bar{\beta}\cos\theta/\beta)}.$$

Hence the probability of producing a π^+ per unit solid angle in the laboratory is

$$\frac{dP}{d\Omega} = \frac{1}{4\pi}\frac{d\cos\theta^*}{d\cos\theta}$$

$$= \frac{p}{4\pi\bar{\gamma}p^*(1 - \bar{\beta}\cos\theta/\beta)}$$

$$= \frac{m_\pi\beta\gamma}{4\pi\bar{\gamma}p^*(1 - \bar{\beta}\cos\theta/\beta)}.$$

(c) The result in (b) shows that $\frac{dP}{d\Omega}$ is singular if $1 - \frac{\bar{\beta}}{\beta}\cos\theta = 0$ which requires $\bar{\beta} > \beta$. When the π^+ goes backward in the center-of-mass frame, $\beta < \bar{\beta}$. Thus there will be an angle θ in the laboratory for which the condition is satisfied. Physically, this is the "turn around" angle, i.e., the maximum possible angle of π^+ emission in the laboratory.

4033

The Q-value (the energy released) of the $He^3(n, p)$ reaction is reported to be 0.770 MeV. From this and the fact that the maximum kinetic energy of β-particles emitted by tritium (H^3) is 0.018 MeV, calculate the mass difference in amu between the neutron and a hydrogen atom (^1H). (1 amu = 931 MeV)

(SUNY, Buffalo)

Solution:

The reaction

$$^3He + n \rightarrow\, ^3H + p$$

has Q-value

$$Q = [M(^3\text{He}) + M(n) - M(^3\text{H}) - M(^1\text{H})] = 0.770 \text{ MeV},$$

whence

$$M(n) - M(^3\text{H}) = 0.770 + M(^1\text{H}) - M(^3\text{He}).$$

As in the decay $^3\text{H} \rightarrow^3 \text{He} + e^- + \bar{\nu}$ the electron has maximum energy

$$E_{\max} = [M(^3\text{H}) - M(^3\text{He})] = 0.018 \text{ MeV},$$

we find

$$M(n) - M(^1\text{H}) = 0.770 + 0.018 = 0.788 \text{ MeV}$$

$$= 8.46 \times 10^{-4} \text{ amu}.$$

4034

Suppose that a slowly moving antiproton is annihilated in a collision with a proton, leading to 2 negative pions and 2 positive pions. ($m_\pi c^2 = 140$ MeV)

(a) What is the average kinetic energy per pion? (MeV)

(b) What is the magnitude of the momentum of a pion with such an energy? (MeV/c)

(c) What is the magnitude of the velocity? (In units of c)

(d) If the annihilation led instead to 2 photons, what would be the wavelength of each? (cm)

(*UC, Berkeley*)

Solution:

(a)

$$p + \bar{p} \rightarrow 2\pi^+ + 2\pi^-,$$

As the incident $\bar{p}$ is slowly moving, we can take $T_{\bar{p}} \approx 0$. Then each pion will have energy $E_\pi \approx \frac{2m_p}{4} = \frac{1}{2}m_p$, and so kinetic energy

$$\bar{T}_\pi \approx \frac{1}{2}m_p - m_\pi = \frac{1}{2}(938 - 2 \times 140) = 329 \text{ MeV}.$$

(b) The momentum of each pion is

$$p = \sqrt{E_\pi^2 - m_\pi^2} \approx \frac{1}{2}\sqrt{m_p^2 - 4m_\pi^2} = 448 \text{ MeV}/c \,.$$

(c) Its velocity is

$$\beta = \frac{p}{E} \approx \frac{2p}{m_p} = 0.955 \,.$$

(d) If the annihilation had led to two photons, the energy of each photon would be

$$E_\gamma = \frac{2m_p}{2} = m_p = 938 \text{ MeV} \,.$$

The wavelength of each photon is

$$\lambda = \frac{c}{\nu} = \frac{2\pi\hbar c}{h\nu} = \frac{2\pi\hbar c}{E_\gamma} = \frac{2\pi \times 197 \times 10^{-13}}{938} = 1.32 \times 10^{-13} \text{ cm} \,.$$

4035

Consider the process of Compton scattering. A photon of wavelength λ is scattered off a free electron initially at rest. Let λ' be the wavelength of the photon scattered in a direction of θ.

(a) Compute λ' in terms of λ, θ and universal parameters.
(b) Compute the kinetic energy of the recoiled electron.

(CUSPEA)

Solution:

(a) Conservation of energy gives (Fig. 4.4)

$$pc + mc^2 = p'c + \sqrt{p_e^2 c^2 + m^2 c^4} \,,$$

or

$$(p - p' + mc)^2 = p_e^2 + m^2 c^2 \,, \tag{1}$$

where m is the mass of electron. Conservation of momentum requires

$$\mathbf{p} = \mathbf{p'} + \mathbf{p}_e$$

or

$$(\mathbf{p} - \mathbf{p'})^2 = \mathbf{p}_e^2 \,. \tag{2}$$

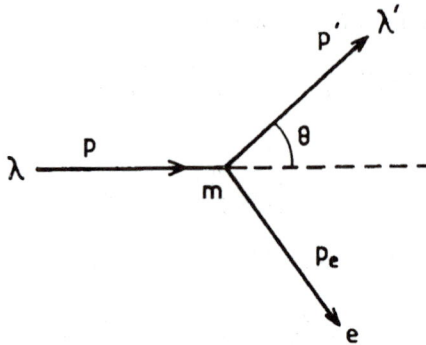

Fig. 4.4

The difference of Eqs. (1) and (2) gives

$$pp'(1 - \cos\theta) = (p - p')mc\,,$$

i.e.,

$$\frac{1}{p'} - \frac{1}{p} = \frac{1}{mc}(1 - \cos\theta)\,,$$

or

$$\frac{h}{p'} - \frac{h}{p} = \frac{h}{mc}(1 - \cos\theta)\,.$$

Hence

$$\lambda' = \lambda + \frac{h}{mc}(1 - \cos\theta)\,.$$

(b) The result of (a) gives

$$p'c = \frac{mc^2}{1 - \cos\theta + \dfrac{mc}{p}}\cdot$$

The kinetic energy of the recoiled electron is

$$T = \sqrt{p_e^2 c^2 + m^2 c^4} - mc^2 = pc - p'c$$

$$= \frac{pc(1 - \cos\theta)}{1 - \cos\theta + \dfrac{mc}{p}}$$

$$= \frac{(1 - \cos\theta)\dfrac{hc}{\lambda}}{1 - \cos\theta + \dfrac{mc\lambda}{h}}\cdot$$

4036

An X-ray photon of initial frequency 3×10^{19} Hz collides with an electron at rest and is scattered through 90°. Find the new frequency of the X-ray. The electron Compton wavelength is 2.4×10^{-12} meters.

(Wisconsin)

Solution:

Suppose that the target electron is free. Then the wavelength of the scattered photon is given by (**Problem 4035(a)**)

$$\lambda' = \lambda_0 + \frac{h}{mc}(1 - \cos\theta),$$

where λ_0 is the wavelength of the incident photon, $h/(mc)$ is the electron's Compton wavelength λ_c. At scattering angle 90° the wavelength of the scattered photon is

$$\lambda' = \lambda_0 + \lambda_c,$$

and the new frequency is

$$\nu' = \frac{c}{\lambda'} = \frac{c}{\dfrac{c}{\nu_0} + \lambda_c} = 2.42 \times 10^{19} \text{ Hz}.$$

4037

Consider Compton scattering of photons colliding head-on with moving electrons. Find the energy of the back-scattered photons ($\theta = 180°$) if the incident photons have an energy $h\nu = 2$ eV and the electrons have a kinetic energy of 1 GeV.

(Wisconsin, MIT, Columbia, Chicago, CCT)

Solution:

Denote the energies and momenta of the electron and photon before and after collision by E_e, p_e, E_γ, p_γ, E'_e, p'_e, E'_γ, p'_γ respectively. Conservation of energy and of momentum give

$$E_\gamma + E_e = E'_\gamma + E'_e,$$

or

$$p_\gamma + E_e = p'_\gamma + E'_e,$$

and

$$-p_\gamma + p_e = p'_\gamma + p'_e.$$

Addition and subtraction of the last two equations give

$$E'_e + p'_e = -2p'_\gamma + E_e + p_e,$$

$$E'_e - p'_e = 2p_\gamma + E_e - p_e,$$

which, after multiplying the respective sides together, give

$$E'^2_e - p'^2_e = E^2_e - p^2_e + 2p_\gamma(E_e + p_e) - 2p'_\gamma(E_e - p_e + 2p_\gamma).$$

With $E'^2_e - p'^2_e = E^2_e - p^2_e = m^2_e$, this becomes

$$p'_\gamma = \frac{p_\gamma(E_e + p_e)}{E_e - p_e + 2p_\gamma} \approx \frac{2p_\gamma E_e}{\dfrac{m^2_e}{2E_e} + 2p_\gamma}$$

$$= \frac{2 \times 2 \times 10^{-6} \times 10^3}{\dfrac{0.511^2}{2 \times 10^3} + 2 \times 2 \times 10^{-6}} = 29.7 \text{ MeV}/c,$$

since $E_e - p_e = E_e - \sqrt{E^2_e - m^2_e} \approx E_e - E_e(1 - \frac{m^2_e}{2E^2_e}) = \frac{m^2_e}{2E_e}$, $E_e + p_e \approx 2E_e$, $E_e \approx T_e$ as $m_e \ll E_e$. Hence the back-scattered photons have energy 29.7 MeV.

4038

(a) Two photons energy ε and E respectively collide head-on. Show that the velocity of the coordinate system in which the momentum is zero is given by

$$\beta = \frac{E - \varepsilon}{E + \varepsilon}$$

(b) If the colliding photons are to produce an electron-positron pair and ε is 1 eV, what must be the minimum value of the energy E?

(Wisconsin)

Solution:

(a) Let $\mathbf{P}$, $\mathbf{p}$ be the momenta of the photons, where $P = E$, $p = \varepsilon$. The total momentum of the system is $|\mathbf{P} + \mathbf{p}|$, and the total energy is $E + \varepsilon$. Hence the system as a whole has velocity

$$\beta = \frac{|\mathbf{P} + \mathbf{p}|}{E + \varepsilon} = \frac{E - \varepsilon}{E + \varepsilon} \,.$$

(b) At threshold the invariant mass squared of the system is

$$S = (E + \varepsilon)^2 - (\mathbf{P} + \mathbf{p})^2 = (2m_e)^2 \,,$$

m_e being the electron mass.

As $(\mathbf{P} + \mathbf{p})^2 = (P - p)^2 = (E - \varepsilon)^2$, the above gives the minimum energy required:

$$E = \frac{m_e^2}{\varepsilon} = 261 \text{ GeV} \,.$$

4039

The universe is filled with black-body microwave radiation. The average photon energy is $E \sim 10^{-3}$ eV. The number density of the photons is ~ 300 cm^{-3}. Very high energy γ-rays make electron-positron-producing collisions with these photons. This pair-production cross section is $\sigma_T/3$, with σ_T being the nonrelativistic electron-photon scattering cross section $\sigma_T = (8\pi/3)r_e^2$, where $r_e = e^2/mc^2$ is the classical radius of electron.

(a) What energy γ-rays would have their lifetimes in the universe limited by this process?

(b) What is the average distance they would *travel* before being converted into e^+e^- pairs?

(c) How does this compare with the size of the universe?

(d) What physical process might limit lifetime of ultra-high-energy protons (energy $\geq 10^{20}$ eV) in this same microwave radiation? (Assume photon-proton scattering to be too small to be important.)

(CUSPEA)

Solution:

(a) Let the energies and momenta of the high energy photon and a microwave photon be E_1, $\mathbf{p}_1$, E_2, $\mathbf{p}_2$ respectively. For e^+e^- production we require

$$(E_1 + E_2)^2 - (\mathbf{p}_1 + \mathbf{p}_2)^2 \geq (2m)^2 \,,$$

where m is the electron mass. As $E_1 = p_1$, $E_2 = p_2$, this becomes

$$2E_1 E_2 - 2\mathbf{p}_2 \cdot \mathbf{p}_2 \geq (2m)^2 \,,$$

or, if the angle between $\mathbf{p}_1$ and $\mathbf{p}_2$ is θ,

$$E_1 E_2 (1 - \cos\theta) \geq 2m^2 \,.$$

Hence

$$E_1 \geq \frac{2m^2}{E_2(1 - \cos\theta)} \,.$$

E_1 is minimum when $\theta = \pi$, i.e., $\cos\theta = -1$. Thus the minimum energy for pair production is

$$E_{\min} = \frac{m^2}{E_2} = \frac{(0.51 \times 10^6)^2}{10^{-3}} = 2.6 \times 10^{14} \text{ eV} \,.$$

Photons of energies above this value would have lieftimes limited by the pair production process.

(b) The mean free path for pair production is

$$l = \frac{1}{\rho\sigma} \approx \frac{1}{\frac{\rho\sigma_T}{3}} = \frac{9}{8\pi\rho r_e^2}$$

$$= \frac{9}{8\pi \times 300 \times (2.8 \times 10^{-13})^2} = 1.5 \times 10^{22} \text{ cm} = 1.6 \times 10^4 \text{ light years} \,.$$

(c) The size of our universe is $R \approx 10^{10}$ light years. Thus

$$l \ll R \,.$$

(d) Suppose the proton collides head-on with a microwave photon. The total energy $\bar{E}$ in the center-of-mass frame is given by the invariant mass squared

$$(E_p + E_\gamma)^2 - (p_p - p_\gamma)^2 = \bar{E}^2 \,,$$

or

$$2E_p E_\gamma + 2p_p p_\gamma + m_p^2 = \bar{E}^2 \,.$$

As $p_\gamma = E_\gamma$, $p_p \approx E_p$,

$$\bar{E} = \sqrt{4E_p E_\gamma + m_p^2}$$

$$= \sqrt{4 \times 10^{20} \times 10^{-3} + (10^9)^2}$$

$$= 1.18 \times 10^9 \text{ eV}.$$

Neglecting $\gamma p \to \gamma p$, we see that, as conservation of baryon number requires baryon number 1 in the products, the possible reactions are the following pion photoproduction

$$\gamma p \to \pi^0 p, \qquad \gamma p \to \pi^+ n.$$

4040

Consider the pion photoproduction reaction

$$\gamma + p \to \pi^0 + p,$$

where the rest energy is 938 MeV for the proton and 135 MeV for the neutral pion.

(a) If the initial proton is at rest in the laboratory find the laboratory threshold gamma-ray energy for this reaction to "go".

(b) The isotropic 3-K cosmic black-body radiation has average photon energy of about 0.001 eV. Consider a head-on collision between a proton and a photon of energy 0.001 eV. Find the minimum proton energy that will allow this pion photoproduction reaction to go.

(c) Speculate briefly on the implications of your result [to part (b)] for the energy spectrum of cosmic ray protons.

(UC, Berkeley)

Solution:

(a) The invariant mass squared of the reaction at threshold is

$$(E_\gamma + m_p)^2 - p_\gamma^2 = (m_p + m_\pi)^2.$$

With $E_\gamma = p_\gamma$, this gives

$$E_\gamma = \frac{(m_p + m_\pi)^2 - m_p^2}{2m_p} = m_\pi + \frac{m_\pi^2}{2m_p} = 145 \text{ MeV}.$$

(b) For head-on collision the invariant mass squared at threshold,

$$S = (E_\gamma + E_p)^2 - (p_\gamma - p_p)^2 = (m_\pi + m_p)^2,$$

gives

$$E_p - p_p = \frac{(m_p + m_\pi)^2 - m_p^2}{2E_\gamma} = 1.36 \times 10^{14} \text{ MeV}.$$

Writing $E_p - p_p = A$, we have

$$p_p^2 = (E_p - A)^2,$$

or

$$m_p^2 - 2AE_p + A^2 = 0,$$

giving the minimum proton energy for the reaction to go

$$E_p = \frac{1}{2A}(A^2 + m_p^2) \approx \frac{A}{2} = 6.8 \times 10^{13} \text{ MeV}.$$

(c) The photon density of 3-K black-body radiation is very large. Protons of energies $> E_p$ in cosmic radiation lose energy by constantly interacting with them. Hence the upper limit of the energy spectrum of cosmic-ray protons is E_p.

4041

The J/ψ particle has a mass of 3.097 GeV/c^2 and a width of 63 keV. A specific J/ψ is made with momentum 100 GeV/c and subsequently decays according to

$$J/\psi \to e^+ + e^-.$$

(a) Find the mean distance traveled by the J/ψ in the laboratory before decaying.

(b) For a symmetric decay (i.e., e^+ and e^- have the same laboratory momenta), find the energy of the decay electron in the laboratory.

(c) Find the laboratory angle of the electron with respect to the direction of the J/ψ.

(Columbia)

Solution:

(a) The total width Γ of J/ψ decay is 63 keV, so its proper lifetime is

$$\tau_0 = \hbar/\Gamma = \frac{6.58 \times 10^{-16}}{63 \times 10^3} = 1.045 \times 10^{-20} \text{ s}.$$

The laboratory lifetime is $\tau = \tau_0 \gamma$, where γ is its Lorentz factor. Hence the mean distance traveled by the J/ψ in the laboratory before decaying is

$$l = \tau\beta c = \tau_0\gamma\beta c = \frac{\tau_0 pc}{m} = 1.045 \times 10^{-20} \times \frac{100}{3.097} \times 3 \times 10^8$$

$$= 1.012 \times 10^{-10} \text{ m}.$$

(b) For symmetric decay, conservation of energy and of momentum give

$$E_J = 2E_e,$$

$$p_J = 2p_e \cos\theta,$$

where θ is the angle the electron makes with the direction of the J/ψ particle. Thus

$$E_e = \frac{1}{2}E_J = \frac{1}{2}\sqrt{p_J^2 + m_J^2} = \frac{1}{2}\sqrt{100^2 + 3.097^2} = 50.024 \text{ GeV}.$$

(c) The equations give

$$\left(\frac{E_J}{2}\right)^2 - \left(\frac{p_J}{2\cos\theta}\right)^2 = E_e^2 - p_e^2 = m_e^2,$$

or

$$\cos\theta = \frac{p_J}{\sqrt{p_J^2 + m_J^2 - 4m_e^2}}$$

$$= \frac{100}{\sqrt{100^2 + 3.097^2 - 4 \times (0.511 \times 10^{-3})^2}} = 0.9995,$$

i.e.

$$\theta = 1.77°.$$

4042

A negative Ξ particle decays into a Λ^0 and a π^-:

$$\Xi^- \to \Lambda^0 + \pi^- .$$

The Ξ^- is moving in the laboratory in the positive x direction and has a momentum of 2 GeV/c. The decay occurs in such a way that in the Ξ^- center-of-mass system the Λ^0 goes at an angle of 30° from the initial Ξ^- direction.

Find the momenta and angles of the Λ^0 and the π^- in the laboratory after the decay.

Rest energies:

$$M_\Xi c^2 = 1.3 \text{ GeV},$$

$$M_\Lambda c^2 = 1.1 \text{ GeV},$$

$$M_\pi c^2 = 0.14 \text{ GeV}.$$

(*Columbia*)

Solution:

The kinematic parameters β, γ and energy E_Ξ for Ξ^- are as follows:

$$E_\Xi = \sqrt{p_\Xi^2 + m_\Xi^2} = 2.385 \text{ GeV},$$

$$\beta_\Xi = \frac{p_\Xi}{E_\Xi} = 0.839,$$

$$\gamma_\Xi = \frac{E_\Xi}{m_\Xi} = 1.835.$$

Denote quantities in the Ξ^- rest frame by a bar. Conservation of momentum and of energy give

$$\bar{\mathbf{p}}_\pi + \bar{\mathbf{p}}_\Lambda = 0,$$

$$\bar{E}_\pi + \bar{E}_\Lambda = m_\Xi.$$

Then

$$\bar{p}_\Lambda = \bar{p}_\pi,$$

and so

$$\bar{E}_\Lambda = \sqrt{\bar{p}_\pi^2 + m_\Lambda^2} = m_\Xi - \bar{E}_\pi .$$

Solving the last equation gives, with $\bar{E}_\pi^2 - \bar{p}_\pi^2 = m_\pi^2$,

$$\bar{E}_\pi = \frac{m_\Xi^2 + m_\pi^2 - m_\Lambda^2}{2m_\Xi} = 0.192 \text{ GeV} ,$$

$$\bar{E}_\Lambda = m_\Xi - \bar{E}_\pi = 1.108 \text{ GeV} ,$$

$$\bar{p}_\Lambda = \bar{p}_\pi = \sqrt{\bar{E}_\pi^2 - m_\pi^2} = 0.132 \text{ GeV}/c .$$

The angle between $\bar{\mathbf{p}}_\Lambda$ and $\mathbf{p}_\Xi$ is $\bar{\theta}_\Lambda = 30°$, and the angle between $\bar{\mathbf{p}}_\pi$ and $\mathbf{p}_\Xi$ is $\bar{\theta}_\pi = 30° + 180° = 210°$.

Lorentz-transforming to the laboratory frame:

For π:

$$p_\pi \sin\theta_\pi = \bar{p}_\pi \sin\bar{\theta}_\pi = 0.132 \times \sin 210^0 = -0.064 \text{ GeV}/c ,$$

$$p_\pi \cos\theta_\pi = \gamma(\bar{p}_\pi \cos\bar{\theta}_\pi + \beta\bar{E}_\pi) = 0.086 \text{ GeV}/c ,$$

giving

$$\tan\theta_\pi = -0.767, \quad \text{or} \quad \theta_\pi = -37.5° ,$$

$$p_\pi = \sqrt{0.086^2 + 0.064^2} = 0.11 \text{ GeV}/c .$$

For Λ:

$$p_\Lambda \sin\theta_\Lambda = \bar{p}_\Lambda \sin\bar{\theta}_\Lambda = 0.132 \times \sin 30° = 0.66 \text{ GeV}/c ,$$

$$p_\Lambda \cos\theta_\Lambda = \gamma(\bar{p}_\Lambda \cos\bar{\theta}_\Lambda + \beta\bar{E}_\Lambda) = 1.92 \text{ GeV}/c ,$$

$$\tan\theta_\Lambda = 0.034, \quad \text{or} \quad \theta_\Lambda = 1.9° ,$$

$$p_\Lambda = \sqrt{1.92^2 + 0.066^2} = 1.92 \text{ GeV}/c .$$

The angle between directions of π and Λ in the laboratory is

$$\theta = \theta_\Lambda - \theta_\pi = 1.9 + 37.5 = 39.4° .$$

4043

A K-meson of rest energy 494 MeV decays into a μ of rest energy 106 MeV and a neutrino of zero rest energy. Find the kinetic energies of the μ and neutrino in a frame in which the K-meson decays at rest.

(UC, Berkeley)

Solution:

Consider the reaction

$$K \to \mu + \nu$$

in the rest frame of K. Conservation of momentum and of energy give

$$\mathbf{p}_\mu + \mathbf{p}_\nu = 0, \quad \text{or} \quad p_\mu = p_\nu,$$

and $E_\mu + E_\nu = m_K$.

We have

$$E_\mu^2 = (m_K - E_\nu)^2 = m_K^2 + E_\nu^2 - 2m_K E_\nu,$$

or, as $E_\nu = p_\nu = p_\mu$ and $E_\mu^2 = p_\mu^2 + m_\mu^2$,

$$p_\mu = \frac{m_K^2 - m_\mu^2}{2m_K} = \frac{494^2 - 106^2}{2 \times 494} = 236 \text{ MeV}/c.$$

The kinetic energies are

$$T_\nu = E_\nu = p_\nu c = p_\mu c = 236 \text{ MeV},$$

$$T_\mu = \sqrt{p_\mu^2 + m_\mu^2} - m_\mu = 152 \text{ MeV}.$$

4044

Pions ($m = 140$ MeV) decay into muons and neutrinos. What is the maximum momentum of the emitted muon in the pion rest frame?

(a) 30 MeV/c.
(b) 70 MeV/c.
(c) 2.7 MeV/c.

(CCT)

Solution:

Denote total energy by E, momentum by $\mathbf{p}$, and consider the reaction $\pi \to \mu + \nu_\mu$ in the pion rest frame. Conservation of energy and of momentum give

$$E_\mu = m_\pi - E_\nu \,,$$

$$\mathbf{p}_\mu + \mathbf{p}_\nu = 0 \,, \qquad \text{or} \qquad p_\mu = p_\nu \,.$$

As for neutrinos $E_\nu = p_\nu$, the first equation becomes, on squaring both sides,

$$p_\mu^2 + m_\mu^2 = (m_\pi - p_\mu)^2 \,,$$

giving

$$p_\mu = \frac{m_\pi^2 - m_\mu^2}{2m_\pi} = 29.9 \text{ MeV}/c \,.$$

Thus the answer is (a).

4045

The η' meson (let M denote its mass) can decay into a ρ^0 meson (mass m) and a photon (mass $= 0$): $\eta' \to \rho^0 + \gamma$. The decay is isotropic in the rest frame of the parent η' meson.

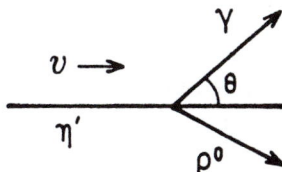

Fig. 4.5

Now suppose that a monoenergetic beam of η' mesons is traveling with speed v in the laboratory and let θ be the angle of the photon relative to the beam, as shown in Fig. 4.5. Let $P(\theta)d(\cos\theta)$ be the normalized probability that $\cos\theta$ lies in the interval $(\cos\theta, \cos\theta + d\cos\theta)$.

(a) Compute $P(\theta)$.

(b) Let $E(\theta)$ be the laboratory energy of the photon coming out at angle θ. Compute $E(\theta)$.

$$(CUSPEA)$$

Solution:

(a) Denote quantities in the rest frame of the η' particle by a bar and consider an emitted photon. Lorentz transformation for the photon,

$$\bar{p}\cos\bar{\theta} = \gamma(p\cos\theta - \beta E),$$

$$\bar{E} = \gamma(E - \beta p\cos\theta),$$

where γ, β are the Lorentz factor and velocity of the decaying η' in the laboratory frame, gives, as for the photon $\bar{p} = \bar{E}$, $p = E$,

$$\cos\bar{\theta} = \frac{\cos\theta - \beta}{1 - \beta\cos\theta},$$

or

$$\frac{d\cos\bar{\theta}}{d\cos\theta} = \frac{1 - \beta^2}{(1 - \beta\cos\theta)^2}.$$

In the rest frame of the η', photon emission is isotropic, i.e., the probability of γ emission per unit solid angle is a constant. Thus

$$dP \propto d\bar{\Omega} = 2\pi\sin\bar{\theta}d\bar{\theta} = 2\pi d\cos\bar{\theta},$$

or

$$\frac{dP}{1} = \frac{2\pi d\cos\bar{\theta}}{4\pi} = \frac{1}{2}d\cos\bar{\theta}.$$

Writing it as $dP = \bar{P}(\bar{\theta})d\cos\bar{\theta}$ we have $\bar{P}(\bar{\theta}) = \frac{1}{2}$. Transforming to the laboratory frame,

$$dP = \bar{P}(\bar{\theta})d\cos\bar{\theta} = P(\theta)d\cos\theta,$$

giving

$$P(\theta) = \frac{1}{2}\frac{d\cos\bar{\theta}}{d\cos\theta} = \frac{1 - \beta^2}{2(1 - \beta\cos\theta)^2}.$$

(b) In the rest frame of the η', conservation laws give

$$\bar{E}_p = M - \bar{E}, \qquad \bar{p}_p = \bar{p},$$

or

$$\bar{E}_p^2 - \bar{p}_p^2 = m^2 = M^2 - 2M\bar{E}.$$

Thus

$$\bar{E} = \frac{M^2 - m^2}{2M}.$$

Lorentz transformation for energy

$$\bar{E} = \gamma E(1 - \beta \cos \theta)$$

gives

$$E = \frac{\bar{E}}{\gamma(1 - \beta \cos \theta)} = \frac{M^2 - m^2}{2(E_\eta - p_\eta \cos \theta)},$$

E_η, p_η being the energy and momentum of the η' in the laboratory.

4046

A K_L^0 meson ($Mc^2 = 498$ MeV) decays into $\pi^+\pi^-$ ($mc^2 = 140$ MeV) in flight. The ratio of the momentum of the K_L^0 to Mc is $p/Mc = 1$. Find the maximum transverse component of momentum that any decay pion can have in the laboratory. Find the maximum longitudinal momentum that a pion can have in the laboratory.

(*Wisconsin*)

Solution:

In the laboratory frame, K_L^0 has velocity

$$\beta_c = \frac{p}{E} = \frac{p}{\sqrt{p^2 + M^2}} = \frac{1}{\sqrt{2}},$$

and hence $\gamma_c = \sqrt{2}$.

Let the energy and momentum of the pions in the rest frame of K_L^0 be $\bar{E}$ and $\bar{p}$ respectively. Energy conservation gives $2\bar{E} = M$, and hence

$$\bar{p} = \sqrt{\bar{E}^2 - m^2} = \frac{1}{2}\sqrt{M^2 - 4m^2} = \frac{1}{2}\sqrt{498^2 - 4 \times 140^2} = 206 \text{ MeV}/c.$$

The transverse component of momentum is not changed by the Lorentz transformation. Hence its maximum value is the same as the maximum value in the rest frame, namely 206 MeV/c.

In the laboratory frame the longitudinal component of momentum of π is

$$p_l = \gamma_c(\bar{p}\cos\bar{\theta} + \beta_c \bar{E}),$$

and has the maximum value $(\cos\bar{\theta} = 1)$

$$p_{l\text{max}} = \gamma_c(\bar{p} + \beta_c \bar{E}) = \gamma_c \left(\bar{p} + \frac{\beta_c M}{2}\right) = \sqrt{2}\left(206 + \frac{498}{2\sqrt{2}}\right)$$

$$= 540 \text{ MeV}/c.$$

4047

(a) A D^0 charmed particle decays in the bubble chamber after traveling a distance of 3 mm. The total energy of the decay products is 20 GeV. The mass of D^0 is 1.86 GeV. What is the time that the particle lived in its own rest frame?

(b) If the decays of many D^0 particles are observed, compare the expected time distributions (in the D^0 rest frame) of the decays into decay mode of branching ratio 1% and the same for a decay mode of branching ratio 40%.

(*Wisconsin*)

Solution:

(a) The total energy of the D^0 before decay is 20 GeV. Hence the Lorentz factor γ of its rest frame is

$$\gamma = \frac{E}{m_0} = \frac{20}{1.86} = 10.75.$$

The velocity of the D^0 (in units of c) is

$$\beta = \sqrt{\frac{\gamma^2 - 1}{\gamma^2}} = 0.996$$

The lifetime of the D^0 in the laboratory is

$$\tau = \frac{l}{\beta c} = \frac{3 \times 10^{-3}}{0.996 \times 3 \times 10^8} = 1.0 \times 10^{-11} \text{ s}$$

and its proper lifetime is

$$\tau_0 = \frac{\tau}{\gamma} = 9.3 \times 10^{-13} \text{ s}.$$

(b) The decay constant of D^0 is $\lambda = \frac{1}{\tau} = 1.07 \times 10^{12} \text{ s}^{-1}$.

In whatever decay mode, the expected time distribution of D^0 decays take the same form $f(t) \approx e^{-\lambda t} = \exp(-1.07 \times 10^{12} \times t)$. In other words, the decay modes of branching ratios 1% and 40% have the same expected time distribution.

4048

The charmed meson D^0 decays into $K^- \pi^+$. The masses of D, K, $\pi = 1.8$, 0.5, 0.15 GeV/c^2 respectively.

(a) What is the momentum of the K-meson in the rest frame of the D^0?

(b) Is the following statement true or false? Explain your answer.

"The production of single K^- mesons by neutrinos (ν_μ) is evidence for D^0 production"

(Wisconsin)

Solution:

In the rest frame of the D^0 meson, momentum conservation gives

$$\mathbf{p}_K + \mathbf{p}_\pi = 0, \quad \text{or} \quad p_K = p_\pi.$$

Energy conservation gives

$$E_K + E_\pi = m_D.$$

i.e.,

$$\sqrt{p_K^2 + m_K^2} + \sqrt{p_K^2 + m_\pi^2} = m_D,$$

leading to

$$p_K = \left[\left(\frac{m_D^2 + m_\pi^2 - m_K^2}{2m_D} \right)^2 - m_\pi^2 \right]^{\frac{1}{2}} = 0.82 \text{ GeV}/c.$$

(b) False. K^- has an s quark. Other particles such as Ξ^*, Ω^-, K^*, which can be produced in neutrino reactions, can also decay into single K^- mesons.

4049

The mean lifetime of a charged π-meson at rest is 2.6×10^{-8} sec. A monoenergetic beam of high-energy pions, produced by an accelerator, travels a distance of 10 meters, and in the process 10% of the pion decay. Find the momentum and kinetic energy of the pions.

(Wisconsin)

Solution:

Suppose the initial number of pions is N_0 and their velocity is β (in units of c). After traveling a distance of l the number becomes

$$N(l) = N_0 \exp\left(\frac{-\lambda l}{\beta c}\right),$$

where λ is the decay constant of pion in the laboratory. As

$$\lambda = \frac{1}{\tau} = \frac{1}{\gamma \tau_0},$$

where $\tau_0 = 2.6 \times 10^{-8}$ s is the proper lifetime of pion, $\gamma = \frac{1}{\sqrt{1-\beta^2}}$, we have

$$\gamma \beta = \frac{l}{\tau_0 c \ln \dfrac{N_0}{N(l)}} = \frac{10}{2.6 \times 10^{-8} \times 3 \times 10^8 \times \ln \dfrac{1}{0.9}} = 12.2\,.$$

The momentum of the pions is

$$p = m\gamma\beta = 0.14 \times 12.2 = 1.71 \text{ GeV}/c\,,$$

and so the kinetic energy is

$$T = \sqrt{\mathbf{p}^2 + m^2} - m \approx 1.58 \text{ GeV}\,.$$

4050

Neutral mesons are produced by a proton beam striking a thin target. The mesons each decay into two γ-rays. The photons emitted in the forward direction with respect to the beam have an energy of 96 MeV, and the photons emitted in the backward direction have an energy of 48 MeV.

(a) Determine $\beta = v/c$ for the mesons.

(b) Determine the (approximate) rest energy of the mesons.

(Wisconsin)

Solution:

(a) In the decay of a π^0 in the laboratory, if one photon is emitted backward, the other must be emitted forward. Let their energies and momenta be E_2, p_2, E_1, p_1 respectively. Conservation of energy gives

$$E = E_1 + E_2 = 96 + 48 = 144 \text{ MeV}.$$

Conservation of momentum gives

$$p = p_1 - p_2 = 96 - 48 = 48 \text{ MeV}/c.$$

Hence the π^0 has velocity

$$\beta = \frac{p}{E} = \frac{48}{144} = \frac{1}{3}.$$

(b) The π^0 rest mass is

$$m = \frac{E}{\gamma} = E\sqrt{1 - \beta^2} = \frac{144}{3} \times \sqrt{8} = 136 \text{ MeV}/c^2.$$

4051

A particle has mass $M = 3$ GeV/c^2 and momentum $p = 4$ GeV/c along the x-axis. It decays into 2 photons with an angular distribution which is isotropic in its rest frame, i.e. $\frac{dP}{d\cos\theta^*} = \frac{1}{2}$. What are the maximum and minimum values of the component of photon momentum along the x-axis? Find the probability dP/dp_x of finding a photon with x component of momentum p_x, as a function of p_x.

(Wisconsin)

Solution:

In the rest frame of the particle, conservation of momentum and of energy require

$$\bar{E}_1 + \bar{E}_2 = M, \qquad \bar{\mathbf{p}}_1 + \bar{\mathbf{p}}_2 = 0.$$

Thus

$$\bar{p}_1 = \bar{p}_2 = \bar{p}, \qquad \bar{E}_1 = \bar{E}_2 = \bar{E} = \frac{M}{2},$$

and the photons have energy

$$\bar{E} = \frac{3}{2} = 1.5 \text{ GeV}$$

and momentum

$$\bar{p} = \bar{E} = 1.5 \text{ GeV}/c.$$

The decaying particle has, in the laboratory,

$$\gamma\beta = \frac{p}{M} = \frac{4}{3},$$

and so

$$\gamma = \sqrt{(\gamma\beta)^2 + 1} = \frac{5}{3}, \qquad \beta = \frac{\gamma\beta}{\gamma} = 0.8.$$

Lorentz transformation gives the x component of photon momentum in the laboratory as

$$p_x = \gamma(\bar{p}\cos\bar{\theta} + \beta\bar{E}) = \gamma\bar{p}(\cos\bar{\theta} + \beta).$$

Hence, p_x is maximum when $\bar{\theta} = 0°$:

$$(p_x)_{\max} = \frac{5}{3} \times 1.5(1 + 0.8) = 4.5 \text{ GeV}/c,$$

p_x is minimum when $\bar{\theta} = 180°$:

$$(p_x)_{\min} = \frac{5}{3} \times 1.5(-1 + 0.8) = -0.5 \text{ GeV}/c.$$

Differentiating the transformation equation we have

$$dp_x = \gamma\bar{p}\,d\cos\bar{\theta}.$$

Hence

$$\frac{dP}{dp_x} = \frac{dP}{d\cos\bar{\theta}} \frac{d\cos\bar{\theta}}{dp_x} = \frac{1}{2} \cdot \frac{1}{\gamma\bar{p}} = 0.2.$$

4052

A neutral pion (π^0) decays into two γ rays. Suppose a π^0 is moving with a total energy E.

(a) What are the energies of the γ-rays if the decay process causes them to be emitted in opposite directions along the pion's original line of motion?

(b) What angle is formed between the two γ's if they are emitted at equal angles to the direction of the pion's motion?

(c) Taking $m_\pi = 135$ MeV and $E = 1$ GeV, give approximate numerical values for your above answers.

(Columbia)

Solution:

(a) Let the momenta and energies of the two γ's be p_{γ_1}, p_{γ_2} and E_{γ_1}, E_{γ_2}, the momentum and energy of the π^0 be p_π, E, respectively. Conservation laws of energy and momentum require

$$E = E_{\gamma_1} + E_{\gamma_2},$$

$$p_\pi = p_{\gamma_1} - p_{\gamma_2}.$$

As

$$E^2 = p_\pi^2 + m_\pi^2, \qquad E_{\gamma_1} = p_{\gamma_1}, \qquad E_{\gamma_2} = p_{\gamma_2},$$

the above equations give

$$m_\pi^2 = 4E_{\gamma_1}E_{\gamma_2} = 4E_{\gamma_1}(E - E_{\gamma_1}).$$

The quadratic equation for E_{γ_1} has two solutions

$$E_{\gamma_1} = \frac{E + \sqrt{E^2 - m_\pi^2}}{2},$$

$$E_{\gamma_2} = \frac{E - \sqrt{E^2 - m_\pi^2}}{2},$$

which are the energies of the two photons.

(b) Let the angles the two photons make with the direction of the pion be θ and $-\theta$. Conservation laws give

$$E = 2E_\gamma,$$

$$p_\pi = 2p_\gamma \cos\theta.$$

Note that, on account of symmetry, the two photons have the same energy and momentum E_γ, p_γ.

The two equations combine to give

$$m_\pi^2 = 4E_\gamma^2 - 4p_\gamma^2 \cos^2 \theta = E^2(1 - \cos^2 \theta) = E^2 \sin^2 \theta \,,$$

or

$$\theta = \pm \arcsin\left(\frac{m_\pi}{E}\right) .$$

Thus the angle between the two photons is

$$\theta_{2\gamma} = 2\theta = 2\arcsin\left(\frac{m_\pi}{E}\right) .$$

(c) Numerically we have

$$E_{\gamma_1} = \frac{10^3 + \sqrt{10^6 - 135^2}}{2} = 995.4 \text{ MeV} \,,$$

$$E_{\gamma_2} = \frac{10^3 - \sqrt{10^6 - 135^2}}{2} = 4.6 \text{ MeV} \,,$$

$$\theta_{2\gamma} = 2\arcsin\left(\frac{135}{1000}\right) = 15.5° .$$

4053

A π^0 meson decays isotropically into two photons in its rest system. Find the angular distribution of the photons in the laboratory as a function of the cosine of the polar angle in the laboratory for a π^0 with momentum $p = 280$ MeV/c. The rest energy of the pion is 140 MeV.

(UC, Berkeley)

Solution:

In the rest frame of the pion, the angular distribution of decay photons is isotropic and satisfies the normalization condition $\int W_0(\cos\theta^*, \phi^*)d\Omega^* = 1$. As a π^0 decays into two photons, $\int W(\cos\theta^*, \phi^*)d\Omega^* = 2$. Note that W is the probability of emitting a photon in the solid angle $d\Omega^*(\theta^*, \phi^*)$ in the decay of a π^0. As W is independent of θ^* and ϕ^*, the integral gives $W \int d\Omega^* = 4\pi W = 2$, or $W(\cos\theta^*, \phi^*) = \frac{1}{2\pi}$. Integrating over ϕ^*, we have

$$\int_0^{2\pi} W(\cos\theta^*)d\varphi^* = W \int_0^{2\pi} d\varphi^* = 1\,,$$

or

$$W(\cos\theta^*) = 1\,.$$

If θ^* corresponds to laboratory angle θ, then

$$W(\cos\theta)d\cos\theta = W(\cos\theta^*)d\cos\theta^*\,.$$

Let γ_0, β_0 be the Lorentz factor and velocity of the decaying π^0. The Lorentz transformation for a photon gives

$$p\cos\theta = \gamma_0(p^*\cos\theta^* + \beta_0 E^*) = \gamma_0 p^*(\cos\theta^* + \beta_0)\,,$$

$$E = p = \gamma_0(E^* + \beta_0 p^*\cos\theta^*) = \gamma_0 p^*(1 + \beta_0\cos\theta^*)\,.$$

Note E^*, p^* are constant since the angular distribution of the photons in the rest frame is isotropic. Differentiating the above equations with respect to $\cos\theta^*$, we have

$$\cos\theta\frac{dp}{d\cos\theta^*} + p\frac{d\cos\theta}{d\cos\theta^*} = \gamma_0 p^*\,,$$

$$\frac{dp}{d\cos\theta^*} = \gamma_0\beta_0 p^*\,,$$

which combine to give

$$\frac{d\cos\theta^*}{d\cos\theta} = \frac{p}{\gamma_0 p^*(1 - \beta_0\cos\theta)} = \frac{1}{\gamma_0^2(1 - \beta_0\cos\theta)^2}\,,$$

use having been made of the transformation equation

$$E^* = \gamma_0(E - \beta_0 p\cos\theta)\,,$$

or

$$p^* = \gamma_0 p(1 - \beta_0\cos\theta)\,.$$

Hence

$$W(\cos\theta) = W(\cos\theta^*)\frac{d\cos\theta^*}{d\cos\theta} = \frac{1}{\gamma_0^2(1-\beta_0\cos\theta)^2}\,.$$

With π^0 of mass 140 MeV/c^2, momentum 280 MeV/c, we have

$$\gamma_0\beta_0 = \frac{280}{140} = 2\,,$$

$$\gamma_0 = \sqrt{(\gamma_0\beta_0)^2 + 1} = \sqrt{5}\,,$$

$$\beta_0 = \frac{\gamma_0\beta_0}{\gamma_0} = \frac{2}{\sqrt{5}}\,,$$

giving the laboratory angular distribution

$$W(\cos\theta) = \frac{1}{(\sqrt{5})^2\left(1 - \dfrac{2}{\sqrt{5}}\cos\theta\right)^2} = \frac{1}{(\sqrt{5} - 2\cos\theta)^2}\,.$$

4054

A neutral pion decays into two γ-rays, $\pi^0 \to \gamma + \gamma$, with a lifetime of about 10^{-16} sec. Neutral pions can be produced in the laboratory by stopping negative pions in hydrogen via the reaction

$$\pi^- + p \to \pi^0 + n\,.$$

The values of the rest masses of these particles are:

$$m(\pi^-) = 140\ \text{MeV}\,, \qquad m(\pi^0) = 135\ \text{MeV}\,, \qquad m(p) = 938\ \text{MeV}\,,$$

$$m(n) = 940\ \text{MeV}\,.$$

(a) What is the velocity of the π^0 emerging from this reaction? Assume that both the π^- and the proton are at rest before the reaction.

(b) What is the kinetic energy of the emerging neutron?

(c) How far does the π^0 travel in the laboratory if it lives for a time of 10^{-16} seconds measured in its own rest frame?

(d) What is the maximum energy in the laboratory frame of the γ-rays from the π^0 decay?

(Columbia)

Solution:

(a) Momentum conservation requires

$$\mathbf{p}_{\pi^0} + \mathbf{p}_n = 0, \quad \text{or} \quad p_{\pi^0} = p_n.$$

Energy conservation requires

$$E_n = m_{\pi^-} + m_p - E_{\pi^0}.$$

With $E^2 - p^2 = m^2$, these equations give

$$E_{\pi^0} = \frac{(m_\pi + m_p)^2 + m_{\pi^0}^2 - m_n^2}{2(m_{\pi^-} + m_p)}$$

$$= 137.62 \text{ MeV}.$$

Hence

$$\gamma = \frac{E_{\pi^0}}{m_{\pi^0}} = 1.019,$$

and

$$\beta = \sqrt{1 - \frac{1}{\gamma^2}} = 0.194.$$

Thus the π^0 has velocity 5.8×10^7 m/s.

(b) The neutron has kinetic energy

$$T_n = m_{\pi^-} + m_p - E_{\pi^0} - m_n$$

$$= 0.38 \text{ MeV}.$$

(c) The lifetime of π^0 in the laboratory is

$$\tau = \tau_0 \gamma = 1.019 \times 10^{-16} \text{ s}.$$

Hence the distance it travels before decaying is

$$l = \tau \beta c = 1.019 \times 10^{-16} \times 5.8 \times 10^7 = 5.9 \times 10^{-9} \text{ m}.$$

(d) The π^0 has $\gamma = 1.019$, $\beta = 0.194$ in the laboratory. In its rest frame, each decay photon has energy

$$E_\gamma^* = \frac{1}{2}m_{\pi^0} = 67.5 \text{ MeV}.$$

Transforming to the laboratory gives

$$E_\gamma = \gamma(E_\gamma^* + \beta p_\gamma^* \cos\theta^*).$$

Maximum E_γ corresponds to $\theta^* = 0$:

$$(E_\gamma)_{\max} = \gamma E_\gamma^*(1+\beta) = 1.019 \times 67.5 \times (1+0.194)$$

$$= 82.1 \text{ MeV}.$$

4055

High energy neutrino beams at Fermilab are made by first forming a monoenergetic π^+ (or K^+) beam and then allowing the pions to decay by

$$\pi^+ \to \mu^+ + \nu.$$

Recall that the mass of the pion is 140 MeV/c^2 and the mass of the muon is 106 MeV/c^2.

(a) Find the energy of the decay neutrino in the rest frame of the π^+.

In the laboratory frame, the energy of the decay neutrino depends on the decay angle θ (see Fig. 4.6). Suppose the π^+ beam has an energy 200 GeV/c^2.

(b) Find the energy of a neutrino produced in the forward direction $(\theta = 0)$.

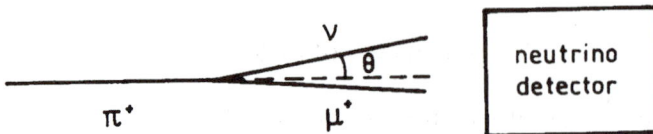

Fig. 4.6

(c) Find the angle θ at which the neutrino's energy has fallen to half of its maximum energy.

(*Chicago*)

Solution:

(a) In the π^+ rest frame conservation laws of energy and momentum require

$$E_\nu + E_\mu = m_\pi \,,$$

$$\mathbf{p}_\nu + \mathbf{p}_\mu = 0 \,, \qquad \text{or} \qquad p_\nu = p_\mu \,.$$

These equations combine to give

$$m_\mu^2 + p_\nu^2 = E_\nu^2 + m_\pi^2 - 2m_\pi E_\nu \,.$$

Assume that neutrino has zero mass. Then $E_\nu = p_\nu$ and the above gives

$$E_\nu = \frac{m_\pi^2 - m_\mu^2}{2m_{\pi^+}} = \frac{140^2 - 106^2}{2 \times 140} = 30 \text{ MeV} \,.$$

(b) For π^+ of energy 200 GeV, $\gamma = \frac{E}{m} = \frac{200}{0.140} = 1429$, $\beta \approx 1$. Lorentz transformation for neutrino

$$E_\nu = \gamma(E_\nu^* + \beta p_\nu^* \cos\theta^*) = \gamma E_\nu^*(1 + \beta\cos\theta^*)$$

gives for $\theta^* = 0$

$$E_\nu = \gamma E_\nu^*(1 + \beta) \approx 1429 \times 30 \times (1 + 1) = 85.7 \text{ GeV} \,.$$

Note $\theta^* = 0$ corresponds to $\theta = 0$ in the laboratory as $p_\nu \sin\theta = p_\nu^* \sin\theta^*$.

(c) The laboratory energy of the neutrino is maximum when $\theta = \theta^* = 0$. Thus

$$(E_\nu)_{\max} = \gamma E_\nu^*(1 + \beta) \,.$$

For $E_\nu = \frac{1}{2}(E_\nu)_{\max}$, we have

$$\gamma E_\nu^*(1 + \beta\cos\theta^*) = \frac{1}{2}\gamma E_\gamma^*(1 + \beta) \,,$$

giving

$$\cos\theta^* = \frac{\beta - 1}{2\beta} \,,$$

which corresponds to

$$\sin \theta^* = \sqrt{1 - \cos^2 \theta^*} = \frac{1}{2\beta} \sqrt{3\beta^2 + 2\beta - 1}.$$

Lorentz transformation equations for neutrino

$$p_\nu \sin \theta = p_\nu^* \sin \theta^*,$$

$$p_\nu \cos \theta = \gamma(p_\nu^* \cos \theta^* + \beta E_\nu^*) = \gamma p_\nu^* (\cos \theta^* + \beta),$$

give

$$\tan \theta = \frac{\sin \theta^*}{\gamma(\cos \theta^* + \beta)}.$$

For $E_\nu = \frac{1}{2}(E_\nu)_{\text{max}}$,

$$\tan \theta_{\frac{1}{2}} = \frac{\sqrt{3\beta^2 + 2\beta - 1}}{\gamma(\beta - 1 + 2\beta^2)} = \frac{1}{\gamma} \cdot \frac{1}{(2\beta - 1)} \sqrt{\frac{3\beta - 1}{\beta + 1}} \approx \frac{1}{\gamma}$$

as $\beta \approx 1$. Hence at half the maximum angle,

$$\theta_{\frac{1}{2}} \approx \frac{1}{\gamma}.$$

4056

One particular interest in particle physics at present is the weak interactions at high energies. These can be investigated by studying high-energy neutrino interactions. One can produce neutrino beams by letting pi and K mesons decay in flight. Suppose a 200-GeV/c pi-meson beam is used to produce neutrinos via the decay $\pi^+ \rightarrow \mu^+ + \nu$. The lifetime of pi-mesons is $\tau_{\pi^\pm} = 2.60 \times 10^{-8}$ sec (in the rest frame of the pion), and its rest energy is 139.6 MeV. The rest energy of the muon is 105.7 MeV, and the neutrino is massless.

(a) Calculate the mean distance traveled by the pions before they decay.

(b) Calculate the maximum angle of the muon (relative to the pion direction) in the laboratory.

(c) Calculate the minimum and maximum momenta the neutrinos can have.

(*UC, Berkeley*)

Solution:

(a) The pions have Lorentz factor

$$\gamma = \frac{E}{m} \approx \frac{p}{m} = \frac{200000}{139.6} = 1433.$$

The lifetime of the pions in the laboratory frame is then $\tau = \gamma \tau_0 = 2.6 \times 10^{-8} \times 1433 = 3.72 \times 10^{-5}$ s.

The speed of the pions is very close to that of light. Thus before decaying the distance traveled is on the average

$$l = c\tau = 3 \times 10^8 \times 3.72 \times 10^{-5} = 1.12 \times 10^4 \text{ m}.$$

(b) Figure 4.7 shows the decay in the laboratory frame Σ and the rest frame Σ^* of the pion.

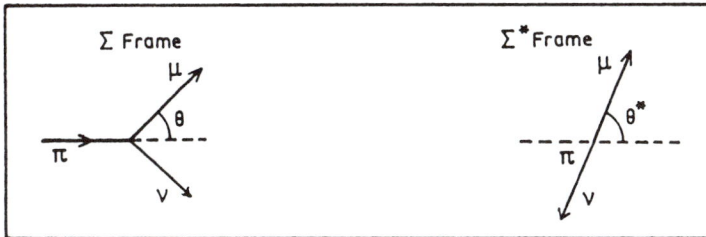

Fig. 4.7

In Σ^*, conservation laws of energy and momentum require

$$E_\nu^* + E_\mu^* = m_\pi,$$

$$\mathbf{p}_\nu^* + \mathbf{p}_\mu^* = 0, \qquad \text{or} \qquad p_\nu^* = p_\mu^*.$$

The above equations combine to give

$$E_\mu^* = \frac{m_\pi^2 + m_\mu^2}{2m_\pi} = 109.8 \text{ MeV}.$$

Lorentz transformation for the muon gives

$$p_\mu \sin \theta = p_\mu^* \sin \theta^* ,$$

$$p_\mu \cos \theta = \gamma(p_\mu^* \cos \theta^* + \beta E_\mu^*) ,$$

where $\gamma = 1433$ is the Lorentz factor of Σ^*, $\beta \approx 1$. Thus

$$\tan \theta = \frac{\sin \theta^*}{\gamma \left(\cos \theta^* + \dfrac{E_\mu^*}{p_\mu^*} \right)} = \frac{\sin \theta^*}{\gamma \left(\cos \theta^* + \dfrac{1}{\beta_\mu^*} \right)} ,$$

where $\beta_\mu^* = \frac{p_\mu^*}{E_\mu^*}$. To find maximum θ, let

$$\frac{d \tan \theta}{d \theta^*} = 0 .$$

This gives $\cos \theta^* = -\beta_\mu^*$, $\sin \theta^* = \sqrt{1 - \beta_\mu^{*2}} = \frac{1}{\gamma_\mu^*}$. Hence

$$(\tan \theta)_{\max} = \frac{1}{\gamma \gamma_\mu^* \left(-\beta_\mu^* + \dfrac{1}{\beta_\mu^*} \right)} = \frac{\beta_\mu^*}{\gamma \gamma_\mu^* (\beta_\mu^{*2} - 1)} = \frac{\gamma_\mu^* \beta_\mu^*}{\gamma} = \frac{\sqrt{\gamma_\mu^{*2} - 1}}{\gamma} .$$

As $\gamma_\mu^* = \frac{E_\mu^*}{m_\mu} = \frac{109.8}{105.7} = 1.039$, $\gamma = 1433$, we have

$$\theta_{\max} = \arctan(\tan \theta)_{\max} \approx \frac{\sqrt{\gamma_\mu^{*2} - 1}}{\gamma} = 1.97 \times 10^{-4} \text{ rad} = 0.011^0 .$$

(c) In the rest frame Σ^*, the neutrino has energy

$$E_\nu^* = m_\pi - E_\mu^* = \frac{m_\pi^2 - m_\mu^2}{2m_\pi} = 29.8 \text{ MeV} ,$$

and hence momentum 29.8 MeV/c. Lorentz transformation gives for the neutrino,

$$p_\nu = E_\nu = \gamma(E_\nu^* + \beta p_\nu^* \cos \theta^*) = \gamma p_\nu^* (1 + \beta \cos \theta^*) .$$

Hence

$$(p_\nu)_{\max} = \gamma p_\nu^* (1 + \beta)$$

$$= 1433 \times 29.8(1+1) = 85.4 \text{ GeV}/c,$$

$$(p_\nu)_{\min} = \gamma p_\nu^* (1 - \beta)$$

$$= [\sqrt{(\gamma\beta)^2 + 1} - \gamma\beta] p_\nu^*$$

$$\approx \frac{p_\nu^*}{2\gamma\beta} = \frac{m_\pi p_\nu^*}{2p_\pi} = \frac{139.6 \times 29.4}{2 \times 200 \times 10^3}$$

$$= 1.04 \times 10^{-2} \text{ MeV}/c.$$

4057

A beam of pions of energy E_0 is incident along the z-axis. Some of these decay to a muon and a neutrino, with the neutrino emerging at an angle θ_ν relative to the z-axis. Assume that the neutrino is massless.

(a) Determine the neutrino energy as a function of θ_ν. Show that if $E_0 \gg m_\pi$ and $\theta_\nu \ll 1$,

$$E_\nu \approx E_0 \frac{1 - \left(\dfrac{m_\mu}{m_\pi}\right)^2}{1 + \left(\dfrac{E_0}{m_\pi}\right)^2 \theta_\nu^2}.$$

(b) The decay is isotropic in the center-of-mass frame. Determine the angle θ_m such that half the neutrinos will have $\theta_\nu < \theta_m$.

(Columbia)

Solution:

(a) Let the emission angle of the muon relative to the z-axis be θ. Conservation of energy and of momentum give

$$E_0 = E_\mu + E_\nu = \sqrt{p_\mu^2 + m_\mu^2} + E_\nu,$$

$$\sqrt{E_0^2 - m_\pi^2} = p_\mu \cos\theta + p_\nu \cos\theta_\nu,$$

$$0 = p_\mu \sin\theta + p_\nu \sin\theta_\nu,$$

As neutrino is assumed massless, $p_\nu = E_\nu$. The momentum equations combine to give

$$p_\mu^2 = E_0^2 - m_\pi^2 + p_\nu^2 - 2\sqrt{E_0^2 - m_\pi^2}\, E_\nu \cos\theta_\nu \,,$$

while the energy equation gives

$$p_\mu^2 = E_0^2 - m_\mu^2 + p_\nu^2 - 2E_0 E_\nu \,.$$

The difference of the last two equations then gives

$$E_\nu = \frac{m_\pi^2 - m_\mu^2}{2(E_0 - \sqrt{E_0^2 - m_\pi^2}\cos\theta_\nu)}$$

$$= \frac{m_\pi^2}{2E_0}\frac{\left[1 - \left(\dfrac{m_\mu}{m_\pi}\right)^2\right]}{\left[1 - \sqrt{1 - \left(\dfrac{m_\pi}{E_0}\right)^2}\cos\theta_\nu\right]}\,.$$

If $E_0 \gg m_\pi$, $\theta_\nu \ll 1$, then

$$\sqrt{1 - \left(\frac{m_\pi}{E_0}\right)^2}\cos\theta_\nu \approx \left[1 - \frac{1}{2}\left(\frac{m_\pi}{E_0}\right)^2\right]\left(1 - \frac{\theta_\nu^2}{2}\right) \approx 1 - \frac{1}{2}\left(\frac{m_\pi}{E_0}\right)^2 - \frac{\theta_\nu^2}{2}\,,$$

and hence

$$E_\nu \approx \frac{m_\pi^2}{E_0} \times \frac{1 - \left(\dfrac{m_\mu}{m_\pi}\right)^2}{\left(\dfrac{m_\pi}{E_0}\right)^2 + \theta_\nu^2} = E_0\frac{1 - \left(\dfrac{m_\mu}{m_\pi}\right)^2}{1 + \left(\dfrac{E_0}{m_\pi}\right)^2\theta_\nu^2}\,.$$

(b) The center-of-mass frame (i.e. rest frame of π) has Lorentz factor and velocity

$$\gamma = \frac{E_0}{m_\pi}\,, \qquad \beta = \sqrt{1 - \frac{1}{\gamma^2}}\,.$$

Denote quantities in the rest frame by a bar. Lorentz transformation for the neutrino

$$p_\nu \sin\theta_\nu = \bar{p}_\nu \sin\bar{\theta}_\nu \,,$$

$$p_\nu \cos\theta_\nu = \gamma(\bar{p}_\nu \cos\bar{\theta}_\nu + \beta\bar{E}_\nu) = \gamma\bar{p}_\nu(\cos\bar{\theta}_\nu + \beta)$$

gives

$$\tan \theta_\nu = \frac{\sin \bar{\theta}_\nu}{\gamma(\beta + \cos \bar{\theta}_\nu)}.$$

As the angular distribution of the neutrinos in the rest frame is isotropic, $\bar{\theta}_m = 90°$. Then

$$\tan \theta_m = \frac{\sin 90°}{\gamma(\beta + \cos 90°)} = \frac{1}{\gamma\beta} = \frac{1}{\sqrt{\gamma^2 - 1}}$$

$$= \frac{1}{\sqrt{\left(\dfrac{E_0}{m_\pi}\right)^2 - 1}} = \frac{m_\pi}{\sqrt{E_0^2 - m_\pi^2}},$$

or

$$\theta_m = \arctan\left(\frac{m_\pi}{\sqrt{E_0^2 - m_\pi^2}}\right).$$

Note that as

$$\frac{d\theta_\nu}{d\bar{\theta}_\nu} = \frac{\cos^2 \theta_\nu}{\gamma} \frac{(1 + \beta \cos \bar{\theta}_\nu)}{(\beta + \cos \bar{\theta}_\nu)^2} \geq 0$$

θ_ν increases monotonically as $\bar{\theta}_\nu$ increases. This means that if $\bar{\theta}_\nu \leq \bar{\theta}_m$ contains half the number of the neutrinos emitted, $\theta_\nu \leq \theta_m$ also contains half the neutrinos.

4058

(a) Calculate the momentum of pions that have the same velocity as protons having momentum 400 GeV/c. This is the most probable momentum that produced-pions have when 400-GeV/c protons strike the target at Fermilab. The pion rest mass is 0.14 GeV/c². The proton rest mass is 0.94 GeV/c².

(b) These pions then travel down a decay pipe of 400 meter length where some of them decay to produce the neutrino beam for the neutrino detector located more than 1 kilometer away. What fraction of the pions decay in the 400 meters? the pions' proper mean lifetime is 2.6×10^{-8} sec.

(c) What is the length of the decay pipe as measured by observers in the pion rest frame?

(d) The pion decays into a muon and a neutrino ($\pi \to \mu + \nu_\mu$, the neutrino has zero rest-mass.) Using the relationship between total relativistic energy and momentum show that the magnitude of the decay fragments' momentum in the pion rest frame is given by $\frac{p}{c} = \frac{M^2 - m^2}{2M}$, where M is the rest mass of pion and m is the rest mass of muon.

(e) The neutrino detectors are, on the average, approximately 1.2 km from the point where the pions decay. How large should the transverse dimension (radius) of the detector be in order to have a chance of detecting all neutrinos that are produced in the forward hemisphere in the pion rest frame?

(UC, Berkeley)

Solution:

(a) The pions and the protons, having the same velocity, have the same γ and hence the same $\gamma\beta$. As

$$p_\pi = m_\pi \gamma\beta, \qquad p_p = m_p \gamma\beta,$$

$$p_\pi = \frac{m_\pi}{m_p} p_p = \frac{0.14}{0.94} \times 400 = 59.6 \text{ Gev}/c.$$

(b) The pions have

$$\gamma\beta = \frac{59.6}{0.14} = 426,$$

and hence $\gamma = \sqrt{(\gamma\beta)^2 + 1} \approx \gamma\beta = 426$.

The pions have proper mean lifetime $\tau_0 = 2.6 \times 10^{-8}$ s and hence mean lifetime $\tau = \gamma\tau_0 = 1.1 \times 10^{-5}$ s in the laboratory. Hence

$$\frac{N}{N_0} = (1 - e^{-\frac{t}{\tau_c}}) = (1 - e^{-0.12}) = 0.114.$$

(c) In the pion rest frame, on account of Fitzgerald contraction the observed length of the decay pipe is

$$\bar{l} = \frac{l}{\gamma} = \frac{400}{426} = 0.94 \text{ m}.$$

(d) In the pion rest frame, energy and momentum conservation laws require

$$E_\mu + E_\nu = m_\pi,$$

$$\mathbf{p}_\mu + \mathbf{p}_\nu = 0, \qquad \text{or} \qquad p_\mu = p_\nu.$$

For a particle, total energy and momentum are related by (taking $c = 1$)

$$E^2 = p^2 + m^2 .$$

For neutrino, as $m = 0$ we have $E_\nu = p_\nu$. The energy equation thus becomes

$$p_\mu^2 + m_\mu^2 = m_\pi^2 - 2p_\nu m_\pi + p_\nu^2 ,$$

or

$$p_\nu = \frac{m_\pi^2 - m_\mu^2}{2m_\pi}$$

i.e.,

$$p = \frac{M^2 - m^2}{2M} .$$

(e) The decay $\pi \to \mu\nu$ is isotropic in the rest frame of the pion. **Problem 4057**(b) gives the neutrinos' 'half-angle' as

$$\theta_{1/2} = \arctan\left(\frac{m_\pi}{\sqrt{E_0^2 - m_\pi^2}}\right) = \arctan\frac{1}{\sqrt{\gamma^2 - 1}} \approx \frac{1}{\gamma} .$$

Thus the diameter of the detector should be larger than

$$L = 2d \tan\theta_{\frac{1}{2}} \approx \frac{2d}{\gamma} = \frac{2 \times 1200}{426} = 5.63 \ m .$$

4059

Consider the decay $K^0 \to \pi^+ + \pi^-$.

Assuming the following transition matrix element

$$T_{if} = \frac{G}{\sqrt{8E_K E_+ E_-}} \frac{P_K(P_+ + P_-)}{m_K} .$$

show that the lifetime of the K^0 meson as measured in its rest system is

$$\tau = \left[\frac{G^2}{8\pi\hbar^4 c}\sqrt{\frac{m_K^2}{4} - \mu^2}\right]^{-1} .$$

(E_K, E_+ and E_- are the relativistic energies of K^0, π^+ and π^- respectively, and P_K, P_+ and P_- are the corresponding 4-momenta. M_K is the K-meson mass and G is the coupling constant. μ is the π-meson mass).

(SUNY, Buffalo)

Solution:

The transition probability per unit time is given by

$$W = \frac{2\pi}{\hbar} |T_{if}|^2 \rho(E) \,.$$

In the rest frame of K^0 meson,

$$E_K = m_K c^2 , \qquad E_+ = E_- = \frac{1}{2} m_K c^2 \,,$$

$$P_K^2 = \frac{E_K^2}{c^2} = m_K^2 c^2 \,,$$

$$(P_+ + P_-)^2 = -(\mathbf{p}_+ + \mathbf{p}_-)^2 + \frac{(E_+ + E_-)^2}{c^2} = m_K^2 c^2 \,.$$

Hence

$$|T_{if}|^2 = \frac{G^2}{8 E_K E_+ E_-} \frac{[P_K (P_+ + P_-)]^2}{m_K^2}$$

$$= \frac{G^2}{8 m_K c^2 \frac{m_K^2}{4} c^4} \frac{m_K^4 c^4}{m_K^2} = \frac{G^2}{2 m_K c^2} \,.$$

For a two-body decay, in the rest frame of the decaying particle,

$$\rho(E) = \frac{1}{(2\pi\hbar)^3} \frac{d}{dE} \int p_1^2 dp_1 d\Omega = \frac{4\pi}{(2\pi\hbar)^3} \frac{d}{dE} \left(\frac{1}{3} p_1^3 \right) \,,$$

assuming the decay to be isotropic.

Noting $\mathbf{p}_1 + \mathbf{p}_2 = 0$, or $p_1^2 = p_2^2$, i.e., $p_1 dp_1 = p_2 dp_2$, and $dE = dE_1 + dE_2$, we find

$$\rho(E) = \frac{4\pi}{(2\pi\hbar)^3} \frac{E_1 E_2 p_1}{E_1 + E_2} = \frac{1}{(2\pi\hbar)^3 c^2} \frac{m_K c^2}{4} \left(\sqrt{\frac{m_K^2}{4} - \mu^2} \right) 4\pi c$$

$$= \frac{m_K c}{8\pi^2 \hbar^3} \sqrt{\frac{m_K^2}{4} - \mu^2} \,,$$

where we have used

$$\frac{d}{dt}\left(\frac{1}{3}p_1^3\right) = \frac{p_1^2 dp_1}{dE_1 + dE_2} = \frac{p_1}{\dfrac{dE_1}{p_1 dp_1} + \dfrac{dE_2}{p_2 dp_2}} = \frac{E_1 E_2 p_1}{E_1 + E_2},$$

for as $E_1^2 = p_1^2 + m_1^2$

$$\frac{dE_1}{p_1 dp_1} = \frac{1}{E_1}, \qquad \text{etc.}$$

Therefore

$$W = \frac{2\pi}{\hbar} \frac{G^2}{2m_K c^2} \frac{m_K c}{8\pi^2 \hbar^3} \sqrt{\frac{m_K^2}{4} - \mu^2}$$

$$= \frac{G^2}{8\pi \hbar^4 c} \sqrt{\frac{m_K^2}{4} - \mu^2},$$

and the lifetime of K^0 is

$$\tau = \left[\frac{G^2}{8\pi \hbar^4 c} \sqrt{\frac{m_K^2}{4} - \mu^2}\right]^{-1}.$$

4060

The possible radioactive decay of the proton is a topic of much current interest. A typical experiment to detect proton decay is to construct a very large reservoir of water and put into it devices to detect Čerenkov radiation produced by the products of proton decay.

(a) Suppose that you have built a reservoir with 10,000 metric tons (1 ton = 1000 kg) of water. If the proton mean life τ_p is 10^{32} years, how many decays would you expect to observe in one year? Assume that your detector is 100% efficient and that protons bound in nuclei and free protons decay at the same rate.

(b) A possible proton decay is $p \to \pi^0 + e^+$. The neutral pion π^0 immediately (in 10^{-16} sec) decays to two photons, $\pi^0 \to \gamma + \gamma$. Calculate the maximum and minimum photon energies to be expected from a proton decaying at rest. The masses: proton $m_p = 938$ MeV, positron $m_{e^+} = 0.51$ MeV, neutral pion $m_{\pi^0} = 135$ MeV.

(CUSPEA)

Solution:

(a) Each H_2O molecule has 10 protons and 8 neutrons and a molecular weight of 18. The number of protons in 10^4 tons of water is then

$$N = \frac{10}{18} \times 10^7 \times 10^3 \times 6.02 \times 10^{23} = 3.34 \times 10^{33} \, ,$$

using Avagadro's number $N_0 = 6.02 \times 10^{23}$ mole^{-1}. The number of expected decays per year is therefore

$$\Delta N = \frac{3 \cdot 34}{\tau_p} \times 10^{33} = \frac{3.34 \times 10^{33}}{10^{32}} = 33.4/\text{year} \, .$$

(b) In the rest frame of the proton, conservation laws of energy and momentum require

$$M_p = E_{\pi^0} + E_{e^+} \, ,$$

$$p_{\pi^0} = p_{e^+} \, .$$

With $E^2 = M^2 + p^2$, these give

$$E_\pi = \frac{M_p^2 + M_\pi^2 - M_e^2}{2M_p}$$

$$= \frac{938^2 + 135^2 - 0.5^2}{2 \times 938} = 479 \text{ MeV} \, .$$

In the rest frame of the π^0 the energy and momentum of each γ are

$$E' = p' = \frac{M_\pi}{2} \, .$$

The π^0 has Lorentz factor and velocity

$$\gamma_\pi = \frac{479}{135} = 3.548 \, ,$$

$$\beta_\pi = \sqrt{1 - \frac{1}{\gamma_\pi^2}} = 0.9595 \, .$$

Lorentz transformation between the π^0 rest frame and the laboratory frame for the photons

$$E_\gamma = \gamma_\pi(E' + \beta_\pi p' \cos\theta') = \frac{M_\pi}{2}\gamma_\pi(1 + \beta_\pi \cos\theta') = \frac{E_\pi}{2}(1 + \beta_\pi \cos\theta')$$

shows that the photons will have in the laboratory maximum energy ($\theta' = 0$)

$$(E_\gamma)_{max} = \frac{E_\pi}{2}(1 + \beta_\pi) = \frac{479}{2}(1 + 0.9595) = 469.3 \text{ MeV},$$

and minimum energy ($\theta' = 180°$)

$$(E_\gamma)_{min} = \frac{E_\pi}{2}(1 - \beta_\pi) = \frac{479}{2}(1 - 0.9595) = 9.7 \text{ MeV}.$$

4061

Consider the decay in flight of a pion of laboratory energy E_π by the mode $\pi \to \mu + \nu_\mu$. In the pion center-of-mass system, the muon has a helicity $h = \frac{\mathbf{s}\cdot\boldsymbol{\beta}}{s\beta}$ of 1, where $\mathbf{s}$ is the muon spin. For a given E_π there is a unique laboratory muon energy $E_\mu^{(0)}$ for which the muon has zero average helicity in the laboratory frame.

(a) Find the relation between E_π and $E_\mu^{(0)}$.

(b) In the nonrelativistic limit, find the minimum value of E_π for which it is possible to have zero-helicity muons in the laboratory.

(Columbia)

Solution:

(a) Consider the spin 4-vector of the muon emitted in the decay $\pi \to \mu + \nu$. In the rest frame of the muon, it is

$$S_\alpha = (\mathbf{S}, iS_0),$$

where $\mathbf{S}$ is the muon spin and $S_0 = 0$.

Now consider the spin 4-vector in the rest frame of the pion, Σ_π. The muon has parameters γ_μ, β_μ in this frame and

$$S'_\alpha = (\mathbf{S}', iS'_0)$$

with

$$\mathbf{S}' = \mathbf{S} + (\gamma_\mu - 1)\mathbf{S} \cdot \hat{\boldsymbol{\beta}}_\mu \hat{\boldsymbol{\beta}}_\mu \,,$$

$$S_0' = \gamma_\mu(S_0 + \mathbf{S} \cdot \boldsymbol{\beta}_\mu) = \gamma_\mu \mathbf{S} \cdot \boldsymbol{\beta}_\mu = \gamma_\mu S \beta_\mu h_\mu \,.$$

In the Σ_π frame,

$$h_\mu = \frac{\mathbf{S} \cdot \boldsymbol{\beta}_\mu}{S\beta_\mu} = 1 \,,$$

and so $\mathbf{S} \cdot \boldsymbol{\beta}_\mu = S\beta_\mu$, i.e. $\mathbf{S}//\boldsymbol{\beta}_\mu$. It follows that

$$\mathbf{S}' = \mathbf{S} + (\gamma_\mu - 1)S\beta_\mu^{-1}\boldsymbol{\beta}_\mu \,,$$

$$S_0' = \gamma_\mu S\beta_\mu \,.$$

Next transform from Σ_π to the laboratory frame Σ, in which the pion has parameters γ_π, β_π, the muon has parameters γ, β. Then

$$S_\alpha^{\mathrm{Lab}} = (\mathbf{S}'', iS_0'') \,,$$

where

$$S_0'' = \gamma_\pi(S_0' + \boldsymbol{\beta}_\pi \cdot \mathbf{S}')$$

$$= \gamma_\pi[\gamma_\mu\beta_\mu S + \boldsymbol{\beta}_\pi \cdot \mathbf{S} + (\gamma_\mu - 1)(\boldsymbol{\beta}_\pi \cdot \boldsymbol{\beta}_\mu)S\beta_\mu^{-1}] \,.$$

As $\mathbf{S}//\boldsymbol{\beta}_\mu$,

$$(\boldsymbol{\beta}_\pi \cdot \boldsymbol{\beta}_\mu)S\beta_\mu^{-1} = (\boldsymbol{\beta}_\pi \cdot \mathbf{S})\beta_\mu\beta_\mu^{-1} = \boldsymbol{\beta}_\pi \cdot \mathbf{S} \,,$$

and

$$S_0'' = \gamma_\pi\gamma_\mu S(\beta_\mu^2 + \boldsymbol{\beta}_\pi \cdot \boldsymbol{\beta}_\mu)\beta_\mu^{-1} = \gamma\beta Sh \,,$$

with

$$h = \gamma_\pi\gamma_\mu\gamma^{-1}\beta^{-1}(\beta_\mu^2 + \boldsymbol{\beta}_\pi \cdot \boldsymbol{\beta}_\mu)\beta_\mu^{-1} \,.$$

At muon energy $E_\mu^{(0)}$, $h = 0$, or

$$\beta_\mu^2 = -\boldsymbol{\beta}_\pi \cdot \boldsymbol{\beta}_\mu \,.$$

Lorentz transformation then gives

$$\gamma = \gamma_\pi\gamma_\mu(1 + \boldsymbol{\beta}_\pi \cdot \boldsymbol{\beta}_\mu)$$

$$= \gamma_\pi\gamma_\mu(1 - \beta_\mu^2) = \frac{\gamma_\pi}{\gamma_\mu} \,.$$

Hence

$$E_\mu^{(0)} = m_\mu \gamma = \frac{m_\mu}{m_\pi} \frac{E_\pi}{\gamma_\mu} .$$

Consider the decay in the rest frame of π. Conservation of momentum and of energy require

$$\mathbf{p}_\nu + \mathbf{p}_\mu = 0, \qquad \text{or} \qquad p_\nu = p_\mu ,$$

$$E_\nu = m_\pi - E_\pi .$$

These combine to give

$$E_\mu = \frac{m_\pi^2 + m_\mu^2}{2m_\pi} ,$$

or

$$\gamma_\mu = \frac{E_\mu}{m_\mu} = \frac{m_\pi^2 + m_\mu^2}{2m_\pi m_\mu} .$$

Hence

$$E_\mu^{(0)} = \frac{m_\mu}{m_\pi} \cdot \frac{2m_\pi m_\mu}{m_\pi^2 + m_\mu^2} E_\pi = \frac{2m_\mu^2}{m_\pi^2 + m_\mu^2} E_\pi .$$

(b) For the average muon helicity $h = 0$ in the laboratory frame, we require

$$\boldsymbol{\beta}_\pi \cdot \boldsymbol{\beta}_\mu = -\beta_\mu^2 ,$$

or

$$\beta_\pi \cos\theta = -\beta_\mu .$$

This means that

$$\beta_\pi \geq \beta_\mu , \qquad \text{or} \qquad \gamma_\pi \geq \gamma_\mu .$$

Hence the minimum pion energy required is

$$(E_\pi)_{\min} = \gamma_\mu m_\pi = \frac{m_\pi^2 + m_\mu^2}{2m_\mu} .$$

2. INTERACTIONS BETWEEN
RADIATION AND MATTER (4062–4085)

4062

The energy loss of an energetic muon in matter is due mainly to collisions with

(a) nucleons.
(b) nuclei.
(c) electrons.

(CCT)

Solution:

A muon loses energy in matter mainly due to collisions with electrons, transferring part of its kinetic energy to the latter, which can either jump to higher energy levels or to be separated from the atoms resulting in their ionization.

So the answer is (c).

4063

A beam of negative muons can be stopped in matter because a muon may be

(a) transformed into an electron by emitting a photon.
(b) absorbed by a proton, which goes into an excited state.
(c) captured by an atom into a bound orbit about the nucleus.

(CCT)

Solution:

A μ^- can be captured into a bound orbit by a nucleus to form a μ-atom. It can also decay into an electron and two neutrinos $(\gamma_\mu, \bar{\nu}_e)$ but not an electron and a photon. So the answer is (c).

4064

After traversing one radiation length, an electron of 1 GeV has lost:

(a) 0.368 GeV

(b) none

(c) 0.632 GeV

of its original energy.

(*CCT*)

Solution:

By definition $E = E_0 e^{-x/\lambda}$, where λ is the radiation length. Thus when $x = \lambda$, $E = E_0 e^{-1} = 0.368$ GeV. The loss of energy is $\Delta E = 1 - 0.368 = 0.632$ GeV, and the answer is (c).

4065

A relativistic proton loses 1.8 MeV when penetrating a 1-cm thick scintillator. What is the most likely mechanism?

(a) Ionization, excitation.

(b) Compton effect.

(c) Pair production.

(*CCT*)

Solution:

When a relativistic proton passes through a medium, energy loss by ionization and excitation comes to $-dE/dx \approx 1\text{–}2$ MeV/g cm^{-2}. The density of the scintilator is $\rho \approx 1$ g cm^{-3}, so $dx = 1$ g cm^{-2}. The energy loss rate $-\frac{dE}{dx} = 1.8$ MeV/g cm^{-2} agrees with ionization loss rate. So the answer is (a).

4066

The mean energy loss of a relativistic charged particle in matter per g/cm^2 is about

(a) 500 eV.

(b) 10 KeV.

(c) 2 MeV.

(*CCT*)

Solution:

As $dE/dx \approx (1 \sim 2)$ MeV/g cm $^{-2}$, the answer is (c).

4067

The critical energy of an electron is the energy at which

(a) the radiation loss equals the ionization loss.
(b) the electron ionizes an atom.
(c) the threshold of nuclear reaction is reached.

(CCT)

Solution:

The critical energy is defined as the energy at which the radiation loss is equal to the ionization loss. The answer is (a).

4068

The straggling of heavy ions at low energy is mostly a consequence of

(a) finite momentum.
(b) fluctuating state of ionization.
(c) multiple scattering.

(CCT)

Solution:

Multiple scattering changes an ion's direction of motion, thus making them straggle. The answer is (c).

4069

The so-called "Fermi plateau" is due to

(a) a density effect.
(b) Lorentz contraction.
(c) relativistic mass increase.

(CCT)

Solution:

At Lorentz factor $\gamma \approx 3$, rate of ionization loss $dE/dx \approx (dE/dx)_{min}$. At $\gamma > 3$, because of its logarithmic relationship with energy, dE/dx increases only slowly with increasing γ. Finally, $\frac{dE}{dz} \approx$ constant when $\gamma > 10$ for a dense medium (solid or liquid), and when $\gamma > 100$ for a dilute medium (gas), because of the effect of electron density. The plateau in the $\frac{dE}{dx}$ vs E curve is known as "Fermi plateau". Thus the answer is (a).

4070

The probability for an energy loss E' in the interval dE' of a charged particle with energy E and velocity v in a single collision is proportional to

(a) $\frac{E'}{E} dE'$.
(b) $E dE'$.
(c) $(\frac{1}{vE'})^2 dE'$.

(CCT)

Solution:

Take collisions with electrons as example. For a single collision, the energy loss of a particle of charge Ze depends only on its velocity v and the impact parameter b : $E' = \frac{2Z^2 e^4}{m_0 v^2 b^2}$, where m_0 is the electron mass. Thus $dE' = -\frac{4Z^2 e^4}{m_0 v^2 b^3} \, db = -A \frac{db}{v^2 b^3}$, where $A = \frac{4Z^2 e^4}{m_0}$ is a constant.

Suppose the electrons are distributed uniformly in the medium. Then the probability of colliding with an electron with impact parameter in the interval between b and $b + db$ is

$$d\sigma = 2\pi b |db| = \frac{2\pi v^2 b^4}{A} dE' = \frac{\pi A dE'}{2(vE')^2} \propto \frac{dE'}{(vE')^2} \, .$$

Hence the answer is (c).

4071

The scattering of an energetic charged particle in matter is due mostly to interactions with (the)

(a) electrons.
(b) nuclei.

(c) quarks.

(CCT)

Solution:

In traversing a medium, a charged particle suffers Coulomb interactions with both electrons and nuclei. However, though collisions with the former are numerous, the momentum transfer in each is very small. Only collisions with the latter will result in appreciable scattering of the traversing particle.

Hence the answer is (b).

4072

The mean scattering angle of a charged particle in matter of a thickness x increases with

(a) x^2.
(b) $x^{1/2}$.
(c) x.

(CCT)

Solution:

The mean scattering angle of a particle of charge Ze in traversing matter of thickness x is $|\bar{\theta}| = \frac{KZ\sqrt{x}}{pv} \propto x^{1/2}$, where K is a constant. Hence the answer is (b).

4073

Consider a 2-cm thick plastic scintillator directly coupled to the surface of a photomultiplier with a gain of 10^6. A 10-GeV particle beam is incident on the scintillator as shown in Fig. 4.8(a).

(a) If the beam particle is a muon, estimate the charge collected at the anode of the photomultiplier.

(b) Suppose one could detect a signal on the anode of as little as 10^{-12} coulomb. If the beam particle is a neutron, estimate what is the smallest laboratory angle that it could scatter elastically from a proton in the scintillator and still be detected?

scintillator

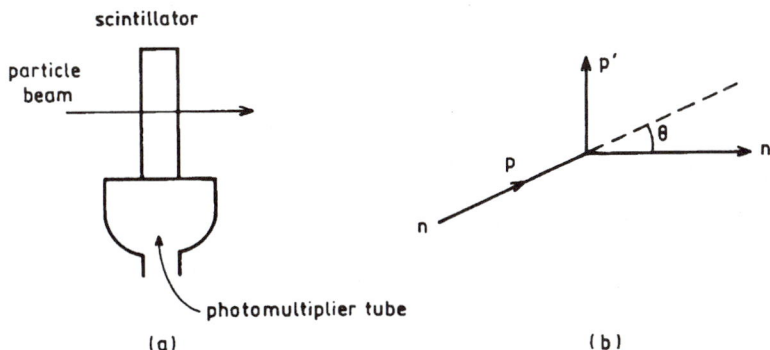

Fig. 4.8

(c) Same as Part (b), but it scatters elastically from a carbon nucleus.

(*Chicago*)

Solution:

(a) From its ionization loss curve, we see that a muon of energy 10 GeV will lose 4 MeV in a plastic scintillator of length 2 cm. Roughly, in a plastic scintillator, producing one photon requires 100 eV of energy. This amount of energy will produce $N_{\rm ph} \approx 4 \times 10^4$ photons in the scintillator. Suppose about 50% of the photons make it to the photomultiplier tube and about 10% of these produce photoelectrons off the cathode. Then the number of photoelectrons emitted is $N_{\rm pe} = 2 \times 10^3$. With a gain of 10^6, the charge collected at the anode of the photomultiplier is $Q = 2 \times 10^9 e = 3.2 \times 10^{-10}$ C.

(b) Figure 4.8(b) shows a neutron scatters by a small angle θ in the laboratory frame. Its momentum is changed by an amount $p\theta$ normal to the direction of motion. This is the momentum of the recoiling nucleus. Then the kinetic energy acquired by it is

$$\frac{p^2\theta^2}{2m} \, ,$$

where m is the mass of the recoiling nucleus. As an energy loss of 4 MeV corresponds to 3.2×10^{-10} C of anode charge, the detection threshold of 10^{-12} C implies that recoil energy as little as 12.5 keV can be detected. Hence the smallest laboratory scattering angle $\theta_{\rm min}$ that can be detected is given by

$$\theta^2_{min} = \frac{2m_p}{p_n^2} \times 12.5 \times 10^3 = \frac{2 \times 10^9}{(10^{10})^2} \times 12.5 \times 10^3 = 2.5 \times 10^{-7} \text{ rad}^2 \,,$$

i.e.

$$\theta_{min} = 5.0 \times 10^{-4} \text{ rad} \,,$$

assuming the recoiling nucleus is a proton.

(a) If the recoiling particle is a carbon nucleus, then

$$\theta^2_{min} = \frac{2m_c}{p_n^2} \times 12.5 \times 10^3 = \frac{2 \times 12 \times 10^9}{(10^{10})^2} \times 12.5 \times 10^3 = 3.0 \times 10^{-6} \text{ rad}^2 \,,$$

i.e.,

$$\theta_{min} = 1.73 \times 10^{-3} \text{ rad} \,.$$

4074

How many visible photons ($\sim$ 5000 Å) does a 100-W bulb with 3% efficiency emit per second?

(a) 10^{19}.
(b) 10^9.
(c) 10^{33}.

(*CCT*)

Solution:

Each photon of $\lambda = 5000$ Å has energy

$$E = h\nu = hc/\lambda = \frac{2\pi \times 197 \times 10^{-7}}{5000 \times 10^{-8}} = 2.5 \text{ eV} \,.$$

So the number of photons is

$$N = \frac{W}{E} = \frac{100 \times 0.03}{2.5 \times 1.6 \times 10^{-19}} = 0.75 \times 10^{19} \approx 10^{19} \,.$$

Hence the answer is (a).

4075

Estimate the attenuation (absorption/scattering) of a beam of 50-keV X-rays in passage through a layer of human tissue (no bones!) one centimeter thick.

(*Columbia*)

Solution:

As the human body is mostly water, we can roughly take its density as that of water, $\rho \approx 1$ g/cm^3. Generally, the absorption coefficient of 50 keV X-rays is about 0.221 cm^2/g. Then the attenuation resulting from the passage through one centimeter of tissue (thickness = 1 cm × 1 g cm^{-3} = 1 g cm^{-2}) is

$$1 - \exp(-0.221 \times 1) = 0.20 = 20\%.$$

4076

Photons of energy 0.3 eV, 3 eV, 3 keV, and 3 MeV strike matter. What interactions would you expect to be important? Match one or more interactions with each energy.

0.3 eV	(a) Pair production	(e) Atomic Ionization
3 eV	(b) Photoelectric effect	(f) Raman Scattering (rotational
3 keV	(c) Compton Scattering	and vibrational excitation)
3 MeV	(d) Rayleigh Scattering	

(Wisconsin)

Solution:

Raman scattering is important in the region of 0.3 eV. Atomic ionization, Rayleigh scattering and Raman scattering are important around 3 eV. Photoelectric effect is important in the region of 3 keV. In the region of 3 MeV, Compton scattering and pair production are dominant.

4077

Discuss the interaction of gamma radiation with matter for photon energies less than 10 MeV. List the types of interaction that are important in this energy range; describe the physics of each interaction and sketch the relative contribution of each type of interaction to the total cross section as a function of energy.

(Columbia)

Solution:

Photons of energies less than 10 MeV interact with matter mainly through photoelectric effect, Compton scattering, and pair production.

(1) *Photoelectric effect*: A single photon gives all its energy to a bound electron in an atom, detaching it completely and giving it a kinetic energy $E_e = E_\gamma - E_b$, where E_γ is the energy of the photon and E_b is the binding energy of the electron. However, conservation of momentum and of energy prevent a free electron from becoming a photoelectron by absorbing all the energy of the photon. In photoelectric effect, conservation of momentum must be satisfied by the recoiling of the nucleus to which the electron was attached. The process generally takes place with the inner electrons of an atom (mostly K- and L-shell electrons). The cross section $\sigma_{p-e} \propto Z^5$, where Z is the nuclear charge of the medium. If $\varepsilon_K < E_\gamma < 0.5$ MeV, $\sigma_{p-e} \propto E_\gamma^{-\frac{7}{2}}$, where ε_K is the binding energy of K-electron. If $E_\gamma > 0.5$ MeV, $\sigma_{p-e} \propto E_\gamma^{-1}$. Thus photoelectric effect is dominant in the low-energy region and in high-Z materials.

(2) *Compton scattering*: A photon is scattered by an electron at rest, the energies of the electron and the scattered photon being determined by conservation of momentum and energy to be respectively

$$E_e = E_\gamma \left[1 + \frac{mc^2}{E_\gamma(1 - \cos\theta)} \right]^{-1} ,$$

$$E_{\gamma'} = E_\gamma \left[1 + \frac{E_\gamma}{mc^2}(1 - \cos\theta) \right]^{-1} ,$$

where m is the electron mass, E_γ is the energy of the incident photon, and θ is the angle the scattered photon makes with the incident direction. The cross section is $\sigma_c \propto Z E_\gamma^{-1} \ln E_\gamma$ (if $E_\gamma > 0.5$ MeV).

(3) *Pair production*: If $E_\gamma > 2m_e c^2$, a photon can produce a positron-electron pair in the field of a nucleus. The kinetic energy of the positron-electron pair is given by $E_{e+} + E_{e-} = E_\gamma - 2m_e c^2$. In low-energy region σ_{e+e-} increases with increasing E_γ, while in high-energy region, it is approximately constant. Figure 4.9 shows the relative cross sections of lead for absorption of γ-rays as a function of E_γ. It is seen that for $E_\gamma \gtrsim 4$ MeV, pair production dominates, while for low energies, photoelectric and Compton effects are important. Compton effect predominates in the energy region from several hundred keV to several MeV.

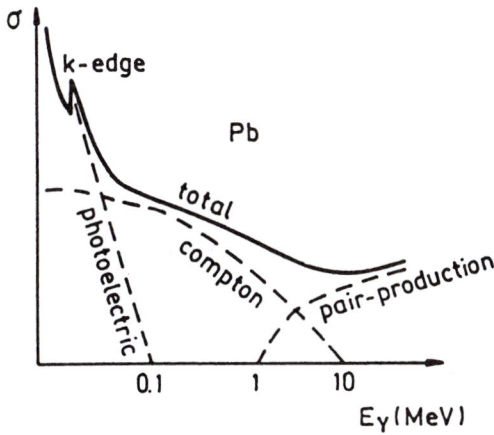

Fig. 4.9

4078

Fast neutrons can be detected by observing scintillations caused by recoil protons in certain (optically transparent) hydrocarbons. Assume that you have a 5 cm thick slab of scintillator containing the same number-density of C and H, namely 4×10^{22} atoms/cm^3 of each kind.

(a) What fraction of ~ 5 MeV neutrons incident normal to the slab will pass through the slab without interacting with either C or H nuclei?

(b) What fraction of the incident neutrons will produce a recoil proton? [Assume $\sigma_H = 1.5$ barns, $\sigma_C = 1.0$ barn. Note 1 barn $= 10^{-24}$ cm^2.]

(Wisconsin)

Solution:

(a) Denote the number of neutrons by N. The number decreases by ΔN after traveling a distance Δx in the scintillator, given by

$$\Delta N = -N(\sigma_H n_H + \sigma_C n_C)\Delta x,$$

where n is the number density of the nuclei of the scintillator. After passing through a distance d, the number of neutrons that have not undergone any interaction is then

$$N = N_0 \exp[-(\sigma_H n_H + \sigma_C n_C)d],$$

giving

$$\eta = N/N_0 = \exp[-(1.5 + 1.0) \times 10^{-24} \times 4 \times 10^{22} \times 5]$$

$$= e^{-0.5} = 60.5\%.$$

(b) The fraction of incident neutrons undergoing at least one interaction is

$$\eta' = 1 - \eta = 39.5\%.$$

Of these only those interacting with protons can produce recoil protons. Thus the fraction of neutrons that produce recoil protons is

$$\eta'' = \eta' \cdot \frac{1.5}{1.5 + 1.0} = \frac{3}{5}\eta' = 23.7\%.$$

4079

The mean free path of fast neutrons in lead is about 5 cm. Find the total neutron cross section of lead (atomic mass number ~ 200, density $\sim 10 \text{ g/cm}^3$).

(Wisconsin)

Solution:

The number of Pb atoms per unit volume is

$$n = \frac{\rho}{A} \times N_0 = \frac{10}{200} \times 6.022 \times 10^{23} = 3.01 \times 10^{22} \text{ cm}^{-3}.$$

The mean free path of neutron in lead is $l = 1/(n\sigma)$, where σ is the interaction cross section between neutron and lead. Hence

$$\sigma = \frac{1}{nl} = \frac{1}{3.01 \times 10^{22} \times 5} = 6.64 \times 10^{-24} \text{ cm}^2 = 6.64 \ b.$$

4080

It is desired to reduce the intensity of a beam of slow neutrons to 5% of its original value by placing into the beam a sheet of Cd (atomic weight 112, density 8.7×10^3 kg/m^3). The absorption cross section of Cd is 2500 barns. Find the required thickness of Cd.

(Wisconsin)

Solution:

The intensity of a neutron beam after passing through a Cd foil of thickness t is given by $I(t) = I_0 e^{-n\sigma t}$, where I_0 is the initial intensity, n is the number density of Cd, and σ is the capture cross section. As

$$n = \frac{\rho N_0}{A} = \frac{8.7}{112} \times 6.022 \times 10^{23} = 4.7 \times 10^{22} \text{ cm}^{-3},$$

the required thickness of Cd foil is

$$t = \frac{1}{n\sigma} \ln \frac{I_0}{I(t)} = \frac{1}{4.7 \times 10^{22} \times 2500 \times 10^{-24}} \ln \frac{1}{0.05} = 0.025 \text{ cm}.$$

4081

A beam of neutrons passes through a hydrogen target (density 4×10^{22} atom/cm^3) and is detected in a counter C as shown in Fig. 4.10. For equal incident beam flux, 5.0×10^5 counts are recorded in C with the target empty, and 4.6×10^5 with the target full of hydrogen. Estimate the total n-p scattering cross section, and its statistical error.

(Wisconsin)

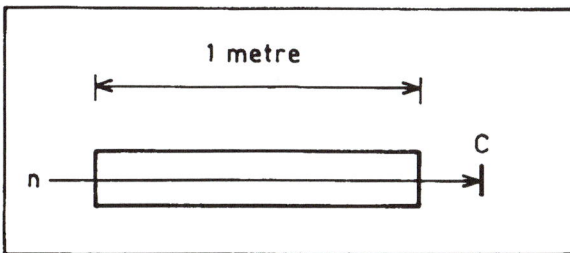

Fig. 4.10

Solution:

Let the total cross section of n-p interaction be σ. After passing through the hydrogen target, the number of neutrons decreases from N_0 to $N_0 e^{-n\sigma t}$, where $n = 4 \times 10^{22}$ cm^{-3} is the atomic concentration of the target. Suppose the numbers of neutrons detected without and with the hydrogen target are N', N'' respectively and η is the neutron-detecting efficiency of C. Then

$$N' = \eta N_0 , \qquad N'' = \eta N_0 e^{-n\sigma t} = N' e^{-n\sigma t} ,$$

and thus

$$N''/N' = e^{-n\sigma t} ,$$

giving the *n-p* scattering cross section as

$$\sigma = \frac{1}{nt} \ln \frac{N'}{N''} = \frac{1}{4 \times 10^{22} \times 100} \ln \frac{5 \times 10^5}{4.6 \times 10^5} = 2.08 \times 10^{-26} \text{ cm}^2 .$$

To estimate the statistical error of σ we note

$$\Delta\sigma = \frac{\partial\sigma}{\partial N'}(\Delta N') + \frac{\partial\sigma}{\partial N'}(\Delta N') ,$$

$$\frac{\partial\sigma}{\partial N'} = \frac{1}{ntN'} ,$$

$$\frac{\partial\sigma}{\partial N''} = -\frac{1}{ntN''} ,$$

$$\Delta N' = \sqrt{N'} , \qquad \Delta N'' = \sqrt{N''} .$$

Hence

$$(\Delta\sigma)^2 = \left(\frac{\partial\sigma}{\partial N'}\right)^2 (\Delta N')^2 + \left(\frac{\partial\sigma}{\partial N''}\right)^2 (\Delta N'')^2 = \frac{1}{(nt)^2}\left(\frac{1}{N'} + \frac{1}{N''}\right) ,$$

or

$$\Delta\sigma = \frac{1}{(nt)}\sqrt{\frac{1}{N'} + \frac{1}{N''}} = \frac{1}{4 \times 10^{22} \times 100}\sqrt{\frac{1}{4.6 \times 10^5} + \frac{1}{5 \times 10^5}}$$

$$\approx 5 \times 10^{-28} \text{ cm}^2 .$$

Therefore

$$\sigma = (2.08 \pm 0.05) \times 10^{-26} \text{ cm}^2 = (20.8 \pm 0.5) \text{ mb} .$$

4082

A beam of energetic neutrons with a broad energy spectrum is incident down the axis of a very long rod of crystalline graphite as shown in Fig. 4.11. It is found that the faster neutrons emerge from the sides of the rod, but only slow neutrons emerge from the end. Explain this very briefly and estimate numerically the maximum velocity of the neutrons which emerge from the end of the rod. Introduce no symbols.

(Columbia)

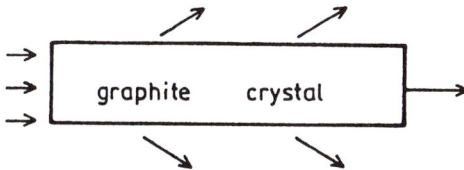

Fig. 4.11

Solution:

Crystalline graphite is a cold neutron filter. High-energy neutrons change directions on elastic scattering with the nuclei in the crystalline graphite and finally go out of the rod. Because of their wave property, if the wavelengths of the neutrons are comparable with the lattice size, interference occurs with the diffraction angle θ satisfying Bragg's law

$$m\lambda = 2d\sin\theta, \qquad \text{with } m = 1, 2, 3\ldots.$$

In particular for $\lambda > 2d$, there is no coherent scattering except for $\theta = 0$. At $\theta = 0$, the neutrons can go through the crystal without deflection. Furthermore, as the neutron absorption cross section of graphite is very small, attenuation is small for the neutrons of $\lambda > 2d$. Graphite is polycrystalline with irregular lattice orientation. The high-energy neutrons change directions by elastic scattering and the hot neutrons change directions by Bragg scattering from microcrystals of different orientations. Finally both leave the rod through the sides. Only the cold neutrons with wavelength $\lambda > 2d$ can go through the rod without hindrance. For graphite, $\lambda > 2d = 6.69$ Å. The maximum velocity of such cold neutrons is

$$v_{\max} = \frac{p}{m} = \frac{h}{m\lambda} = \frac{2\pi\hbar c^2}{\lambda m c^2} = \frac{2\pi \times 197 \times 10^{-13} \times 3 \times 10^{10}}{6.69 \times 10^{-8} \times 940}$$

$$= 0.59 \times 10^5 \ cm/s = 590 \ m/s \,.$$

4083

Mean free path for 3-MeV electron-neutrinos in matter is

$$10, 10^7, 10^{17}, 10^{27} \ g/cm^2 \,.$$

(Columbia)

Solution:

The interaction cross section between neutrino and matter is $\sigma \approx 10^{-41}$ cm^2, and typically the atomic number-density of matter $n \approx 10^{23}$ cm^{-3}, density of matter $\rho \approx 1$ g/cm^3. Hence the mean free path of neutrino in matter is $1 = \rho/n\sigma \approx 10^{18}$ g/cm^2. The third answer is correct.

4084

Čerenkov radiation is emitted by a high-energy charged particle which moves through a medium with a velocity greater than the velocity of electro-magnetic-wave propagation in the medium.

(a) Derive the relationship among the particle velocity $v = \beta c$, the index of refraction n of the medium, and the angle θ at which the Čerenkov radiation is emitted relative to the line of flight of the particle.

(b) Hydrogen gas at one atmosphere and 20°C has an index of refraction $n = 1 + 1.35 \times 10^{-4}$. What is the minimum kinetic energy in MeV which an electron (of mass 0.5 MeV/c^2) would need in order to emit Čerenkov radiation in traversing a medium of hydrogen gas at 20°C and one atmosphere?

(c) A Čerenkov-radiation particle detector is made by fitting a long pipe of one atmosphere, 20°C hydrogen gas with an optical system capable of detecting the emitted light and of measuring the angle of emission θ to an accuracy of $\delta\theta = 10^{-3}$ radian. A beam of charged particles with momentum

100 GeV/c are passed through the counter. Since the momentum is known, measurement of the Čerenkov angle is, in effect, a measurement of the rest mass m_0. For a particle with m_0 near 1 GeV/c^2, and to first order in small quantities, what is the fractional error (i.e., $\delta m_0/m_0$) in the determination of m_0 with the Čerenkov counter?

(CUSPEA)

Solution:

(a) Figure 4.12 shows the cross section of a typical Čerenkov wavefront. Suppose the particle travels from O to A in t seconds. The radiation sent out while it is at O forms a spherical surface with center at O and radius $R = ct/n$. The Čerenkov radiation wavefront which is tangent to all such spherical surfaces is a conic surface. In the triangle AOB, OB = $R = ct/n$, OA = $vt = \beta ct$, and so $\cos\theta = \text{OB}/\text{OA} = 1/(n\beta)$.

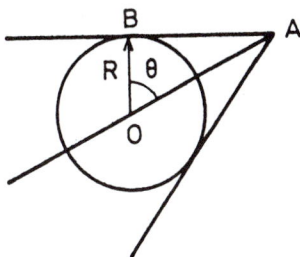

Fig. 4.12

(b) As $\cos\theta = \frac{1}{n\beta}$, we require

$$\beta \geq \frac{1}{n}.$$

Thus

$$\beta_{\min} = \frac{1}{n} = \frac{1}{1 + 1.35 \times 10^{-4}} \approx 1 - 1.35 \times 10^{-4},$$

and so

$$\gamma_{\min} = \frac{1}{\sqrt{(1+\beta)(1-\beta)}} \approx \frac{1}{\sqrt{2 \times 1.35 \times 10^{-4}}} = 60.86.$$

The minimum kinetic energy required by an electron is therefore

$$T = (\gamma - 1)mc^2 = 59.86 \times 0.5 = 29.9 \text{ MeV}.$$

(c) The rest mass $m_0 c^2$ is calculated from (taking $c = 1$)

$$m_0^2 = \frac{p^2}{(\gamma\beta)^2} = \frac{p^2(1 - \beta^2)}{\beta^2} = \frac{p^2}{\beta^2} - p^2$$

$$= p^2 n^2 \cos^2\theta - p^2.$$

Differentiating with respect to θ gives

$$2m_0 dm_0 = -2p^2 n^2 \cos\theta \sin\theta d\theta.$$

Hence

$$\delta m_0 = \frac{p^2 n^2}{2m_0} \sin 2\theta \delta\theta.$$

With $m_0 \approx 1 \text{ GeV}/c^2$, $p = 100 \text{ GeV}/c$,

$$\gamma = \frac{\sqrt{p^2 + m_0^2}}{m_0} = \sqrt{10^4 + 1}.$$

Thus

$$\cos\theta = \frac{1}{n\beta} = \frac{\gamma}{n\sqrt{\gamma^2 - 1}}$$

$$= \frac{\sqrt{10^4 + 1}}{(1 + 1.35 \times 10^{-4}) \times 10^2}$$

$$\approx \frac{1 + 0.5 \times 10^{-4}}{1 + 1.35 \times 10^{-4}}$$

$$\approx 1 - 0.85 \times 10^{-4}$$

$$\approx 1 - \frac{\theta^2}{2},$$

and hence

$$\theta^2 \approx 1.7 \times 10^{-4}$$

or

$$\theta \approx 1.3 \times 10^{-2} \text{ rad}.$$

As θ is small, $\sin 2\theta \approx 2\theta$, and

$$\frac{\delta m_0}{m_0} = \frac{p^2 n^2 \theta}{m_0^2} \delta\theta$$

$$\approx 10^4 \times 1.3 \times 10^{-2} \times 10^{-3} = 0.13.$$

4085

A proton with a momentum of 1.0 GeV/c is passing through a gas at high pressure. The index of refraction of the gas can be changed by changing the pressure.

(a) What is the minimum index of refraction at which the proton will emit Čerenkov radiation?

(b) At what angle will the Čerenkov radiation be emitted when the index of refraction of the gas is 1.6? (Take rest mass of proton as 0.94 GeV/c^2.)

(Columbia)

Solution:

(a) The proton has Lorentz factor

$$\gamma = \frac{\sqrt{p^2 + m^2}}{m} = \frac{\sqrt{1 + 0.94^2}}{0.94} = 1.46$$

and hence

$$\beta = \sqrt{1 - \frac{1}{\gamma^2}} = 0.729.$$

For the proton to emit Čerenkov radiation we require

$$\frac{1}{n\beta} \leq 1,$$

or

$$n \geq \frac{1}{\beta} = \frac{1}{0.729} = 1.37.$$

(b)

$$\cos\theta = \frac{1}{n\beta} = \frac{1}{1.6 \times 0.729} = 0.86,$$

giving

$$\theta = 31°.$$

3. DETECTION TECHNIQUES AND
EXPERIMENTAL METHODS (4086–4105)

4086

The mean energy for production of a free ion pair in gases by radiation is

(a) equal to the ionization potential.
(b) between $20 \sim 40$ eV.
(c) in good approximation 11.5Z.

$$(CCT)$$

Solution:

The average energy needed to produce a pair of free ions is larger than the ionization potential, as part of the energy goes to provide for the kinetic energy of the ions. The answer is (b).

4087

At low E/p the drift velocity of electrons in gases, v_{Dr}, follows precisely the relation $v_{Dr} \propto E/p$. This can be explained by the fact that

(a) the electrons each gains an energy $\varepsilon = eE \int ds$.
(b) the electrons thermalize completely in inelastic encounters with the gas molecules.
(c) the cross section is independent of electron velocity.

$$(CCT)$$

Solution:

The electrons acquire an average velocity $v_{Dr} = \frac{p}{2m_e} = \frac{eE\tau}{2m_e}$, in the electric field E, where τ is the average time-interval between two consecutive collisions. As $\tau = \frac{l}{v_{Dr}} \propto \frac{1}{\sigma v_{Dr}}$, where l is the mean free path of the electrons in the gas and σ is the interaction cross section, we have

$$v_{Dr} \propto \frac{E}{\sigma p} \propto \frac{E}{p}$$

if σ is independent of velocity. If σ is dependent on velocity, the relationship would be much more complicated. Hence the answer is (c).

4088

The mean ionization potential is a mean over energies of different

(a) atomic excitation levels.
(b) molecular binding energies.
(c) electronic shell energies.

(CCT)

Solution:

The mean ionization potential is defined as the average energy needed to produce a pair of positive and negative ions, which is the average of the molecular binding energies. The answer is (b).

4089

The efficiency of a proportional counter for charged particles is ultimately limited by

(a) signal-to-noise ratio.
(b) total ionization.
(c) primary ionization.

(CCT)

Solution:

If the mean primary ionization of a charged particle is very small, there is a finite probability that the charged particle may not produce sufficient primary ionization for its observation because of statistical fluctuation. Hence the answer is (c).

4090

Spectra of monoenergetic X-rays often show two peaks in proportional counters. This is due to

(a) escape of fluorescent radiation.
(b) Auger effect.
(c) Compton scattering.

(CCT)

Solution:

The escape of fluorescent radiation causes the spectrum to have two peaks. The larger peak is the total energy peak of the X-rays, while the smaller one is due to the fluorescent X-rays escaping from the detector. The answer is (a).

4091

A Geiger counter consists of a 10 mm diameter grounded tube with a wire of 50 μm diameter at +2000 V in the center. What is the electrical field at the wire?

(a) 200^2 V/cm.

(b) 150 kV/cm.

(c) 1.5×10^9 V/cm.

(CCT)

Solution:

With $R_0 = 0.5 \times 10^{-2}$ m, $R_i = 75 \times 10^{-6}$ m, $V = 200$ V, $\gamma = 25 \times 10^{-6}$ m,

$$E(r) = \frac{V}{r \ln \dfrac{R_0}{R_i}} = 1.51 \times 10^7 \text{ V}/m$$

$$= 151 \text{ kV/cm}.$$

Hence the answer is (b).

4092

For Question 4091, the electrical field at the tube wall is

(a) 0 V/cm.

(b) 377 V/cm.

(c) 754 V/cm.

(CCT)

Solution:

Same as for **Problem 4091** but with $r = 0.5 \times 10^{-2}$ m:

$$E(r) = 7.55 \times 10^4 \text{ V/m} = 755 \text{ V/cm}.$$

The answer is (c).

4093

What limits the time resolution of a proportional counter?

(a) Signal-to-noise ratio of the amplifier.

(b) Slow signal formation at the anode (slow rise time).

(c) Random location of the ionization and therefore variable drift time.

$$(CCT)$$

Solution:

Randomness of the location of the primary ionization causes the time it takes for the initial ionization electrons to reach the anode to vary. The anode signals are produced mainly by the avalanche of the electrons which reach the anode first. Thus large fluctuation results, making the resolution poor. The answer is (c).

4094

What is the mechanism of discharge propagation in a self-quenched Geiger counter?

(a) Emission of secondary electrons from the cathode by UV quanta.

(b) Ionization of the gas near the anode by UV quanta.

(c) Production of metastable states and subsequent de-excitation.

$$(CCT)$$

Solution:

The answer is (b).

4095

Does very pure NaI work as a good scintillator?

(a) No.
(b) Only at low temperatures.
(c) Yes.

(*CCT*)

Solution:

The answer is (b).

4096

What is the advantage of binary scintillators?

(a) They are faster.
(b) They give more amplitude in the photodetector.
(c) They are cheaper.

(*CCT*)

Solution:

The advantage of binary scintillators is their ability to restrain the Compton and escape peaks, and so to increase the total energy-peak amplitudes in the photodetector. The answer is (b).

4097

A charged particle crosses a NaI(TI)-scintillator and suffers an energy loss per track length dE/dx. The light output dL/dx

(a) is proportional to dE/dx.
(b) shows saturation at high dE/dx.
(c) shows saturation at high dE/dx and deficiency at low dE/dx.

(*CCT*)

Solution:

NaI(T1) is not a strictly linear detector. Its photon output depends on both the type of the traversing particle and its energy loss. When the energy loss is very small the departure from nonlinearity of the photon output is large, while when dE/dX is very large it becomes saturated. The answer is (c).

4098

Monoenergetic γ-rays are detected in a NaI detector. The events between the Compton edge and the photopeak occur

(a) predominantly in thin detectors.
(b) predominantly in thick detectors.
(c) never.

(*CCT*)

Solution:

In general, the number of events in the region between the Compton edge and the photopeak is smaller than in other regions. In the spectrum, such events appear as a valley. In neither a thin detector or a thick detector can they become dominant. The answer is (c).

4099

The light emission in organic scintillators is caused by transitions between

(a) levels of delocalized σ electrons.
(b) vibrational levels.
(c) rotational levels.

(*CCT*)

Solution:

Actually the fast component of the emitted light from an organic scintillator is produced in the transition between the $^0S_1$ level and the delocalized $^1S_0$ level. The answer is (a).

4100

A proton with total energy 1.4 GeV transverses two scintillation counters 10 m apart. What is the time of flight?

(a) 300 ns.

(b) 48 ns.

(c) 33 ns.

(CCT)

Solution:

The proton has rest mass $m_p = 0.938$ GeV and hence

$$\gamma = \frac{E}{m_p} = \frac{1.4}{0.938} = 1.49 \,,$$

$$\beta = \sqrt{1 - \frac{1}{\gamma^2}} = 0.74 \,.$$

The time of flight is therefore

$$t = \frac{10}{0.74 \times 3 \times 10^8} = 4.5 \times 10^{-8} \; s = 45 \; ns \,.$$

The answer is (b).

4101

What is the time of flight if the particle in Question 4100 is an electron?

(a) 330 ns.

(b) 66 ns.

(c) 33 ns.

(CCT)

Solution:

An electron with energy 1.4 GeV $\gg m_e c^2 = 0.51$ MeV has $\beta \approx 1$. Hence the time of flight is

$$t \approx \frac{10}{3 \times 10^8} = 3.3 \times 10^{-8} \; s = 33 \; ns \,.$$

Thus the answer is (c).

4102

How would you detect 500 MeV γ-rays? With

(a) hydrogen bubble chamber.
(b) shower counter (BGO).
(c) Geiger counter.

(CCT)

Solution:

As 500 MeV γ-rays will cause cascade showers in a medium, we need a total-absorption electromagnetic shower counter for their detection. The BGO shower counter makes a good choice because of its short radiation length and high efficiency. Hence the answer is (b).

4103

How would one measure the mean lifetime of the following particles?

(1) $U^{238} : \tau = 4.5 \times 10^9$ years,
(2) Λ^0 hyperon : $\tau = 2.5 \times 10^{-10}$ sec,
(3) ρ^0 meson : $\tau \approx 10^{-22}$ sec.

(Wisconsin)

Solution:

(1) The lifetime of ^{238}U can be deduced from its radioactivity $-dN/dt = \lambda N$, where the decay rate is determined directly by measuring the counting rate. Given the number of the nuclei, λ can be worked out and $\tau = 1/\lambda$ calculated.

(2) The lifetime of Λ^0 hyperon can be deduced from the length of its trajectory before decaying according to $\Lambda^0 \to p^+\pi^-$ in a strong magnetic field in a bubble chamber. From the opening angle and curvatures of the tracks of p and π^-, we can determine the momentum of the Λ^0, which is the sum of the momenta of p and π^-. Given the rest mass of Λ^0, its mean lifetime can be calculated from the path length of Λ^0 (**Problem 3033**).

(3) The lifetime of ρ^0 meson can be estimated from the invariant mass spectrum. From the natural width ΔE of its mass in the spectrum, its lifetime can be estimated using the uncertainty principle $\Delta E \Delta \tau \approx \hbar$.

4104

The "charmed" particles observed in e^+e^- storage rings have not yet been seen in hadron-hadron interactions. One possible means for detecting such particles is the observation of muons resulting from their leptonic decays. For example, consider a charmed particle c with decay mode

$$c \to \mu\nu .$$

Unfortunately, the experimental situation is complicated by the presence of muons from π decays.

Consider an experiment at Fermilab in which 400 GeV protons strike a thick iron target (beam dump) as depicted in Fig. 4.13.

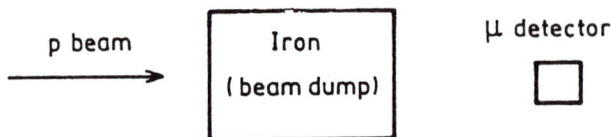

Fig. 4.13

Some of the muons entering the detector will be from π decays and some from c decays (ignore other processes). Calculate the ratio of muons from c decays to those from π decays under the following assumptions:

(a) the pions that have suffered interaction in the dump completely disappear from the beam,

(b) the energy spectra of both π and c are flat from minimum up to the maximum possible energy,

(c) the mass of the c is 2 GeV/c^2 and its lifetime is $\ll 10^{-10}$ sec,

(d) one can ignore muon energy loss in the iron,

(e) one can ignore any complications due to the geometry of the muon detector,

(f) the p–p inelastic cross section is 30 mb, and the mean charged pion multiplicity in inelastic interactions is 8.

Be specific. State any additional assumptions. Give a numerical value for the ratio at $E_\mu = 100$ GeV assuming the total production cross section for c to be 10 μb per Fe nucleus and that it decays to $\mu\nu$ 10% of the time.

(*Princeton*)

Solution:

In addition to the assumptions listed in the question, we also assume the charge independence of nucleon interactions so that $\sigma_{pp} = \sigma_{pn}$ and the mean charged-pion multiplicities are the same for pp and pn collisions.

For ^{56}Fe, the number densities of protons and neutrons are the same, being

$$N_p = N_n = \frac{28}{56} \times 7.8 \times 6.02 \times 10^{23} = 2.35 \times 10^{24} \text{ cm}^{-3} \, .$$

Denote the flux of protons in the incident beam by $\phi(x)$, where x is the target thickness from the surface of incidence. As

$$\frac{d\phi}{dx} = -(\sigma_{pp}N_p + \sigma_{pn}N_n)\phi = -2\sigma_{pp}N_p\phi \, ,$$

$$\phi = \phi_0 e^{-2\sigma_{pp}N_p x} \, .$$

If the target is sufficiently thick, say $x = 10$ m $= 10^3$ cm, then

$$\phi = \phi_0 \exp(-2 \times 30 \times 10^{-27} \times 2.35 \times 10^{24} \times 10^3) = 5.8 \times 10^{-62}\phi_0 \, ,$$

at the exit surface, showing that the beam of protons is completely dissipated in the target. This will be assumed in the following.

Consider first the c quarks produced in p–Fe interactions in the target. From the given data $\sigma_{p\text{Fe}}(c) = 10 \ \mu b$, $\sigma_{pp} = 30$ mb, we find the number of c quarks so produced as

$$N_c = \int N_{\text{Fe}}\sigma(c)d\phi \approx N_{\text{Fe}}\sigma(c)\phi_0 \int_0^\infty e^{-2\sigma_{pp}N_p x}dx$$

$$= \frac{N_{\text{Fe}}}{2N_{pp}} \frac{\sigma(c)}{\sigma_{pp}}\phi_0 = \frac{1}{56} \times \frac{10^{-5}}{30 \times 10^{-3}}\phi_0 = 5.95 \times 10^{-6}\phi_0 \, .$$

As the c quarks have lifetime $\ll 10^{-10}$ s, all those produced in p–Fe interactions will decay in the target, giving rise to muons 10% of the times. Thus

$$N_{\mu c} = 0.1N_c = 5.95 \times 10^{-7}\phi_0 \, .$$

Next consider the muons arising from the decay of charged pions produced in p-nucleon interactions. After emission the pions may interact with the nucleons of the target and disappear from the beam, as assumed, or decay in flight giving rise to muons. For the former case we assume $\sigma_{\pi p} = \sigma_{\pi n} \approx \frac{2}{3}\sigma_{pp} = 20$ mb at high energies. For the latter case the lifetime of the charged pions in the laboratory is γ_π/λ, where λ is the decay constant and $\gamma_\pi = (1 - \beta_\pi^2)^{-\frac{1}{2}}$, $\beta_\pi c$ being the mean velocity of the pions. Then the change of N_π per unit interval of x is

$$\frac{dN_\pi}{dx} = 8(\sigma_{pp}N_p + \sigma_{pn}N_n)\phi(x) - \left(\frac{\lambda}{\gamma_\pi\beta_\pi c} + \sigma_{\pi p}N_p + \sigma_{\pi n}N_n\right)N_\pi$$

$$= 16\sigma_{pp}N_p\varphi_0 e^{-2\sigma_{pp}N_p x} - \left(\frac{\lambda}{\gamma_\pi\beta_\pi c} + 2\sigma_{\pi p}N_p\right)N_\pi$$

$$= 8B\phi_0 e^{-Bx} - B'N_\pi\,,$$

where $B = 2\sigma_{pp}N_p$, $B' = 2\sigma_{\pi p}N_p + \lambda'$, $\lambda' = \frac{\lambda}{\gamma_\pi\beta_\pi c}$. The solution of the differential equation is

$$N_\pi = \frac{8B}{B' - B}(e^{-Bx} - e^{-B'x})\phi_0\,.$$

Hence the number of charged pions which decay in the target per unit interval of x is

$$\frac{dN_\pi(\lambda)}{dx} = \frac{\lambda}{\gamma_\pi\beta_\pi c}N_\pi(\lambda) = \frac{8B\lambda'}{B' - B}(e^{-Bx} - e^{-B'x})\phi_0\,.$$

Integration from $x = 0$ to $x = \infty$ gives

$$N_\pi(\lambda) = \frac{8B\lambda'}{B' - B}\left(\frac{1}{B} - \frac{1}{B'}\right)\phi_0 = \frac{8\lambda'\phi_0}{B'}\,.$$

The branching ratio for $\pi \to \mu\nu \approx 100\%$, so that $N_{\mu\pi} \approx N_\pi(\lambda)$. This means that the energy spectrum of muons is also flat (though in actual fact high-energy muons are more likely than low-energy ones), making the comparison with $N_{\mu c}$ much simpler.

Take for example $E_\mu \sim 100$ GeV. Then $E_\pi \gtrsim 100$ GeV, $\beta_\pi \approx 1$, $\gamma_\pi \gtrsim 714$, and so

$$\lambda' = \frac{\lambda}{\gamma_\pi\beta_\pi c} = \frac{1}{2.6 \times 10^{-8} \times 714 \times 3 \times 10^{10}} = 1.8 \times 10^{-6}\ \text{cm}^{-1}\,.$$

As

$$\sigma_{\pi p} N_p = 20 \times 10^{-27} \times 2.34 \times 10^{24} = 4.7 \times 10^{-2} \text{ cm}^{-1} \gg \lambda'.$$

$$N_\pi(\lambda) \simeq \frac{8\lambda'\phi_0}{2\sigma_{\pi p}N_p} = \frac{8 \times 1.8 \times 10^{-6}\phi_0}{2 \times 4.7 \times 10^{-2}} = 1.5 \times 10^{-4}\phi_0.$$

Hence

$$\frac{N_{\mu c}}{N_{\mu \pi}} = \frac{5.95 \times 10^{-7}}{1.5 \times 10^{-4}} = 4 \times 10^{-3}.$$

4105

An experiment has been proposed to study narrow hadronic states that might be produced in $p\bar{p}$ annihilation. Antiprotons stored inside a ring would collide with a gas jet of hydrogen injected into the ring perpendicular to the beam. By adjusting the momentum of the beam in the storage ring the dependence of the $p\bar{p}$ cross section on the center-of-mass energy can be studied. A resonance would show up as a peak in the cross section to some final state.

Assume that there exists a hadron that can be produced in this channel with a mass of 3 GeV and a total width of 100 keV.

(a) What beam momentum should be used to produce this state?

(b) One of the motivations for this experiment is to search for charmonium states (bound states of a charmed quark-antiquark pair) that cannot be seen directly as resonance in e^+e^- annihilation. Which spin-parity states of charmonium would you expect to be visible as resonance in this experiment but not in e^+e^- annihilation?

Rough answers are O.K. for the remaining questions.

(c) Assume that the beam momentum spread is 1%. If the state shows up as a peak in the total cross section vs. center-of-mass energy plot, how wide would it appear to be?

(d) How wide would the state appear to be if oxygen were used in the gas jet instead of hydrogen?

(e) Assume that the jet is of thickness 1 mm and of density 10^{-9} gram/cm^3, and that there are 10^{11} circulating antiprotons in a ring of diameter 100 m. How many events per second occur per cm^2 of cross section? (In other words, what is the luminosity?) How many $p\bar{p}$ annihilations would occur per second?

(f) If the state (whose total width is 100 keV) has a branching ratio of 10% to $p\bar{p}$, what is the value of the total cross section expected at the peak (assuming the target jet is hydrogen)?

(*Princeton*)

Solution:

(a) In the laboratory frame, the velocity of the gas jet is very small and the target protons can be considered as approximately at rest. At threshold, the invariant mass squared is

$$S = (E_p + m_p)^2 - p_p^2 = M^2 \,.$$

With $E_p^2 = m_p^2 + p_p^2$, $M = 3$ GeV, this gives

$$E_p = \frac{M^2 - 2m_p^2}{2m_p} = \frac{3^2 - 2 \times 0.938^2}{2 \times 0.938} = 3.86 \text{ GeV} \,,$$

and hence the threshold momentum

$$p_p = \sqrt{E_p^2 - m_p^2} = 3.74 \text{ GeV}/c \,.$$

(b) In e^+e^- collisions, as e^+e^- annihilation gives rise to a virtual photon whose J^P is 1^-, only the resonance state of $J^P = 1^-$ can be produced. But for $p\bar{p}$ reaction, many states can be created, e.g.,

$$\text{for } S = 0, \; l = 0, \; J^P = 0^+ \,;$$

$$S = 1, \; l = 0, \; J^P = 1^- \,;$$

$$S = 1, \; l = 1, \; J^P = 0^-, \; 1^-, \; 2^- \,;$$

$$l = 2, \; J^P = 1^+, \; 2^+, \; 3^+ \,.$$

Therefore, besides the state $J^P = 1^-$, other resonance states with $J^P = 0^-, 0^+, 1^+, 2^-, 2^+, 3^+ \cdots$ can also be produced in $p\bar{p}$ annihilation.

(c) At threshold

$$p^2 = E^2 - m^2 = \frac{M^4}{4m_p^2} - M^2 \,.$$

Differentiating we have

$$2p\Delta p = M^3 \frac{\Delta M}{m_p^2} - 2M\Delta M \,,$$

or

$$\Delta M = \frac{2m_p^2 p^2 \frac{\Delta p}{p}}{M^3 - 2m_p^2 M} \,.$$

With $\frac{\Delta p}{p} = 0.01$, this gives

$$\Delta M = \frac{2 \times 0.938^2 \times 3.74^2 \times 0.01}{3^3 - 2 \times 0.938^2 \times 3}$$

$$= 1.13 \times 10^{-2} \text{ GeV} \,.$$

Since $\Delta M \gg \Gamma$, the observed linewidth is due mainly to Δp.

(d) If oxygen was used instead of hydrogen, the proton that interacts with the incident antiproton is inside the oxygen nucleus and has a certain kinetic energy known as the Fermi energy. The Fermi motion can be in any direction, thus broadening the resonance peak. For a proton in an oxygen nucleus, the maximum Fermi momentum is

$$p_F \approx \frac{\hbar}{R_0} \left(\frac{9\pi Z}{4A} \right)^{1/3} = \frac{\hbar c}{R_0 c} \left(\frac{9\pi}{8} \right)^{1/3}$$

$$= \frac{197 \times 10^{-13}}{1.4 \times 10^{-13} c} \left(\frac{9\pi}{8} \right)^{1/3} = 210 \text{ MeV}/c \,,$$

where the nuclear radius is taken to be $R = R_0 A^{1/3}$, which is much larger than the spread of momentum ($\Delta p = 3.47$ MeV/c). This would make the resonance peak much too wide for observation. Hence it is not practicable to use oxygen instead of hydrogen in the experiment.

(e) The antiprotons have velocity βc, where

$$\beta = \frac{p_p}{E_p} = \frac{3.74}{3.86} = 0.97 \,.$$

The number of times they circulate the ring per second is

$$\frac{\beta c}{100\pi}$$

and so the number of encounters of $p\bar{p}$ per second per cm^2 of cross section is

$$B = 10^{11} \times \frac{0.97 \times 3 \times 10^{10}}{100 \times 10^2 \times \pi} \times 0.1 \times 10^{-9} \times 6.023 \times 10^{23}$$

$$= 5.6 \times 10^{30} \ cm^{-2} \ s^{-1} \ .$$

Suppose $\sigma_{p\bar{p}} \approx 30$ mb. The number of $p\bar{p}$ annihilation expected per second is

$$\sigma_{p\bar{p}}B = 30 \times 10^{-27} \times 5.6 \times 10^{30}$$

$$= 1.68 \times 10^5 \ s^{-1} \ .$$

(f) The cross section at the resonance peak is given by

$$\sigma = \frac{(2J+1)}{(2J_p+1)(2J_{\bar{p}}+1)} \frac{\pi \lambda^2 \Gamma_{p\bar{p}}\Gamma}{(E-M)^2 + \dfrac{\Gamma^2}{4}} \ .$$

At resonance $E = M$. Suppose the spin of the resonance state is zero. Then as $J_p = J_{\bar{p}} = \frac{1}{2}$,

$$\sigma(J=0) = \pi \lambda^2 \frac{\Gamma_{p\bar{p}}}{\Gamma} \ .$$

With $\lambda = \frac{\hbar}{p_p}$, $\frac{\Gamma_{p\bar{p}}}{\Gamma} = 0.1$, we have

$$\sigma = \pi \times \left(\frac{\hbar c}{p_p c}\right)^2 \times 0.1 = \pi \times \left(\frac{197 \times 10^{-13}}{3740}\right)^2 \times 0.1$$

$$= 8.7 \times 10^{-30} \ cm^2$$

$$= 8.7 \ \mu b \ .$$

4. ERROR ESTIMATION
   AND STATISTICS (4106–4118)

4106

Number of significant figures to which α is known: 4, 8, 12, 20

(Columbia)

Solution:

$$\alpha = \frac{e^2}{\hbar c} = \frac{1}{137.03604(11)},$$

the answer is 8.

4107

If the average number of counts in a second from a radioactive source is 4, what is the probability of recording 8 counts in one second?

(Columbia)

Solution:

The count rate follows Poisson distribution. Hence

$$P(8) = 4^8 e^{-4}/8! = 0.03.$$

4108

Suppose it is intended to measure the uniformity of the thickness of an aluminium filter placed perpendicular to an X-ray beam. Using an X-ray detector and source, equal-exposure transmission measurements are taken at various points on the filter. The number of counts, N, obtained in 6 trials were 1.00×10^4, 1.02×10^4, 1.04×10^4, 1.06×10^4, 1.08×10^4, 1.1×10^4.

(a) Calculate the standard deviation associated with these measurements.

(b) What do the measurements tell you about the uniformity of the filter?

(c) Given that $N = N_0 e^{-\mu t}$, how is a fractional uncertainty in N related to a fractional uncertainty in t?

(d) For a given number of counts at the detector, would the fractional error in t be larger for small t or large t?

(Wisconsin)

Solution:

(a) The mean of the counts is $\bar{N} = \frac{1}{n}\sum_1^n N_i = 1.05 \times 10^4$. The standard deviation of a reading is

$$\sigma = \sqrt{\frac{1}{n-1}\sum_{i}^{n}(N_i - \bar{N})^2} = 0.037 \times 10^4 \,.$$

(b) If the Al foil is uniform, the counts taken at various locations should follow the Poisson distribution with a standard deviation

$$\Delta N = \sqrt{N} \approx \sqrt{1.05 \times 10^4} = 0.01 \times 10^4 \,.$$

Since the standard deviation of the readings (0.037×10^4) is more than three times ΔN, the foil cannot be considered uniform.

(c) Write $N = N_0 e^{-\mu t}$ as $\ln N = \ln N_0 - \mu t$. As $\frac{dN}{N} = -\mu dt$, we have

$$\frac{\Delta N}{N} = \mu \Delta t \,,$$

or

$$\frac{\Delta N}{N} = \mu t \left(\frac{\Delta t}{t}\right) \,.$$

(d) As

$$\frac{\Delta t}{t} = \frac{1}{\mu t}\frac{\Delta N}{N} \,,$$

for a given set of data, the smaller t is, the larger is the fractional error of t.

4109

You have measured 25 events $J \to e^+ e^-$ by reconstructing the mass of the $e^+ e^-$ pairs. The apparatus measures with $\Delta m/m = 1\%$ accuracy. The average mass is 3.100 GeV. What is the error?

(a) 6.2 MeV
(b) 1.6 MeV
(c) 44 MeV.

(CCT)

Solution:

As Δm is the error in a single measurement, the standard deviation is

$$\sigma = \sqrt{\frac{1}{25-1}\sum(\Delta m)^2} = \sqrt{\frac{25}{24}}\Delta m \approx \Delta m = 31 \text{ MeV} \,.$$

Hence the standard deviation of the mean, or the standard error, is

$$e = \frac{\sigma}{\sqrt{25}} = 6.2 \text{ MeV}.$$

Thus the answer is (a).

4110

In a cloud chamber filled with air at atmospheric pressure, 5 MeV alpha particles make tracks about 4 cm long. Approximately how many such tracks must one observe to have a good chance of finding one with a distinct sharp bend resulting from a nuclear encounter?

(Columbia)

Solution:

As the nuclear radius is $R = r_0 A^{1/3}$, where $r_0 = 1.2$ fm and $A = 14.7$ for the average air nucleus, the nuclear cross section σ is

$$\sigma \approx \pi R^2 = \pi \times (1.2 \times 10^{-13} \times 14.7^{1/3})^2 = 2.7 \times 10^{-25} \text{ cm}^2.$$

The number density of nuclei in the cloud chamber is

$$n = \frac{\rho N_A}{A} = \frac{0.001293 \times 6.023 \times 10^{23}}{14.7} = 5.3 \times 10^{19} \text{ cm}^{-3}.$$

Hence the mean free path is $\lambda = \frac{1}{n\sigma} = 7.0 \times 10^4$ cm.

Therefore, for a good chance of finding a large-angle scattering one should observe about $7 \times 10^4 / 4 \approx 20000$ events.

4111

The positive muon (μ^+) decays into a positron and two neutrinos,

$$\mu^+ \rightarrow e^+ + \nu_e + \bar{\nu}_\mu,$$

with a mean lifetime of about 2 microseconds. Consider muons at rest polarized along the z-axis of a coordinate system with a degree of polarization

P, and confine our observations to the highest-energy positrons from muon decays. These positrons are emitted with an angular distribution

$$I(\cos\theta)d\Omega = (1 + P\cos\theta)\frac{d\Omega}{4\pi},$$

where θ is the angle between the positron direction and z-axis, and $d\Omega$ is the solid angle element into which the positron is emitted.

(a) Assume $P = +1$. What is the probability that for the first six positrons observed, three are in the forward hemisphere ($\cos\theta > 0$) and three are in the backward hemisphere ($\cos\theta < 0$)?

(b) Assume that P is in the neighborhood of 1, but not accurately known. You wish to determine P by comparing the numbers of observed forward (N_f) and backward (N_b) decay positrons. How many muon decays, N ($N = N_f + N_b$), must you observe to determine P to an accuracy of $\pm 1\%$?

(CUSPEA)

Solution:

(a) As $d\Omega = 2\pi d\cos\theta$, the probability of a forward decay is

$$P_f = 2\pi \int_0^1 \frac{(1 + P\cos\theta)d\cos\theta}{4\pi} = \frac{1}{2}\left(1 + \frac{P}{2}\right),$$

and the probability of a backward decay is

$$P_b = 2\pi \int_{-1}^0 \frac{(1 + P\cos\theta)d\cos\theta}{4\pi} = \frac{1}{2}\left(1 - \frac{P}{2}\right).$$

If we observe N positrons, the probability of finding N_f positrons in the forward and N_b positrons in the backward hemisphere, where $N = N_f + N_b$, is according to binomial distribution

$$W = \frac{N!}{N_f!N_b!}(P_f)^{N_f}(P_b)^{N_b}.$$

For $P = 1$, the above give $P_f = 3/4$, $P_b = 1/4$. With $N = 6$, $N_f = N_b = 3$, the probability is

$$W = \frac{6!}{3!3!}\left(\frac{3}{4}\right)^3\left(\frac{1}{4}\right)^3 = 0.132.$$

(b) P can be determined from

$$P_f - P_b = \frac{P}{2},$$

i.e.,

$$P = 2(P_f - P_b) = 2(2P_f - 1),$$

where $P_f = \frac{N_f}{N}$, $P_b = \frac{N_b}{N}$ are to be obtained from experimental observations. With N events observed, the standard deviation of N_f is

$$\Delta N_f = \sqrt{NP_f(1 - P_f)}.$$

So

$$\Delta P_f = \frac{\Delta N_f}{N} = \sqrt{\frac{P_f(1 - P_f)}{N}}.$$

Hence

$$\Delta P = 4\Delta P_f = 4\sqrt{\frac{P_f(1 - P_f)}{N}},$$

or

$$N = \frac{16P_f(1 - P_f)}{(\Delta P)^2}.$$

With $P \approx 1$, $\Delta P \approx 0.01P = 0.01$, $P_f \approx \frac{3}{4}$, N must be at least

$$N_{\min} = \frac{16 \cdot \dfrac{3}{4} \cdot \dfrac{1}{4}}{(10^{-2})^2} = 30000.$$

4112

Carbon dioxide in the atmosphere contains a nearly steady-state concentration of radioactive ^{14}C which is continually produced by secondary cosmic rays interacting with atmosphere nitrogen. When a living organism dies, its carbon contains ^{14}C at the atmospheric concentration, but as time passes the fraction of ^{14}C decreases due to radioactive decay. This is the basis for the technique of radiocarbon dating.

In the following you may assume that the atmospheric value for the ratio $^{14}C/^{12}C$ is 10^{-12} and that the half life for the ^{14}C β-decay is 5730 years.

(a) It is desired to use radiocarbon dating to determine the age of a carbon sample. How many grams of a sample are needed to measure the age to a precision of ± 50 years (standard deviation of 50 years)? Assume that the sample is actually 5000 years old, that the radioactivity is counted for one hour with a 100% efficient detector, and that there is no background.

(b) Repeat part (a), but now assume that there is a background counting rate in the detector (due to radioactivity in the detector itself, cosmic rays, etc.) whose average value is accurately known to be 4000 counts per hour.

(*CUSPEA*)

Solution:

(a) ^{14}C decays according to

$$N = N_0 e^{-\lambda t}.$$

Its counting rate is thus

$$A = -dN/dt = \lambda N_0 e^{-\lambda t} = \lambda N.$$

Differentiating we have

$$\frac{dA}{dt} = -\lambda^2 N_0 e^{-\lambda t} = -\lambda A,$$

and hence

$$\Delta A/A = \lambda \Delta t.$$

The decay constant is $\lambda = \frac{\ln 2}{T_{1/2}} = \frac{\ln 2}{5730} = 1.21 \times 10^{-4}$ yr^{-1}. As the counting rate per hour A follows the Poisson distribution,

$$\frac{\Delta A}{A} = \frac{\sqrt{A}}{A} = \frac{1}{\sqrt{A}} = 50\lambda,$$

giving

$$A = \left(\frac{1}{50 \times 1.21 \times 10^{-4}} \right)^2 = 2.73 \times 10^4 \ h^{-1}.$$

Let the mass of carbon required be x grams. Then

$$A = \frac{\lambda x N_A}{12} \times 10^{-12} \times \exp(-5000\lambda) \,,$$

giving

$$x = \frac{12A \times 10^{12} \times e^{5000\lambda}}{N_A \lambda}$$

$$= \frac{12 \times 2.73 \times 10^4 \times 365 \times 24}{6.023 \times 10^{23} \times 1.21 \times 10^{-4}} \times 10^{12} \times e^{5000 \times 1.21 \times 10^{-4}}$$

$$= 72.1 \ g \,.$$

(b) With a background counting rate of A_B, the total rate is $A + A_B \pm \sqrt{A + A_B}$. As A_B is known precisely, $\Delta A_B = 0$. Hence

$$\Delta(A + A_B) = \Delta A = \sqrt{A + A_B} \,,$$

or

$$\frac{\Delta A}{A} = \sqrt{\frac{1}{A} + \frac{A_B}{A^2}} \,.$$

With $\frac{\Delta A}{A} = \lambda \Delta t = C$, say, the above becomes

$$C^2 A^2 - A - A_B = 0 \,.$$

Hence

$$A = \frac{1}{2C^2}(1 + \sqrt{1 + 4C^2 A_B})$$

$$= \frac{1}{2 \times (1.21 \times 10^{-4} \times 50)^2} \left[1 + \sqrt{1 + 4 \times (1.21 \times 10^{-4} \times 50)^2 \times 4000} \right]$$

$$= 3.09 \times 10^4 \ h^{-1} \,,$$

and the mass of sample required is

$$m = \frac{3.09 \times 10^4}{2.73 \times 10^4} \times 72.1 = 81.6 \ g \,.$$

4113

A Čerenkov counter produces 20 photons/particle. The cathode of the photomultiplier converts photons with 10% efficiency into photoelectrons. One photoelectron in the multiplier will produce a signal. Of 1000 particles, how many passes unobserved?

(a) none
(b) 3
(c) 130

(CCT)

Solution:

Consider the passage of a particle. It produces 20 photons, each of which has a probability $P = 0.1$ of producing a photoelectron and so being detected. The particle will not be observed if none of the 20 photons produces photoelectrons. The probability of this happening is

$$P(0) = \frac{20!}{0!20!}(0.1)^0(0.9)^{20}$$

$$= 0.122 \, .$$

Hence of the 1000 incident particles, it is expected that 122 will not be observed. Thus the answer is (c).

4114

A radioactive source is emitting two types of radiation A and B, and is observed by means of a counter that can distinguish between the two. In a given interval, 1000 counts of type A and 2000 of type B are observed. Assuming the processes producing A and B are independent, what is the statistical error on the measured ratio $r = \frac{N_A}{N_B}$?

(Wisconsin)

Solution:

Writing the equation as

$$\ln r = \ln N_a - \ln N_B$$

and differentiating both sides, we have

$$\frac{dr}{r} = \frac{dN_A}{N_A} - \frac{dN_B}{N_B} \, .$$

As N_A and N_B are independent of each other,

$$\left(\frac{\Delta r}{r}\right)^2 = \left(\frac{\Delta N_A}{N_A}\right)^2 + \left(\frac{\Delta N_B}{N_B}\right)^2 \, .$$

Now N_A and N_B follow Poisson's distribution. So $\Delta N_A = \sqrt{N_A}$, $\Delta N_B = \sqrt{N_B}$, and hence

$$\frac{\Delta r}{r} = \sqrt{\frac{1}{N_A} + \frac{1}{N_B}} = \sqrt{\frac{1}{1000} + \frac{1}{2000}} = 3.9\% \, ,$$

or

$$\Delta r = \frac{1000}{2000} \times 0.039 = 0.020 \, ,$$

which is the standard error of the ratio r.

4115

A sample of β-radioactive isotope is studied with the aid of a scintillation counter which is able to detect the decay electrons and accurately determine the individual decay times.

(a) Let τ denote the mean decay lifetime. The sample contains a large number N of atoms, and the detection probability per decay is ε. Calculate the average counting rate in the scintillator. You may assume τ to be much longer than any period of time over which measurements are made. In a measurement of τ, 10,000 counts are collected over a period of precisely one hour. The detection efficiency of the scintillator is independently determined to be 0.4 and N is determined to be 10^{23}. What is the measured value of τ? What is the statistical error in this determination of τ (standard deviation)?

(b) Let $P(t)dt$ be the probability that two successive counts in the scintillator are at t and $t + dt$. Compute $P(t)$ in terms of t, ε, N, τ.

(*CUSPEA*)

Solution:

(a) As $\tau \gg$ time of measurement, N can be considered constant and the average counting rate is

$$R = \frac{\varepsilon N}{\tau}.$$

Hence

$$\tau = \frac{\varepsilon N}{R} = \frac{0.4 \times 10^{23}}{10^4} = 0.4 \times 10^{19} \text{ h} = 4.6 \times 10^{14} \text{ yr}.$$

The statistical error of R is $\sqrt{R}$ as counting rates follow Poisson's distribution. Then

$$\frac{\Delta \tau}{\tau} = \frac{\Delta R}{R} = \frac{1}{\sqrt{R}} = \frac{1}{\sqrt{10^4}} = 0.01,$$

or

$$\Delta \tau = 4.6 \times 10^{12} \text{ yr}.$$

(b) The first count occurs at time t. This means that no count occurs in the time interval 0 to t. As the expected mean number of counts for the interval is $m = Rt$, the probability of this happening is

$$\frac{e^{-m} m^0}{0!} = e^{-m} = e^{-Rt}.$$

The second count can be taken to occur in the time dt. As $m' = Rdt$, the probability is

$$\frac{e^{-m'} m'}{1!} = e^{-Rdt} Rdt \approx Rdt.$$

Hence

$$P(t)dt = Re^{-Rt} dt$$

or

$$P(t) = \frac{\varepsilon N}{\tau} \exp\left(-\frac{\varepsilon N t}{\tau}\right).$$

4116

A minimum-ionizing charged particle traverses about 1 mg/cm^2 of gas. The energy loss shows fluctuations. The full width at half maximum (fwhm) divided by the most probable energy loss (the relative fwhm) is about

(a) 100%.

(b) 10%.

(c) 1%.

(CCT)

Solution:

The energy loss of a minimum-ionizing charged particle when it transverses about 1 mg/cm^2 of gas is about 2 keV. The average ionization energy for a gas molecule is about 30 eV. The relative fwhm is then about

$$\eta = 2.354 \left(\frac{\varepsilon F}{E_0}\right)^{1/2} = 2.354 \left(\frac{30F}{2000}\right)^{1/2} = 29(F)^{1/2}\%,$$

where $F < 1$ is the Fanor factor. The answer is (b).

4117

An X-ray of energy ε is absorbed in a proportional counter and produces in the mean $\bar{n}$ ion pairs. The rms fluctuation σ of this number is given by

(a) $\sqrt{\bar{n}}$.

(b) $\sqrt{F\bar{n}}$, with $F < 1$.

(c) $\pi \ln \bar{n}$.

(CCT)

Solution:

The answer is (b).

4118

A 1 cm thick scintillator produces 1 visible photon/100 eV of energy loss. It is connected by a light guide with 10% transmission to a photomultiplier (10% efficient) converting the light into photoelectrons. What is the variation σ in pulse height for the proton in Problem 4065?

(a) 21.2%

(b) 7.7%

(c) 2.8%

(CCT)

Solution:

The energy loss of the proton in the scintillator is $\Delta E = 1.8$ MeV $= 1.8 \times 10^6$ eV per cm path length. Then the mean number of photons produced in the scintillator is

$$\bar{n} = \frac{1.8 \times 10^6}{100} = 1.8 \times 10^4 .$$

With a transmission efficiency of 10% and a conversion efficiency of 10%, the number of observed photoelectrons is $\bar{N} = 1.8 \times 10^4 \times 0.1 \times 0.1 = 180$. The percentage standard deviation is therefore

$$\sigma = \frac{\sqrt{\bar{N}}}{\bar{N}} = \frac{1}{\sqrt{\bar{N}}} = \frac{1}{\sqrt{180}} = 7.5\% .$$

Hence the answer is (b).

5. PARTICLE BEAMS
AND ACCELERATORS (4119–4131)

4119

(a) Discuss the basic principles of operation of cyclotrons, synchrocyclotrons and synchrotrons. What are the essential differences among them? What limits the maximum energy obtainable from each?

(b) Discuss the basic principles of operation of linear accelerators such as the one at SLAC. What are the advantages and disadvantages of linear accelerators as compared to circular types?

(c) For what reason have colliding-beam accelerators ("intersecting storage ring") been constructed in recent years? What are their advantages and disadvantages as compared to conventional fixed-target accelerators?

(Columbia)

Solution:

(a) The cyclotron basically consists of two hollow, semicircular metal boxes — the dees — separated along their straight edges by a small gap. An ion source at the center of the gap injects particles of charge Ze into one of the dees. A uniform and constant magnetic field is applied perpendicular to the dees, causing the particles to orbit in circular paths of radius r given by

$$\frac{mv^2}{r} = ZevB.$$

The particles are accelerated each time it crosses the gap by a radio-frequency electric field applied across the gap of angular frequency $\omega_r = \frac{ZeB}{m} = w_p$, the angular frequency of revolution of the particles. As w_p is independent of the orbit radius r, the particles always take the same time to cover the distance between two successive crossings, arriving at the gap each time at the proper phase to be accelerated.

An upper limit in the energy attainable in the cyclotron is imposed by the relativistic increase of mass accompanying increase of energy, which causes them to reach the accelerating gap progressively later, to finally fall out of resonance with the rf field and be no longer accelerated.

In the synchrocyclotron, this basic limitation on the maximum energy attainable is overcome by varying the frequency of the rf field, reducing it step by step in keeping with the decrease of w_p due to relativistic mass change. While in principle there is no limit to the attainable energy in the synchrocyclotron, the magnet required to provide the magnetic field, which covers the entire area of the orbits, has a weight proportional to the third power of the maximum energy. The weight and cost of the magnet in practice limit the maximum attainable energy.

In the synchrotron the particles are kept in an almost circular orbit of a fixed radius between the poles of a magnet annular in shape, which provides a magnetic field increasing in step with the momentum of the particles. Accelerating fields are provided by one or more rf stations at points on the magnetic ring, the rf frequency increasing in step with the increasing velocity of the particles. The highest energy attainable is limited by the radiation loss of the particles, which on account of the centripetal acceleration radiate electromagnetic radiation at a rate proportional to the fourth power of energy.

Comparing the three types of accelerators, we note that for the cyclotron both the magnitude of the magnetic field and the frequency of the rf field are constant. For the synchrocyclotron, the magnitude of the magnetic field is constant while the frequency of the rf field changes synchronously with the particle energy, and the orbit of a particle is still a spiral. For the synchrotron, both the magnitude of the magnetic field and the frequency of the rf field are to be tuned to keep the particles in a fixed orbit.

(b) In a linear accelerator such as SLAC, charged particles travel in a straight line along the axis of a cylindrical pipe that acts as a waveguide, which has a rf electromagnetic field pattern with an axial electric field component to provide the accelerating force. Compared to ring-shaped accelerators, the linear accelerator has many advantages. As the particles move along a straight line they are easily injected and do not need extraction. In addition, as there is no centripetal acceleration radiation loss is neglectable. It is especially suited for acclerating electrons to very high energies. Another advantage is its flexibility in construction. It can be lengthened in steps. Its downside is its great length and high cost as compared to a ring accelerator of equal energy.

(c) In the collision of a particle of mass m and energy E with a stationary particle of equal mass the effective energy for interaction is $\sqrt{2mE}$, while for a head-on collision between colliding beams of energy E the effective energy is $2E$. It is clear then that the higher the energy E, the smaller will be the fraction of the total energy available for interaction in the former case. As it is difficult and costly to increase the energy attainable by an accelerated particle, many colliding-beam machines have been constructed in recent years. However, because of their lower beam intensity and particle density, the luminosity of colliding-beam machines is much lower than that of stationary-target machines.

4120

(a) Briefly describe the cyclotron and the synchrotron, contrasting them. Tell why one does not use:

(b) cyclotrons to accelerate protons to 2 GeV?

(c) synchrotrons to accelerate electrons to 30 GeV?

(Columbia)

Solution:

(a) In the cyclotron, a charged particle is kept in nearly circular orbits by a uniform magnetic field and accelerated by a radio frequency electric field which reverses phase each time the particle crosses the gap between the two D-shape electrodes. However, as its mass increases accompanying the increase of energy, the cyclotron radius of the particle $r = \frac{mv}{eB}$ increases, and the cyclotron frequency $w = \frac{eB}{m}$ decreases. Hence the relative phase of particle revolution relative to the rf field changes constantly. In the synchrotron the bending magnetic field is not constant, but changes with the energy of the particle, causing it to move in a fixed orbit. Particles are accelerated by resonant high frequency field at one or several points on the orbit, continually increasing the energy (cf. **Problem 4119(a)**).

(b) In the cyclotron, as the energy of the particle increases, the radius of its orbit also increases and the accelerating phase of the particle changes constantly. When the kinetic energy of the particle is near to its rest energy, the accumulated phase difference can be quite large, and finally the particle will fall in the decelerating range of the radio frequency field when it reaches the gap between the D-shaped electrorodes. Then the energy of the particle cannot be further increased. The rest mass of the proton is ~ 1 GeV. To accelerate it to 2 GeV with a cyclotron, we have to accomplish this before it falls in the decelerating range. The voltage required is too high in practice.

(c) In the synchrotron the phase-shift probem does not arise, so the particle can be accelerated to a much higher energy. However at high energies, on account of the large centripetal acceleration the particle will radiate electromagnetic radiation, the synchrotron radiation, and lose energy, making the increase in energy per cycle negative. The higher the energy and the smaller the rest mass of a particle, the more intense is the synchrotron radiation. Obviously, when the loss of energy by synchrotron radiation is equal to the energy acquired from the accelerating field in the same interval of time, further acceleration is not possible. As the rest mass of electron is only 0.511 MeV, to accelerate an electron to 30 GeV, we must increase the radius of the accelerator, or the accelerating voltage, or both to very large values, which are difficult and costly in practice. For example, a 45 GeV e^+e^- colliding-beams facility available at CERN has a circumference of 27 km.

4121

Radius of 500 GeV accelerator at Batavia is 10^2, 10^3, 10^4, 10^5 m.

(Columbia)

Solution:

In a magnetic field of induction B, the radius of the orbit of a proton is

$$R = \frac{m\gamma\beta c}{eB} = \frac{m\gamma\beta c^2}{eBc}.$$

For a proton of energy 500 GeV, $\beta \approx 1$, $m\gamma c^2 = 500$ GeV. Hence, if $B \sim 1\,T$ as is generally the case,

$$R = \frac{500 \times 10^9 \times 1.6 \times 10^{-19}}{1.6 \times 10^{-19} \times 1 \times 3 \times 10^8} = \frac{5}{3} \times 10^3 \text{ m}.$$

Thus the answer is 10^3 m.

4122

In a modern proton synchrotron (particle accelerator) the stability of the protons near the equilibrium orbit is provided by the fact that the magnetic field B required to keep the particles in the equilibrium orbit (of radius R) is nonuniform, independent of θ, and can often be parametrized as

$$B_z = B_0 \left(\frac{R}{r}\right)^n,$$

where z is the coordinate perpendicular to the plane of the equilibrium orbit (i.e., the vertical direction) with $z = 0$ at the equilibrium orbit, B_0 is a constant field required to keep the particles in the equilibrium orbit of radius R, r is the actual radial position of the particle (i.e. $\rho = r - R$ is the horizontal displacement of the particle from the equilibrium orbit), and n is some constant. Derive the frequencies of the vertical and horizontal betatron oscillations for a particular value of n. For what range of values of n will the particles undergo stable oscillations in both the vertical and horizontal directions around the equilibrium orbit?

(Columbia)

Solution:

Using the cylindrical coordinates (r, θ, z), we can write the equation of motion of the particle

$$\frac{d}{dt}(m\mathbf{v}) = e\mathbf{E} + e\mathbf{v} \times \mathbf{B}$$

as

$$\frac{d}{dt}\left(m\frac{dr}{dt}\right) - mr\left(\frac{d\theta}{dt}\right)^2 = eE_r + eB_z r\frac{d\theta}{dt} - eB_\theta\frac{dz}{dt},$$

$$\frac{1}{r}\frac{d}{dt}\left(mr^2\frac{d\theta}{dt}\right) = eE_\theta + eB_r\frac{dz}{dt} - eB_z\frac{dr}{dt},$$

$$\frac{d}{dt}\left(m\frac{dz}{dt}\right) = eE_z + eB_\theta\frac{dz}{dt} - eB_r r\frac{d\theta}{dt}.$$

On the orbit of the particle the electric field is zero and the magnetic field is independent of θ, i.e.,

$$E_\theta = E_r = E_z = B_\theta = 0.$$

The first and third of the above equations reduce to

$$\frac{d}{dt}\left(m\frac{dr}{dt}\right) - mr\left(\frac{d\theta}{dt}\right)^2 = eB_z r\frac{d\theta}{dt}, \tag{1}$$

$$\frac{d}{dt}\left(m\frac{dz}{dt}\right) = -eB_r r\frac{d\theta}{dt}. \tag{2}$$

On the equilibrium orbit, $r = R$ and Eq. (1) becomes

$$mR\left(\frac{d\theta}{dt}\right)^2 = -eB_0 R\left(\frac{d\theta}{dt}\right),$$

or

$$\frac{d\theta}{dt} = -\frac{eB_0}{m} = -\omega_0, \qquad \text{say}.$$

ω_0 is the angular velocity of revolution of the particle, i.e., its angular frequency.

The actual orbit fluctuates about the equilibrium orbit. Writing $r = R + \rho$, where ρ is a first order small quantity, same as z, and retaining only first order small quantities we have, near the equilibrium orbit,

$$B_z(r, z) \approx B_0\left(\frac{B}{r}\right)^n \approx B_0\left(1 + \frac{\rho}{R}\right)^{-n} \approx B_0\left(1 - \frac{n\rho}{R}\right).$$

As
$$\nabla \times \mathbf{B} = 0,$$
considering the θ component of the curl we have
$$\frac{\partial B_r}{\partial z} = \frac{\partial B_z}{\partial r},$$
from which follows
$$B_r(\rho, z) \approx B_r(\rho, 0) + \left(\frac{\partial B_r}{\partial z}\right)_{z=0} z = 0 + \left(\frac{\partial B_z}{\partial r}\right)_{z=0} z$$

$$= -\left(\frac{nB_z}{r}\right)_{z=0} z = -\frac{nB_0}{R} z,$$

since $B = B_z = B_0$ for $\rho = 0$.

To consider oscillations about R, let $r = R + \rho$. On using the approximate expressions for B_z and B_r and keeping only first order small quantities, Eqs. (1) and (2) reduce to

$$\frac{d^2\rho}{dt^2} = -\omega_0^2(1 - n)\rho,$$

$$\frac{d^2 z}{dt^2} = -\omega_0^2 nz.$$

Hence if $n < 1$, there will be stable oscillations in the radial direction with frequency

$$\omega_\rho = \sqrt{1 - n}\,\omega_0 = \frac{\sqrt{1 - n}\,eB_0}{m}.$$

If $n > 0$, there will be stable oscillations in the vertical direction with frequency

$$\omega_z = \sqrt{n}\,\omega_0 = \frac{\sqrt{n}\,eB_0}{m}.$$

Thus only when the condition $0 < n < 1$ is satisfied can the particle undergo stable oscillations about the equilibrium orbit in both the horizontal and vertical directions.

4123

A modern accelerator produces two counter-rotating proton beams which collide head-on. Each beam has 30 GeV protons.

(a) What is the total energy of collision in the center-of-mass system?

(b) What would be the required energy of a conventional proton accelerator in which protons strike a stationary hydrogen target to give the same center-of-mass energy?

(c) If the proton-proton collision rate in this new machine is 10^4/sec, estimate the required vacuum in the system such that the collision rate of protons with residual gas be of this same order of magnitude in 5 m of pipe. Take 1000 m as the accelerator circumference, $\sigma_{p-\text{air}} = 10^{-25}$ cm^2, and the area of the beam as 1 mm^2.

(Columbia)

Solution:

(a) The center-of-mass system is defined as the frame in which the total momentum of the colliding particles is zero. Thus for the colliding beams, the center-of-mass system (c.m.s.) is identical with the laboratory system. It follows that the total energy of collision in c.m.s. is $2E_p = 2 \times 30 = 60$ GeV.

(b) If a conventional accelerator and a stationary target are used, the invariant mass squared is

$$S = (E_p + m_p)^2 - p_p^2$$
$$= E_p^2 - p_p^2 + 2E_p m_p + m_p^2$$
$$= 2E_p m_p + 2m_p^2 .$$

In c.m.s.

$$S = (60)^2 = 3600 \text{ GeV}^2 .$$

As S is invariant under Lorentz transformation we have

$$2E_p m_p + 2m_p^2 = 3600 ,$$

or

$$E_p = \frac{1800 - 0.938^2}{0.938} = 1918 \text{ GeV} ,$$

as the required incident proton energy.

(c) Let n, s be the number density of protons and cross sectional area of each colliding beam, L be the circumference of the beam orbit, l be the length of the pipe of residual air with density ρ. The number of collisions per unit time in the colliding beam machine is

$$r = \frac{N}{\Delta t} = \frac{N_p N_p \sigma_{pp}}{\left(\dfrac{L}{c}\right)} = \frac{(nsL)^2 c\sigma_{pp}}{L} = n^2 s^2 Lc\sigma_{pp} .$$

The number of collisions per unit time in the air pipe is

$$r' = \frac{N'}{\Delta t'} = \frac{N_p N_a \sigma_{pa}}{\left(\dfrac{L+l}{c}\right)} \approx (nsL)\left(\frac{\rho s_l N_A}{A}\right)\frac{c\sigma_{pa}}{L} ,$$

where A is the molecular weight of air and N_A is Avodagro's number.

If $r' = r$, the above give

$$\rho = \frac{A}{N_A}\frac{L}{l}\frac{\sigma_{pp}}{\sigma_{pa}}n .$$

As $r = 10^4$ s^{-1}, we have

$$n = \left(\frac{10^4}{s^2 Lc\sigma_{pp}}\right)^{\frac{1}{2}} = \left(\frac{10^4}{10^{-4} \times 10^5 \times 3 \times 10^{10} \times 3 \times 10^{-26}}\right)^{\frac{1}{2}}$$

$$= 1.8 \times 10^9 \text{ cm}^{-3} ,$$

taking $\sigma_{pp} = 30$ mb$= 3 \times 10^{-26}$ cm^2. Hence

$$\rho = \frac{29}{6.02 \times 10^{23}}\left(\frac{1000}{5}\right)\left(\frac{3 \times 10^{-26}}{10^{-25}}\right) \times 1.8 \times 10^9$$

$$= 5.3 \times 10^{-12} \text{ g cm}^{-3} ,$$

The pressure P of the residual air is given by

$$\frac{5.3 \times 10^{-12}}{1.3 \times 10^{-3}} = \frac{P}{1} ,$$

i.e., $P = 4 \times 10^{-9}$ atm.

4124

Suppose you are able to produce a beam of protons of energy E in the laboratory (where $E \gg m_p c^2$) and that you have your choice of making a single-beam machine in which this beam strikes a stationary target, or dividing the beam into two parts (each of energy E) to make a colliding-beam machine.

(a) Discuss the relative merits of these two alternatives from the following points of view:

(1) the threshold energy for particle production,

(2) the event rate,

(3) the angular distribution of particles produced and its consequences for detector design.

(b) Consider the production of the Z^0 particle ($Mc^2 \approx 90$ GeV) at threshold in a $p + p$ collision. What is the energy E required for each type of machine?

(c) At beam energy E, what is the maximum energy of a π meson produced in each machine?

(CUSPEA)

Solution:

(a) (i) The invariant mass squared is the same before and after reaction:

$$S = -(p_1 + p_2)^2 = -(p'_1 + p'_2 + p)^2 ,$$

where p_1, p_2 are the initial 4-momenta of the two protons, p'_1, p'_2 are their final 4-momenta, respectively, and p is the 4-momentum of the new particle of rest mass M.

Then for one proton being stationary initially, $p_1 = (\mathbf{p}_1, E_p)$, $p_2 = (0, m_p)$ and so

$$S = (E_1 + m_p)^2 - \mathbf{p}_1^2$$

$$= (E_1^2 - \mathbf{p}_1^2) + m_p^2 + 2E_1 m_p$$

$$= 2m_p^2 + 2E_1 m_p .$$

At threshold the final state has

$$p'_1 = p'_2 = (0, m_p), \qquad p = (0, M) \qquad \text{and so}$$

$$S' = (2m_p + M)^2 .$$

For the reaction to proceed we require

$$S \geq S',$$

or

$$E_1 \geq m_p + 2M + \frac{M^2}{2m_p}.$$

For colliding beams, we have $p_1 = (\mathbf{p}_c, E_c)$, $p_2 = (-\mathbf{p}_c, E_c)$ and the invariant mass squared

$$S'' = (2E_c)^2 - (\mathbf{p}_c - \mathbf{p}_c)^2 = 4E_c^2.$$

The requirement $S'' \geq S'$ then gives

$$E_c \geq m_p + \frac{M}{2}.$$

Note that $E_1 \gg E_c$ if $M \gg m_p$. Hence colliding-beam machine is able to produce the same new particle with particles of much lower energies.

(ii) Since a fixed target provides an abundance of target protons which exist in its nuclei, the event rate is much higher for a stationary-target machine.

(iii) With a stationary-target machine, most of the final particles are collimated in the forward direction of the beam in the laboratory. Detection of new particles must deal with this highly directional geometry of particle distribution and may have difficulty in separating them from the background of beam particles.

With a colliding-beam machine the produced particles will be more uniformly distributed in the laboratory since the total momentum of the colliding system is zero. In this case the detectors must cover most of the 4π solid angle.

(b) Using the formulas in (a) (i) we find, with $m_p = 0.94$ GeV, $M = 90$ GeV, the threshold energies for a fixed-target machine,

$$E_1 = m_p + 2M + \frac{M^2}{2m_p} = 0.94 + 2 \times 90 + \frac{90^2}{2 \times 0.94} = 4489 \text{ GeV},$$

and for a colliding-beam machine,

$$E_c = m_P + \frac{M}{2} = 0.94 + \frac{90}{2} = 45.94 \text{ GeV}.$$

(c) *Colliding-beam machine*

Let p_1, p_2 be the momenta of the protons in the final state, and p_π be the momentum of the pion produced. Conservation of energy requires

$$2E = \sqrt{m_p^2 + p_1^2} + \sqrt{m_p^2 + p_2^2} + \sqrt{m_\pi^2 + p_\pi^2}.$$

Conservation of momentum requires

$$\mathbf{p}_1 + \mathbf{p}_2 + \mathbf{p}_\pi = 0,$$

or

$$p_\pi^2 = p_1^2 + p_2^2 + 2p_1 p_2 \cos \alpha.$$

This means that for p_π to have the maximum value, the angle α between $\mathbf{p}_1$, $\mathbf{p}_2$ must be zero, since $\frac{\partial p_\pi}{\partial \alpha} = \frac{p_1 p_2}{p_\pi} \sin \alpha$. Thus at maximum p_π, the three final particles must move in the same line. Write

$$\mathbf{p}_2 = -(\mathbf{p}_1 + \mathbf{p}_\pi).$$

The energy equation becomes

$$2E = \sqrt{m_p^2 + (p_\pi + p_1)^2} + \sqrt{m_\pi^2 + p_\pi^2} + \sqrt{m_p^2 + p_1^2}.$$

Differentiating we have

$$0 = \frac{(p_\pi + p_1) d(p_\pi + p_1)}{\sqrt{m_p^2 + (p_\pi + p_1)^2}} + \frac{p_\pi dp_\pi}{\sqrt{m_\pi^2 + p_\pi^2}} + \frac{p_1 dp_1}{\sqrt{m_p^2 + p_1^2}}.$$

Letting $dp_\pi/dp_1 = 0$, we find

$$-\frac{p_1}{\sqrt{m_p^2 + p_1^2}} = \frac{(p_\pi + p_1)}{\sqrt{m_p^2 + (p_\pi + p_1)^2}}.$$

Hence

$$p_\pi = -2p_1, \quad p_2 = p_1.$$

Thus at maximum E_π,

$$2E = 2\sqrt{m_p^2 + p_1^2} + \sqrt{m_\pi^2 + (2p_1)^2},$$

or
$$4E^2 - 4EE_{\pi\,max} + m_\pi^2 + 4p_1^2 = 4m_p^2 + 4p_1^2\,,$$

giving the maximum pion energy
$$E_{\pi\,max} = \frac{4E^2 + m_\pi^2 - 4m_p^2}{4E} \approx E\,,$$

as $E \gg m_p$.

Stationary-target machine: When E_π is maximum, the two final-state protons are stationary and the pion takes away the momentum of the incident proton. Thus
$$E_\pi + 2m_p = E + m_p\,,$$

or
$$E_\pi = E - m_p \approx E \qquad \text{as} \qquad E \gg m_p\,.$$

4125

An electron (mass m, charge e) moves in a plane perpendicular to a uniform magnetic field. If energy loss by radiation is neglected the orbit is a circle of some radius R. Let E be the total electron energy, allowing for relativistic kinematics so that $E \gg mc^2$.

(a) Explain the needed field strength B analytically in terms of the above parameters. Compute B numerically, in gauss, for the case where $R = 30$ meters, $E = 2.5 \times 10^9$ electron volts. For this part of the problem you will have to recall some universal constants.

(b) Actually, the electron radiates electromagnetic energy because it is being accelerated by the B field. However, suppose that the energy loss per revolution ΔE is small compared to E. Explain the ratio $\Delta E/E$ analytically in terms of the parameters. Then evaluate this ratio numerically for the particular value of R given above.

(*CUSPEA*)

Solution:

(a) Let $\mathbf{v}$ be the velocity of the electron. Its momentum is $\mathbf{p} = m\gamma\mathbf{v}$, where $\gamma = (1 - \frac{v^2}{c^2})^{-\frac{1}{2}}$. Newton's second law of motion gives
$$\frac{d\mathbf{p}}{dt} = m\gamma\frac{d\mathbf{v}}{dt} = e\mathbf{v} \times \mathbf{B}\,,$$

as $|\mathbf{v}|$ and hence γ are constant since $\mathbf{v} \perp \mathbf{B}$, or

$$\left|\frac{d\mathbf{v}}{dt}\right| = \frac{evB}{m\gamma}.$$

As

$$\left|\frac{d\mathbf{v}}{dt}\right| = \frac{v^2}{R},$$

where R is the radius of curvature of the electron orbit,

$$B = \frac{m\gamma v}{eR},$$

or

$$B = \frac{pc}{eRc} = \frac{\sqrt{E^2 - m^2c^4}}{eRc} \approx \frac{E}{eRc}$$

$$= \frac{2.5 \times 10^9 \times 1.6 \times 10^{-19}}{1.6 \times 10^{-19} \times 30 \times 3 \times 10^8} = 2.8 \times 10^{-1} \text{ T}$$

$$= 2.8 \times 10^3 \text{ Gs}.$$

(b) The power radiated by the electron is

$$P = \frac{e^2}{6\pi\varepsilon_0 c^3}\gamma^6\left[\dot{v}^2 - \left(\frac{\mathbf{v} \times \dot{\mathbf{v}}}{c}\right)^2\right]$$

$$= \frac{e^2\dot{v}^2}{6\pi\varepsilon_0 c^3}\gamma^4$$

$$= \frac{e^2 v^4}{6\pi\varepsilon_0 c^3}\frac{\gamma^4}{R^2},$$

as $\dot{\mathbf{v}} \perp \mathbf{v}$. The energy loss per revolution is then

$$\Delta E = \frac{2\pi R P}{v} = \frac{4\pi}{3}\left(\frac{e^2}{4\pi\varepsilon_0 mc^2 R}\right)(\gamma\beta)^3\gamma mc^2$$

$$= \frac{4\pi}{3}\left(\frac{r_0}{R}\right)(\gamma\beta)^3 E = \frac{4\pi}{3}\left(\frac{r_0}{R}\right)(\gamma^2 - 1)^{\frac{3}{2}}E,$$

where $r_0 = 2.8 \times 10^{-15}$ m is the classical radius of electron and $\beta = \frac{v}{c}$.
With $\gamma = \frac{2.5 \times 10^9}{0.51 \times 10^6} = 4.9 \times 10^3$,

$$\frac{\Delta E}{E} \approx \frac{4\pi}{3} \times \frac{2.8 \times 10^{-15}}{30} \times (4.9 \times 10^3)^3$$

$$= 4.6 \times 10^{-5}.$$

The results can also be obtained using the relevant formulas as follows.

(a)
$$p(\text{GeV}/c) = 0.3B(\text{T})R(\text{m})$$

giving
$$B = \frac{p}{0.3R} = \frac{2.5}{0.3 \times 30} \approx 0.28 \text{ T}.$$

(b)
$$\Delta E(\text{keV}) \approx 88E(\text{GeV})^4/R(\text{m})$$

giving
$$\frac{\Delta E}{E} = 88E^3 \times 10^{-6}/R$$

$$= 88 \times 2.5^3 \times 10^{-6}/30$$

$$= 4.6 \times 10^{-5}.$$

4126

Draw a simple, functional cyclotron magnet in cross section, showing pole pieces of 1 m diameter, yoke and windings. Estimate the number of ampere-turns required for the coils if the spacing between the pole pieces is 10 cm and the required field is 2 $T(= 20 \text{ kgauss})$. $\mu_0 = 4\pi \times 10^{-7}$ J/A². m.

(Columbia)

Solution:

Figure 4.14 shows the cross section of a cyclotron magnet. The magnetic flux ϕ crossing the gap betwen the pole pieces is

$$\phi = \frac{NI}{R},$$

where
$$R = \frac{d}{\mu_0 S},$$

d being the gap spacing and S the area of each pole piece, is the reluctance. By definition the magnetic induction is $B = \frac{\phi}{S}$. Thus

$$NI = \phi R = \frac{Bd}{\mu_0} = \frac{2 \times 10 \times 10^{-2}}{4\pi \times 10^{-7}} = 1.59 \times 10^5 \text{ A-turns}.$$

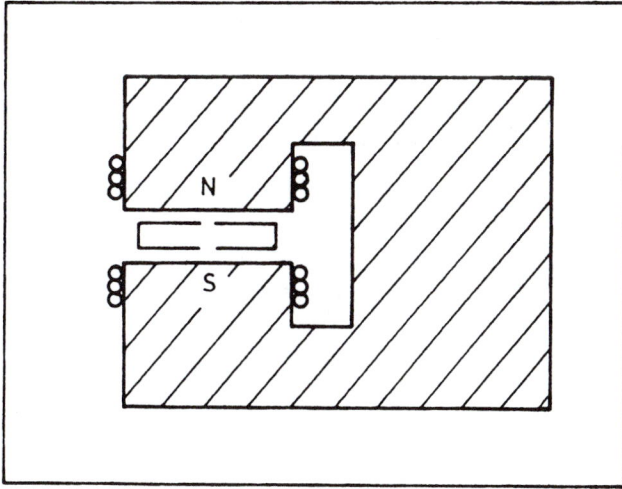

Fig. 4.14

4127

In general, when one produces a beam of ions or electrons, the space charge within the beam causes a potential difference between the axis and the surface of the beam. A 10-mA beam of 50-keV protons ($v = 3 \times 10^6$ m/sec) travels along the axis of an evacuated beam pipe. The beam has a circular cross section of 1-cm diameter. Calculate the potential difference between the axis and the surface of the beam, assuming that the current density is uniform over the beam diameter.

(Wisconsin)

Solution:

The beam carries a current

$$ I = \int \mathbf{j} \cdot d\mathbf{S} = \int_0^R j 2\pi r dr = \pi R^2 j = \pi R^2 \rho v \,, $$

where j and ρ are the current and charge densities respectively. Thus

$$ \rho = \frac{I}{\pi R^2 v} \,. $$

At a distance r from the axis, Gauss' flux theorem

$$2\pi r l E = \pi r^2 l \rho / \varepsilon_0$$

gives the electric field intensity as

$$E = \frac{r\rho}{2\varepsilon_0} = \frac{r}{2\pi\varepsilon_0} \frac{I}{vR^2} \, .$$

As $E = -\frac{dV}{dr}$, the potential difference is

$$\Delta V = \int_0^R E(r) dr = \frac{I}{2\pi\varepsilon_0 v R^2} \int r \, dr = \frac{I}{4\pi\varepsilon_0 v}$$

$$= \frac{9 \times 10^9 \times 10 \times 10^{-3}}{3 \times 10^6} = 30 \text{ V} \, .$$

4128

Cosmic ray flux at ground level is 1/year, 1/min, 1/ms, 1/μs, cm^{-2} sterad^{-1}.

(Columbia)

Solution:

The answer is $1/(\text{min} \cdot \text{cm}^2 \cdot \text{sterad})$. At ground level, the total cosmic ray flux is $1.1 \times 10^2/(\text{m}^2 \cdot \text{s} \cdot \text{sterad})$, which consists of a hard component of $0.8 \times 10^2/(\text{m}^2 \cdot \text{s} \cdot \text{sterad})$ and a soft component of $0.3 \times 10^2/(\text{m}^2 \cdot \text{s} \cdot \text{sterad})$.

4129

Particle flux in a giant accelerator is 10^4, 10^8, 10^{13}, 10^{18} per pulse.

(Columbia)

Solution:

A typical particle flux in a proton accelerator is 10^{13}/pulse.

4130

Which particle emits the most synchrotron radiation light when bent in a magnetic field?

(a) Proton.

(b) Muon.

(c) Electron.

(CCT)

Solution:

The synchrotron radiation is emitted when the trajectory of a charged particle is bent by a magnetic field. **Problem 4125** gives the energy loss per revolution as

$$\Delta E = \left(\frac{4\pi}{3}\right)\left(\frac{e^2}{4\pi\varepsilon_0}\right)\frac{1}{R}\beta^3\gamma^4,$$

where R, the radius of curvature of the trajectory, is given by

$$R = \frac{m\gamma\beta c}{eB}.$$

Thus for particles of the same charge and γ, $\Delta E \propto m^{-1}$. Hence the answer is (c).

4131

The magnetic bending radius of a 400 GeV particle in 15 kgauss is:

(a) 8.8 km.

(b) 97 m.

(c) 880 m.

(CCT)

Solution:

The formula

$$p(\text{GeV}/c) = 0.3B(T)R(m)$$

gives

$$R = \frac{p}{0.3B} = \frac{400}{0.3 \times 1.5} = 880 \text{ m}.$$

Or, from first principles one can obtain

$$R = \frac{m\gamma\beta c}{eB} \approx \frac{m\gamma c^2}{eBc} = \frac{400 \times 10^9 \times 1.6 \times 10^{-19}}{1.6 \times 10^{-19} \times 1.5 \times 3 \times 10^8} = 880 \text{ m},$$

as $B = 15$ kGs $= 1.5$ T.

Hence the answer is (c).

INDEX TO PROBLEMS

Abnormal magnetic moment of μ 3009

Absorption spectrum of HCl 1135

Accelerators 4119, 4120, 4121, 4122, 4123, 4124

α-decay 2033, 2035, 2107

α-spectrum measurement 2086

Allowed and forbidden transitions of Mg 1086

Angular momentum quantization 1035

Associated production of strange particles 3009

Atom formed by

    e with μ^+ 1061

    μ^- with nucleus 1059, 1060, 1062, 1063, 1064, 1066

    μ^- with π^+ 1065

    Ω^- with Pb nucleus 1058

    spin-1 'electrons' with He nucleus 1026

Atomic clock 1057

Atomic model of

    Bohr 1042, 1049

    Thomas-Fermi 1013

    Thomson 1045

Atomic transitions 1038, 1043, 1044, 1068, 1069, 1070, 1071, 1086

Auger effect 1008

Bag model of hadron 3080, 3081

β-decay 1039, 1047, 2006, 2085

    of Fermi and Gamow-Teller types 2088

β-decay, Fermi's theory of 3009

β^+-decay 2087, 2088

    vs K-capture 2084

Binding energy of electron in atom 1106

Black-body radiation 1002, 1077

Bohr orbit 1033

Bohr radius 1003

Bohr-Sommerfeld quantization 1036

Bragg reflection from NaCl crystal 1101

Carbon dating 2106, 4112

Čerenkov radiation 4084, 4085, 4125

Charged particle in magnetic field 4002, 4007, 4131

Charmed particle detection by μ 4104

Charmonium from $p\bar{p}/e^-e^+$ annihilations 4105

Clebsch-Gordan coefficients 3026

Colliding-beam kinematics 4001

Colliding-beam machine vs single-beam machine 4123, 4124

Color quantum number 3072, 3078, 3079

Compton effect 1034

Compton wavelength 4010, 4035, 4036, 4037

Continuous electron-spin resonance spectroscope 1111

Cosmic black-body radiation 4039

Cosmic-ray μ in geomagnetic field 4004, 4005

Coulomb barrier penetration 2007

Count rate statistics 4107, 4108, 4111, 4113, 4114, 4115, 4118

Cyclotron magnet 4126

Cyclotron vs synchrotron 4120

D particle 3087, 3088

Decay, angular distribution of products of 3054, 3055

    conservation laws in 3016, 3017, 3018, 3021, 3049

    relative rates of 3027, 3055

Decay of

    hyperon 3017

    K 3021, 3056, 4059

    Λ 3053, 3054, 3055

    ^{8}Li 2080

    μ 3022, 4026

    n 3014

    π 3013, 3050

    π, μ 3049, 3050

    p 3066, 4060

    Σ 3024

Decays 3012, 3013, 3029

    leptonic 3038, 3039, 3040

    nonleptonic weak (hyperon) 3035

Density of nuclear matter 2008

Detailed balancing 3036

Detection of particles by scintillator 4073

Deuterium 'molecule' (dqd) 1142
Deuteron
    photodistintegration 2049
    represented by square well 2050
    states 2053, 2056
    theory 2058
Diatomic molecule 1124, 1129
    modeled as dumbbell 1128, 1133
    represented by harmonic oscillator 1124
Dissociation energy of hydrogen molecule 1125
Doublet structure of sodium line 1035, 1092
Drell-Yan quark annihilation 3079
Effect of external potential on atomic energy levels 1012
Electric dipole moment of n 3009
Electric dipole transition 1040, 1079, 1080, 1081, 1093, 1096, 2092
Electric field needed to ionize atom 1004
Electric multipole transition 1038, 2093
Electric polarizability of atom 1076
Electron configuration of atom 1039, 1071, 1075, 1081, 1082, 1083
    1085, 1087, 1088, 1090, 1093, 1095, 1098, 1116
Electron in nucleus, argument against 2001
Electrostatic energy of speherical charge 2009
Energy levels in
    atom with spin-3/2 'electrons' 1067
    He atom 1071, 1072, 1074, 1100
    H-like atom 1044
    Mg atom 1086
    molecule 1127, 1139, 1140, 1141
    system of heavy quark-antiquark pair 3090
Energy level corrections 1048
    due to finite size of nucleus 1050, 1051
    due to presence of electric field 1121, 1122
    due to presence of magnetic field 1057, 1080, 1116, 1117, 1118
    1119, 1120
Error estimation 4109
    η particle production and decay 3032
Exchange force 1132

Excitation by bombardment of
    atom with electron 1011
    molecule with neutron 1131
Excitation energies of mirror nuclei 2011
Experiments important in history of atomic physics 1113
Fermi plateau 4069
Fermi transition between isospin multiplets 2091
Fine structure of atomic levels 1009, 1028, 1052, 1054, 1055, 1099
Fission 2029, 2030, 2035, 2037, 2041, 2043, 2117
Franck-Hertz experiment 1034
Frequency shift of photon falling through gravity 2099
Fusion 2006, 2044, 2045, 2046
    γ-ray absorption 2094, 2095
    γ-ray emission 2096, 2097
Geiger-Nuttall law of α-decays 2076, 2078
Gell-Mann–Nishijima relation 3074
Gluons 3082
Gluons, evidence for 3072
G-parity operator 3005
Half life of
    particle 4002
    radioactive nucleus 2040, 2066, 2077, 2079
Hall effect 1112
Heavy neutrino 3048
Helicity 3043, 3050
Helicity of μ from π decay 4061
Higgs boson 3064
Hund's rule 1008, 1078, 1082, 1095
Hyperfine structure of atomic levels 1029, 1030, 1041, 1052, 1053
    1054, 1099
Inertness of noble gas 1077
Intermediate boson 3059, 3061
Ionization energy 1007, 1009, 1046, 1076
Isobaric analog states 2014, 2069
Isobaric nuclei 2081, 2082
Isospin assignment 2014
Isospin muliplet 2012, 2013

JJ coupling 1094

Josephson effect 1112

J/ψ particle 3009

K lifetime 4059

    production 3058

    regeneration 3009

Kinematics of

    collision 4013, 4019, 4020, 4021, 4022, 4024, 4025, 4027, 4028

        4029, 4030, 4031, 4033, 4034, 4038, 4040, 4054, 4055

    decay 4006, 4023, 4041, 4042, 4043, 4044, 4045, 4046, 4047, 4048

        4049, 4050, 4051, 4052, 4053, 4056, 4057, 4058

    relativistic particle 4003, 4004, 4006, 4008, 4009, 4030

K_S/K_L ratio 3056, 3057

Lamb-Rutherford experiment 1034

Lamb shift 1008, 1032, 1037

Λ particle production in πp scattering 3021

Landé g-factor 1083, 1091, 1109

Landé interval rule 1008

Lepton number conservation 3011

Lepton types 3011

Lifetime measurements 4103

Lifetimes against different types of interaction 3018

LS coupling 1079, 1080, 1081, 1083, 1088, 1089, 1091, 1094, 1097

Lyman alpha-line 1009

Magnetic moment of

    atom 1077

    deuteron 2057

    electron 1009

    nucleus 2015, 2070

Magnetic monopole 2071

Meson of charge 2, argument against 3021

Metal as free electrons in potential well 1014

Molecule, homonuclear 1130

    hydrogen 1132

    H_2^+ 1123

Mössbauer spectroscopy 1111

Multiple-choice questions on
    accelerators 4129, 4130
    atomic physics 1005, 1018, 1020, 1021, 1023, 1027, 1030, 4069, 4088
       1126, 1127
    cosmic rays 4128
    elementary interactions 3002, 3004, 3007, 3045, 4011, 4070, 4071
       4083, 4116
    experimental errors 4106, 4117
    experimental methodology 1001, 3046, 4089, 4090, 4091, 4092, 4093
       4094, 4095, 4096, 4097, 4098, 4099, 4102
    nuclear physics 2008, 2040, 2108, 3020
    particle kinematics 4100, 4101
    particle physics 3008, 3010, 3051, 4062, 4063, 4064, 4065, 4066
       4067, 4068, 4074, 4075, 4086, 4087
Neutrino
    capture by isotopes 2089
    from the sun 2046
    interaction cross section 3045, 3046
    interaction with matter 3047
    mass 3044
    oscillation 3068
    properties 3042, 3043
    types 3042, 3067
Neutron-antineutron oscillation 3069, 3070
Neutron decay modes 3014
Neutron density in uranium 2039
Neutron interaction in scintillator 4078
Neutron irradiation of
    gold 2101
    Li 2103
    nuclei 2104
Neutron passage through graphite rod 4082
Neutron scattering cross sections 2118, 2119, 4079, 4081
Neutron star 2047
Noble gas atomic structure 1077, 1083
Nuclear
    binding energy 2025, 2026, 2027, 2028

excitation energy 2108, 2111
    ground state 2035
    reaction 2109, 2110, 2112, 2115, 2116, 2120
Nuclear precession in magnetic field 2005
Nuclear radius
    determination 2002, 2035
    from mirror nuclei 2009, 2010
Nuclear reactor of
    breeder type 2042
    fission type 2043
Nuclear shell model 2065, 2067, 2068, 2072
    magic numbers 2060, 2071
    single-particle levels 2061, 2062, 2064, 2069
Nucleon form factor 2021
Nucleon-nucleon interactions 2048, 2090
Nucleus, double-magic 2066
    effect of deformation of 2004, 2073
    magnetic moment of 2006
    models of 2059
Nucleus represented by potential box 2063
N/Z ratio for stable nuclei 2031, 2032
Pairing force 2075
Para- and ortho-states 1134
    of He atom 1073, 1077
    of hydrogen molecule 1133
Parity of atomic level 1097
Parity of π^0 3021
Parity operator 3052
Parity violation in ep scattering 3065
Particle interactions 3013, 3017, 3019, 3037
    angular distribution in 4032
    conservation laws in 3001, 3006, 3015, 3016, 3025
    cross sections for 3037
    relative cross sections for 3025, 3026
    relative strengths of 3001
    threshold for 3019, 3036, 4012, 4014, 4015, 4016, 4017, 4018, 4031
      4032

Particle interactions between

 e^+, e^- 3077

 e, ν 3059, 3060

 e^+, p and e^-, p 3021

 K^0, e 3062

 N, N 2090, 3017, 3019

 ν, q 3063

 π, d 3030, 3019

 p, $\bar{p}$ 3013

Particle tracks in emulsion/bubble chamber 3033, 3069

Particle types 3003

Particle with magnetic moment in magnetic field 1025

Photoexcitation of atom 1010, 1019

Photon interactions in matter 4076, 4077

π quantum numbers and properties 3023

Potential difference across particle beam 4127

Pressure exerted by electron on cavity walls 1024

Proton-radioactivity 2034

Ψ particle 3084, 3085, 3086

Pulsed nuclear magnetic resonance spectroscope 1111

Quantum chromodynamics 3082, 3083

Quantum numbers of hadron 3074, 3075

Quark model of hadron 3072, 3073, 3074, 3075

Radioactive capture $p + n \rightarrow d + \gamma$ 2051, 2052

Radioactivity series 2100, 2102

Raman spectrum 1136, 1137

Recombination of split neutron beams 1027

Relative population in energy level 1090

Resonance particles 3034

Resonance states in e^+e^- annihilation 3089

s wave scattering 1016

Scattering by atom of p 2018

Scattering by hard sphere 2016

Scattering by nucleus of

 α 2113

 e 2021

 p 2023

Scattering cross section calculations involving
    Born approximation 2017
    known total cross section 2022
    phase shift 2019, 2020
    Rutherford formula 1017
Semi-emperical nuclear mass formula 2024, 2036
Separation energy of neutron from nucleus 2071
Σ particle 3028
Singlet and triplet states of hydrogen molecule 1041
Spectral line broadening 1006, 1055
    by Doppler effect 1021, 1022
Spectral line intensity 1077, 1084
Spectroscopic notation for atomic levels 1069, 1070, 1071, 1075
    1078, 1085, 1089, 1090, 1093, 1116
Spin echo experiment 1110
Spin of free proton 1009
Spin-orbit interaction 1031, 1056
Spontaneous transition, lifetime for 1039
Stern-Gerlach experiment 1015, 1034, 1077, 1114, 1115
SU(3) multiplets 3076, 3078
Synchrotron 4122
System of
    bosons 2074
    nucleons 2073
    two nucleons 2048, 2054, 2055
Time-reversal operator 3052
Transition between molecular levels 1135, 1138, 1140, 1141
Transmission spectrum of HCl 1138
Two-neutrino experiments 3009
Van de Graaff generator experiment 2114
X-ray
    absorption spectrum 1103, 1108
    emission spectrum determination 1107
    K-lines 1102, 1104, 1105
Zeeman effect 1120
Zeeman effect, anomalous 1008